Current Advances in Mechanical Engineering

Current Advances in Mechanical Engineering

Edited by
Rene Sava

CWILLFORD PRESS
www.willfordpress.com

Published by Willford Press,
118-35 Queens Blvd., Suite 400,
Forest Hills, NY 11375, USA

ISBN: 978-1-68285-570-6

Cataloging-in-Publication Data

Current advances in mechanical engineering / edited by Rene Sava.
    p. cm.
Includes bibliographical references and index.
ISBN 978-1-68285-570-6
1. Mechanical engineering. 2. Engineering. I. Sava, Rene.
TJ145 .C87 2019
621--dc21

For information on all Willford Press publications
visit our website at www.willfordpress.com

WILLFORD PRESS

Contents

Permissions

List of Contributors

Index

Preface

This book strives to provide a fair idea about mechanical engineering and the latest advances within the field. As a field of study, mechanical engineering is concerned with the study of mechanical structures by analyzing their construction and design.It integrates the principles and concepts of physics, material science and engineering. Thermodynamics, structural analysis, electricity are some of the focus areas of mechanical engineering. The chapters compiled in this book bring forth some of the most innovative concepts and elucidates the unexplored aspects of mechanical engineering. The topics introduced herein elucidate new techniques and methods that have been adopted over the years due to technological progress. Comprehensive design and easy to understand language, make this book an ideal reference text for both students and experts. It will also help new researchers by foregrounding their knowledge in this discipline.

After months of intensive research and writing, this book is the end result of all who devoted their time and efforts in the initiation and progress of this book. It will surely be a source of reference in enhancing the required knowledge of the new developments in the area. During the course of developing this book, certain measures such as accuracy, authenticity and research focused analytical studies were given preference in order to produce a comprehensive book in the area of study.

This book would not have been possible without the efforts of the authors and the publisher. I extend my sincere thanks to them. Secondly, I express my gratitude to my family and well-wishers. And most importantly, I thank my students for constantly expressing their willingness and curiosity in enhancing their knowledge in the field, which encourages me to take up further research projects for the advancement of the area.

Editor

Strength and Behavior of Bolted Ultra-High Performance Concrete Panel Joint with Geometric Parameter

Yang-Hee Kwon[1], Soo-Hyung Chung[2] and Sung-Gul Hong[1]*

[1]*Department of Architecture and Architectural Engineering, Seoul National University, 1 Gwanak-Ro, Gwanak-Gu, Seoul 08826, Republic of Korea*
[2]*SENSE B/D, 6 Beodeunaru-ro 19-gil, Youngdeungpo-gu, Seoul 07226, Republic of Korea*

Abstract

One of the most reasonable ways to connect thin ultra-high performance concrete (UHPC) facades to building structures is bolted joints. To design the joints for the facades economically and safely, a clear investigation of the structural behavior is needed. In this study, the joint strength, failure modes and strain concentration phenomena of the bolted UHPC panels are investigated by direct tensile test. The main experimental variables are the geometric parameters, which are the width of the specimen, its thickness, and the distance from the center of the hole to the edge. Experimental results show that ductility of the joint depends on the failure mode. In addition, it is shown that the increase in the material cost of the panel, such as size and thickness, does not necessarily lead to an increase in the joint strength.

Keywords: Ultra-High Performance Concrete (UHPC); Facade; Panel; Bolted joint; Cleavage failure; Net-tension failure

Introduction

The development of information technology has led to the development of freeform design technology in the field of architecture, which is attracting much public attention. Metal or concrete has been used as the most suitable material for manufacturing such a preform facade. However, fabricating the preform facades using these materials can lead to some problems such as design complexity and high cost.

UHPC can be a reasonable alternative to the freeform facade, because it has outstanding mechanical performance, self-compacting property and durability [1,2]. Although the material cost of UHPC itself is much higher than that of normal concrete [2,3], the UHPC facade manufactured by precast method can be assembled by bolt fastening, so the construction cost can be reduced and the whole construction cost can be reduced. Nevertheless, the study on the bolt joints of UHPC panels is still lacking. Therefore, this study was conducted to investigate the structural performance of the bolt joint for connecting UHPC panels.

Experimental Procedure

Test parameters

The experimental program was designed to investigate the effect of panel geometry on the behavior of the panel. Based on this, three basic geometric variables, which can affect the strength and failure mode of the joint, were considered. These parameters are shown in Figure 1 and also can be summarized as follows.

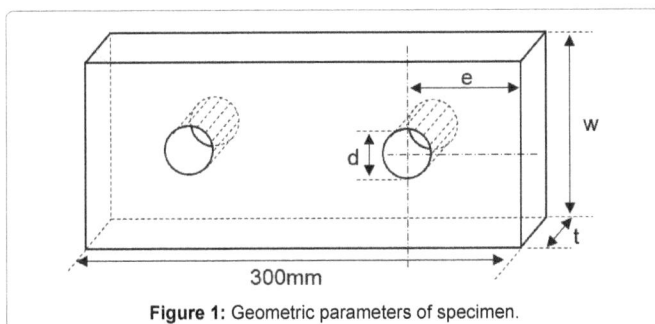

Figure 1: Geometric parameters of specimen.

Cement	Silica fume	Quartz sand	Silica flour	Water	Super-plasticizer	Steel fiber*
1	0.25	1.1	0.35	0.22	0.03	2%

* Volumetric ratio of UHPC

Table 1: Mix proportion of ultra-high performance concrete (by wt% of cement).

In Figure 1, e/d is the ratio between the end distances (e) to the hole diameter (d). The three ratios were considered as e/d=2, 3 and 4 or e=48 mm, 72 mm and 96 mm at d=24 mm, respectively. In addition, w/d is the ratio between the plate width (w) and the hole diameter (d). The three ratios were planned as w/d=4, 6 and 8 when the plate width is 96 mm, 144 mm and 192 mm, respectively. Lastly, to investigate the effect of panel thickness on the behavior of the joint, it was divided into two types of 20 mm and 30 mm.

Preparation of specimens

Total 18 UHPC panels were fabricated according to the raw materials, mix proportion (Table 1) and procedures of our previous studies [4-6]. When all the mixing was completed, the prepared mold which was placed on a flat table was filled with fresh UHPC by the self-compacting property. After 24 hrs, the mold was demolded and the UHPC was cured at a temperature of 90°C and 95% for 48 hrs. Then, the specimen was cured at a room temperature of 20°C and 60% until the test.

Test setup and instrumentation

Figure 2 shows an experiment prepared to investigate the force and strain transmitted to the hole of a UHPC panel through a bolt and its behavior. The panel and the loading device, universal testing machine

***Corresponding author:** Sung-Gul Hong, Department of Architecture and Architectural Engineering, Seoul National University, Gwanak-Ro, Gwanak-Gu, Seoul 08826, Republic of Korea, E-mail: sglhong@snu.ac.kr

(UTM) was fixed by bolts inserted into the holes. The tensile force was applied to the fixed specimen by the displacement control method at a speed of 0.005 ± 0.0015 mm/sec.

Linear variable differential transducers (LVDTs) were installed on both sides of UTM to measure the relative displacement between the panel and the UTM, and the load was measured by the load cell. A total of six strain gauges were attached to estimate the stress distribution around the hole in the panel. Its arrangement is shown in Figure 3.

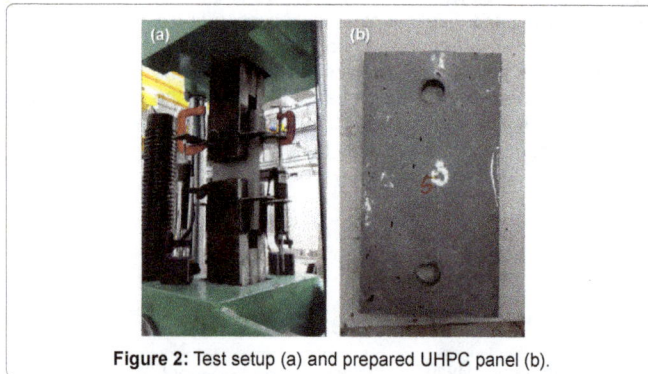

Figure 2: Test setup (a) and prepared UHPC panel (b).

Specimen number	d [mm]	e/d	w/d	t [mm]	Max. load [kN]	Failure mode
1	24	2	4	20	8.3	Cleavage
2	24	2	4	30	11.6	Cleavage
3	24	2	6	20	9.3	Cleavage
4	24	2	6	30	13.5	Cleavage
5	24	2	8	20	10.8	Cleavage
6	24	2	8	30	11.7	Cleavage
7	24	3	4	20	10.7	Net tension
8	24	3	4	30	16.7	Net tension
9	24	3	6	20	13.3	Cleavage
10	24	3	6	30	25.0	Cleavage
11	24	3	8	20	16.5	Cleavage
12	24	3	8	30	23.6	Cleavage
13	24	4	4	20	13.1	Net tension
14	24	4	4	30	15.5	Net tension
15	24	4	6	20	18.6	Net tension
16	24	4	6	30	29.5	Net tension
17	24	4	8	20	19.9	Cleavage
18	24	4	8	30	29.1	Cleavage

Table 2: Test results.

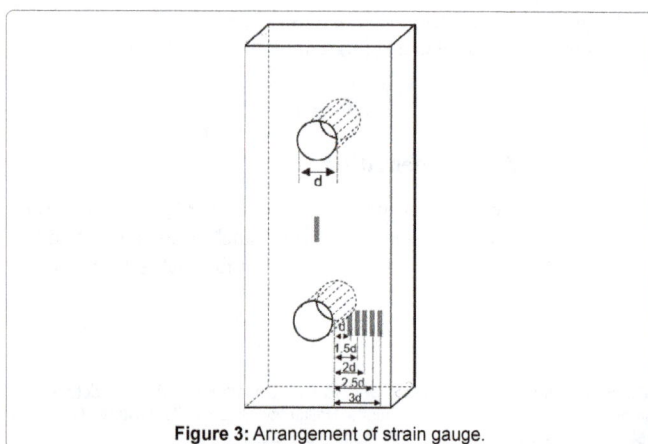

Figure 3: Arrangement of strain gauge.

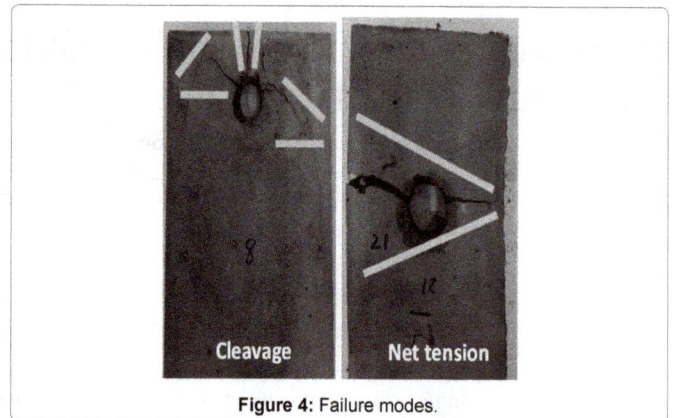

Figure 4: Failure modes.

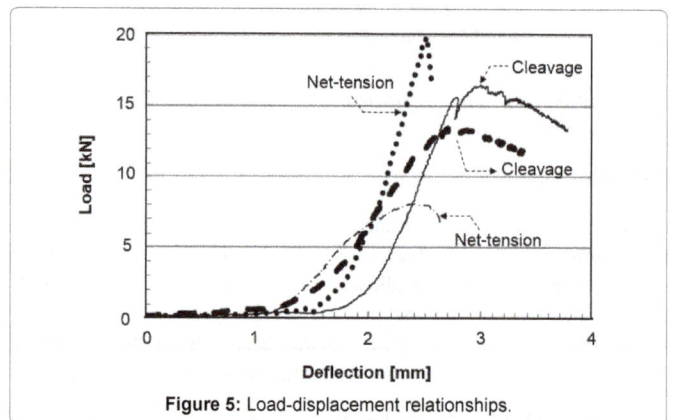

Figure 5: Load-displacement relationships.

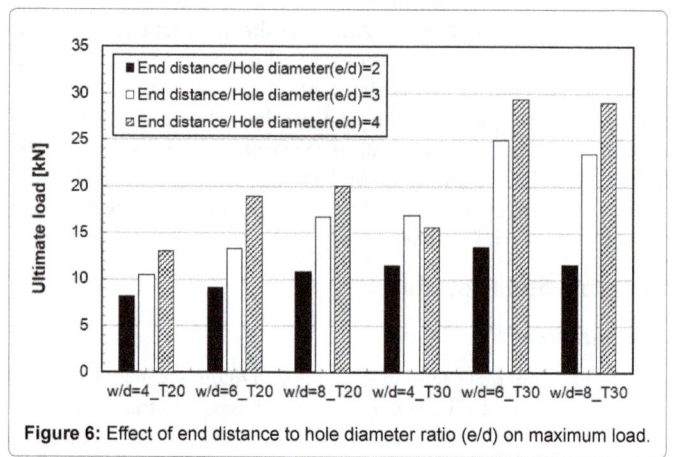

Figure 6: Effect of end distance to hole diameter ratio (e/d) on maximum load.

Results and Discussion

The specimen information and test results are summarized in Table 2. The test results are analyzed by divide into three categories, failure modes, load-displacement characteristics, and Strain concentration factor.

Failure modes

UHPC panels with bolted connections were found to exhibit two modes of failure, such as cleavage and net-tension failures, due to the tensile loading (Figure 4). The cleavage failure was characterized by large cracks from the hole to the nearest side of the panel; then two additional cracks occur in parallel, which improve the ductility of the panel. The widths and lengths of these cracks increase until the maximum load is reached. On the other hand, the net-tension failure was caused by cracks occurring perpendicular to the loading direction through the hole.

Load-displacement characteristics

Figure 5 shows typical load displacement relationships of this study. The displacement of each panel was determined as the average values measured from the two LVDTs. The initial displacement (about 0 mm to 1-1.5 mm) occurred without load resistance because the load was not transferred to the panel through the bolt, where the bolt slipped off the plates. Compared with net-tension failure, cleavage failure showed more ductile behavior, i.e., the load decreased more slowly after the peak load. On the other hand, in the case of net-tension failure, the load

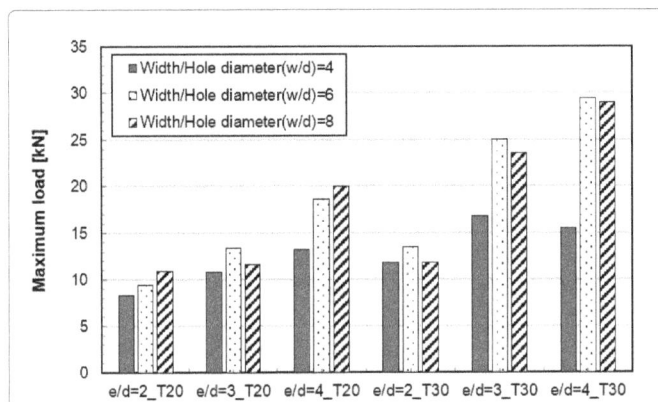

Figure 7: Effect of panel width to hole diameter ratio (w/d) on maximum load.

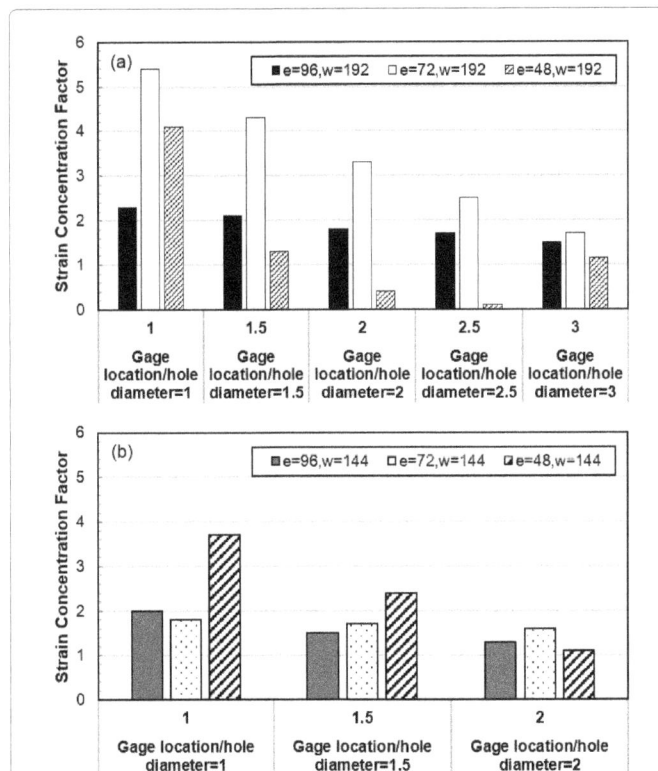

Figure 8: Effect of end distance to hole diameter ratio (e/d) on strain concentration factor.

suddenly dropped, which means the brittle failure. To quantitatively evaluate the load redistribution capability of each failure mode, the ductility ratio was estimated. This was determined as the displacement at 80% of the maximum load in the post peak state to the displacement at the maximum load. Based on this, it was found that the ductility ratio is higher when the UHPC panel shows cleavage failure than when it shows net-tension failure; especially, in the case of net-tension failure, this ratio is less sensitive depending on specimen geometry compared with cleavage failure.

Strength of panel joint based on geometric parameter

Figure 6 shows the effect of e/d on the maximum load of the connection of bolted UHPC panels. Only results for the same width and thickness were presented in this figure. As expected, e/d was found to be a factor affecting failure mode. One notable result is that when w/d was 4 at t=30 mm, the failure mode was changed from cleavage failure to net-tension failure as e/d increases. This change in failure mode was also directly related to the sudden change in the maximum load and the strength. Without change of the failure mode, the strength was increased with increasing e/d.

The maximum load as a function of w/d is shown in Figure 7 which includes only the results of the specimens with the same end distance. Changes in panel width at the bolted joints were found to be the factors that affect the failure mode as well as its structural performance. It is confirmed that the final load increases as w/d ratio increases. In addition, it is also confirmed that as the e/d increases, the failure mode changes from net-tension fracture to cleavage fracture when the end distance is short (w/d=4 at t=20).

Based on Figures 6 and 7, it is confirmed that increasing the panel thickness in the selected thickness range (20-30 mm) is not directly linked to the increase in joint strength of the UHPC facade. In practice, this result is especially important because the structural performance of a facade is often determined by its connections. In other words, the method of increasing the panel thickness for a conservative design is not always guaranteed to increase its structural performance, but rather it is desirable to change the geometrical parameters, as presented in this study. These experimental results can help to design the joints of UHPC facades economically and efficiently.

Strain concentration around hole

The strain distribution of the selected six panels subjected to tension was analyzed. The purpose was to verify the strain concentration around the bolt hole. The results are presented in Figures 8 and 9 on the basis of geometric parameters which were set in this study. In these figures, the strain concentration factor is defined as the ratio of the strain near the hole to the strain at a distance where the concentration does not occur. In all specimens, actually the concentration phenomenon was found, i.e., the closer location from the hole, the higher the strain concentration factor is.

It is also confirmed that this factor increases when the width of the panel increases (compare Figures 8a and 8b). The change in edge distance also affected the strain concentration factor (compare Figures 9a and 9b). However, the degree of the change was not as critical as the case of panel width. Overall, by analyzing the strain distribution as a function of the geometric parameters, it was confirmed that the strain distribution and thus stress concentration of the connection of UHPC panel depend on sensitively the panel width.

Conclusion

The strength and failure mode characteristics of the bolted joint

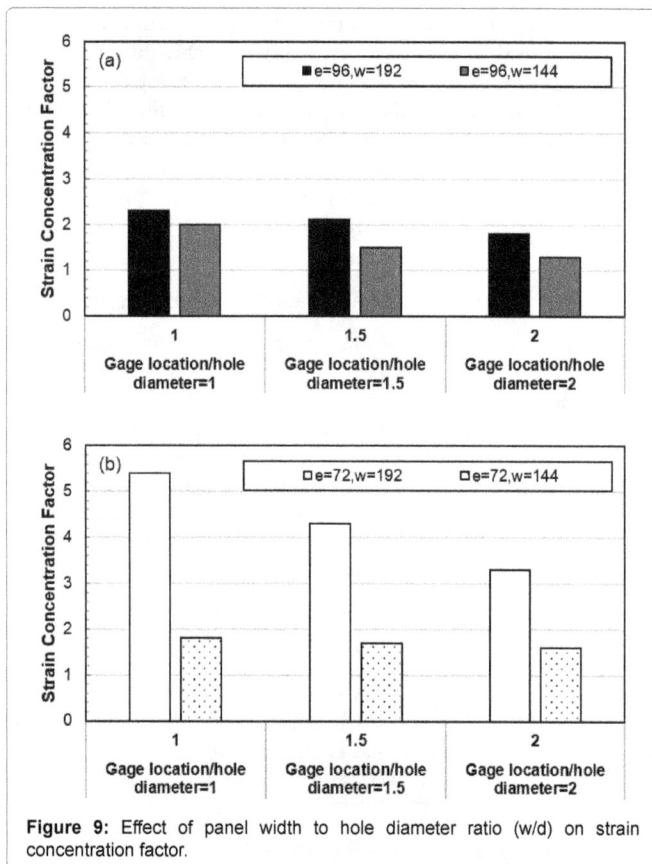

Figure 9: Effect of panel width to hole diameter ratio (w/d) on strain concentration factor.

of UHPC panel were investigated based on geometric parameters. Above all, to design the panel safely, it is necessary to understand the complex material and structure behavior of UHPC, not simply increase the thickness of the panel. The major findings of this study are summarized as follows.

The panel joints that were subjected to direct tensile load via bolt showed two types of failure modes. They behaved more ductile when showed cleavage failure than when showed net-tension failure. Therefore, it is desirable to intend cleavage failure to increase the ductility. Increasing the edge distance, width, and thickness can be effective in increasing the joint strength. However, this increase should be preconditioned that there is no change in failure mode (e.g., cleavage failure to net-tension failure). The change of failure mode was also affected by the geometric parameters. Lastly, the stain concentration around the bolt hole was confirmed by the attached strain gauge and changed sensitively with the panel width.

Acknowledgment

The Institute of Engineering Research in Seoul National University provided research facilities for this work.

References

1. Tayeh BA, Bakar BA, Johari MM (2013) Characterization of the interfacial bond between old concrete substrate and ultra-high performance fiber concrete repair composite. Materials and Structures 46: 743-753.

2. Brühwiler E, Denarié E (2008) Rehabilitation of concrete structures using ultra-high performance fibre reinforced concrete. Proceedings of the 2nd International Symposium on UHPC: 5-7th March, Germany, p: 895-902.

3. Habel K, Denarié E, Brühwiler E (2006) Structural response of elements combining ultra high-performance fiber-reinforced concretes and reinforced concrete. J Structural Engineering 132: 1793-1800.

4. Kang SH, Gyephel T, Hong SG, Moon J (2015) Effect of water-entraining admixtures on the hydro-mechanical properties of ultra-high performance concrete. 14th International Congress on the Chemistry of Cement.

5. Kang SH, Hong SG, Moon J (2016) Influence of internal curing on autogenous and drying shrinkages of ultra-high performance concrete considering heat treatment. FIB Symposium, Maastricht, The Netherlands.

6. Kang SH, Hong SG, Kwon YH (2017) Effect of permanent formwork using ultra-high performance concrete on structural behaviour of reinforced concrete beam subjected to bending as a function of reinforcement parameter. J Applied Mechanical Engineering.

Fabrication and Experimental Analysis on L/D Ratio of Vortex Tube

Taparia N*, Ritesh Kumar C, Kanwar L and Verma D
Department of Mechanical engineering, SRM University, India

Abstract

Now-a-days, the first and foremost important quality of any research or development is its eco-friendly nature. As we know environment safety has become an important aspect of the industries and people in common. Vortex tube is also known as non-conventional cooling device that will produce cold air and hot air when we passed compressed air from compressor without causing any harm to the nature. In vortex tube, when a compressed air from compressor is tangentially passed into vortex chamber a free and forced vortex flow will be generated which will be divide into two air streams i.e., one has lower temperature than inlet temperature and other one has higher temperature than inlet temperature. It can be used for any type of spot cooling application. In this research an attempt is made to fabricate and test a simple vortex tube. The effect of change in length and diameter of vortex tube i.e., (L/D) ratio is investigated and presented in this paper.

Keywords: Vortex tube; Orifice; Inlet pressure; Tangential nozzle

Introduction

General refrigeration systems use very harmful refrigerants, as they are the major cause for depleting ozone layer. Major research work is focussing on to eliminate the use of very harmful refrigerant but side by side we are focussing on their efficiency or c.o.p.

The vortex tube is also known as non-conventional cooling device which was invented by scientist named RANQUE so it is also called as ranque tube, which is a remarkably a simple device, reliable and produces 2 air stream i.e., one has higher temperature than inlet temperature (hot air) and other one has lower temperature than inlet temperature (cold air) simultaneously from the compressed gas. It can be used for spot cooling applications like drilling, welding etc. The working mechanism of vortex tube can be observed physically, but tough to explain, as there is no perfect explanation to explain the phenomenon. In vortex tube compressed air is sent at an angle (tangentially) through the nozzle. The circular or vortex motion of compressed air inside the vortex tube is produced by swirl generators. As the vortex drives inside the tube, 2 temperature streams are formed. The high temperature stream moves along the tube circumference while low temperature stream is in the inner core axis. The higher temperature stream is then allowed to release through the control valve located at one end of the tube and the lower temperature stream is then reflected back from the control valve and released through another end of the tube. This results temperature distribution inside the vortex tube [1-6].

Vortex Tube

The Vortex tube which is a mechanical device working as one of the non-conventional refrigerating or cooling machine without any movable part, by dividing a compressed air into a low temperature region than inlet temperature region and a high temperature region than inlet temperature region. Division of 2 different temperatures is known as temperature separation effect. The air coming from the hot end can be around 190°C, and the air coming from the "cold end" can be around -50°C. When compressed gas is passed at an angle (tangentially) into the vortex tube chamber through the nozzle, a circular flow is produced inside this chamber. When the air circulates inside the centre of this chamber, it is aggrandized and cooled. In this chamber, part of the air circulates to the hot end, and another part coming out through the cold side. Component of the air in the vortex tube reverses from the control valve and moves from the higher temperature end to the lower temperature end. And hence at the control valve side the air released with a higher temperature as compared to inlet temperature, while at another end, the low temperature air is released (Figure 1).

Working

When compressed air is passed from compressor through the nozzle we have made holes at an angle (tangentially) and then it passes through that hole with high velocity and hence, air moves towards the control valve through the hot side pipe. Then 2 types of vortex flow is created in this chamber i.e., free and forced one and air travels in circular like motion along the circumference of hot side. As the air travels in the hot side pipe towards the control valve by using the opening of this valve the flow can be restricted. When the pressure inside the hot side tube near to the valve is greater than outside atmospheric pressure by using the valve, then reversed flow starts through the central portion of

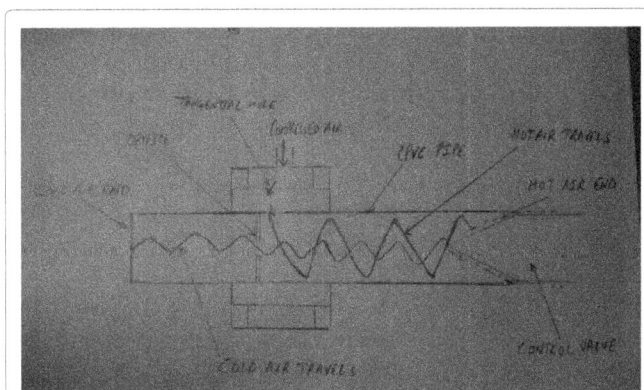

Figure 1: Vortex tube.

*Corresponding author: Taparia N, Assistant Professor, Department of Mechanical engineering, SRM University, India
E-mail:nikhiltaparia123@gmail.com

hot side pipe from high pressure region to low pressure region. In this process, energy interaction takes place between these 2 streams (i.e., reversed and forward) and therefore air stream through the central portion has lower temperature than the inlet whereas the air stream in forward direction has higher temperature than inlet temperature. Then, the air stream which gets back towards the diaphragm and when it passes through the hole present in this diaphragm hole the air gets cooled and comes out from the cold side pipe, while another air stream is passed through the opening of the valve. With the change in control valve openings, we can get various values of temperature difference at cold end and hot end [6-8].

Phenomenon of Energy Transfer

As we know many researchers has a proposed theory for the energy exchange phenomenon between the hot stream and cold air stream in vortex tube. But no theory has been able to explain this unconventional energy interaction.

Van Deemter gave an explanation that wen compressed air passes into the tube it creates 2 type of vortex flow i.e., free and forced vortex motion. Centrifugal acceleration helps the vortex to move along the circumference. Now as this vortex move towards the end where we have placed the control valve, rotation of the air almost stops. So there is the generation of stagnation conditions in the tube. Due to this pressure inside the tube starts to rise and when it becomes greater than the outside pressure i.e., atmospheric pressure, a flow in reversed direction is generated. This flow moves to the opposite direction of original flow from vortex chamber. This reversed flow is also known as forced vortex flow [9,10].

The velocity of the free vortex decreases as it moves from nozzle to valve at end, which causes the sliding between 2 adjacent planes, which are moving towards the valve.

An another hypothesis was given by Prof. Parulekar which states that turbulent mixing in centrifugal fluids cause the energy transfer from low pressure to high pressure.

When compressed air is passed to the vortex chamber it will produce a free vortex motion of air. Now this air will flow toward the end of tube where we have a control valve in the shape of conical. As per theory given by Van Deemeter, stagnation condition is created at the end. Now when pressure near the valve rises with more than outside atmospheric pressure, a reversed flow is generated. The free vortex which moves in forward direction will influence this reversed flow. Then, it will create a reversed flow which is known as forced vortex flow. Energy is needed to create this forced vortex which is provided by free vortex. This amount of energy transferred by free vortex is less than the amount of energy transferred from inner core to the circumference. Hence, we will get the air with less temperature than inlet temperature at one end and the air with more temperature than inlet temperature at another end where we have placed control valve [11,12].

These are the 2 hypothesis by which we can understand how energy interacts and how we will get the cold air and hot air in 2 different ends.

Parts of Vortex Tube

Nozzle

It is a mechanical device that is used to increase the velocity of any stream by decreasing its area at the end and it is placed at the end of any pipe or hose.

Diaphragm

It is a circular in cross-section with very small thickness having a small hole of specific diameter at the centre. Air stream travelling through the core of hot side is emitted to the diaphragm hole and hence it gets expanded and gets cooled.

Valve

Valve is used to obstruct the flow of air at the hot side so that by controlling it we can get cold air on the other end of tube and it is also used to check the quantity of hot air through vortex tube.

Hot air side

In the Vortex tube, this is cylindrical in cross section and it can be varied by taking different lengths as per experiment.

Cold air side

In the Vortex tube, this is also in cylindrical cross-section through which cold air is passed and its length is depending on the length of hot side.

Fabrication and Analysis

Chlorinated polyvinyl chloride (CPVC): It is a thermoplastic which can be made by chlorination of polyvinyl chloride i.e., PVC resin. It can be used for common replacement for metal piping because its strength, easy installation and low cost. CPVC which is a thermoplastic material i.e., it can be moulded into different shapes to create pipes, valves and other liquid handling supplies. CPVC can withstand a wider range of temperatures. Temperature over 60 degree Celsius can softening of material and weakening of joints but CPVC can handle temperature up to 94 degree Celsius. The more resisting properties of CPVC make it useful for various applications. CPVC exhibits fire-retardant properties (Figure 2).

Constructions Details

We have designed 2 different vortex tube by taking different length for hot and cold side and different diameter of pipes so that we will get 2 length and diameter ratio i.e., L/D (Table 1).

Experimental Procedure

At first we run the compressor for 15 mins to get stable pressure. Throughout our experiment we have maintained the input conditions constant. As we pass the pressurized air into the vortex chamber through the holes that are made at an angle i.e., tangentially then, in this chamber first air circulates and enter through the holes that are made in the hot side pipe and air moves towards the hot end where we have setup of control valve. As the air reaches near the control valve and when the pressure of the air becomes greater than the outside

Figure 2: CPVC exhibits fire-retardant properties.

atmospheric pressure by partially closing the control valve, a reversal axial flow takes place through the core of hot side which starts from high pressure region to low pressure region and hence, heat interaction takes place between reversed stream and forward stream. Therefore, air stream passing through the core gets lower temperature than the inlet temperature while air stream which is moving towards the control valve gets higher temperature than inlet temperature. With the change in control valve openings, we can get various values of temperature difference at cold end and hot end.

Analysis

For given L/D ratio of pipe A and pipe B, here are the temperature and pressure variation given as follows (Table 2a-2d):

For pipe A: L/D is 19.68 mm

For pipe B: L/D is 31.49 mm

Result and Conclusion

We have taken 2 pipes i.e., pipe A and pipe B and their L/D ratios are 19.68 and 31.49 respectively. As from the graph no 1 and 2 we can

see that the cooling effect depends upon L/D ratio. As we increases L/D ratio the pipe delivers more cooling effect and we also notice that as we increases the pressure the cooling effect increases (Figures 3-6).

In graph no 3 and 4 we have seen the change in temperature from initial stage at both the end for an individual pipe separately by varying the pressure.

$$\Delta T_C = T_I - T_C$$

$$\Delta T_H = T_H - T_I$$

Where,

$T_I =$ Inlet temperature

$T_C =$ Cold end temperature

Parameter	Pipe A	Pipe B
Length of hot side (MM)	400	450
Length of cold side (MM)	100	150
Diameter of hot side (MM)	25.4	19.05
Diameter of cold side (MM)	25.4	19.05
L/D Ratio (MM)	19.68	31.49
Diameter of orifice (MM)	12.7	9.525

Table 1: Hot and cold side and different diameter of pipes.

S.No	Pressure (BAR)	Temperature at cold end (deg. celsius)	Temperature at hot end (deg. celsius)
1	4.0	24.8	38.1
2	4.5	23.2	39.8
3	5.0	22.1	40.1
4	5.5	21.4	41.9

(a)

S.No	Pressure (BAR)	$\Delta T_C = T_i - T_c$ (DEG. Celsius)	$\Delta T_H = T_h - T_i$ (DEG. Celsius)
1	4.0	7.2	6.1
2	4.5	8.8	7.8
3	5.0	9.9	8.1
4	5.5	10.6	9.8

(b)

S.No	Pressure (BAR)	Temperature at cold end (deg. celsius)	Temperature at hot end (deg. celsius)
1	4.0	21.7	41.5
2	4.5	20.3	43.2
3	5.0	19.1	45.8
4	5.5	18.5	47.1

(c)

S.No	Pressure (BAR)	$\Delta T_C = T_i - T_c$ (DEG. Celsius)	$\Delta T_H = T_h - T_i$ (DEG. Celsius)
1	4.0	10.3	9.5
2	4.5	11.7	11.2
3	5.0	12.9	13.8
4	5.5	13.5	15.1

(d)

Table 2: Analysis for given L/D ratio of pipe A and pipe B are the temperature and pressure variation.

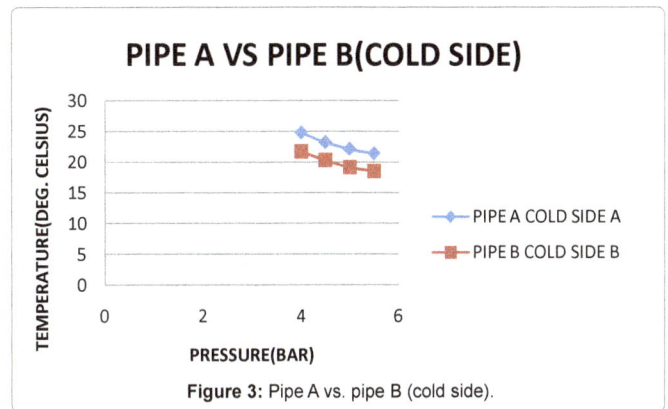

Figure 3: Pipe A vs. pipe B (cold side).

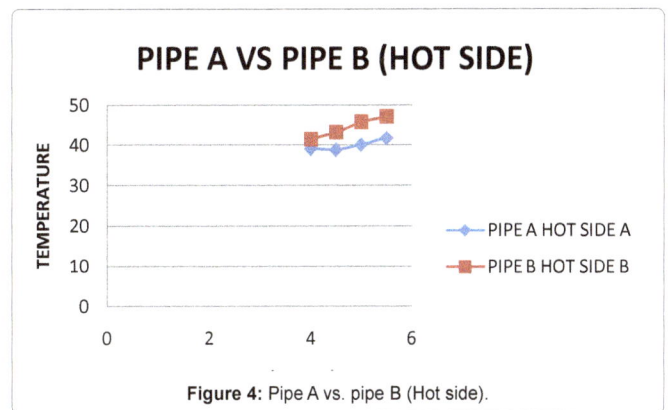

Figure 4: Pipe A vs. pipe B (Hot side).

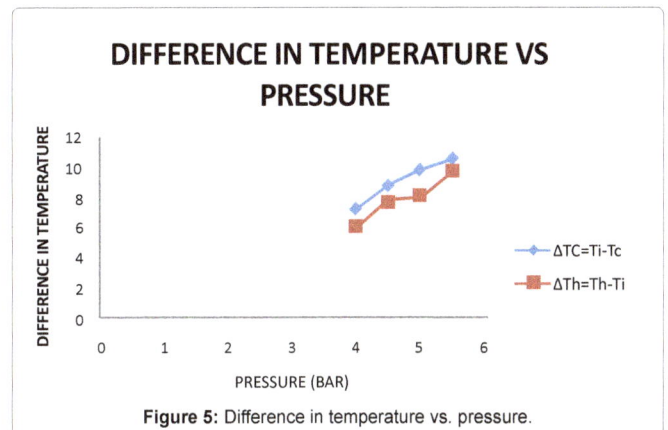

Figure 5: Difference in temperature vs. pressure.

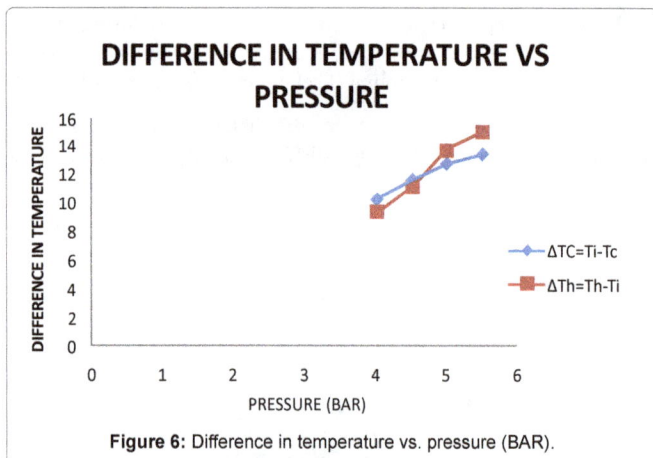

Figure 6: Difference in temperature vs. pressure (BAR).

T_H = Hot end temperature

As we increase the pressure the ΔT_C and ΔT_H gets increase.

Conclusion

From the above analysis we can conclude that as we increases the L/D ratio, we can get more temperature difference between cold end and hot end side. The maximum temperature difference in pipe A between cold end and hot end with initial temperature of the air is 10.6 and 9.8 respectively. The maximum temperature difference in pipe B between cold end and hot end with initial temperature of the air is 13.5 and 15.1 respectively.

References

1. Prabakaran J, Vaidyanathan S (2010) Effect of orifice and pressure of counter flow vortex tube. Indian J Sci technology.

2. Gao (2005) Experimental study on a simple Ranque Hilsch vortex tube. Cryogenics 45: 173-183.

3. Stephan K, Lin S, Durst M, Huang F, Seher D (1983) An investigation of energy separation in a vortex tube. Int J Heat and Mass Transfer 26: 341-348.

4. Ranque GJ, Domkundwar (2014) Experiments in expansion in a vortex tube with simultaneous exhaust of hot air and cold air refrigeration and air-conditioning.

5. Kargaran M, Farzaneh M, Hagighat-Hosini SJ, Arabkoohsar A (2007-2012) Experimental investigation the effects of orifice diameter and tube length on a vortex tube performance.

6. Pourmahmoud N, Bramo AZ (2011) The effect of l/d ratio on the temperature separation in the counter flow vortex tube.

7. Yilmaz M, Kaya M, Karagoz S, Erdogan S (2009) A review on design criteria for vortex tubes. Heat and Mass Transfer 45: 613-632.

8. Desai PS (2004) Refrigeration and air conditioning. Khanna Publishers.

9. Thipse SS (2007) Refrigeration and air conditioning. Jaico Publishing House.

10. Chang Q, Li Q, Zhou G (2011) Experimental investigation of vortex tube refrigerator with a divergent hot tube. Int J refrigeration 34: 322-327.

11. Newton R, Shackelford JS (2012) Hilsch vortex tube cooling/heating .

12. Takahama H (1965) Studies on vortex tubes. Bulletin of JSME 8: 433-440.

3

Concrete Stress-Strain Characterization by Digital Image Correlation

Saldaña HA, [1], Márquez Aguilar PA[2]* and Molina OA[2]

[1]*School of Chemical Sciences and Engineering is also from the Autonomous University of the State of Morelos*

[2]*Research Center for Engineering and Autonomous University of the State of Morelos Av. University # 1001, Col. Chamilpa. Cp. 62209. Cuernavaca Morelos, Mexico*

Abstract

In mechanics of materials it is important to know the stress-strain relation of each material in order to understand their behaviour under different loads. Concrete is one of the most used materials in structural mechanics and they are always under axial loads. This work is implemented using one beam and the speckles created by its reflection. Strain field measurement with noninvasive techniques is needed in order to sense rough-like materials. We present an experimental approach that describes the mechanical behavior of structural materials under compression tests, which are done in a universal testing machine. In this work we show an evaluation of the in-field measurements obtained by digital image correlation allowing us to evaluate the heterogeneous strain evolution observed during these test.

Introduction

Stress-strain diagrams are very important in understanding the behaviour of materials under load [1]; they show the elastic, plastic and rupture part of materials. There are two methods to obtain the diagram: the invasive methods in which mechanics take advantage doing physical tests as the test-tube in which is placed a small piece of the probe and it is applied a specific load to it, then is measured the deformation in displacement [2]. Optical methods are also used as an invasive way to determine residual stress, in-field displacements and strain, in which the most used technique, is the hole-drilling method, created in 1930 by Mathar [3], nowadays this technique is standardized by ASTM [4]. When a laser light felt on a matt surface such as paper or rubber-like material, a speckle pattern is formed and a high-contrast grainy pattern will be reflected. This effect was called egranularity by Rigden and Gordon [5]. The simplest image matching procedure is cross-correlation (CC) that can be performed either in physical space [6,7]; in Fourier space [8] by using Fast Fourier Transforms (FFT) one can evaluate the CC function very quickly. Anuta [9] took advantage of the high speed of FFT algorithm doing digital multispectral and multi-temporal statistical pattern-recognition. Kuglin and Hines [10] observed that information about the displacement of one image with respect to another is included in the phase component of the cross-power spectrum of the images. Correlation technique was applied analysing the strain evolution of a rubber-like material under multi-axial stresses [11]; also for strain measurements using a Vic 2D system in order to obtain the strain diagram of a compound material [12] and it is also applied in resistance heating tensile test [13].

Concrete is one of the most used materials in structural mechanics, in 1988 Rots and Borst loaded a piece of concrete in a tension test and they got around 3 N/mm² and 0.02% stress-strain respectively [14], this material was used for the development of techniques in modelling the seismic response [15]; Wasantha found a relation between stress-strain for concrete under heavy vehicle loadings [16].

The development of non-destructive testing methods is the main challenge for the assessment of structural elements in existing constructions. In this work we measure in-plane strain of a common structural material subjected to a continuum axial load. A compression test is done to analyse the strain evolution of our sample and Digital Image Correlation (DIC) combined with CC method are done in order to obtain in-plane strain deformation, in which we use one laser beam impinging our material in its cross-section during the compression test.

From these experimental tests, we correlate the stress-strain diagram with in-field strain measurements, also it is included the accuracy and mean errors of the tests.

Developments

Speckles statistics

The random intensity distribution is called speckle pattern, which is formed when coherent light is reflected from a rough surface or when light is propagated through a medium with random refractive index fluctuations [17]. In general the statistical properties of speckle patterns depend both on the coherence of the incident light and the detailed properties of the random surface or medium.

The surfaces of most materials are extremely rough on the scale of an optical wavelength ($\lambda \approx 5x10^{-7}$ meters). When monochromatic light is reflected from such a surface, the optical wave resulting at any distant point consists of many coherent components; the interference of all the components is then called the speckle pattern.

Let a speckle signal be $u(x,y,z;t)$ at an observation point (x,y,z) and time t with u as a fully monochromatic wave, the analytic signal takes the form:

$$u(x,y,z;t) = A(x,y,z)exp^{(i2\pi v t)} \qquad (1)$$

where v is the optical frequency and A represents the phasor amplitude of the field, which is a function of space [18]. When a speckle pattern arises by free-space propagation, the amplitude of the electric field at a given observation point (x,y) consist in every de-phased contributions from all different scattering regions of the rough surface, then the

*Corresponding author: Márquez Aguilar PA, Research Center for Engineering and Applied Science University of the State of Morelos Av. University # 1001, Col. Chamilpa. Cp. 62209. Cuernavaca Morelos, Mexico
E-mail: pmarquez@uaem.mx

amplitude $A(x,y,z)$ is represented as a sum of every contribution $k = 1, 2....., N$:

$$A(x,y,z) = \sum_{k=1}^{\infty} \frac{1}{\sqrt{N}} a_k(x,y,z) = \sum_{k=1}^{\infty} \frac{1}{\sqrt{N}} a_k exp^{(i\varphi_k)} \quad (2)$$

Equation (2) must follow some statistical properties [17,18]: The amplitude $a_k / \sqrt{N}$ and the phase φ_k of each element are statistically independent of each other. The phases are uniformly distributed between ($\pi, -\pi$).

In order to study the intensity distribution of the resultant field is necessary to take the real and imaginary part of Eq. (1):

$$A^{(r)} = \frac{1}{\sqrt{N}} \sum_{k=1}^{\infty} |a_k| \cos \varphi_k \quad (3)$$

$$A^{(i)} = \frac{1}{\sqrt{N}} \sum_{k=1}^{\infty} |a_k| \sin \varphi_k \quad (4)$$

When assumptions of the statistical properties are considered it is possible to stablish:

$$[A^{(r)}]^2 = \frac{1}{N} \sum_{k=1}^{\infty} \sum_{m=1}^{\infty} |a_k| |a_m| (\cos \varphi_k \cos \varphi_m = \frac{1}{N} \sum_{k=1}^{\infty} \frac{[|a_k|]^2}{2} \quad (5)$$

$$[A^{(i)}]^2 = \frac{1}{N} \sum_{k=1}^{\infty} \sum_{m=1}^{\infty} |a_k| |a_m| (\sin \varphi_k \sin \varphi_m = \frac{1}{N} \sum_{k=1}^{\infty} \frac{[|a_k|]^2}{2} \quad (6)$$

thus is seen that real and imaginary parts of complex field have identical variances and are uncorrelated. Generally N is a number extremely large so real and imaginary parts can be expressed by (3, 4) as sums of a very large number of independent random variables. It follows from the central limit theorem [17] that as $N \rightarrow \infty$, A^r and A^i are asymptotically Gaussian. Coupling this fact with the results of (5, 6), the joint probability density function of the real and imaginary parts of the field is:

$$P_{r,i}(A^{(i)}, A^{(i)}) = \frac{1}{2\pi\sigma^2} exp[\frac{([A^r]^2 + [A^i]^2)}{2\sigma^2}] \quad (7)$$

where

$$\sigma^2 = \frac{1}{N} \sum_{k=1}^{\infty} \frac{[|a_k|]^2}{2} \quad (8)$$

This kind of density function is commonly known as a circular Gaussian density function. One of the most important parameters measured in optics is the intensity, thus according from the known statistics of the amplitude is necessary to find the corresponding statistical properties of the intensity. Therefore the intensity I and the phase θ of the field are related to the real and imaginary parts of the amplitude [19]:

$$A^{(r)} = \sqrt{I} \cos \theta \quad (9)$$

$$A^{(i)} = \sqrt{I} \sin \theta \quad (10)$$

Substituting equations (9) and (10) in (7) we can observe that the probability distribution of intensity would be:

$$P_{i,\theta}(I,\theta) = \frac{1}{4\pi\sigma^2} exp[\frac{-I}{2\sigma^2}] \quad (11)$$

Image correlation

Digital image correlation is an optical method that uses a mathematical correlation analysis to examine digital image data taken while samples are in mechanical tests. This technique consists on capture consecutive images with a digital camera during the deformation period in order to evaluate the change in surface characteristics and understand the behavior of the specimen while it is subjected to an increasing load.

It is well known that for cross correlation technique sequenced images are needed. We consider a plane covered by a speckle pattern as the image intensity at time t_0 defined by $f(x,y)$, where the two dimensional FFT is:

$$F[f(x,y)] = F(u,v) = \iint_{-\infty}^{\infty} f(x,y)exp^{-i2\pi(ux+vy)} dxdy \quad (12)$$

A second image is considered where the intensity is in a time $t_1 = t_0 + \Delta t$ defined by $g(x,y)$; assuming $g(x,y)$ as a translation of $f(x,y)$ we have: $g(x,y) = f(x-x_0, y-y_0)$. Cross-spectrum $I(u,v)$ can be defined by the multiplication of two signals, considering F the Fourier transform of the first and G the Fourier transform of the second and * the FFT complex conjugate we get:

$$I(u,v) = F(u,v) \bullet G^*(u,v) = F(u,v) \bullet F(u,v)exp^{i2\pi(ux_0+vy_0)} \quad (13)$$

in Eq. 13 it can be seen that the displacement can be expressed according to two components corresponding to two directions; therefore the phase can be decomposed into two expressions with a variation between $-\pi$ and $+\pi$ [13].

In order to get cross correlation we need to take the inverse Fourier transform of the cross spectrum. Lets call U and V a pair of 2-D images, where U represents the reference image and V the deformed image, it is taken the FFT form both images and cross-correlation is defined by:

$$CC = F^{-1}[F(U) \bullet F^*(V)] \quad (14)$$

where F implies 2-D Fourier transform, F^{-1} implies 2-D inverse Fourier transform. The use of FFT requires that images U and V are the same size and have dimensions that are powers of 2. In the present work, we analyze $2^{10} x 2^{10}$ images in order to map a bigger area of the whole image and the shift $\delta x (= \delta y)$ between two consecutive images is 128 pixels in order to get a better result. These two parameters define the mesh formed by the images used to describe the displacement field [11].

Compresion tests

Physical tests are used in order to know mechanical properties of materials and compression test is one of these tests which enable the user to understand the behavior of a material under a continuous axial load; from this test we obtain the stress-strain diagram [2]. In this work we made a mixture of sand-cement with a ratio of 3 × 1 respectively with dimensions of 5 × 5 × 4.5 cm. and they undergo into compression test according to ASTM E-9 [20]. The tests were performed with a speed ratio of 0.5 mm/s up to 3 GPa approx. Since we are working on the elastic part of the diagram, therefore we can apply Hooke's law:

$$\sigma = E \bullet \varepsilon \quad (15)$$

where σ is the stress, E is the Young modulus of the material: 2.2 GPa experimentally obtained by the universal machine and ε is the dimensionless deformation.

In DIC procedure, when the same reference picture is used, it is not possible to measure large displacements in a sequence of pictures, but when they remain small enough it is possible to assume the first image as the reference for the whole analysis. In the present work, we work with deformations less than two millimeters. It is well-known that the infinitesimal strain tensor ε_c is well adapted to small displacements and it can be evaluated as:

$$\varepsilon_c = \frac{1}{2}(R + R^T) - 1 \quad (16)$$

where ε_c is the strain, R is an orthogonal second rank tensor and R^T implies transpose.

Theoretical analysis

As it is well-known, laser beams propagates a unidirectional wave with some divergence and with a finite cross-section; the most common output for such a beam is Gaussian in the TEM^{00} mode, which its intensity distribution is:

$$I(r) = I_0 exp[\frac{-2r^2}{w^2}] \tag{17}$$

where $r = (x^2 + y^2)^{1/2}$ and w is the spot size and depends on the z-coordinate [21]. The intensity profile is a Gaussian and it maintains its profile while it propagates. The Gaussian beam equation is deduced from Helmholtz equation and is represented by [22]:

$$E(r,z) = A\frac{w_0}{w(z)}exp[\frac{-r^2}{[w(z)]^2} + \frac{kr^2}{2R(z)} + kz - \eta(z)] \tag{18}$$

where w_0 is the beam waist, $w(z)$ is the beam spot size, $R(z)$ is the curvature radius of the spherical waves and $\eta(z)$ is the beam phase angle [23]. And the intensity distribution of a Gaussian Beam is:

$$I(r,z) = I_0 \frac{w_0^2}{[w(z)]^2}exp[\frac{-2r^2}{[w(z)]^2}] \tag{19}$$

Gaussian beams are able to pass through different media; the light reflection occurs when it arrives to the boundary separating two media of different optical densities and some of the energy is reflected back into the first medium [24], taking this outset, our laser-beam strikes a rough material and the reflection can be studied as a speckle pattern. The ratio between the intensity of the reflected beam and the incident beam is called reflectivity and is expressed by:

$$R = \frac{I_r}{I_i} \tag{20}$$

where I_r and I_i are the reflected and incident beams respectively and when a beam pass through the media, there exist transmissivity and according to the conservation law of energy [25]:

$$T + R = 1 \tag{21}$$

In the present work, as we are working with solid materials, the transmissivity is zero. Therefore, we can assume that:

$$I_i \bullet R = I_r \tag{22}$$

for R ≤ 1 and according to cross-correlation analysis Eq. (14), taking U and V as speckle patterns and using Eq. (11), expanding formally and changing coordinates we obtain

$$F(u,v) = \iint\limits_{-\infty}^{\infty} \frac{1}{4\pi\sigma^2}exp[\frac{-(x^2+y^2)}{2\sigma^2}]exp^{-i2\pi(ux+vy)}dxdy \tag{23}$$

$$U(\rho:t) = \frac{1}{4\sigma^2}exp[\frac{\pi^2\sigma^2}{2}\rho^2] \tag{24}$$

Being $U(\rho:t)$ the first image Fourier transform at an initial time t_0; the same process is done for $V(\rho:\Delta t)$, where V is the second image Fourier transform at a time t_1 and cross correlation is defined by:

$$CC = [\frac{1}{4\sigma^2}]^2 exp^{i2\pi(\Delta x+\Delta y)} \tag{25}$$

Finally it is taken the correlation phase from the exponential and infinitesimal strain tensor takes place obtaining in-field measurements. From Eq. (25) is possible to see that the relation will keep a Gaussian form, taking this outset, it is assumed that this problem is similar to a common reflection, where the scattered area will decrease and such decrement will be proportional to the deformation.

Experimental Setup and Data Processing

Three probes were made by adding 3×1 sand-cement mixture and they were placed in the universal testing machine in order to begin the compression tests. A diagram of this method is shown schematically in Figure 1a. The output of a Diode-Pumped Solid-State laser (L) with a wavelength λ=532 nm and power of 200 mW [26], is propagated through a positive lens (l) which is placed in front of the sample (Sa) in order to irradiate the cross-section face and the scattering reflection impacts a screen (Sc) which is placed aside the laser beam; the material is first completely flat and it is compressed by the machine (M). During the compression tests, the speed of the compression load was 0.5 mm/s with duration of 20 minutes approximately and while they were taking place the speckle reflection was recorded with a high resolution video camera (Vc) (Figure 1b).

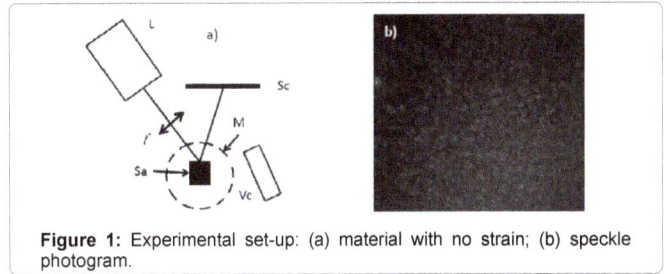

Figure 1: Experimental set-up: (a) material with no strain; (b) speckle photogram.

Once the video is recorded, it is divided into thousands of photograms in order to load each image and process it. A program is written in Matlab for Digital Image Correlation (DIC). The cross-correlation calculation is described below: in Figure 2a it is shown how the code loads the image and it is converted into a gray-scale image as it is shown in Figure 2b; the image is crop into a 1024×1024 in order to begin the FFT analysis of all the images, taken the first as the reference; from the second image and forward, they are considered as deformed images.

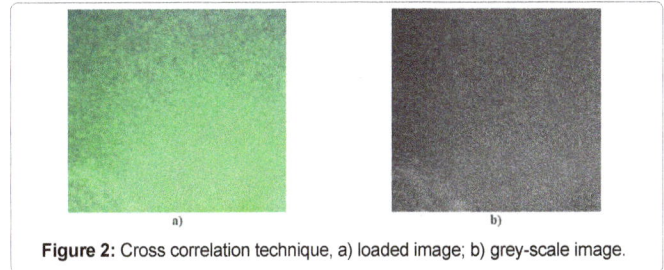

Figure 2: Cross correlation technique, a) loaded image; b) grey-scale image.

Once it is done the cross correlation technique we obtain the phase component as Kuglin and Hines show [10]. Therefore we apply a shift between each interval $\delta \times = \delta y$=128 pixels as is shown in Figure 3.

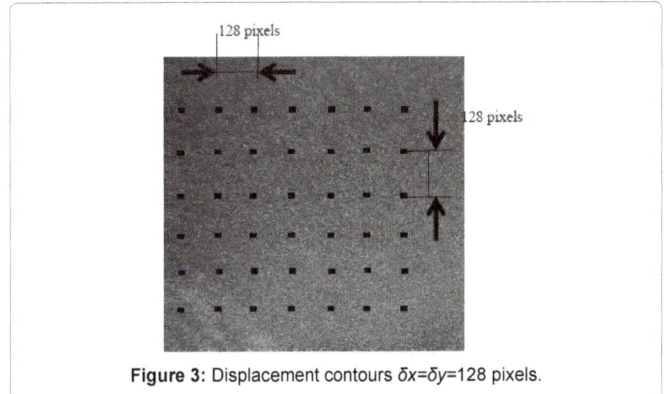

Figure 3: Displacement contours δx=δy=128 pixels.

The first strain measured by the universal machine is 0.002 while cross correlation technique allows us to measure 0.0019349; the shear component is equal to or less than 6.1×10^{-5}. It has been reported that for pure rotation measurements, this technique had 2×10^{-4} sensitivity [13].

Results and Discussion

In the section above, we mentioned that three samples were prepared for compression test, for each one has been obtained two graphs of interest, the strain diagram is shown in Figure 4 and the DIC plot which is shown in Figure 5, both graphs correspond to the first sample.

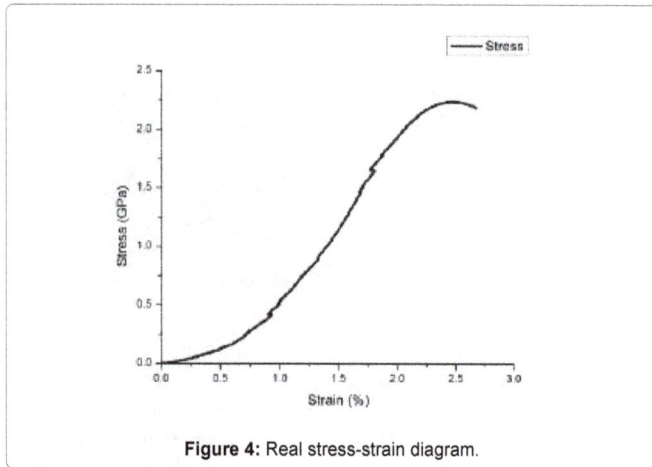

Figure 4: Real stress-strain diagram.

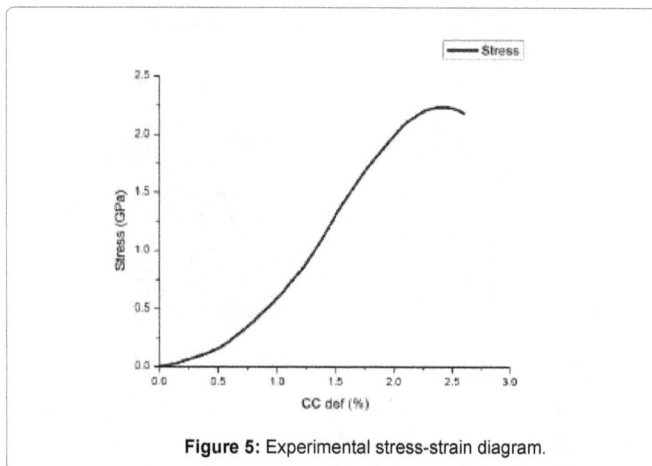

Figure 5: Experimental stress-strain diagram.

In Figure 4 we plot the strain diagram obtained from the machine and in Figure 5 we plot the stress-strain diagram from the experimental results obtained by Eq. (16). For both graphs we see a linear behavior.

In the next figure we compare the real deformation and the experimental data obtained, also some statistical results are shown in Table 1.

In Figure 6 we plot both graphs of interest, stress-strain from the machine and cross-correlation deformation from Eq. (15) (black) and from Eq. (16) (red) respectively. These results are shown for each sample; in Table 1 it is shown the statiscis of the first test, it was observed that there are such a good correlation obtaining 6.4% error from the real strain measurement.

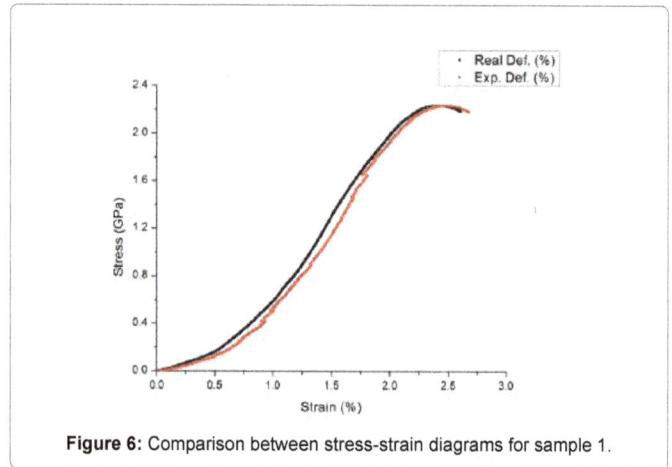

Figure 6: Comparison between stress-strain diagrams for sample 1.

1ˢᵗ Sample	
Difference from real E	2.17 ± 0.8 Gpa
Mean relative error ε	6.40%
Standard deviation	1.45

Table 1: Statistical results for sample 1 during the compression tests and cross-correlation analysis.

In Figure 7 it is plot both graphs of interest and Table 2 summarises showing the statiscis of this sample, it was observed a better correlation than sample 1 obtaining 5% error from the real strain measurement.

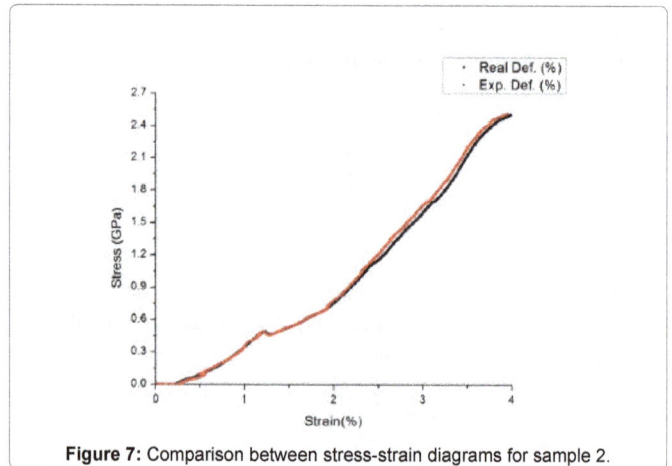

Figure 7: Comparison between stress-strain diagrams for sample 2.

2ⁿᵈ Sample	
Difference from real E	2.19 ± 0.5 Gpa
Mean relative error ε	5.00%
Standard deviation	1.9

Table 2: Statistical results for sample 2 during the compression tests and cross-correlation analysis

In Figure 8 it is plot both graphs of interest and Table 3 summarises showing the statiscis of this sample, it was observed a simmilar behaviour than sample 1 obtaining 6% error from the real strain measurement.

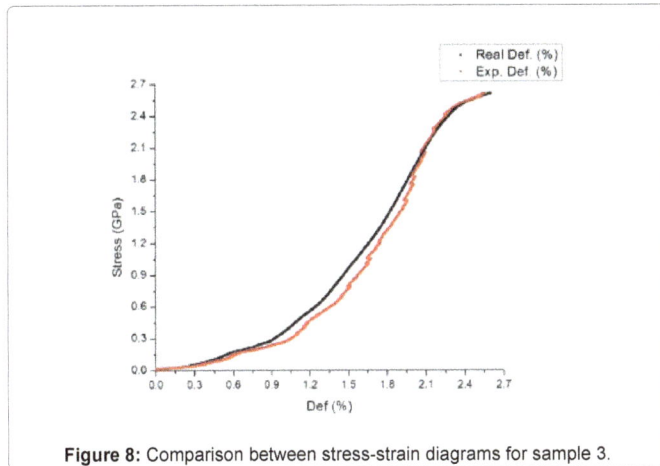

Figure 8: Comparison between stress-strain diagrams for sample 3.

3ʳᵈ Sample	
Difference from real E	2.17 ± 0.8 Gpa
Mean relative error ε	6.10%
Standard deviation	1.45

Table 3: Statistical results for sample 3 during the compression tests and cross-correlation analysis.

For samples 1 and 3 the behaviour of the experimental graph is below of the real stress-strain diagram but for sample 2 it is more correlated and behaves upper than real measurement. As they are all samples made by the same mixture, all of them behave approximately the same but the speckle reflection differs because of their surface, that's why we have got these results.

In Table 4 we show the accuracy and mean error of each measurement, so we can assume that the mean accuracy of our method is 94.4%.

Probe	Accuracy	Error
1	94%	6%
2	95%	5%
3	94%	6%

Table 4: Accuracy and error of the test.

By comparing the average strain determined for three different specimens with the compression stress values measured with the load cell, it was demonstrated that the present technique can measure the relative strain with an average uncertainty of 5.6%.

Conclusion

It was measured the strain of a rough-like material under a compression test using one laser beam, it's speckle reflection and digital image correlation treatment, obtaining 94.4% accuracy. We obtain 6.1×10^{-5} sensitivity, which is less than the reported 2×10^{-4}. We demonstrate that cross-correlation plus speckle metrology could be a reliable technique for measuring strain such as standardized compression tests.

Acknowledgement

Alonso Saldaña Heredia wants to thank CONACYT for the grant No. 360140.

References

1. Beasley F (2011) Theory and Design for Mechanical Measurements. 3ʳᵈ edn. USA John Wiley and Sons.

2. Gere J, Goodno B (2009) Mecánica de Materiales 7ᵗʰ edn. México Cengage Learning.

3. Mathar J (1934) Determination of Initial Stresses by Measuring the Deformation Around Drilled Holes. Transactions ASME 56: 249-254.

4. (2008) ASTM Determining Residual Stresses by the Hole-Drilling Strain Gage Method. ASTM Standard Test Method E837-08 American Society for Testing and Materials West Conshohocken PA.

5. Rigden JD, Gordon EI (1962) The granularity of scattered optical laser light Proceedings of the Institute of Radio Engineers 50: 2367-2368.

6. Peters WH, Ranson WF(1982) Digital image techniques in experimental stress analysis. Opt Eng 21: 427-441.

7. Sutton MA, Mc Neill SR, Helm JP, Chao YJ (2000) Advances in two dimensional and three dimensional computer vision. Photomechanics Topics in Applied Physics 77 Springer Berlin. 323-372.

8. Chen DJ, Chiang FP, Tan YS, Don HS (1993) Digital speckle-displacement measurement using a complex spectrum method. Appl Opt 32:1839-1852.

9. Anuta PE (1970) Spatial registration of multispectral and multitemporal digital imagery using fast Fourier transform techniques. IEEE Trans Geosci Electron 8: 353-368.

10. Kuglin CD, Hines DC (1975) The phase correlation image alignment method. In Proc. Int'l Conf Cybernetics and Society: 163-165.

11. Chevalier L, Calloch S, Hild F, Marco Y (2001) Digital image correlation used to analyze the multiaxial behavior of rubber-like materials. Eur J mech A/solids 20: 169-187.

12. Cintrón R, Saouma V (2008) Strain measurements with digital image correlation system Vic-2D. The George E Brown Jr. Network for Earthquake Engineering Simulation CU-NEES-08-06.

13. Pradille C, Bellet M, Chastel Y (2010) A Laser speckle method for measuring displacement field Application to resistance heating tensile test on steel. Applied Mechanics and Materials 24: 135-140.

14. Rots J, Borst R (1989) Analysis of concrete fracture in "direct" tension. Int J Solids Structures 25: 1381-1394.

15. Gerin M, Adebar P (2004) Accounting for shear in seismic analysis of concrete structures. 13ᵗʰ World Conference on Earthquake Engineering Vancouver B.C, Canada Paper No. 1747.

16. Wasantha MA (2005) Analysis and verification of stresses and strains and their relationship to failure in concrete pavements under heavy vehicle simulator loading. Ph.D. Thesis University of Florida USA.

17. Dainty JC (1975) Laser speckle and related phenomena. Springer-Verlag Berlin Heidelberg 9.

18. Goodman JW (1976) Some fundamental properties of speckle J Opt Soc Am 66: 1145-1150.

19. Fontenelle H (2009) Laser speckle imaging: spatio-temporal image enhancement. Ph.D Thesis University of Patras Greece.

20. ASTM E-9 ICS Number Code 77.040.10 (Mechanical testing of metals).

21. Sirohi RS (2009) Optical methods of measurement. Wholefield techniques 2ⁿᵈ Edition Taylor and Francis Group. USA.

22. Alda J (2003) Laser and Gaussian Beam Propagation and transformation. Encyclopaedia of optical engineering.

23. Yariv A (1990) Quantum Electronics 3ʳᵈ edn. USA John Wiley & Sons.

24. Wood R (1988) Physical Optics 3ʳᵈ edn. Washington DC Optical Science of America.

25. Born M, Wolf E (1970) Principle of optics 4ᵗʰ ed. Pergamon Press UK

26. (2015) Laserglow Technologies LCS-0532 Low-Cost DPSS Laser System Laserglow Part Number: C53200XSX Laser Product Datasheet generated.

4

Arc Energy Characteristics Analysis of AC Square Wave Submerged Arc Welding using WVD

Si-Wen X[1,3], Wang C[1,2], Zhi-Peng Z[1,2] and Kuan-Fang H[1,2],*

[1]College of Mechanical and Electrical Engineering, Hunan University of Science and Technology, Xiantan 411201, PR China
[2]Hunan Provincial Key Laboratory of Health Maintenance for Mechanical Equipment, Hunan University of Science and Technology, Xiantan 411201, PR China
[3]Hunan Provincial Key Laboratory of High efficiency precision cutting for Difficult-to-cut Material, Hunan University of Science and Technology, Xiangtan 411201, PR China

Abstract

In process of AC square wave submerged arc welding, electric signal waveform determines the arc energy characteristics in the time and frequency domain, which also reflects the arc stability of welding process and quality indirectly. Using Wigner-Ville distribution (WVD), time frequency analysis is conducted to the current signal of AC square wave submerged arc welding. The Pseudo Wigner-Ville and smoothed Wigner-Ville distribution are discussed separately to suppress the cross-term interference. The numerical results indicate that the Choi-Williams kernel function has much superiority for time frequency analysis of the submerged arc welding electric signal. It can effectively suppress cross-term and eliminate noise in the Wigner-Ville distribution of welding electric signal, as well as owing the ability to portray the local feature of arc energy in AC square wave submerged arc welding.

Keywords: AC square wave submerged arc welding; Arc energy characteristics; WVD; Choi-williams kernel

Introduction

The AC square wave submerged arc welding is suitable for the ring seam welding, multi-arc welding and other special occasions because of the advantages of quick speed of crossing arc zero, non-magnetic blow and high welding deposition rate [1,2]. In the actual welding process, the arc space and the electrode surface temperature change versus time during the alternating positive and negative polarity of arc [3,4], the arc resistance is not constant and changes as the arc current changes, and the welding power supply source easily interfered by the electric characteristic and the external factors. The presence of such phenomenon results in the irregular distortion of the actual outputting arc current and voltage waveform, especially in the process of high-speed and high current arc welding, which has direct influence on the arc stability and welding formation quality, and also reflects the arc characteristic information in the time domain and frequency domain. It is one of the effective ways to achieve detection of the arc stability and welding quality by the way of arc signal analysis and arc information extraction [5-7].

Submerged arc welding process is a complex and dynamic process of interaction of multiple factors. The actual monitoring of electrical arc current and voltage are non-stationary signals. Wigner-Ville time-frequency distribution (WVD) is a basic and most widely used time-frequency distribution. WVD is secondary time-frequency distribution that can obtain the instantaneous energy, frequency and power spectral density of signal, which is a powerful tool for analysis of non-stationary and time-varying signal. WVD is a two-dimensional joint function as a time-frequency energy density function in two-dimensional plane. WVD is given by the Heisenberg uncertainty principle when time-bandwidth product reaches the lower bound, which greatly improves the feature extraction accuracy of the welding arc signal with characteristics of high energy accumulation and good time-frequency resolution [8,9]. Based on WVD, arc electrical signals of the submerged arc welding process are analyzed, and the electrical energy distribution characteristics in time-frequency plane are obtained to assess the arc stability and welding quality. It is directive significance for the study of the arc stability and the welding seam formation in AC square wave submerged arc welding process.

Secondary Time-Frequency Analysis

WVD satisfies most of the mathematical properties, which is the most basic and applied time-frequency distribution. The Fourier transform of the signal $x(t)$ is $X(j\Omega)$, and then the signal $x(t)$ is defined as Wigner

$$W_x(t,\Omega) = \int_{-\infty}^{\infty} x(t+\tau/2)x^*(t-\tau/2)e^{-j\Omega\tau}d\tau \tag{1}$$

W_x is Wigner-Ville distribution of signal $x(t)$. WVD can be understood as a spectrum corresponding to each time based on this time as the center, which conducts the Fourier transform to the results from the signal multiplied by the right and left of all parts. The advantage of the WVD is the good focusing ability of time-frequency.

Since the arc electric signal of submerged arc welding is a multi-component signal, it has a serious cross-term on WVD time-frequency analysis. So it is necessary to take effective measures to eliminate the interference of cross-terms. The suppression of cross-terms can effectively by adding kernel function. The common adding kernel functions of Wigner-Ville distribution as below:

Pseudo Wigner-Ville Distribution

The oscillation characteristics of the cross-terms can be eliminated by the smoothing of WVD, that is, a smooth window function is added in time domain, and then the pseudo Wigner-Ville distribution is obtained:

$$PW_x(t,\Omega) = \int_{-\infty}^{+\infty} h(\tau)x\left(t+\frac{\tau}{2}\right)x^*\left(t-\frac{\tau}{2}\right)e^{-j\Omega\tau}d\tau \tag{2}$$

*****Corresponding author:** Kuan-fang H, College of Mechanical and Electrical Engineering, Hunan University of Science and Technology, Xiantan 411201, PR China
E-mail: hkf791113@163.com

Where $h(\tau)$ is a window function, the windowing function in the time domain is shorter, the smoothing effect on frequency domain is more obvious, the eliminating effect of the cross-term is also better. However, the elimination of the cross-term may result in the reduction of resolution, that is, the smoothing of Pseudo Wigner-Ville in the frequency direction may make the frequency resolution of the signal worse.

Smoothed WVD

In general, let $G(t, \Omega)$ be a time-frequency distribution of a window function, then the smoothed WVD is defined by convolution of $G(t, \Omega)$ and $W_x(t, \Omega)$ in both t and Ω, marking $SW_x(t, \Omega)$:

$$SW_x(t, \Omega) = \frac{1}{2\pi} \iint W_x(u, \xi) G(t - u, \Omega - \xi) \, du \, d\xi \qquad (3)$$

Where the effect of $G(t, \Omega)$ on $W_x(t, \Omega)$ depends on the shape of $G(t, \Omega)$. In fact, the left part of (3) is one of the general forms of the Cohen class. The time-frequency distribution of Cohen class has high time-frequency resolution, but it is difficult to eliminate the cross-term. Ref [9] used Choi-Williams distribution (CWD) to describe the time frequency energy distribution of arc signal of CO_2 welding, which has effectively suppressed the cross-term interference.

Figure 1 is an arc current signal x(t) of AC square wave submerged arc welding by experiment, the specific welding parameters and experimental phenomena is shown in Table 1. It can be seen from the Figure 1, the current waveform is an irregular square wave of 50 Hz alternating positive and negative, which contains interference signals of other frequency components in each waveform peaks and valleys and impacts at the crossing zero position. The sampling frequency is 2.5 kHz, 2500 number signals are selected to be calculated and analyzed.

The WVD of the signal is shown in Figure 2. It can be seen from Figure 2, the energy distribution of the signal is mainly concentrated in the contour of 50 Hz, and there contain other frequency contour lines of mutations and interference components, which is fully consistent with current outputting parameters of the actual AC square wave submerged arc welding process. It shows the WVD can effectively extract the current signal characteristics of submerged arc welding. However, the cross-terms are easily mixed with the principal and other

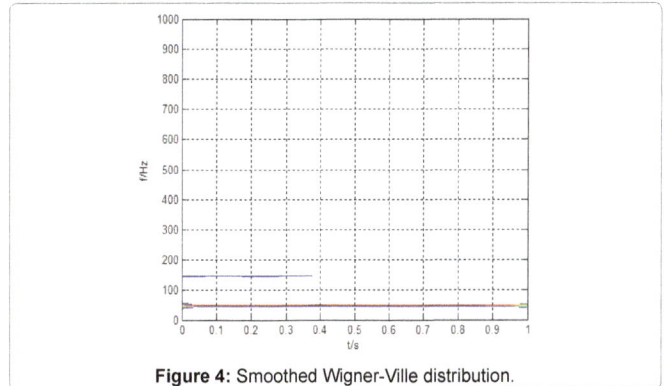

Figure 2: WVD of current signal.

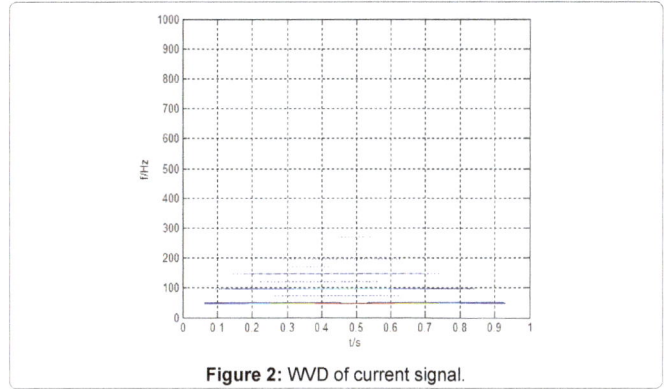

Figure 3: Pseudo Wigner-Ville distribution.

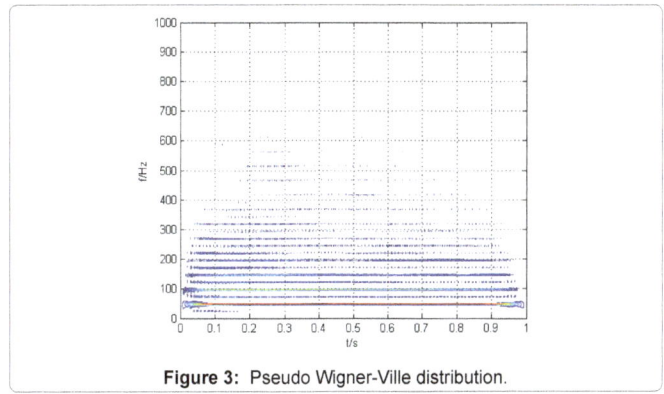

Figure 4: Smoothed Wigner-Ville distribution.

Experiment number	Currents (A)	Voltage (V)	Speed (m/min)	Frequency (Hz)	Duty ratio	Welding cases
1	630	40	0.6	50	0.5	No short-circuit and breaking arc, arc stability, good welding formation

Table 1: Welding parameters and experimental phenomena.

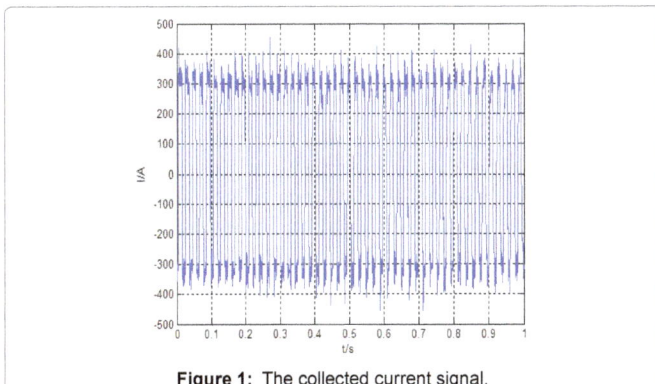

Figure 1: The collected current signal.

items, which disturb the interpretation of WVD. So, the cross-terms are easily seemed as the studying object for the interpretation of arc energy information.

According to the above method of cross-terms suppression, the pseudo Wigner-Ville distribution and smoothed Wigner-Ville distribution are used to reduce the cross-term interference. Two steps are generally taken to eliminate different frequency components, the first step is calculation of the analytical signal instead of actual signal, and the second step is to suppress cross-term interference by time-frequency analysis of pseudo Wigner-Ville distribution and smoothed Wigner-Ville distribution. The calculation results are shown in Figure 3 and Figure 4.

Figure 3 is Pseudo Wigner-Ville distribution, Figure 4 is smoothed Wigner-Ville distribution. It can be seem form Figures 3 and 4 that the main component of 50 Hz can seen in the Wigner-Ville distribution with other interference components and cross-term interference. Since these cross-terms fluctuate along the time axis, therefore it is difficult to distinguish the characteristics of welding arc energy distribution. The pseudo Wigner-Ville distribution has eliminated the cross-terms at a certain degree. The smoothed Wigner-Ville distribution has greatly reduced the influence of the cross-terms due to the time domain smooth.

It can be further seen from the Figures 3 and 4, the pseudo Wigner-Ville distribution and smoothed Wigner-Ville distribution can suppress the cross-terms effect at a certain degree, improve the application effect of the WVD, and can clearly distinguish the energy distribution characteristics of the main current waveform, but it cannot observe the local features of impact crossing zero and mutations of AC square wave current waveform, that is, it cannot effectively portray local features of arc energy of AC square wave submerged arc welding.

The signal of Figure 1 is still considered as the studying object, taking two steps to suppress the cross-term interference and eliminate the noise signal of different frequency components. The first step is to calculate the analytical signal instead of actual signal; the second step is taking the Choi-Williams kernel as time-frequency analysis [9]. The time frequency results of the signal by WVD with Choi-Williams kernel is shown in Figure 5.

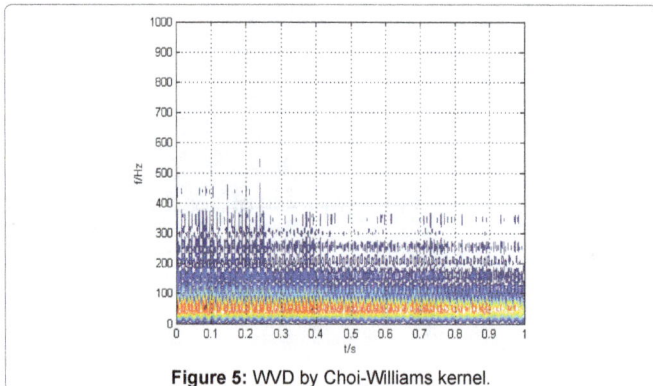

Figure 5: WVD by Choi-Williams kernel.

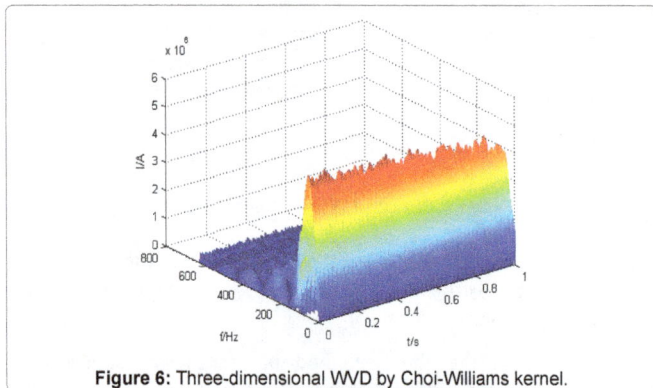

Figure 6: Three-dimensional WVD by Choi-Williams kernel.

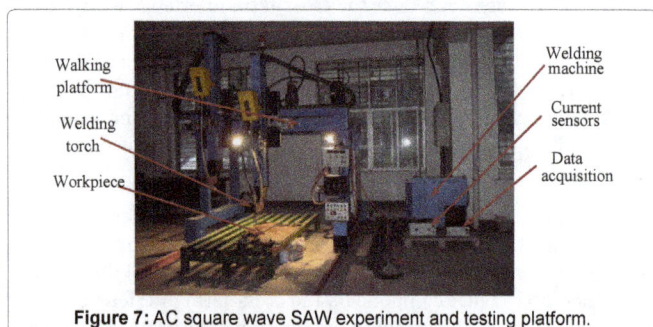

Figure 7: AC square wave SAW experiment and testing platform.

Experiment sequence	Currents (A)	Voltage (V)	Welding speed (m/min)	Frequency (Hz)	Duty ratio	Welding conditions
1	630	40	1.2	50	0.5	Breaking arc, arc instability, poor welding formation
2	630	40	1.2	80	0.5	No short-circuit and breaking arc, arc stability, good welding formation
3	630	40	1.2	100	0.5	No short-circuit and breaking arc, arc stability, good welding formation

Table 2: Welding parameters and results.

Comparing Figure 5 with Figure 6, using the Choi-Williams kernel as time frequency analysis, it effectively suppresses the cross-term influences of the main component and other interference components. Although the noise interference is not completely eliminated, the influence of the noise to WVD can be negligible. The energy distribution of arc signal retains the components of impact and mutation, which means a valid portrayed capacity of local features for AC square wave submerged arc welding.

From the above calculation analysis, WVD can effectively extract the characteristics of the welding arc signal, which can provide an intuitive and accurate criterion for the arc stability and welding quality. The Choi-Williams kernel based WVD analysis can effectively suppress cross interference term, eliminate the influence of noise to the welding arc electrical signal, and highlight the local characteristics of arc energy.

Experiential Results and Analysis

The measurement instruments for arc current and voltage signals of AC square wave submerged arc welding are consists of the Hall sensor, Ethernet data acquisition, industrial computer and other parts, and the experimental platform is shown in Figure 7. The collected welding current, voltage signal are transported to industrial control computer by cable transmission. The collected signal are analyzed and processed by Matlab. Experiments are done by alternating current square wave submerged arc welding machine MZE1000. The material of work piece is low carbon steel Q235 with slab thickness of 20 mm, the welding wire trademark is H08A with diameter of 4.0 mm, and welding flux is HJ431. Under the conditions of the different welding parameters such as voltage, current and welding speed, the welding experiment of submerged arc welding is carried out, and the current and voltage signals are acquired corresponding with the welding parameters. Welding parameters and testing phenomena are shown in Table 2. The Choi-Williams kernel based WVD is adopted to analyze the welding current signals in Table 2, the welding current signals and corresponding results of time-frequency distribution are shown in Figures 8-10.

The current waveform and WVD distribution of experiment sequence 3 (Figure 10).

It can be seen from Figures 8-10, the joint distribution of the amplitude of the welding current signal on the time and frequency plane can be clearly observed. The energy of each group signal mainly concentrates on the 50 Hz, 80 Hz and 100 Hz, with some other irregular frequency components fluctuating along the time axis. The irregular frequency components are actual outputting current waveform distortion of the welding power source with random distribution. The current waveform distortion and the size range directly affect the distribution of arc energy, and then affect the welding formation.

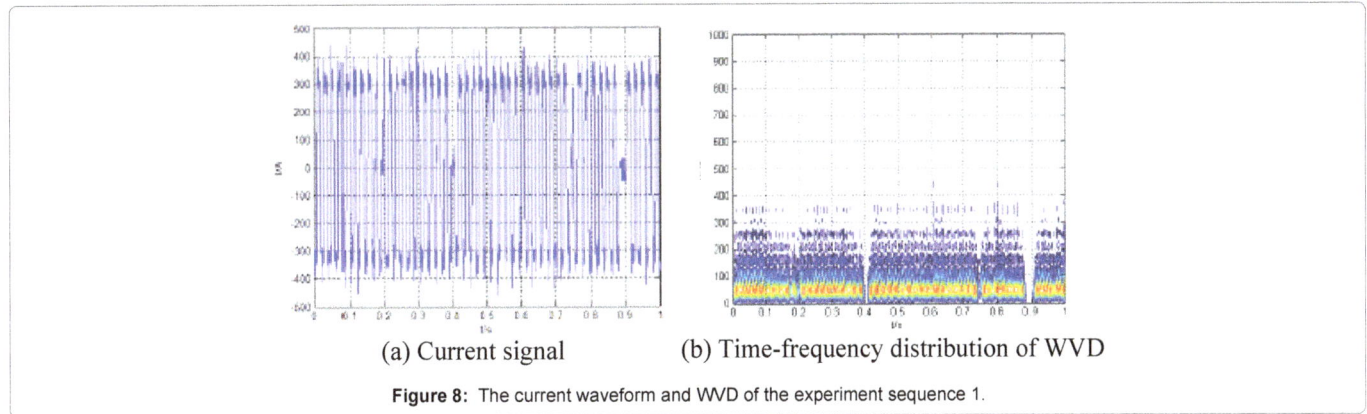

(a) Current signal   (b) Time-frequency distribution of WVD

Figure 8: The current waveform and WVD of the experiment sequence 1.

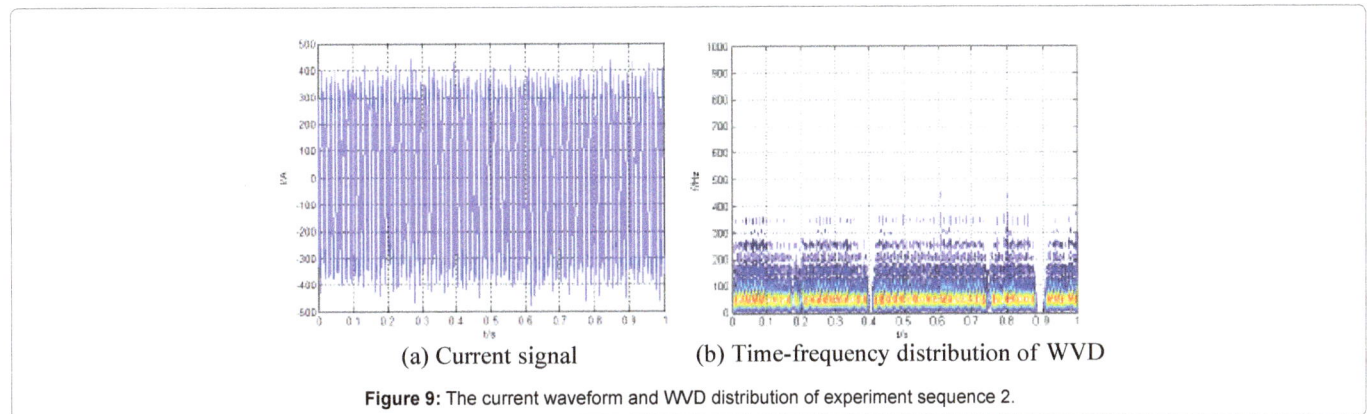

(a) Current signal   (b) Time-frequency distribution of WVD

Figure 9: The current waveform and WVD distribution of experiment sequence 2.

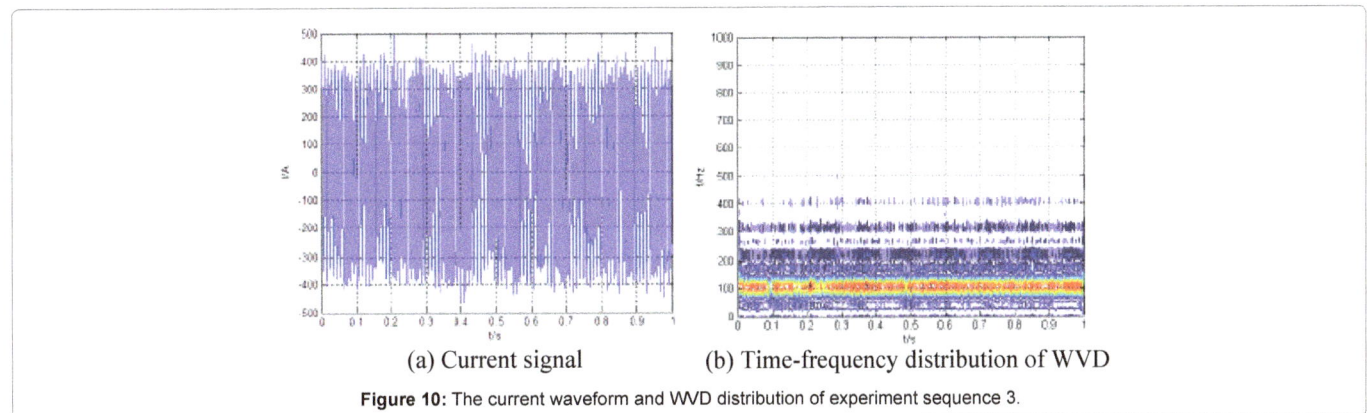

(a) Current signal   (b) Time-frequency distribution of WVD

Figure 10: The current waveform and WVD distribution of experiment sequence 3.

Figures 8b, 9b and 10b are the time-frequency spectrum calculated by welding current signal under the same duty ratio and different frequencies. The three group signals of time-frequency spectrum of the main frequency components are 50 Hz, 80 Hz and 100 Hz. The amplitude of the current signal is not much difference in frequency domain, the difference among the three group signals are the temporal change in the energy. The arc energy is different versus time seen from the three groups of time-frequency distribution. The energy of the unit time is changed frequently and relatively concentrated as an increasing of frequency.

It can be seen from Figures 9 and 10, the time and frequency contour of arc energy distribution is regular alternation in stable arc burning condition. Figure 8 is the current waveform and time frequency distribution at the condition of breaking arc at a time due to some factors. The arc energy reduces suddenly, which is abnormal phenomena in the process of welding and interferes the arc stability and arc energy distribution. At end of the broking arc, the arc continue burning steadily, the arc energy distribution becomes regular in time and frequency contour.

Conclusion

Time frequency analysis of WVD is conducted to the current signal of AC square wave submerged arc welding. The Pseudo Wigner-Ville and smoothed Wigner-Ville distribution are discussed separately to suppress the cross-term interference. The Choi-Williams kernel function based WVD can effectively suppress cross-term and highlight local characteristics of arc energy, which provides an intuitive, accurate criterion for assessment of welding arc stability and welding quality. By applying WVD to the arc current signals collected under different working conditions, the WVD is indicated as a direct, effective and practical time frequency exaction method for arc energy characteristic.

Acknowledgment

This work is supported by National Natural Science Foundation of China (51475159, 51005073), is gratefully acknowledged.

References

1. Tusek J (2000) Mathematical modeling of melting rate in twin-wire welding. J Mater Process Tech 100: 250-256.

2. Kuan-fang H, Shi-sheng H, De-yi S (2008) Development of AC Square-wave inverter for High-power submerged arc welding. J South China University Tech 36: 79-82.

3. Hirohira A (1979) Welding arc phenomena, Beijing. China Machine Press.

4. Shisheng H (2006) Arc welding power source and the digital control. Beijing. China Machine Press.

5. Jiaxiang X, Zhiping Y (2003) Wavelet packet analysis to electronic signal of arc welding process. Chinese J Mech Eng 39: 128-130.

6. Yiqing Z, Jiaxiang X, Kuanfang H (2011) A denoising method with exponential decay threshold for square arc signals in submerged arc welding. Transactions of the China Welding Institution 32: 5-8.

7. Yiqing Z, Zhenmin W, Jiaxiang X (2012) Wavelet detection and analysis for fault signal of welding arc. Electric Welding Machine 42: 47-49.

8. Yi L (2007) Application of joint time-frequency analysis to electrical signals of CO2 arc welding. Transactions of the China Welding Institution 28: 75-78.

9. Luo Yi, Guangfeng W, Chuntian L (2008) Application of Choi-Williams distribution to electrical signals detection in CO2 arc welding. Transactions of the China Welding Institution 29: 101-103.

5

Evolution of Japanese Automobile Manufacturing Strategy Using New JIT: Developing QCD Studies Employing New SCM Model

Amasaka K*

Department of Mechanical Engineering, Aoyama Gakuin University, Japan

Abstract

This paper introduces New JIT that contributes to the evolution of Japanese automobile manufacturing strategy. We believe that the key to successful global manufacturing is joint task team activities between the manufacturer and affiliated/non-affiliated suppliers employing Strategic Stratified Task Team Model. To realize this, we create the New SCM Model for strengthen of SCM. Here, we introduce typical research examples of how this model improved the bottleneck problems of worldwide automobile manufactures.

Keywords: New JIT strategy; QCD studies for automobile manufacturing; New SCM model

Introduction

Toda's management challenge of manufacturer is to provide high QCD (Quality, Cost, and Delivery) products ahead of competitors through "Market creating" activities, with priority given to customers. To accomplish manufacturing that places top priority on customers with a good QCD and in a rapidly changing technical environment, it is important to develop a new production technology and establish new process management to enable global production. For the automobile manufacturing industry, the key to success in global production is systematizing its management methods when modeling strategic SCM (supply chain management).

This paper analyzes and proves the significance of strategically employing New JIT, new management technology principle at Japanese automobile manufacturing Amasaka [1-3] In the implementation stage, automobile manufacturers endeavoring to become global companies are required to collaborate with not only affiliated companies, but also with non-affiliated companies to achieve harmonious coexistence among them based on cooperation and competition called "Japan Supply System" Amasaka [4,5]. To realize manufacturing of excellent quality for customer, we employ the Strategic Stratified Task Team Model" utilizing two core models: Structured Model of Stratified Joint Task Teams and Strategic Cooperative Creation Team Model for strengthen of SCM strategy [6].

Concretely, we create the New SCM Model consisting of three core models: Strategic Task Team Model, Global Partnering Model, and Simultaneous Fulfillment of QCD Approach Model. In typical research examples, we verify the effectiveness of New SCM Model by Total Task Management Team activities between automobile assembly maker and suppliers through the solution of the bottleneck problems of the world wide automobile manufactures Amasaka [7-9].

Importance of Strategic QCD Studies

Progress of Japanese automobile manufacturing industry

The leading Japanese automobile management technology that contributed most to worldwide manufacturing from the second half of the 20th century was the Japanese Production System. This is typified by the Toyota Production System (TPS) Ohno [10]. This system has been further developed as production systems known as Just in Time (JIT) and Lean Systems (Doos et al; Hayes and Wheelwright [11]; Womack and Jones [12]; Taylor and Brunt [13]. The history of

such development is shown in Figure 1, "Transition in Management Technology," (Amasaka [14,15].

As seen in the diagram, the Japanese manufacturing represented by Toyota Production System constitutes the basis for the manufacturing carried out worldwide today. Among the main administrative management technologies that have contributed to the development of Japanese manufacturing are Industrial Engineering, Operations Research, Quality Control, Administrative Management, Marketing Research, Production Control, and Information Technology. These are shown on the vertical axis of the diagram. On the horizontal axis, a variety of elemental technologies, management methods, and scientific methodologies are arranged in chronological order. Conventional Japanese manufacturing has developed from in-house production to cooperative relationships with suppliers although, since the beginning administrative management technology has become increasingly complicated.

Therefore, the current task of today's manufacturing sector is to succeed in global production. A key to this is the deployment of supply chain management on a global scale that encompasses cooperative business operations even with overseas suppliers, and the ever growing need for the systemization of such operation methods. In particular, during the implementation stage the organically combined use of partnering and digital engineering (CAE, CAD, CAM), and SCM will become necessary as they are essential for the deployment of the main components of Toyota Production System, called Just in Time (JIT) and Total Quality Management (TQM). Therefore, in-depth study of the kind of administrative management technology that will be effective even for next-generation business operations is also urgently needed.

In recent years, however, both developed Western nations and developing nations have advanced the study of Toyota Production System and TQM and re-acknowledged the importance of the quality of administrative management technology. They have also promoted

---

*Corresponding author: Amasaka K, Department of Mechanical Engineering, Aoyama Gakuin University, Japan, E-mail: amasaka@hn.catv.ne.jp

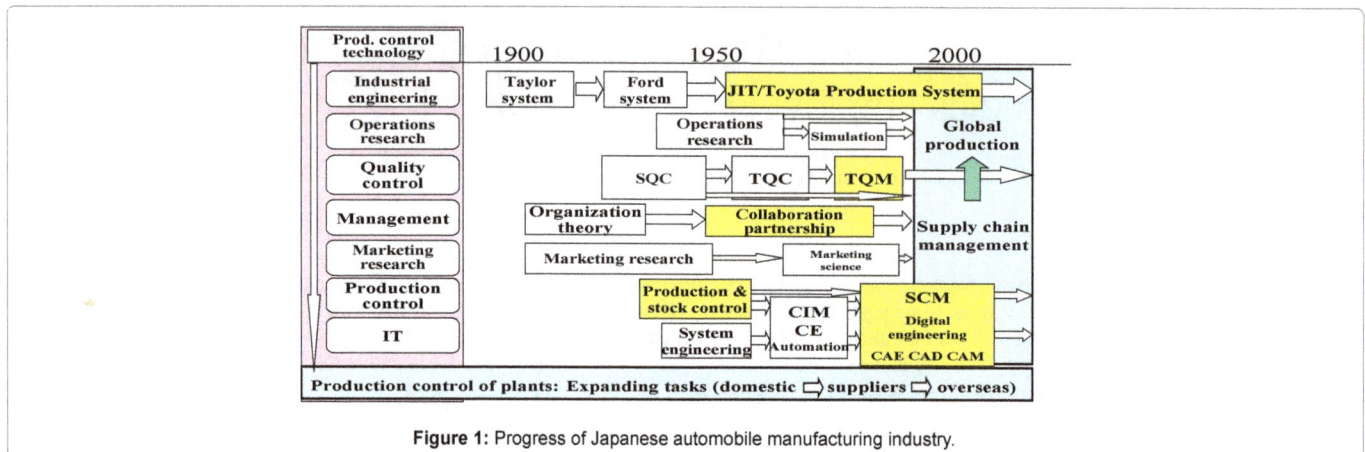

Figure 1: Progress of Japanese automobile manufacturing industry.

the reinforcement of quality in manufacturing on a national level Nezu [16].

As a result of such efforts, the superior quality of Japanese products has been rapidly compromised. One distinctive example of this is shown in a comparison of the quality of automobiles sold in the United States. Although Toyota, still a leading Japanese car manufacturer, can be seen to have achieved steady improvements in the quality of its automobiles (IQS, Initial Quality Study) up to now, GM of the United States and Hyundai of Korea have also promoted quality improvements and achieved even more dramatic results (Amasaka, 2007b, 2008a). The observations above indicate that in order for Japanese manufacturers to continue to play the leading role in the world, it is urgent to reform their administrative management technology from a fresh standpoint, rather than simply clinging to the successful experiences they have enjoyed up to now.

Needs for the reform of Japanese-style management technology

Looking at the recent automobile recall problems, we see a rapidly increasing number of manufacturing quality issues with their roots in technological product design and product (Joiner [17]; Nihon Keizai Shinbun [18]; Amasaka [19,20]. If we are to turn the tide, we cannot be content with simply resolving individual technical issues. Instead, we must evolve core technologies that result in the overhaul of every business process from development and production to SCM, and establish and systematically apply a new management technology model that intelligently links them together Amasaka [21,22]. The top priority issue of the industrial field today is the "new deployment of global marketing" for surviving the era of global quality competition Kotler [23]. The pressing management issue particularly for Japanese manufacturers to survive in the global market is the "uniform quality worldwide and production at optimum locations" which is the prerequisite for successful global production.

To realize manufacturing that places top priority on customers with a good QCD and in a rapidly changing technical environment, it is essential to create a core principle capable of changing the technical development work processes of development and design divisions Amasaka [24]. Furthermore, a new quality management technology principle linked with overall activities for higher work process quality in all divisions is necessary for an enterprise to survive (Burke and Trahant [25]; Amasaka [26]. The creation of attractive products requires each of the sales, engineering/design, and production departments to be able to carry out management that forms linkages throughout the whole

organization (Seuring et al. eds [27]; Amasaka [26]. From this point of view, the reform of Japanese-style management technology is desired once again. In this need for improvements, Toyota is no exception Goto [28]; Amasaka [26].

Similarly, it is important to develop a new production technology principle and establish new process management principles to enable global production. Furthermore, new marketing activities independent from past experience are required for sales and service divisions to achieve firmer relationships with customers. In addition, a new quality management technology principle linked with overall activities for higher work process quality in all divisions is necessary for an enterprise to survive Amasaka [29].

Importance of strategic QCD studies with affiliated and non-affiliated suppliers

IT development has led to a market environment where customers can promptly acquire the latest information from around the world with ease. In this age, customers select products that meet their lifestyle and have a sense of value on the basis of a value standard that justifies the cost. Thus the concept of "Quality" has expanded from being product quality, which is oriented to business quality, to becoming corporate management quality-oriented. Customers are strict in demanding the reliability of enterprises through the utility values (quality, reliability) of their products (Evans and Dean [30]; Amasaka [26]. Advanced companies in countries all over the world, including Japan, are shifting to global production. The purpose of global production is to realize "uniform quality worldwide and production at optimum locations" in order to ensure company's survival am idst fierce competition (Doz and Hamel [31]; Amasaka [26].

For the manufacturing industry, the key to success in global production is systematizing its management methods when modeling strategic SCM for its domestic and overseas suppliers. In-depth studies of the Toyota Production System called JIT and Lean Production System, TQM, partnering, and digital engineering will be needed when these methods are implemented in the future. Above all, manufacturers endeavoring to become global companies are required to collaborate not only with affiliated companies, but also with non-affiliated companies to achieve harmonious coexistence among them based on cooperation and competition. In other words, a so-called "feder ation of companies" is needed (Hamel and Prahalad [32].

New JIT Strategy, Surpassing JIT

Traditional Japanese production system and quality management: JIT

One of the greatest contributions that Japan made to the world is JIT. JIT is a production system that enables provision of what customers desire when they desire it. JIT is also introduced in a number of enterprises in the United States and Europeans a key management technology (Taylor and Brunt [13]; Amasaka [1]. The Japanese-style production system represented by the current Toyota Production System (JIT) is a production system which has been developed by Toyota. Implementing TQM in the production process, this production system aims to achieve the simultaneous of quality and productivity in pursuit of maximum rationalization while recognizing the principle of cost reduction.

This is the essential concept of JIT and therefore, these have been positioned as a core part of Toyota's management technology and often likened to the wheels on both sides of a vehicle Amasaka [33,34]. However, JIT which is representing the Japanese-style production system today has already been developed as an internationally shared system, known as a Lean System and is no longer an exclusive technology of Toyota in Japan. In the Western countries also, the importance of quality control has been recognized through the studies on the Japanese TQM. As a result, TQM activities have been increasingly popular. Therefore, the superiority in quality of Japanese products assured by the Japanese-style quality control has been gradually undermined in recent years [35].

New JIT, New Management Technology Principle

Having said the above, it is the author's conjecture that it is clearly impossible to lead the next-generation by merely maintaining the two Toyota management technology principles, JIT and TQM. To overcome this issue, it is essential to renovate not only JIT, which is the core principle of the production process, but also to establish core principles for marketing, design and development, production and other departments.

The next-generation management technology model surpassing JIT, New JIT, which the author has proposed through theoretical and systematic analyses as shown in Figure 2, is the Just in Time system

for not only manufacturing, but also for customer relations, sales and marketing, product planning, R&D, design, production engineering, logistics, procurement, administration and management, for enhancing business process innovation and introduction of new concepts and procedures.

New JIT contains hardware and software systems as the next generation technical principles for accelerating the optimization (high linkage) of work process cycles of all the divisions. The first item, the hardware system, consists of the TMS, TDS, and TPS, which are the three core elements required for establishing new management technology principles for sales, R&D, design, engineering, and production, among others.

The expectations and role of the first principle TMS include the following: (i) Market creation through the gathering and use of customer information, (ii) Improvement of product value by understanding the elements essential to raising merchandize value, (iii) Establishment of hardware and software marketing system to form ties with customers and (iv) Realization on the necessary elements for adopting a corporate attitude (behavioral norm) of enhancing customer value and developing customer satisfaction (CS), customer delight (CD) customer retention (CR), and networks.

The expectations and role of the second principle TDS are the systemization of design management method which is capable of clarifying the following: (i) Collection and analysis of updated internal and external information that emphasizes the importance of design philosophy, (ii) Development design process, (iii) Design method that incorporates enhanced design technology for obtaining general solutions, and (iv) Design guideline for designer development (theory, action and decision-making).

The expectations and new role of the third principle TPS comprise the following: (i) Customer-oriented production control systems that place the priority on internal and external quality information, (ii) Creation and management of a rational production process organization, (iii) QCD activities using advanced production technology and (iv) Creation of active workshop capable of implementing partnership. For the second item, the strategic quality management system, the author is establishing a new principle of quality control; Science TQM Amasaka [36] called TQM-S (TQM by utilizing Science SQC, New Quality Control Principle (Amasaka, 2004c) as a software system for

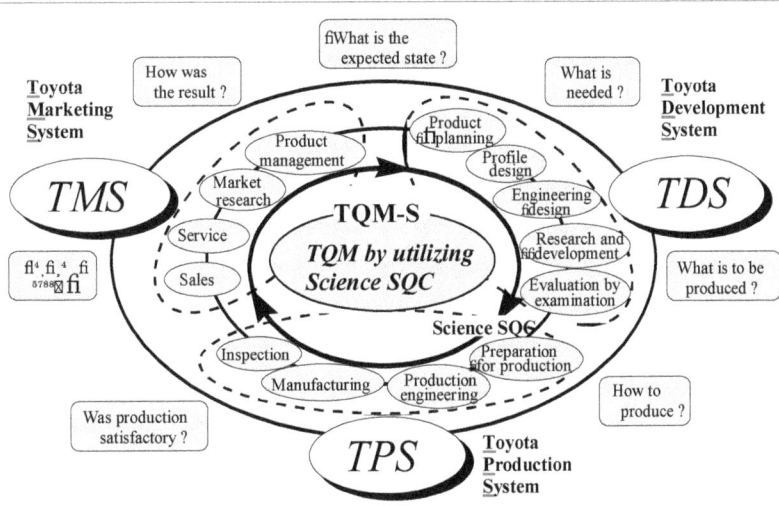

Figure 2: New JIT, New management technology principle.

innovating the business process quality of the 13 departments shown in the diagram (Figure 2).

Strategic Stratified Task Team Model

Japan supply system-partnering chains as the platform-type

As an example, Toyota group consists of a total 14 enterprises Toyota [37] Twelve enterprises, which branched from Toyota Automatic Loom Works, Ltd., form the nucleus of the primary group parts manufacturers (called "Kyohokai") that supply parts directly to Toyota, to which join Hino Motors, Ltd. and Daihatsu Motor Co., Ltd. Each of the group companies is closely linked to Toyota in a wide and solid supplier-assembler relation "Toyota Supply System" called "Japan Supply System" as shown in Figure 3.

An automobile is assembled with some 20,000 parts. Since it is not economical for the assembler to manufacture all the parts in-house, considerable portion of the parts are normally purchased from outside suppliers (parts manufacturers). Therefore, to years, and the situation still remains unchanged for it. If parts purchased from the supplier (parts manufacturer) have low dependability, vehicles assembled with them have also low dependability naturally. This is exactly the reason why "performance of a vehicle almost depends on the parts" (Amasaka, 2000). In this sense, the assembly makers (automobile manufactures) and parts manufacturers (suppliers) are the inhabitant of the same fate-sharing community. Actual supplier-assembler relation is generally quite complex and many-sided.

Relationship between Toyota and its suppliers is unique in many points compared with those of other assemblers. There is the saying that "Toyota wrings a towel even when it is dry", indicating Toyota's strict demand to supplier s for their prices and quality. At the same time, no other assemblers are as enthusiastic as Toyota in raising strong suppliers through education and training. As early as in 1939, Toyota established its basic concept of the purchasing activities for promoting coexistence and co-prosperity.

To realize this, Toyota strengthened its suppliers by making it a rule to continue transaction forever once started. No other assemblers have their supplier groups as powerful as those of Toyota. The 14 Toyota group companies form the nucleus of the powerful supplier system. As thus far stated, close relation between Toyota and its group members is quite cooperative in one sense while simultaneously very competitive in another. This represents the supplier- assembler relation unparalleled elsewhere.

There is no denying that the strength of Toyota that can keep on supplying popular vehicle model such as "Lexus" of high dependability originates from within Toyota's own. But it is also the fact that part of Toyota's strength comes from the strength of its supply system consisting of Toyota group and thousands of other suppliers or Toyota's skill in managing such powerful supply system. In the following, the authors intend to zero in on Toyota's quality management activities in exploring the secret of the strength of Toyota that continues to manufacture vehicles of high dependability. This is because it is quality management activities themselves that provide important bonds to cooperation or partnering between Toyota and its group companies.

Platform-type partnering chains by structured model of stratified joint task teams

Concretely speaking, the author believes that a company has to (i) join forces with domestic suppliers to enhance the intellectual productivity of plant divisions, and (ii) succeed in "global production" to promote overseas operations and develop local production. In the implementation stage, first, (A) the quality management theory of "Science SQC" Amasaka [9] will be applied, as the figure shows, as the methodology for scientifically solving problems through the strategic linkage of these 3 core elements.

Second, as Figure 4 shows, (B) the structured model of stratified joint task teams formed from partnering linkages will be developed systematically and organizationally to promote the strategic development of New JIT This model will consist of Task 1 to Task 8 teams involving the group, department, division, field, whole company, affiliated companies, non-affiliated companies, and overseas affiliates.

As indicated in the figure, the level of problem-solving technology rises strategically to product development strategy I and II through joint task teams of intra-company departments and divisions (Task-1 to Task-5, Task team, Task management team, and Total task management team) in proportion with the improvement of the stratified task level. This technology is further expanded to quality management strategy I to II through the domestic joint task teams of affiliated and non-affiliated companies and overseas counterparts (foreign groups: affiliated/non-affiliated) (Task-6 to Task-8, Joint A to Joint C).

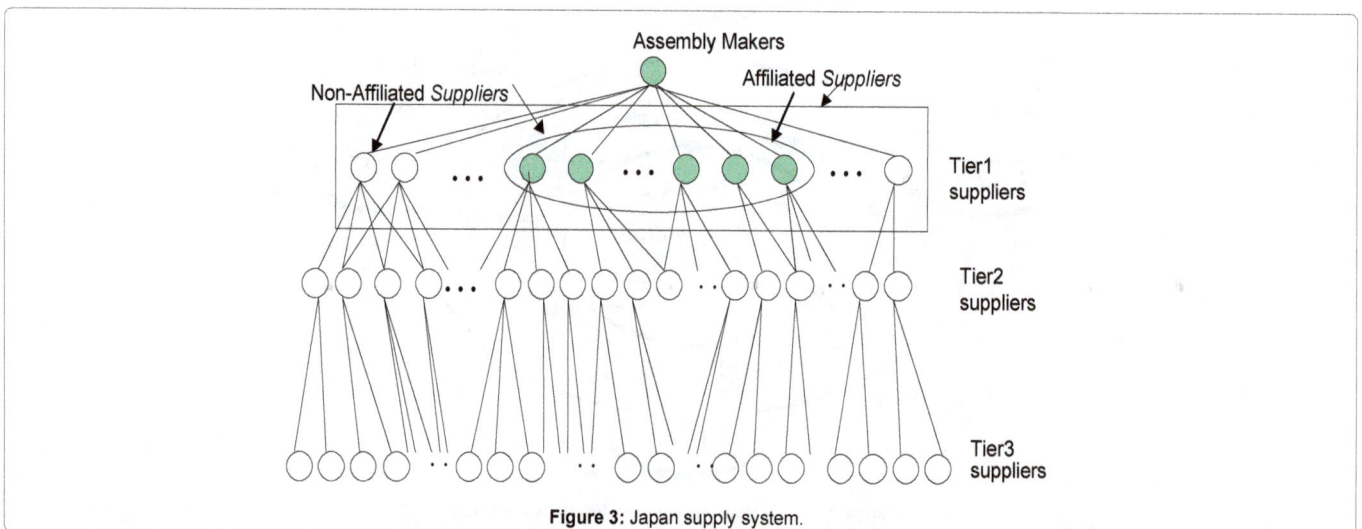

Figure 3: Japan supply system.

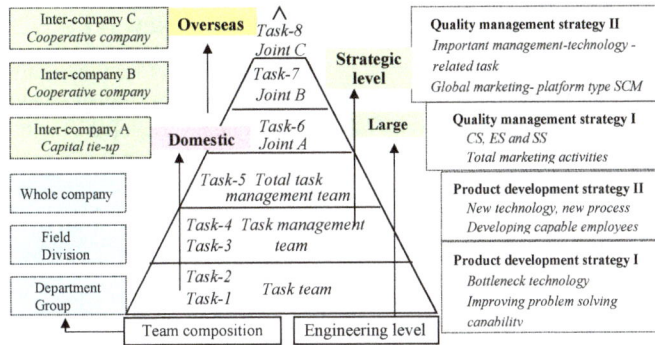

Figure 4: Structured model of strategic stratified joint task teams.

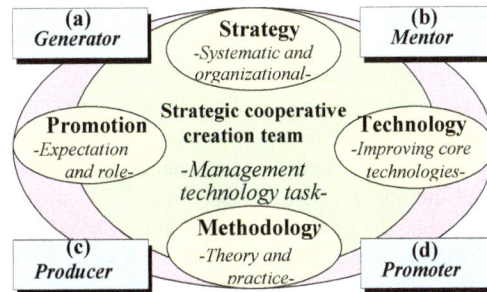

Figure 5: Strategic activity frame of the cooperative creation team model.

Task 6 (Joint A) is aimed at establishing a collaboration with the group suppliers with whom inter-company A has a capital tie-up and Task 7 (Joint 7) is aimed at a collaboration with suppliers that are not within its group. The mission of Task 8 (Joint C) is to strengthen cooperation with overseas suppliers as a strategic alliance.

Construction of the strategic cooperative creation team model

In order for the strategic stratified task team created in Figure 4 to perform New JIT and solve issues of management technology, formation of the Strategic Cooperative Creation Team Model indicated by Figure 5 is essential Yamaji [38]. To empower the team, team members should collectively have the capabilities of (1) "Strategy" for systematic and organizational activities, (2) "Technology" to improve core technologies, (3) "Methodology" to practically identify the gap between theories and the actual, and (4) "Promotion" to fulfill the expectations and roles of the team. If the task team tackles a strategic issue requiring high technologies, the members have to be ingenious enough as (a) "Generators" and at the same time they have to be able to perform strategic analysis as (b) "Mentors".

In addition, to infuse effective drive force in the team activities, creativity as (c) "Producers" and leadership to orchestrate all members' ideas as (d) "Promoters" toward target achievement is important. As the key to the successful team activities, the team leader (Administrator) should select the members who have at least one of the capabilities for (a) to (d), commission authority and responsibilities to the members, and have himself/herself concentrate on risk management. For this reason, as the leader, a person who has an experience of clearing business obstacles should be appointed, so that the leader is capable enough to lead the team overcoming difficulties.

Creation of New SCM Model for QCD Studies Employing New JIT

Today's management challenge is to provide high QCD products ahead of their competitors through "Market Creating" activities, with priority given to customers. This is the mission of New JIT. To realize manufacturing that provides excellent quality to the customer, the author has created New SCM Model with three core elements as follows; (i) "Strategic Task Team Model" between the manufacturer and affiliated/non- affiliated suppliers, (ii) "Global Partnering Model" for strategically implementing New JIT, and (iii) "Simultaneous QCD Fulfillment Approach Model" for developing New JIT.

Strategic Task Team Model" between the Manufacturer and Affiliated/Non-affiliated Suppliers

We believe that the key to successful global production is joint task activities between the manufacturer and affiliated/non-affiliated suppliers as stated above. In other words, it is important for the companies involved to work hard together in world markets under the principle of "harmonious coexistence through cooperation and mutual competition" in order to establish improved management technologies.

An example of concrete measures for development is "Strategic Task Team Model between the manufacturer and affiliated/ non-affiliated suppliers" as shown in Figure 6. To purchase the necessary parts, it will be important for the manufacturer to mutually cooperate with (a) Supplier I (in-house parts maker (own company)), (b) Supplier II, affiliated manufacturer (capital participation), (c) Supplier III, non-affiliated manufacturer, and (d) Supplier IV, manufacturer with foreign capital. In the stage of actual implementation, it is important to strategically organize the stratified task teams from the following viewpoints and by setting the objective to be continual improvement of management technologies: (i) Product strategy, (ii) Engineering strategy, (iii) Quality strategy, (iv) QCD effect, (v) Value of the task teams, and (vi) Human resource strategy.

After solving the most important management technology challenges at the beginning, the important job for the manufacturer's general administrator is to select jointly from his own company and suppliers: (1) "Generators" gifted with a special capacity for creating ideas, (2) "Mentors" having the ability to give guidance and a dvice, (3) "Producers" with the capability to achieve and execute, and (4) "Promoters" capable of implementing things as an organization.

Global Partnering Model" for Strategically Implementing New JIT

Understanding the need for strategically implementing New JIT by applying the aforementioned Strategic Task Team Model between the manufacturer and affiliated/non-affiliated suppliers, we create the 4-core structured "Global Partnering Model (GPM)" in Figure 7 that implements Science SQC. This principle has been proven effective in strategically solving management technology problems in this author's previous studies. As shown in the figure, GPM is composed of four cores, namely (1) Stratified joint task team (GPM-HT, Task-1 to Task-8 by New JIT) in mutual cooperation with affiliated and non-affiliated suppliers, (2) Stratified education training for improving the skills of staff and managers (GPM-HE, the Hierarchical Education by New JIT), (3) Stratified leader training (GPM-HL, the Hierarchical Leaders Growth by New JIT) and (4) Overseas study system (GPM-SA, the Studying Abroad System by New JIT).

To render the created "Global Partnering Model" effective in the implementation stage, it is important to adopt the hardware system with three core elements (TMS, TDS and TPS), and the software system (TQM-S).

Simultaneous QCD Fulfillment Approach Model Developing New JIT

In recent years, leading manufacturers in Japan have been deploying a new production strategy called "globally consistent levels of quality and simultaneous global launch (production at optimal locations)" in order to get ahead in the "worldwide quality competition", and "high quality assurance in manufacturing - simultaneous achievement of QCD" Amasaka and Sakai [39,40]. This is the key to successful global production, and has become a prerequisite for its accomplishment developing New JIT.

However, it has been observed that, despite the fact that overseas plants have the relevant production systems, facilities, and materials equivalent to those that have made Japan the world leader in manufacturing, the "building up of quality - assuring of process capability (Cp)" has not reached a sufficient level due to the lack of skills of the production operators at the manufacturing sites. Under such a circumstance, there are many studies abroad for globalization (Lagrosen [41]: Ljungström [42]; Burke et al., [43]; Hoogervorst et al. [44]).

In order to realize the key to global production, considering the importance of scientific quality management, and on the basis of the "New JI T" which has verified the effectiveness of the "stratified task team", the authors create the "Simultaneous QCD Fulfillment Approach Model" for developing New JIT as shown in Figure 8 by Yamaji [45].

The function of the (i) Quality Assurance (QA) and TQM promotion as corporate environment factors for succeeding in "global production are (1) Customer Satisfaction (CS), ES (Employee Satisfaction), and SS (Social Satisfaction), (2) High Quality assurance, (3) simultaneously achieve QCD, (4) success in global partnering. Intellectual productivity and (5) Evolution of quality management.

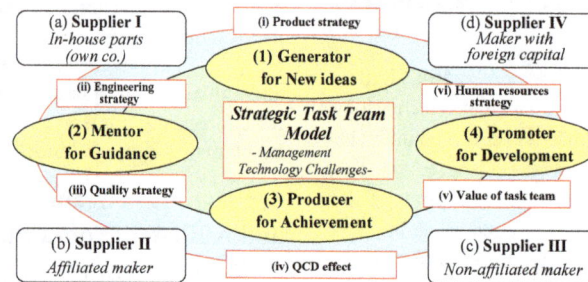

Figure 6: Strategic task team model between maker and affiliated/non-affiliated suppliers.

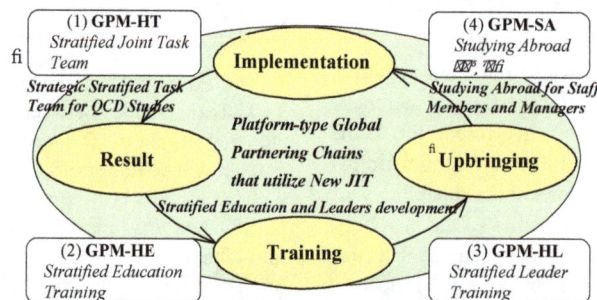

Figure 7: Global partnering model for strategically implementing new JIT.

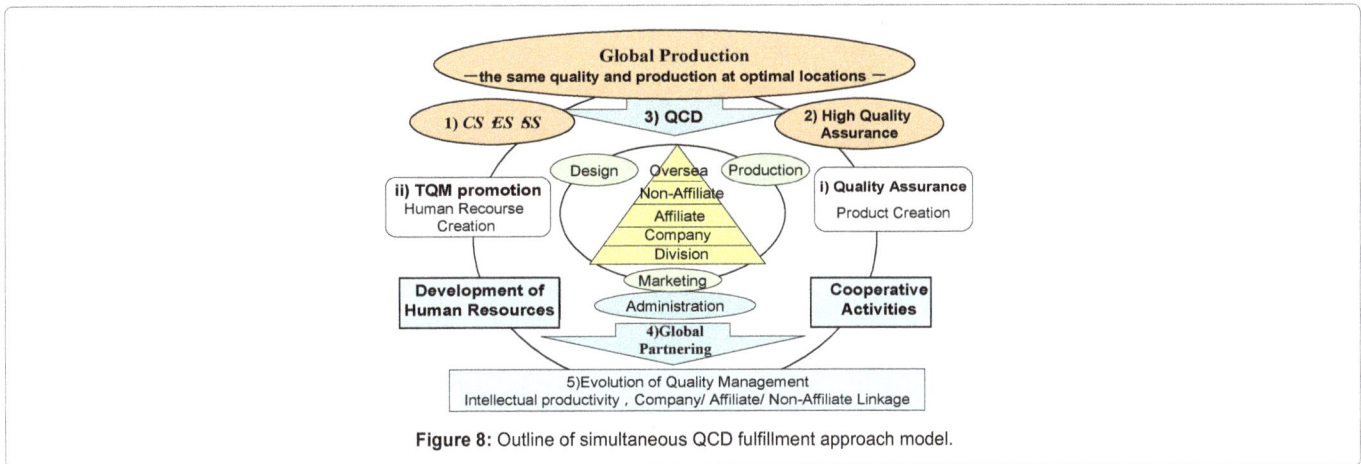

Figure 8: Outline of simultaneous QCD fulfillment approach model.

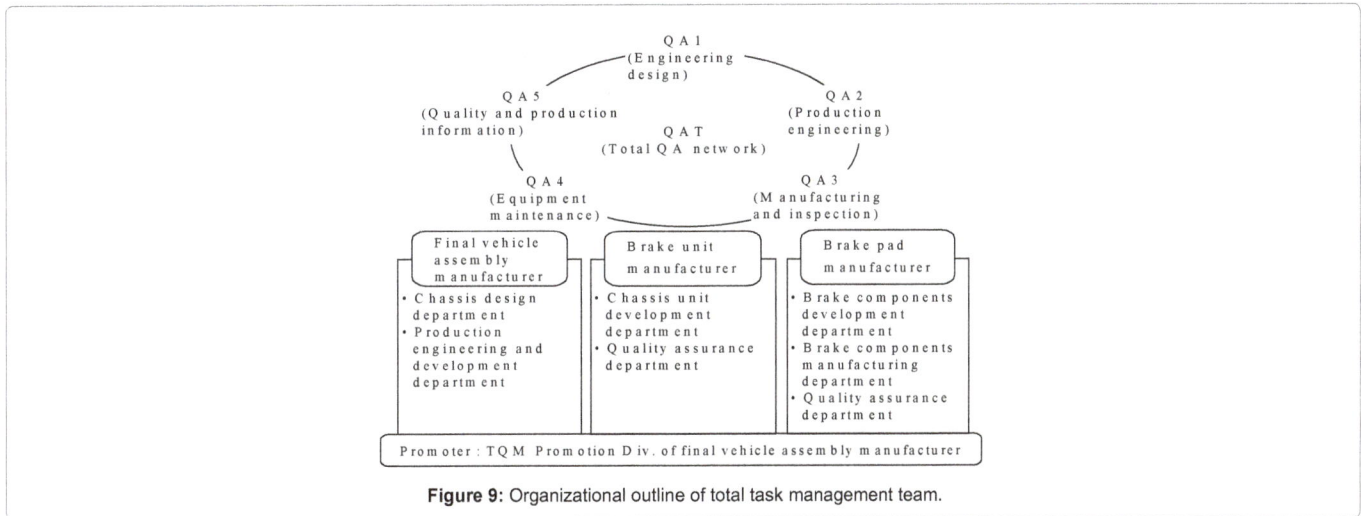

Figure 9: Organizational outline of total task management team.

More specifically, the (i) QA division needs to promote manufacturing of high reliable manufacturing, and cooperative activity across the organization is indispensable to achieve that. In the (ii) TQM promotion division, it is important to cultivate human resources that have even higher skills, knowledge, and creativity. Therefore, the value of intelligent human resources must be promoted in an effort to improve the productivity of white collar workers.

In order to realize the above, global partnering which enables a strategic cooperation among divisions, such as designing, production, marketing and administration as well as the entire company, affiliated companies, non-affiliated companies, and overseas corporations, must be achieved. Improvement of intellectual productivity is simultaneous achievement of QCD utilizing this model.

Application examples employing new SCM model

In this section, we present examples showing the results of research and effectiveness where the "New SCM Model" employing Strategic Stratified Task Team Model mentioned above has been applied in current bottleneck problems, braking performance, transaxle oil seal leakage, and others.

Brake Pad Quality Assurance

Disk brakes work on a principle that the pressing of pads to the rotors by the calipers generate braking force. Brake noise is generated when the pad and the rotor are in contact with each other in a delicately unstable condition. The condition that allows noise or abnormal sound to be generated with ease and the response quality of brake are items contrary to each other. Therefore, it is important to analyze the properties of pads and the response quality of noises and brake, and reduce dispersions in the properties of pads that affect the generation of noises and the braking effect Miller [46].

The key point of this activity lies in establishing technologies that make sensitivity analysis of factors on the braking effect and noise in the aspect of design engineering, and to make these mutually contrary braking performances compatible with each other through bi-directional interaction with the manufacturing engineering.

Total Task Management Team Using Three Pillars of Total QA Network

To establish an organizational system that allows the vehicle design, parts design, production process design, manufacture, inspection, maintenance, sale (service, marketing quality) divisions including these of automotive manufacturer to share know how and information supports engineers' conception. This links to the improvement of technology and improves the quality of business process. Here, this organization is realized in the form of a "Total Task Management Team" by final vehicle assembly manufacturer, brake unit manufacturer, and brake pad manufacturer as shown in Figure 9.

Five teams were formed, which consisted of QA1 (Engineering design), QA2 (Production engineering), QA3 (Manufacture and Inspection), QA4 (Facility Maintenance) and QA5 (Quality and Production information).

These five teams mutually cooperated in developing Total QA Network (QAT) (Amasaka, 2012). As a key technology of quality assurance activities, we create here three pillars of management namely "Total Technical Management (TM)", "Total Production Management (PM)" and "Total Information Management (IM)". QAT activities are implemented by fusing the three total management in Figure 10 into one.

Total technical management

Makes a sensitivity analysis of dispersion in both raw materials and the process condition for the brake response quality and the brake noise, and reviews the engineering method and clarifies process condition control items.

Total product management

To realize the process conditions and control items, divisions

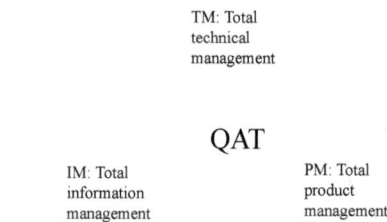

Figure 10: Three total QA network management activities.

such as development, production design, production engineering, manufacturing, maintenance, and quality assurance all combined carry on production activities using the "QA Network" table Kojima [47] of manufacturing based on the matrix diagram method and the process FMEA (Failure Mode and Effect Analysis), and FTA (Fault Tree Analysis), etc. utilizing QFD (Quality Function Deployment).

Total information management

Establishes a system for making a timely feedback of the market quality information (dealers), following process information (completed vehicle manufacturers) and self-process information (parts manufacturers) to the process.

Development of Total QA Network using SQC Technical Methods

To promptly optimize the Total QA Network systematically and organically, the "SQC Technical Methods" which is popularly used as the "Mountain-Climbing for Problem-Solving" utilizing Statistical Science as shown in Figure 11 is applied. Various types of arrows in the figure represent team activities of QA1 through QA5 respectively.

Total technical management activities

Makes analysis on each material (Figure 11, TM1), conducts a market survey (Figure 11, TM2), and makes factorial analysis on the basis of the above results (sensitivity analysis), boiling down the material to a short term. For example, in the factorial analysis I (Figure 11, TM3), it is found, by using the principal component analysis and other, that in the raw material characteristics, stratified mineral particle size and the diameter of inorganic fiber are related with the noise and friction characteristics as shown in Figure 12. Area "a" represents an area where the noise and friction are in considerably bad condition, while area "b" representing an area where the effect remains. We found area "c" where the both characteristics are not contrary to each other.

We have found through the factorial analysis I that variation of the

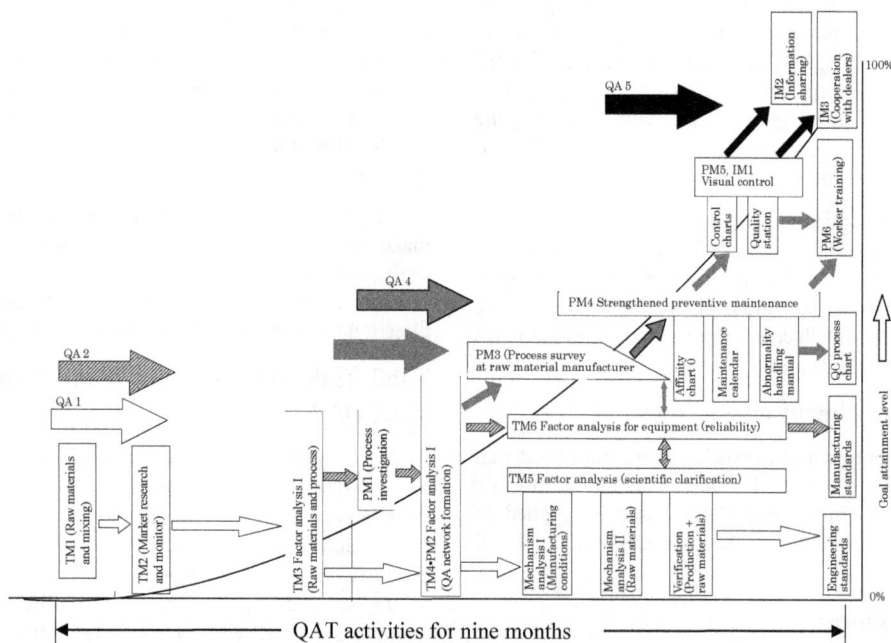

Figure 11: Development of total QA network using SQC technical methods.

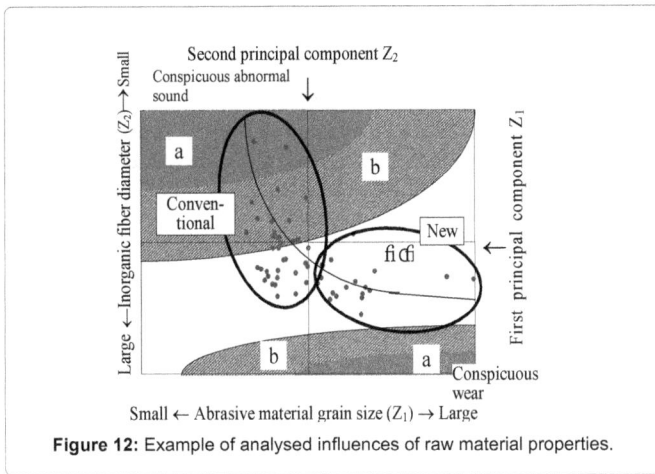

Figure 12: Example of analysed influences of raw material properties.

process and manufacturing conditions (Figure 11, PM1) is important. With regard to ground components (inorganic fibers and hard fine particles), we verified the factorial effect in combination with a technical analysis of a single material and the state of dispersion in the pad through the electron microscope observation. The authors were thus able to clarify the mechanism of variation of the braking effect caused by manufacturing dispersion of inorganic fibers. This led to a successful improvement in cooperation with the material manufacturers.

On the stage of production preparation, the drawings of the product were created using the QA Network table of engineering (Figure 11, TM4) according to the market quality. The drawings of the equipment were created using the QA Network table of manufacturing (Figure 11, PM2) on the basis of quality of conformance. On this stage, we boiled factors down to important factors, which are related to the acceptance of raw materials and the management of the condition and state of manufacturing process.

In addition, we then analyzed phenomena on respective factors selected (Figure 11, TM5). We grasped the equipment condition quantitatively to optimize tolerance in the product drawing scientifically. For example, by using unbalanced regression plotting, we came to understand that the thermoforming temperature and the substitute characteristics of the braking noise are in causal relation as shown in Figure 13.

Each area "a" represents an area where there is the lack of strength or an area not suitable for the forming while area "b" representing an area having residual effect of the lack of strength or unsuitableness for forming. Considering the strength and formability, we decided on Area "c" for the management condition of thermo-forming process. To retain the management condition of area "c", we implemented a factorial analysis on the side of the equipment (Figure 11, TM6). For example, we made a factorial analysis of dispersion in the temperature of the forming dies. This enabled us to provide the uniform pad forming temperature.

Total product management

To realize the process conditions found through "Total Technical Management", factors for non-conformance were investigated with the process survey (Figure 11, PM1) and association

With quality characteristics was determined with the "QA Network" of manufacture (Figure 11, PM2). Moreover, to complete the "QA Network", a brake unit manufacturer, a brake pad manufacturer

and raw material manufacturer formed a task team, for instance, to carry on quality review mutually (Figure 11, PM3).

As the result, present process capability became known. This developed to the prevention of defect and flow-out of defectives, then to the strengthening of preventive maintenance (Figure 11, PM4: Preventive maintenance calendar), visual management (Figure 11, PM5: In line SQC), Worker Training (Figure 11, PM6: Manual for measures of abnormal quality).

Total information management

We quality check station (Figure 5, IM1), which readily provides process information as the result of the "Total Product Management". Moreover, we established routes by which market information (DAS; Dynamic Assurance System) Sasaki [48] held by the completed vehicle manufacturer can be shared by parts manufacturers (Figure 5, IM2) and the route by which actual parts information can be acquired from dealers (Figure 5, IM3).

At present, we are summarizing the market information, process information, engineering information and other quality information of QA1 through QA5. We promote to upgrade the pad performance and quality by analyzing the above-mentioned information. In addition, we wish to state that the scheduled development of wide QCD studying activities by both companies realized given results where market claim was reduced to one-sixth and under.

Effect

Through practical use of Total Task Management Team, the effect of the QCD research activities is remarkable as follows;

*Estimated market claim ratio reduction to 1/6 (from 2.6% to 0.4%) *In-process fraction defective: down 60% (from 0.5% to 0.2%) *Short convergence of initial failures: (from 9 months to 3 months) *Cost reduction: 9.4% (156 Yen/unit)

High Reliability Assurance of the Transaxle Oil Seal Leakage

The oil seal for the drive system works to seal lubricant inside the transaxle unit. The cause and effect relationship between the oil seal design parameters and sealing performance is not necessarily fully clarified. As a result, oil leakage from the oil seal is not completely eliminated, presenting a continual engineering problem (Lopez et al. [49]. So far, oil seal quality improvement has been made as follows; a development designer having empirical engineering capability

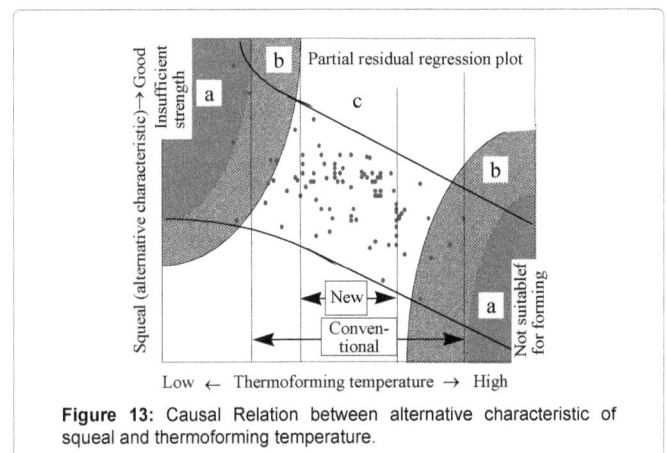

Figure 13: Causal Relation between alternative characteristic of squeal and thermoforming temperature.

recovers the leaking oil seal parts from the market, analyzes the cause of leakage with proper technology, and incorporates countermeasures into the design.

Many of the recent leaking parts, however, exhibit no apparent problem and the cause of the leakage is often undetectable. This makes it difficult to map out permanent measures to eliminate the leakage; Nozawa et al. [50].

Dual total task management team

Effective solution of technical problems requires the formation of teams and understanding of the essence of problems by the teams as a whole. To ensure high reliability of product design and quality assurance, a Total Task Management Team involving final vehicle assembly manufacture with transaxle unit (T company), and oil seal manufacturer (N company) personnel was created to transform the implicit knowledge (relating to product and processes in both organization) into explicit knowledge and to create new technology of interest to both organizations. The Dual Total Task Management Team named "DOS-Q" (Drive-train Oil Seal-Quality Assurance Team: T DOS-Q5 and N DOS-Q8) is shown in Figure 14.

"T DOS-Q5" constituting teams comprise Q1 and Q2 in charge of investigation into the cause of the "oil leakage" and Q3-Q5, which handled manufacturing problems relating to drive shafts, vehicles and transaxles. Similarly, "N DOS-Q8" formed teams Q1 through Q8. Q1 and Q2 at T Company interacted closely with their counterparts at N company to improve the reliability of the oil seal as a single unit and, likewise, Q3-Q8 handled the manufacturing problems for quality assurance.

Accordingly, the teams shared their individual knowledge (relating to empirical techniques and other technical information) to apply them to solving the problems under consideration. Each team had a general manager and the joint team was led by T Company's TQM promotion general manager for the vehicle reliability assurance. The methodology of TDS-D (Total Design System for Drive-train Development) involving TM (Technology Management), Pm (Production Management), and IM (Information Management) was used by utilizing "Three Total QA Network Management Activities" as the above Figure 10.

Moreover, to realize optimization of the "DOS-Q5" business process, problem solving is formulated using the "Development of

Total QA Network using SQC Technical Methods" by using the same approach as the above Figure 11.

Fault analysis and factorial analysis

In the conventional cases of oil seal unitary sample collection process, it was from time to time observed that there was available no information on the mating part or the collected sample was attached with foreign matters which hampered the determination of the cause of leakage. To prevent this, recovery process was improved along with the acquisition of the background or the history of the market recovery. Thus Weibull analysis could be made in Figure 15 with credibility based on the comparison result with the actual parts by recovering non-defectives as well as defectives or by recovering whole of the transaxle unit in order to reproduce oil leakage.

As the result, we have obtained a new knowledge that the failure type of "oil leakage" is a mixed model of three failures of the "Decreasing failure rate", "Constant failure rate" and "Increasing failure rate" which was not found in th e rule of thumb. Moreover, we could understand correct particle size distribution and composition of foreign matters attached to the oil leaking parts by recovering them in a special recovery case. It was also newly discovered that two oil seals of apparently similar degree of wear produce difference in pumping quantity depending on the mileage.

For the factorial analysis to researching the oil leakage phenomenon and mechanism of the failure type, not only defective parts (oil leaking parts) but non-defectives (non-oil leaking parts) were used. As the result, the difference of two parts was detected through the discriminating analysis as shown in Figure 16. For example, It is known that the factor analysis to the design factor of hardness of oil seal rubber has the largest influence among explanatory factors evaluated. The result agrees with the experience technology.

Such survey and analysis could not be made with the conventional separate investigation activities of the vehicle manufacturer or the parts manufacturer. This was accomplished through the "Total Task Management Team" activities between the companies of T and N.

Clarification of oil Leakage mechanism by visualization

We have arranged the information obtained from the above-mentioned activities, which have boiled down to the following new

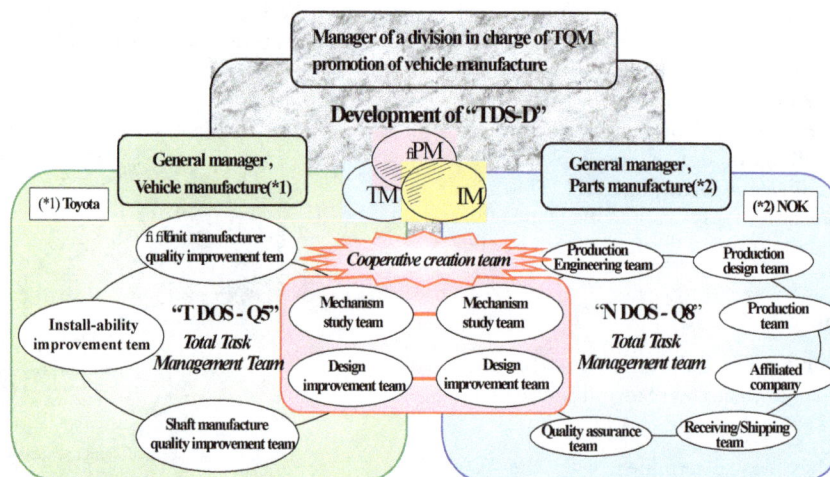

Figure 14: Configuration of cooperative creation team (Total task management team).

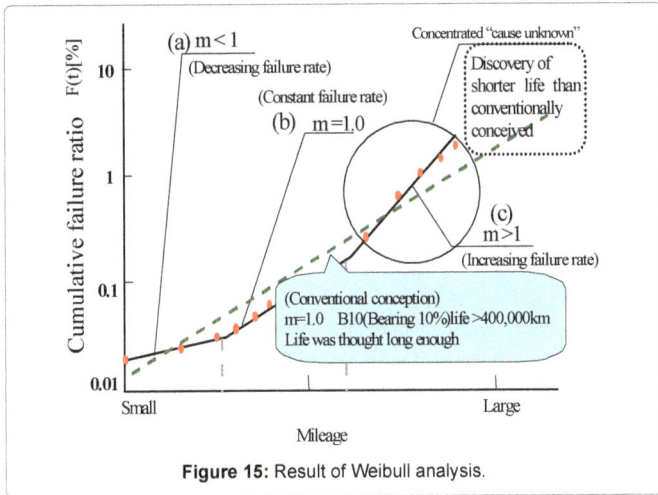

Figure 15: Result of Weibull analysis.

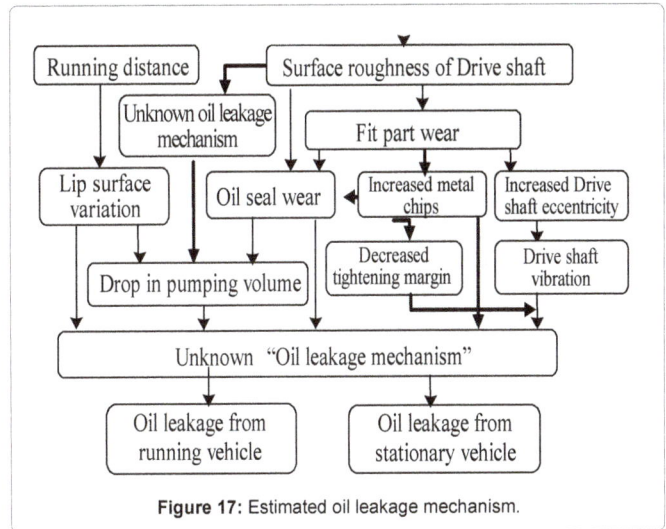

Figure 16: Influential effect of each factor.

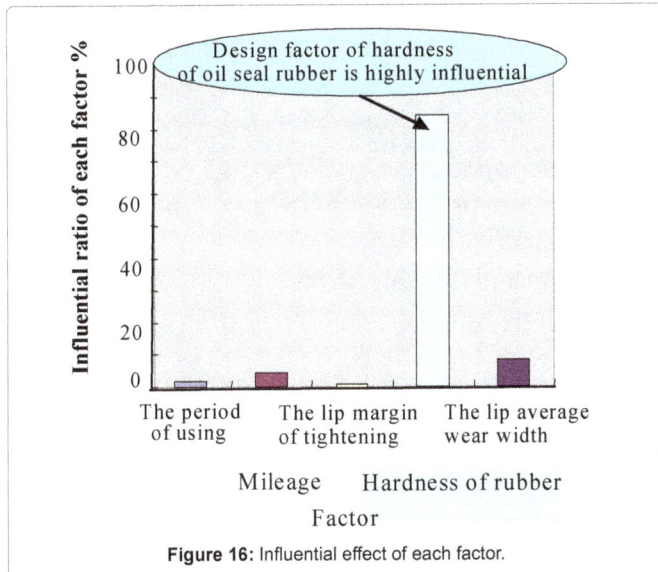

Figure 17: Estimated oil leakage mechanism.

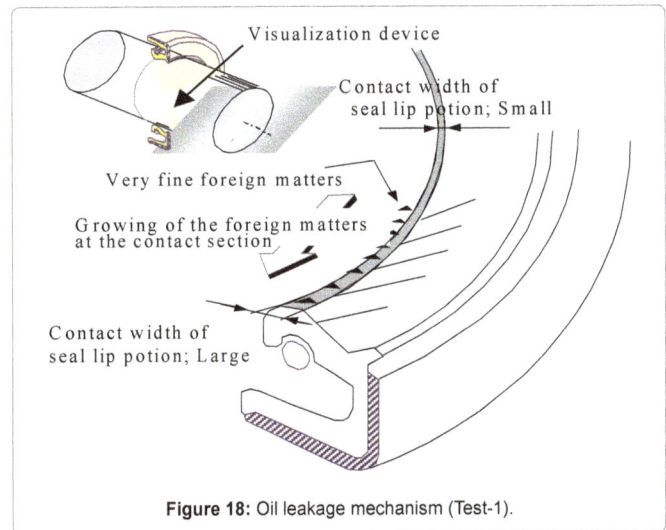

Figure 18: Oil leakage mechanism (Test-1).

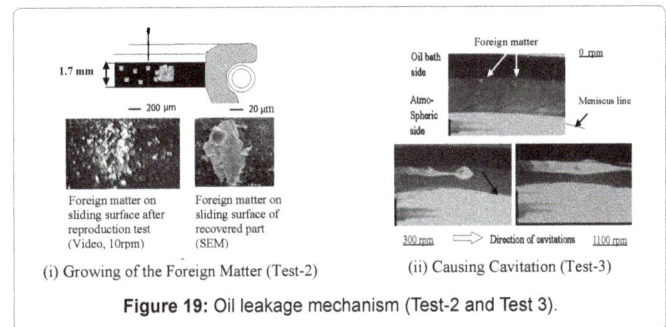

Figure 19: Oil leakage mechanism (Test-2 and Test 3).

facts : (1) When compared the recovery parts with the forced wear parts, sealing performance differs because of the difference in the lip surface condition, (2) Sealing performance drops if the axis has greater surface roughness. On the basis of such findings and the quantitative analysis data, the team generated another affinity and association diagrams concerning oil leakage and estimated the oil leakage mechanism to discover the presence of "unknown mechanism" as shown in Figure 17.

Accordingly, a device was developed to visualize the dynamic behavior of the oil seal lip to turn this "unknown mechanism" into explicit knowledge as shown in Figure 18 and Figure 19. As the result of observation of the recovered parts from the market with this visualization device, a process was observed by which very fine foreign matters which were conventionally thought not to affect the oil leakage grow at the contact section (Test-1 and Test-2). As a result of the component analysis, it is confirmed that the fine foreign matter is the powder produced during engagement of gears inside the transaxle gear box.

These fine foreign matters are piled on the microscopic irregularities on the lip sliding surface to bring changes to the microscopic pressure distribution eventually degrading the sealing performance. Also the presence of a mechanism was confirmed from a separate observation

result that foreign matters that had bitten into the lip sliding surface caused the aeration (cavitations) to be generated to the oil flow on the lip sliding surface to deteriorate the sealing performance (Test-3). As far as the authors know, such knowledge was not given consideration conventionally and only discovered through the current team activities.

Applying optimal CAE design approach model

We address the technological problem of oil seal leakage in automotive drive trains as a way to construct an "Optimal CAE Design Approach Model" for quality assurance. The model is used to explain cavitation caused by the metal particles (foreign matter) generated through transaxle wear, a pressing issue in the automobile industry.

Oil Simulator Using Highly Reliable CAE Analysis Technology Component Model

We used the knowledge obtained from the visualization experiment to logically outline the faulty as the above "(3) Clarification of Oil Leakage Mechanism by Visualization". This was done in order to capture the problem using the Highly Reliable CAE Analysis Technology Component model as shown in Figure 20. Using this process, the author was able to arrive at a hypothesis for why the cavitation was occurring; namely, factors like low pump volume and seal damage had compromised the tightness of the seal and lead to oil leaks.

As the figure indicates, the designs are optimized by integrating several aspects of the calculation process, including problem (root cause) identification, conceptualizing the problem logically, and calculation methods (precision of calculators). Once the root causes of the problem are identified, it is critical that there is no discrepancy between the mechanism described and the results of prototype evaluations.

The visualization experiment revealed that cavitation was occurring due to a weakening of the oil seal in areas (surfaces) that were in contact with the rotating drive shaft. This weakening was causing oil seal leaks. The Rayleigh Plesset Model for controlling steam and condensation was used as a CAE analysis model that could explain the problem.

The finite element method and non-stationary analyses were used as convenient algorithms. The Reynolds-averaged Navier-Stokes equation, Bernoulli's principle, and lubrication theory were appropriate theoretical formulas. Accuracy was ensured, and the time integration method was used to perform calculations in a realistic timeframe. Each of the above elements was used to construct the Oil Seal Simulator.

CAE Analysis Examples

A cavitation is generated at the following steps; Oil collides with a foreign substance - the flow velocity rise near a foreign substance - the fall of pressure - decreased pressure is carried out to below saturated vapor pressure - emasculation of oil - generating of a cavitation as shown in Figure 21. This analysis results at a rotation speed of 1100 rpm.

(i) The fluid speed analysis was then conducted in order to look more closely at the mechanism causing cavitation. The analysis revealed that rapid changes in fluid speed were occurring in the vicinity of foreign particles, and that fluid speed drops immediately before the oil collides with foreign matter. This led to the conclusion that the presence of foreign particle was having an effect on oil flow.

(ii) Comparing cavitation and the fluid speed analysis results against the results of the pressure analysis reveals that in areas of reduced pressure, oil was disappearing inside the cavities being formed meaning that drops in pressure e were likely being caused by these concave areas.

(iii) Cavitation analysis confirmed the cavitation occurring around foreign matter, thus replicating the results of the visualization experiment. At the same time, the finding that cavitation becomes more significant as the rotation speed of the drive shaft increases was similarly replicated.

Verification and Consideration

The above CAE analysis allowed, we to clarify the faulty mechanism causing cavitation; namely, the presence of metal foreign particles was affecting the strength of the oil flow, causing drops in pressure in areas

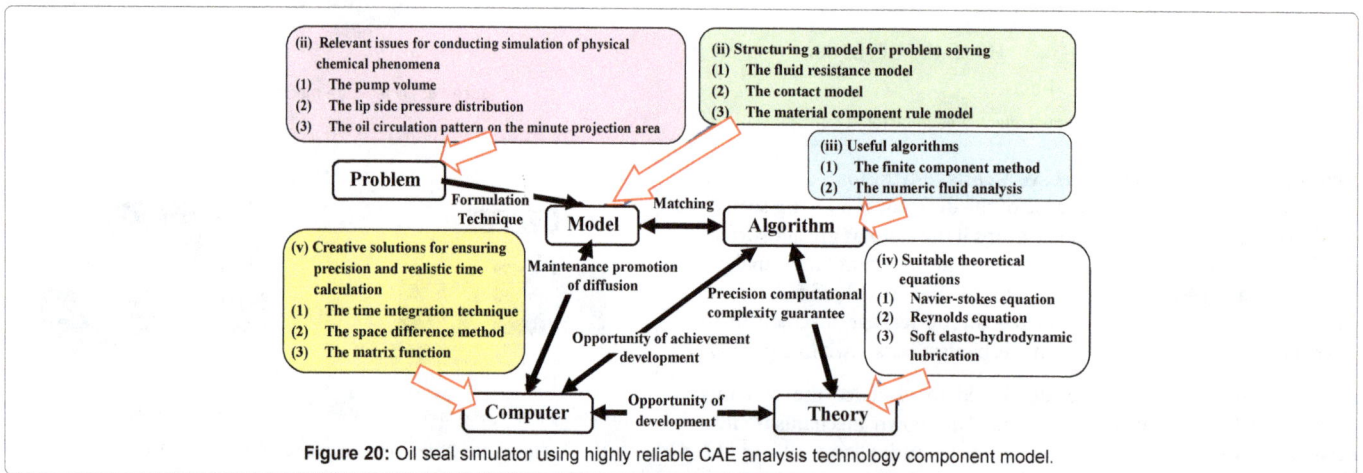

Figure 20: Oil seal simulator using highly reliable CAE analysis technology component model.

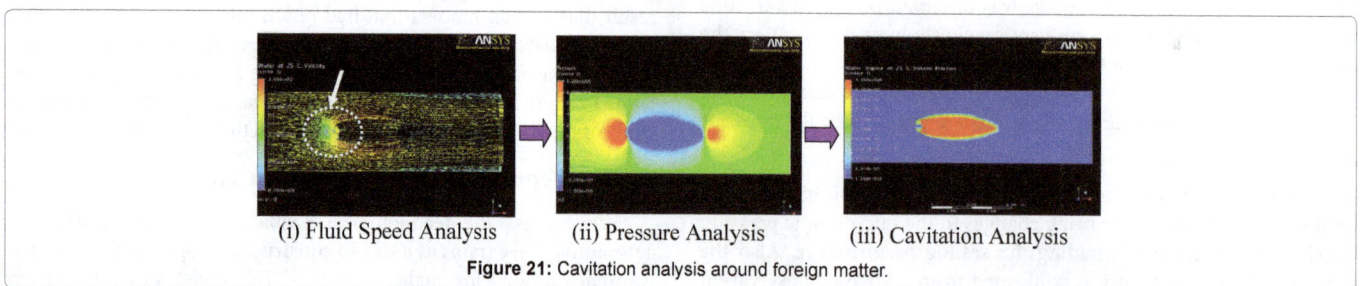

(i) Fluid Speed Analysis   (ii) Pressure Analysis   (iii) Cavitation Analysis

Figure 21: Cavitation analysis around foreign matter.

with faster oil flow and creating cavities. In addition, a similar analysis of changes in the shape and size of the foreign particles revealed that these changes were also causing changes in cavitation. These CAE analysis results indicate a close link between particle size/shape and cavitation.

Preproduction and testing/evaluation of prototypes add a significant amount of time and cost to the development process. However, precise CAE allowed manufacturers to eliminate preproduction (as well as prototype testing/evaluation) and still predict the mechanism causing cavitation and oil leaks. Though gaps such as minute surface variations caused by foreign particles and the shape of the oil film model exist, the CAE analysis allowed the authors to recreate the changes in flow speed and pressure around the foreign metal particles that were causing cavitation - changes which typically cannot be identified.

The deviation between the CAE analysis results and the results of the prototype testing were less than 5%, attesting to the usefulness of precise CAE analysis in certain cases.

Design Changes and Process Control for Improving Reliability

These results led to two measures to improve design quality (shape and materials): (1) strengthen gear surfaces to prevent occurrence of foreign matter even after the B10 life (L10 Bearing to MTBF) to over 400,000 km (improve quality of materials and heat treatments), and (2) formulate a design plan to scientifically ensure optimum lubrication of the surface layer of the oil seal lip where it rotates in contact with the drive shaft.

The result of these countermeasures was a reduction in oil seal leaks (market complaints) to less than 1/20th their original incidence as shown in Figure 22. We believe that this research result will contribute to period shortening of the development design and simultaneous fulfillment of QCD greatly from now on.

Application to Similar QCD Studies

We were able to apply the New SCM MODEL to critical QCD studies of automobile manufacturing using New JIT strategy, including predicting and controlling the special characteristics employing Strategic Stratified Task Team as follows; (i) Preventing vehicles' rusting (ii) Intelligence production operating system (iii) Simultaneous fulfillment of QCD for improving automotive chassis

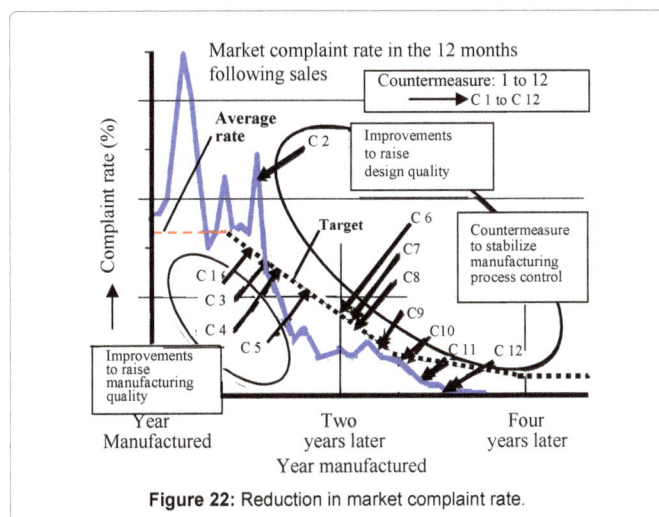

Figure 22: Reduction in market complaint rate.

(paint corrosion resistance and welding process), (iv) Automobile Body Color Development Approach Model Muto et al. [51] (v) prevention of bolt looseness (Hashimoto et al., [52], (vi) New Global Partnering Production Model, (vii) New Turkish Production System and New Vietnam Production Model (Siang et al., [53]; Miyashita [54] and (viii) joint task team activities between Toyota Motor Corporation and Toyota Motor Thailand.

Conclusion

Today's management challenge of manufacturer is to provide high QCD products ahead of competitors for the market creating. The current task of today's Japanese automobile manufacturing is to succeed in global production by evolution of manufacturing and SCM between the assembly makers and affiliated/non-affiliated suppliers. To realize this, we developed the Strategic Stratified Task Team Model with the Structured Model of Strategic Stratified Joint Task Teams, and Strategic Cooperative Creation Team Model by employing New JIT.

Concretely, we created the New SCM Model consisting of the Strategic Task Team Model, Global Partnering Model, and Simultaneous QCD Fulfillment Approach Model developing Strategic Stratified Task Team Model. By applying the created New SCM Model, we could illustrate the effectiveness of strategic QCD studies on current bottleneck problems, braking performance, transaxle oil seal leakage, and many others.

In the future, New JIT will be positioned as "new principle of next-generation manufacturing" of worldwide production management technology strategy, and applied to solving various practical problems utilizing New SCM Model.

Acknowledgement

The author would like to acknowledge the generous support received from the following researchers. All these at Toyota Motor Corporation, many affiliated/nonaffiliated suppliers, and those connected with the Amasaka laboratory at Aoyama Gakuin University that assisted with author's research.

References

1. Amasaka K (2002) New JIT a new management technology principle at Toyota. Int J Production Economics 80: 135-144.

2. Amasaka K (2007a) High linkage model advanced TDS TPS & TMS: strategic development of new JIT at Toyota. Int J Operations and Quantitative Management 13:101-121.

3. Amasaka K. (2014a) New JIT new management technology principle. J Advanced Manufacturing Systems 13: 197-222.

4. Amasaka K (2000) Partnering chains as the platform for quality management in Toyota. Proceedings of the 1st world conference on production and operations management sevilla spain: 1-13.

5. Amasaka K (2008a) Simultaneous fulfillment of QCD - strategic collaboration with affiliated and non-affiliated suppliers. New Theory of Manufacturing Surpassing JIT: Evolution of Just-in-Time : 199-208.

6. Amasaka K (2008b) Strategic QCD studies with affiliated and non-affiliated suppliers utilizing new JIT. Encyclopedia of Networked and Virtual Organizations Information Science Reference Hershey New York 3: 1516-1527.

7. Amasaka K (2012) Science TQM new quality management principle the quality management strategy of toyota. Bentham Science Publishers

8. Amasaka K (2015) Keynote speaker global manufacturing strategy of new JIT: Surpassing JIT. International Conference on Information Science and Management Engineering Phuket Thailand: 1-12.

9. Amasaka K (2014b) New JIT new management technology principle. Taylor & Francis CRC Press

10. Ohno T (1977) Toyota Production System Diamond-Sha.

11. Hayes RH, Wheelwright SC (1984) Restoring our competitive edge competing through manufacturing.Wiley New York.

12. Womack JP, Jones DT (1994) From lean production to the lean enterprise. Harvard Business Review: 93-103.

13. Taylor D, Brunt D (2001) Manufacturing operations and supply chain management - Lean Approach Thomson Learning.

14. Amasaka K (2005) New japan production model an innovative production management principle strategic implementation of new JIT. Proceedings of the 16th Annual Conference of the Production and Operations Management Society Michigan Chicago IL: 1-17

15. Amasaka K (2007b) New japan production model an advanced production management principle: key to strategic implementation of new JIT. The International Business & Economics Research Journal 6: 67-79.

16. Nezu K (1995) Scenario of the jump of US - Manufacturing Industry Based on CALS. Industrial Research Institute.

17. Joiner BL (1994) Fourth generation management: the new Business consciousness joiner associates Inc McGraw-Hill.

18. Nihon Keizai Shinbun (2000) Worst Record: 40% increase of vehicle recalls (July 6, 2000) Risky quality apart from production increase it is re call rapid increase (February 8, 2006).

19. Amasaka K, (2008c) Science TQM a new quality management principle: The quality management strategy of Toyota. The J Management & Engineering Integration 1: 7-22.

20. Amasaka K (2007c) Highly reliable CAE model the key to strategic development of advanced TDS. J Advanced Manufacturing Systems 6: 159-176.

21. Amasaka K (2004a) Applying new JIT A management technology strategy model at toyota - strategic QCD studies with affiliated and non-affiliated suppliers. Proceedings of the 2nd World Conference on Production and Operations Management Society Cancun Mexico: 1-22.

22. Amasaka K (2004b) Development of science TQM a new principle of quality management effectiveness of strategic stratified task team at Toyota. Int J Production Research 42: 3691-3706.

23. Kotler F (1999) Kotler marketing the free press a division of Simon & Schuster Inc.

24. Amasaka K (2005) 4.2 Analysis of sources of variation for preventing vehicles rusting science SQC new quality control principle The quality strategy of Toyota. Springer: 23-31.

25. Burke W, Trahant W (2000) Business climate shift oxford butterworth-heinemann.

26. Amasaka K (2004b) Science SQC New quality control principle. The quality strategy of Toyota Springer.

27. Seuring S, Muller M, Goldbach M, Schneidewind U (2003) Strategy and organization in supply chains Heiderberg Physica.

28. Goto T (1999) Forgotten origin of management quality taught by GHQ CSS management lecture productivity publications.

29. Amasaka K (2007) New japan model science TQM: theory and practice for strategic quality management study group of the ideal situation the quality management of the manufacturing industry maruzen

30. Evans JR, Dean JW (2003) Total quality management organization and strategy. Thomson South-Western.

31. Doz YL, Hamel G (1998) Alliance advantage. Boston MA harvard business school press.

32. Hamel G, Prahalad CK (1994) Competing for the future. Harvard Business School Press Boston.

33. Amasaka K (1988) Concept and progress of Toyota production system (Plenary lecture) co-sponsorship The Japan Society of Precision Engineering Japan.

34. Amasaka K (1999) TQM at toyota toyota's TQM activities to create better car. A Training of Trainer's Course on Evidence based on participatory quality improvement international health program (TOT Course on EPQI) 1-17 Touhoku University School of Medicine (WHO Collaboration Center) Sendai-city Miyagi Japan.

35. Amasaka K (2014c) New JIT new management technology principle: surpassing JIT. J Procedia Technology Special Issues 16: 1135-1145.

36. Amasaka K (2008c) An integrated intelligence development design CAE model utilizing new JIT Application to Automotive High Reliability Assurance. J Advanced Manufacturing Systems 7: 221-241.

37. Toyota Motor Corp (1987) Creation unlimited 50 Years History of Toyota Motor Corporation.

38. Yamaji M, Amasaka K (2009) Strategic productivity improvement model for white-collar workers employing science TQM. The J Japanese Operations Management and Strategy 1: 30-46.

39. Amasaka K, Sakai H (2010) Evolution of TPS fundamentals utilizing new JIT strategy proposal and validity of advanced TPS at toyota. J Advanced Manufacturing Systems 9: 85-99.

40. Amasaka K, Sakai H (2011) The new japan global production model NJ-GPM: strategic development of advanced TPS. The J Japanese Operations Management and Strategy 2: 1-15.

41. Lagrosen S (2004) Quality management in global firms. The TQM Magazine 16: 396-402.

42. Ljungström M (2005) A model for starting up and implementing continuous improvements and work development in practice. The TQM Magazine 17: 385-405.

43. Burke RJ (2005) Effects of Reengineerin g on the Employee Satisfaction - Customer Satisfaction Relationship. The TQM Magazine 17: 358-363.

44. Hoogervorst JAP et al (2005) Total quality management: the need for an employee-centred coherent approach. The TQM Magazine 17: 92-106.

45. Ebioka K, Sakai H, Yamaji M, Amasaka K (2007) A new global partnering production model NGP-PM Utilizing Advanced TPS. J Business & Economics Research 5: 1-8.

46. Miller N (1978) An analysis of disk brake squeal. SAE Technical Paper 780332.

47. Kojima T, Amasaka K (2011) The total quality assurance networking model for preventing defects: building an effective quality assurance system using a total QA network. Int J Management & Information Systems 15: 1-10.

48. Sasaki S (1972) Collection and analysis of reliability information in automotive industries. The 2nd reliability and maintainability symposium Union of Japanese Scientist and Engineers: 385-405.

49. Lopez AM, Nakamura K, Seki K (1997) A study on the sealing characteristics of lip seals with helical ribs. Proceedings of the 15th International Conference of British Hydromechanics Research Group Ltd Fluid Sealing :1-11.

50. Nozawa Y, Ito T, Amasaka K (2013) High precision CAE analysis of automotive transaxle oil seal leakage. China-USA Business Review 12: 363-374.

51. Muto M, Miyake R, Amasaka K (2011) Constructing an automobile body color development approach model. J Management Science 2: 175-183.

52. Hashimoto K, Onodera T, Amasaka K (2014) Developing a highly reliable CAE analysis model of the mechanisms that cause bolt loosening in automobiles. American J Engineering Research 10: 178-187.

53. Siang YY, Sakalsiz MM, Amasaka K (2010) Proposal of new turkish production system (NTPS): Integration and Evolution of Japanese and Turkish Production System. J Business Case Study 6: 69-76.

54. Miyashita S, Amasa, K. (2014) Proposal of a new Vietnam production model NVPM: A new integrated production system of Japan and Vietnam. IOSR Journal of Business and Management 16:18-25.

6

Effect of Permanent Formwork using Ultra-High Performance Concrete on Structural Behaviour of Reinforced Concrete Beam Subjected to Bending as a Function of Reinforcement Parameter

Sung-Hoon Kang, Sung-Gul Hong and Yang-Hee Kwon*

Department of Architecture and Architectural Engineering, Seoul National University, 1 Gwanak-Ro, Gwanak-Gu, Seoul 08826, Republic of Korea

Abstract

In order to use ultra-high performance concrete (UHPC) as a permanent formworks for a reinforced concrete (RC) slab in a construction field, the structural performance of the RC-UHPC composite beams subjected to bending are experimentally investigated. The main parameters are the rebar location and the UHPC thickness. The experimental results show that the crack patterns of the composite specimen is different compared to the reference RC specimen, because of the crack localization phenomenon of UHPC. Within the post peak state of the load-deflection relationship, the composite specimen behaves similar to the RC specimen. However, the reinforcement of deformed bars in the UHPC section shows a synergy effect on structural performance; both load and deformation capacity is significantly increased. Nevertheless, it is required to reinforce the bars in normal concrete section for preventing debonding failure as well as for retaining the load resisting capacity at the post peak state. The best structural performance of the composite specimen is found when the reinforcement ratios are the same in the both normal concrete and UHPC sections. The results of this study help to apply and design the thin UHPC panel as a permanent formwork for retrofitting existing structures.

Keywords: Ultra-High Performance Concrete (UHPC); Permanent formwork; Composite beam

Introduction

Ultra-High Performance Concrete (UHPC) has outstanding mechanical properties (e.g., compressive strength>150 MPa) and durability, compared with conventional concrete [1,2]. This has provided new opportunities to develop new concrete technology. However, the unit cost of this material is very high [2,3]; thus, it should be carefully considered for the use of this material in practice. Utilizing a thin UHPC layer as a permanent formwork (Figure 1) for concrete slab or deck is a promising and effective method because it can improve structural performance (e.g., as crack-resisting capacity, reduction of deflection), durability and watertightness. As a first step, it is necessary to investigate the structural performance of a composite beam which is composed of thin UHPC and normal concrete. Therefore, this study experimentally investigated the flexural behaviour of the composite beam by varying reinforcing steel parameters: location and thickness of UHPC layer. In particular the significance of the location of the rebar (deformed bar) was discussed.

Experimental Procedure

Materials properties and mix proportions

The UHPC was prepared according to the mix proportion listed in Table 1 [4,5]. By using a pan type mixer, ordinary Portland cement type I, undensified silica fume, quartz sand and silica flour were firstly blended for 10 min, and then water and superplasticizer were added into the powder and mixed. Lastly, steel fiber was added and mixed for another 3 min. Once the mixing was finished, the slump flow was measured according to ASTM C1611 [6]. The slump flow value was determined as 700 ± 50 mm. Based on the ASTM C39 [7], the compressive strength was measured as 143 MPa at 28d. The specimen was cured under ambient curing condition of temperature of 20 ± 2°C and R.H of 60 ± 5% [8,9]. A commercial ready-mix concrete with the maximum aggregate size of 25 mm was used for the composite beam which was intended for a concrete slab production. The compressive strength of the concrete was 18 MPa, which was measured at 28d. As a main experimental parameter, deformed bar was used. Its diameter, yield strength and ultimate strength are 10 mm, 465 MPa and 600 MPa, respectively. The location and reinforce ratio are different per each specimen.

Test parameters

Total eight beams were prepared for the experiment. They are composed of two reinforced concrete beams and six composite beams as presented in Table 2. The dimensions were determined based on typical concrete slabs in residential buildings. As mentioned in the Section 1, the main parameters were set as the reinforcement ratio, location of rebar and UHPC thickness. The specimen names in Table 2 indicate the parameters. For instance, NC4+UC0 indicate that 4-D10 deformed bars are reinforced in the NC section of the specimen, but no bars are reinforced in the UHPC section. Basically, all UHPC sections were designed as thin as possible; thus, the thickness was determined

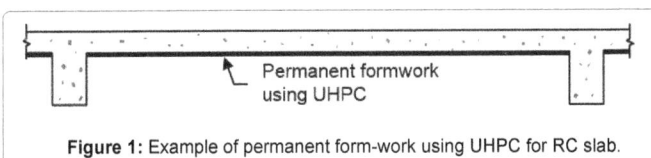

Figure 1: Example of permanent form-work using UHPC for RC slab.

*Corresponding author: Yang-Hee Kwon, Department of Architecture and Architectural Engineering, Seoul National University, Gwanak-Ro, Gwanak-Gu, Seoul 08826, Republic of Korea, E-mail: yanga1126@snu.ac.kr

Cement	Silica fume	Quartz sand	Silica flour	Water	Super-plasticizer	Steel fiber*
1	0.25	1.1	0.35	0.22	0.03	2%
* Volumetric ratio of UHPC						

Table 1: Mix proportion of ultra-high performance concrete (by wt% of cement).

Specimen	NC4	NC4+UC0	NC0+UC4	NC2+UC2
Cross section	180 / 25 / 5@60	210 / 55 / 5@60	210 / 15 / 5@60	210 / 55 / 15 / 3@100
Sher span to depth ratio	4.84	3.85	3.85	3.85
Rebar in — Normal concrete	4-D10	4-D10	-	2-D10
Rebar in — UHPC	-	-	4-D10	2-D10

Specimen	NC2	NC0+UC0	NC0+UC4*	NC0+UC2
Cross section	210 / 25 / 3@100	210 / 300	220 / 25 / 5@60	210 / 15 / 3@100
Sher span to depth ratio	4.05	3.85	3.85	3.85
Rebar in — Normal concrete	2-D10	-	-	-
Rebar in — UHPC	-	-	2-D10	4-D10

Table 2: Beam sections and parameters.

as 30 mm, considering the 10 mm of cover thickness of the UHPC layer. However, one specimen, NC0+UC4*, has 10 mm thicker UHPC section than NC0+UC4, to investigate the effect of the stiffness of the UHPC on the structural performance. All specimens were designed to show the tension control failures. In other words, the reinforcement ratios of the specimens are ranged between 0.25% and 0.62%, which are safely lower than the balance reinforcement ratio of 1.57%.

Fabrication of RC-UHPC composite specimen

To fabricate the composite specimen, thin UHPC panel was firstly prepared, and then normal concrete (NC) was casted on the panel after 2 days. All specimens were cured under ambient temperature of 20 ± 2°C and R.H of 60 ± 5%. It is common practice to apply heat treatment for precast UHPC production. However, in this study, heat was not applied because of the deterioration in the bond strength between the two different concretes and reflecting more realistic condition of retrofit application. Specifically, we conducted the preliminary test to

investigate the effect of the heat treatment of the UHPC on the flexural behaviour of the composite beam (150 mm × 50 mm × 550 mm size). One of two UHPC panels (20 mm-thickness) was exposed to 80°C for 48 hrs (28 d compressive strength: 180 MPa), but the other one was cured at the ambient condition without the heat treatment. As shown in Figure 2, the 3-point bending test revealed that the brittle debonding failure occurred in the specimen only when the UHPC panel subjected to heat treatment before overlaid by NC. This is because the cement hydration reaction had almost terminated after the heat treatment, which makes impossible to gain chemical bond strength for a composite action.

Regarding structural performance of UHPC, fiber orientation in UHPC panel is a crucial factor [10]. To align their direction with the panel especially middle part of the specimen, fresh UHPC was firstly poured on the one sloped end of the mould; then, the mould filled with fresh UHPC was flatted and lastly the UHPC was flatten (Figure 3). At 2nd day, NC was cast on the UHPC panel and the composite specimens were cured until the test day.

Test program

The test setup is presented in Figure 4. By using a hydraulic actuator (maximum capacity of 500 kN), displacement was controlled (1 mm/min) for static 4-point bending test. The load and deflection that are

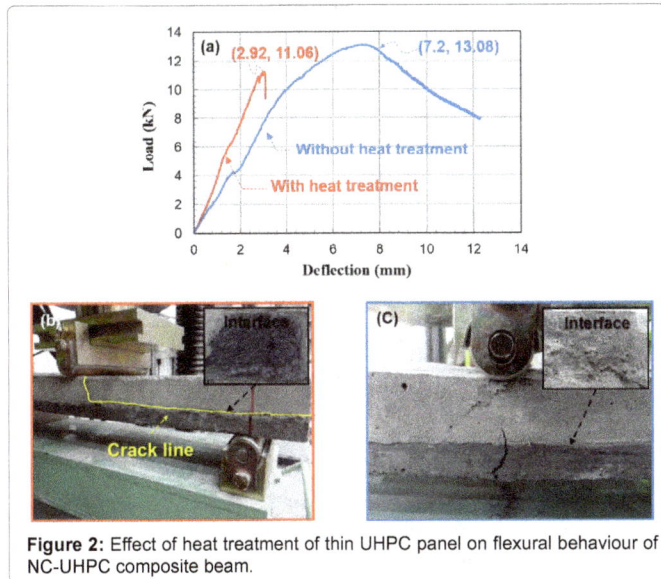

Figure 2: Effect of heat treatment of thin UHPC panel on flexural behaviour of NC-UHPC composite beam.

pointed at the Figure 4 were measured by load cells and linear variable differential transformers (LVDTs), respectively. In addition, the strains of concrete and rebar were measured by the attached strain gauges (orange lines).

Results

Crack pattern and failure mode

Figure 5 shows the cracked specimens at the end of the test. Above of all, there are several flexural cracks in NC specimens between the loading points (Figures 5a and 5c). Moreover, these flexural cracks were localized in the composite specimens (NC2+UC2, NC4+UC0) as shown in Figures 5b and 5d. In other words, if the UC2 (or UC0) is used for the permanent formwork and NC2 (or NC4) is casted on to the formwork, crack pattern will be changed compared to a traditional formwork which has to be removed. This change can be explained by the crack localization phenomenon of the permanent UHPC panel, which was observed in this experiment. Multiple micro cracks simultaneously occurred in the UHPC that was subjected to tensile stress due to the loading; finally, one single crack expanded and the UHPC panel was detached. The detached both sides of panels retain straight shape, but the RC maintained curved shaped. After that, it was observed that the contribution of the panel to the structural performance was disappeared. As a result, the behaviour of the composite beams should be changed to that of RC beams. Thus, the localized crack was changed to several flexural cracks. These flexural cracks in the composite beam

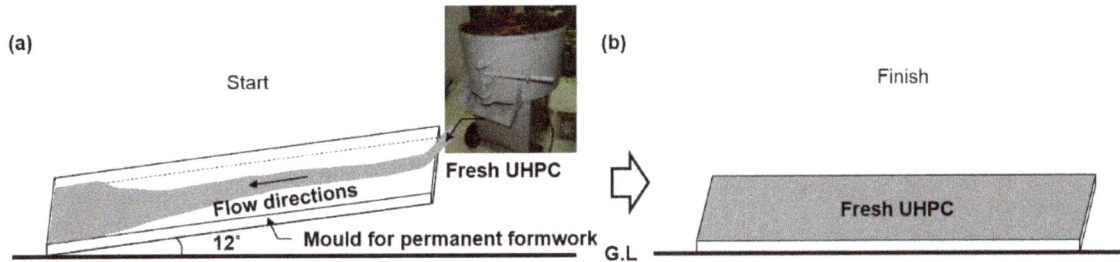

Figure 3: Preparation of thin UHPC permanent formwork considering uniform fiber orientation (i.e., horizontal direction to the surface).

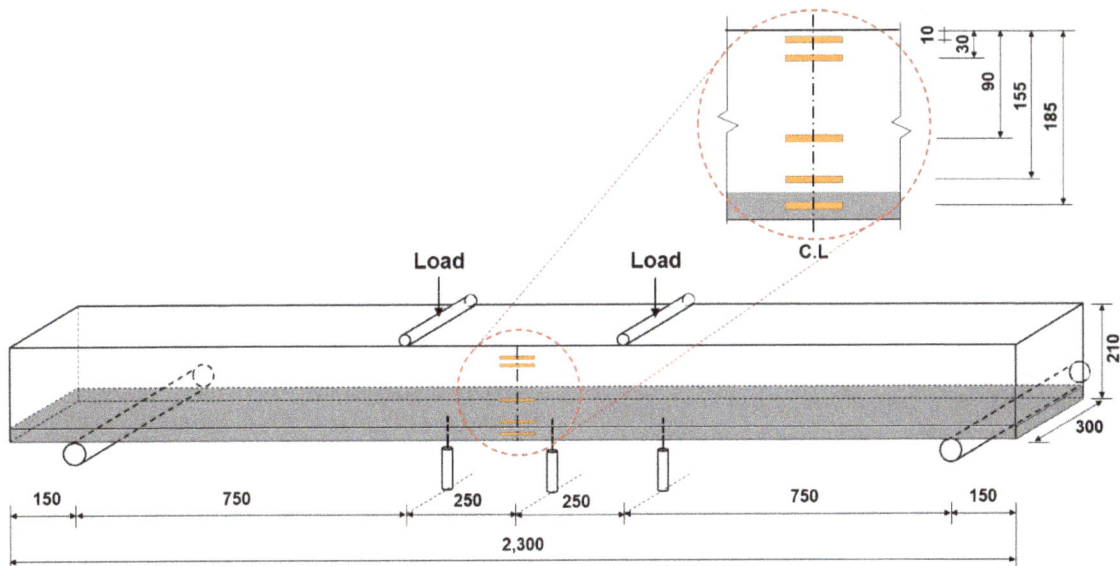

Figure 4: Setup for 4-point bending test.

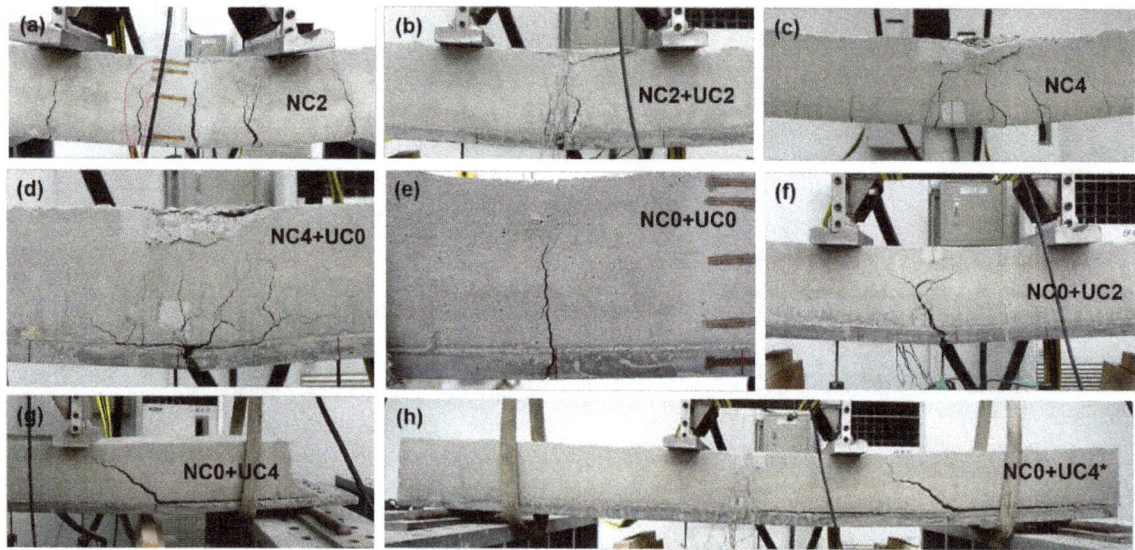

Figure 5: Cracked specimens after test.

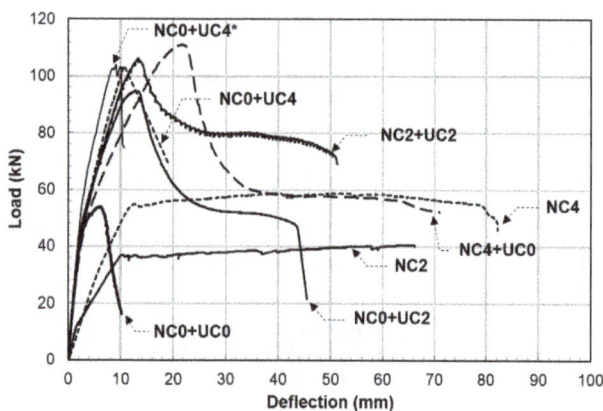

Figure 6: Load-midspan deflection relationship.

hardening behaviour by the fiber bridging action in the UHPC panel, the specimen suddenly collapsed. In other word, since all UHPC section was cracked after peak load, the entire load resistance capacity disappeared without leaving residual resistance. This brittle failure was also observed during the tests of NC0+UC4 and NC0+UC4*, in which deformed bars were reinforced. In these specimens, the debonding failure which was explained in Section 3.1 occurred even before the macro cracks were developed in a concrete and the bars yielded. Moreover, increasing the thickness of UHPC by 10 mm accelerated the debonding failure (compare NC0+UC4 and NC0+UC4*). Rather, reducing reinforcement ratio by 50% led to certainly better result in this case, i.e., the specimen NC0+UC2 showed ductile flexural failure, and UHPC and rebar reached their tensile strength. After the failure of UHPC, the specimen still had the residual resistance due to the embedded rebar.

Among the eight specimens, the best performance regarding both the load and deformation capacity was shown in NC4+UC0 and NC2+UC2. They satisfied the condition that rebar is reinforced in the NC section and its reinforcement ratio is higher than the UHPC section. The specimen NC4+UC0 marked the highest peak load among the all specimens. However, after the peak load its behaviour was almost identical to that of NC4 due to the failure of UHPC. On the other hands, NC2+UC2 retained residual load resistance after the peak load, which can be apparent by comparing with NC2. Especially, the structural performance of NC2+UC2 is outstanding in both pre and post peak load in terms of the initial stiffness as well as the residual load resistance.

Discussion

Figure 7 explains the effect of each parameter of this study on the structural behavior of the composite beam. By reinforcing small portion of rebar (close to the minimum reinforcement ratio of concrete slab) in the UHPC section, the structural performance of the beam was significantly increased in terms of both the load and deformation capacity (Figure 7a). These remarkable increases were possible due to the intensified tension stiffening effect in the UHPC section which was caused by the fiber bridging action [11]. Thus, it is reasonable to use

occurred when rebar was reinforced. On the other hand, brittle flexural failure occurred in other case as shown in Figures 5e and 5f.

On the other hand, two composite specimens which have 4-D10 deformed bars in UHPC section showed brittle debonding failure (Figures 5g and 5h); the upper part of the interface between NC and UHPC was suddenly detached and slipped.

Load-deflection relationship

The relationships between load and mid-span deflection are presented in Figure 6. Within the pre peak state, the load-deflection curve of the composite specimens sharply increased than that of RC specimens. This indicates significantly increased initial stiffness due to the contribution of the UHPC. However, the increased load in the composite specimen steadily drop again after once peak load reached; the decrease rate depended on the location of rebar and thickness of the UHPC section.

Table 3 summarizes the test results including the data and information obtained from load cell, LVDT and strain gauges. The peak load of the specimen NC0+UC0 was 35% higher than that of NC2. Although this composite specimen showed pseudo strain

Specimen	At rebar yield		At peak load					Failure mode	Note
	P_y (kN)	Δ_y (mm)	P_{max} (kN)	Δ_{pmax} (mm)	ε_s	ε_c	ε_u		
NC4	55	12.3	59	37	$\varepsilon_s > \varepsilon_y$	$\varepsilon_c < \varepsilon_o$	-	Flexure	Rebar fracture
NC4+UC0	102	16.4	111	22	$\varepsilon_s > \varepsilon_y$	$\varepsilon_o < \varepsilon_c < \varepsilon_{cu}$	$\varepsilon_u > \varepsilon_{u,t}$	Flexure	Concrete crushing
NC0+UC4	-	-	104	11	$\varepsilon_s = \varepsilon_y$	$\varepsilon_c < \varepsilon_o$	$\varepsilon_u < \varepsilon_{u,t}$	Debonding	Slip of interface
NC0+U4C'	-	-	103	10	$\varepsilon_s < \varepsilon_y$	$\varepsilon_c < \varepsilon_o$	$\varepsilon_u < \varepsilon_{u,t}$	Debonding	Slip of interface
NC2+UC2	104	12.7	106	13	$\varepsilon_s > \varepsilon_y$	$\varepsilon_c < \varepsilon_o$	$\varepsilon_u = \varepsilon_{u,t}$	Flexure	Rebar fracture
NC2	37	9.9	40	66	$\varepsilon_s > \varepsilon_y$	$\varepsilon_c < \varepsilon_o$	-	Flexure	Rebar fracture
NC0+UC0	-	-	54	6	-	$\varepsilon_c < \varepsilon_o$	$\varepsilon_u = \varepsilon_{u,t}$	Flexure	UHPFRC fracture
NC0+UC2	90	10.23	94	12	$\varepsilon_s > \varepsilon_y$	$\varepsilon_c < \varepsilon_o$	$\varepsilon_u > \varepsilon_{u,t}$	Flexure	Rebar fracture

P_y = Load when reinforcements yields.
P_{max} = Peak load.
Δ_y = Mid span deflection at P_y
Δ_{pmax} = Mid span deflection at P_{max}.
ε_c = Strain of concrete.
ε_{cu} = Ultimate strain of concrete.
ε_o = Strain of concrete at maximum stress.
ε_s = Tensile strain of steel.
ε_u = Tensile strain of UHPC.
$\varepsilon_{u,t}$ = Tensile strain of UHPC at maximum stress.
ε_y = Strain of steel at yield stress.

Table 3: Summary of test results.

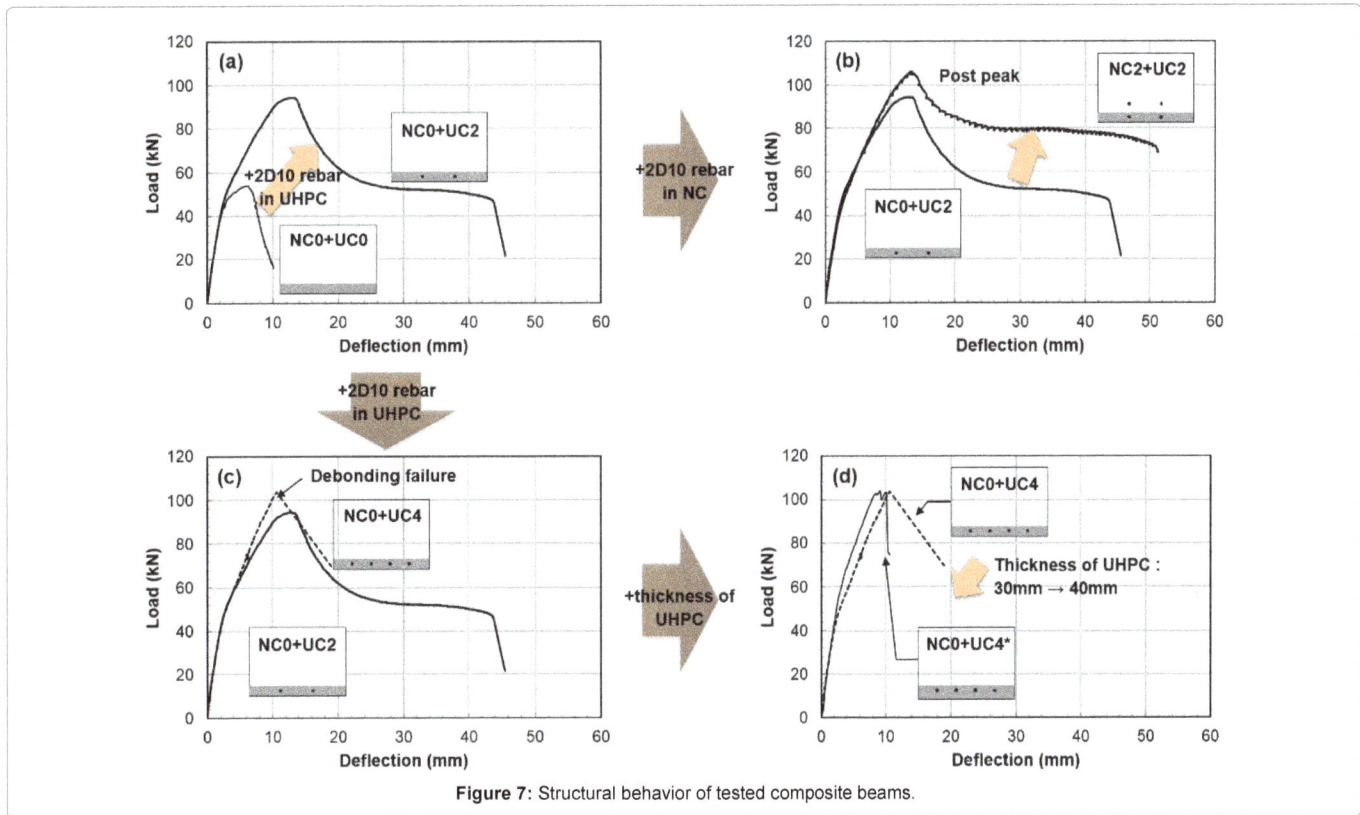

Figure 7: Structural behavior of tested composite beams.

deformed bars for structural UHPC, because of the intensified tension stiffening effect as well as fiber bridging action.

When the same ratio of rebar was additionally reinforced in the concrete section, the post peak load was maintained highly (Figure 7b). Thus, it is recommended to use additional rebar in concrete section for structural safety. In addition, it is interesting to note that the fibers in UHPC contributed to increase of the safety, because the teeth shape in the load-deflection curve can be formed when the fibers pull out from the matrix of UHPC while resisting against loading.

However, when the same rebar was additionally reinforced in the UHPC section, the failure mode was changed negatively (Figure 7c). As the gap of stiffness between NC and UHPC increased, relatively weak part of concrete suddenly failed. Figure 7d also evidently shows the same trend of the result. It reveals the importance of the balance in the stiffness when designing the permanent formwork of UHPC.

Conclusion

This study investigated the structural performance of the RC-UHPC composite beam subjected to bending, in order to utilize the

UHPC panel as a permanent formwork for concrete slab or deck. In particular, the load-deflection behavior of the beam was observed based on reinforcement ratio, location of rebar and thickness of UHPC.

The load resisting capacity of the RC beam could be significantly increased by using the UHPC permanent formwork. When rebar was not included in the UHPC section, the behavior of this composite beam was almost identical to that of the RC beam because the resisting capacity of UHPC disappeared after the peak load. The reinforcement of rebar in UHPC section can further increase the structural behavior because of the intensified tension stiffening effect. However, in this case, there is a prerequisite for the structural safety of balancing the stiffness between the two concretes. If not, the interfacial bond failure can occur. Therefore, the same reinforcement ratios of rebar in concrete and UHPC sections marked the best result in terms of both load and deformation capacity.

Acknowledgment

The Institute of Engineering Research in Seoul National University provided research facilities for this work.

References

1. Tayeh BA, Bakar BA, Johari MM (2013) Characterization of the interfacial bond between old concrete substrate and ultra-high performance fiber concrete repair composite. Materials and Structures 46: 743-753.

2. Brühwiler E, Denarié E (2008) Rehabilitation of concrete structures using ultra-high performance fibre reinforced concrete. Proceedings of the 2nd International Symposium on UHPC: 5-7th March, Germany, p: 895-902.

3. Habel K, Denarié E, Brühwiler E (2006) Structural response of elements combining ultra high-performance fiber-reinforced concretes and reinforced concrete. J Structural Engineering 132: 1793-1800.

4. Kang SH, Gyephel T, Hong SG, Moon J (2015) Effect of water-entraining admixtures on the hydro-mechanical properties of ultra-high performance concrete. 14th International Congress on the Chemistry of Cement during 13-16th October in Beijing, China.

5. Kang SH, Hong SG, Moon J (2016) Influence of internal curing on autogenous and drying shrinkages of ultra-high performance concrete considering heat treatment. FIB Symposium, Maastricht, The Netherlands.

6. ASTM C1611 (2014) Standard test method for slump flow of self-consolidating concrete. ASTM International.

7. ASTM C39 (2016) Standard test method for compressive strength of cylindrical concrete specimens. ASTM International.

8. Association Française de Génie Civil (AFGC) (2013) Ultra high performance fibre-reinforced concrete-Recommendations (Revised edition) Paris, France.

9. Richard P, Cheyrezy M (1995) Composition of reactive powder concretes. Cement and Concrete Research 25: 1501-1511.

10. Markovic I (2006) High-performance hybrid-fibre concrete: Development and utilisation. Delft University of Technology, The Netherlands.

11. Bischoff PH (2003) Tension stiffening and cracking of steel fiber-reinforced concrete. J Materials in Civil Engineering 15: 174-182.

7

Design and Analysis of a Tension Leg Platform's Column for Arabian Sea

Latif U * and Naeem Shah A

Department of Mechanical Engineering, University of Engineering and Technology, Lahore 54000, Pakistan

Abstract

Offshore structures such as boats, ships, oil rigs etc. are under continuous wave loading. These waves exert pressure on the structure which in turn produces stresses in them. These waves can strike structure from any direction and it is very difficult to calculate the pressure exerted by the waves analytically as the shape of the offshore structures is rather complex. The pressures and associated stresses and buckling caused by the striking waves are harmful to the structure, and may lead to their failure. Keeping in view the above scenarios, the column of an offshore oil platform was designed on the basis of hurricane's history of Arabian Sea. The normal and worst sea environment conditions were kept in mind while designing the column and its subsequent analysis was made on the basis of stress and buckling in this study. After analyzing the structure, it is known that the offshore column design is safe for particular sea state. The factor of safety (fos) was 1.90 and 1.77 for the normal and worst conditions, respectively. The pressure was decreased with the increase in wave frequency, while increased with the increase in wave amplitude. Further, buckling modes of extraction were in the range of 1.23 to 1.59.

Keywords: Buckling; Offshore; Pressure; Sea state; Stress; TLP

Introduction

According to latest surveys made for the exploration and exploitation of the oil reserves, it is estimated that deep water oil reserves are in abundance as compared to the existing land reserves Mallory [1]. Therefore, an intensive study along with practical application has been focused on offshore structures in recent times. Offshore structures remain under continuous wave loading during their service or operational life. The life time span of an offshore oil rig is commonly between 25 to 40 years Arvid and Moan [2].

Offshore structures are used by oil and gas industries like Shell Pvt. Ltd. for extraction of natural resources in the marine environment. The most suitable type of offshore structures for intermediate and higher depths is Tension Leg Platform (TLP) Chakrabarti [3]. This Type of offshore structure mostly selected due to its higher stability in harsh sea environment as compared to other offshore structures.

A critical component of the TLP is its column which needs to be modeled with care so that platform may perform its operation smoothly and maintain its stability in the harsh sea environment. Different Types of waves like regular, irregular, long crested and short crested waves strike with the offshore oil rigs column and causes the loading/pressure on it.

The environment for which offshore engineers are designing a column can be hostile, so the constraints and safety measures which govern the design are very crucial parameters. In current study, a TLP column was designed for the particular Sea State on the basis of the history of cyclones which have been encountered in the Arabian Sea so far. The main focus of the study is on the design and analysis of the column of offshore oil rig/platform against stresses and buckling. Two different failure modes are addressed in this paper.

Factors/Data Involved in the Design and Analysis

As the strong evidence of the existence of hydrocarbon/Oil and Gas resources in Pakistani Sea and therefore some research activities related to this field have been reported in the near past, so the TLP column was designed on the basis of hurricanes data available in the literature. This is expected to be helpful in the exploration of the Oil and Gas from the specified location like Costal Area of Makran in Pakistan.

Hydrocarbon resources availability

Offshore regions of Pakistan have thick marine sedimentary reservoir rocks with potential traps both structural and stratigraphic Nuzhat [4]. All the prerequisites of oil and gas occurrence i.e. source; reservoir and traps are amply present in the offshore basins of Pakistan. Many seismically delineated anticlinal traps are available for exploration. Deltas, their fans and cones are favorable locations for oil and gas accumulation. The Indus delta and Indus offshore areas have proved to be a major hydrocarbon producing area. In the deltaic region, exploration activities have identified many promising fields Nuzhat [4]. A large number of untested anticlinal structures have been seismically delineated in the Balochistan offshore area. Surface features such as oil seepages and mud volcanoes show the generation of hydrocarbons in the area. Gas deposits (mostly methane) are common along the Makran coast Nuzhat [4]. Offshore Makran has shown good quality reservoir rock. From the Makran continental margin gas hydrates can be extracted as another source of energy that has been discovered recently Nuzhat [4].

Hurricanes data (encountered in Arabian Sea)

The Indian Ocean is considered among the largest oceans of the world and it is at number three in the list of largest oceans of the world. Arabian Sea is the northwestern part of this ocean. Indian Ocean is included in the oceans of the world which have the hot environment or higher temperature as compared to the other oceans of the world. Gulfs and countries which are situated along with the Arabian Sea are Gulf of Aden and Gulf of Oman and Pakistan, India, Maldives, Sri Lanka and

*Corresponding author: Latif U, Department of Mechanical Engineering, University of Engineering and Technology, Lahore 54000, Pakistan
E-mail: usman_mechanical@yahoo.com

Somalia. The Arabian Sea has the ability to generate cyclones but the storms are very infrequent in this part of the ocean. Most prominent examples of cyclones are "Gonu" and "Phet" Hussain [5]. Total 50 numbers of cyclones has been observed in this sea since 1890 to 2013 Hussain [5].

Description of the Model

The TLP column is made of HY-80 Alloy Steel. The reason behind the selection of this material is its higher resistance power against the corrosion. We know that the sea water is brackish water due to which corrosion rate of every material increases in sea water as compared to land use. The corrosion rate of HY-80 alloy steel is 2-3MPY Dexter [6]. The detailed description of the column is given in the Table 1.

Ideally, the model should be as large as it can be, so that the scaling effect and errors can be reduced or kept to a minimum level. Scaling effect is commonly involved when it is required to make the prototype of the model to test it in the lab. But in this case, authors performed the analysis with the help of different softwares. Authors have used the true dimensions of the column and thus calculated the true and complete effects/impacts of the wave loading on the column (Figure 1).

Hydrodynamic Analysis in Frequency Domain Using ANSYS AQWA

ANSYS AQWA software addresses the vast majority of analysis requirements associated with hydrodynamic assessment of all types of offshore and marine structures like SPARs, Ships, FPSOs, Semi-submersibles, Break Waters and Tension Leg Platform [I]. ANSYS AQWA Diffraction provides an integrated facility for developing primary hydrodynamic parameters required to undertake complex motions and response analysis. Hydrodynamic analysis results, such as motions and pressures, can be transferred to ANSYS structural mechanics products, ANSYS ASAS or generically defined FE models for subsequent structural analysis [I].

ANSYS AQWA Suite extends ANSYS AQWA Diffraction to include analysis capabilities for global performance of moored and connected systems subject to random sea states. Simulations may be static or dynamic in frequency and in time domain [I].

Pressure mapping

The pressure mapping is a process or method which enables to define the parameters in one specific software and then import the results or text into software. The pressure mapping was performed by using ANSYS ASAS, ANSYS AQWA WAVE and Load Case File. ANSYS AQWA Wave software may be used to investigate the effects of currents, waves and wind on fixed or floating offshore and marine structures including Floating Production Storage and Offloading unit, Spars, TLP, Semi-Submersible units, ships etc. ANSYS Inc. [7].

Von misses criteria for the investigation of stresses

Von Misses criterion is used to check whether the design is safe or not against the applied stresses. A complex three dimensional system of stresses is developed by applying load (pressure) on a body in three dimensions. Magnitude and direction of stresses are changed at every point within a body. The failure of a structure at a given point thus can be calculated by using Von Misses criterion given as follows [II]:

$$(S1-S2)^2 + (S2-S3)^2 + (S3-S1)^2 = 2S_e^2 \qquad (2)$$

Where S1, S2 and S3 are the principal stresses and Se is the equivalent stress or "Von Misses Stress".

Block LANCZOS method for buckling analysis

Block LANCZOS method is an iterative method used to find the different modes of extraction of eigen buckling in ANSYS APDL Mechanical. As it is an iterative method so it performs iterations till the exact values are obtained.

Results and Discussion

Effect of wave frequency on pressure

Figure 2 shows the effect of different wave frequencies on the pressure applied on the column. It is clear that there is an inverse relationship between frequency and applied pressure on the column. As the value of frequency of the wave increases, the applied pressure decreases.

As we know that, frequency is the number of cycles per unit time. So, if the value of frequency increases it means that number of cycles per unit time also increases. As the number of cycles increases in a unit time then the height of the wave decreases and a wave of smaller amplitude strikes with the structure which in turn produces smaller pressure on the structure. This is the reason for the decrease in pressure as the frequency is increased.

Effect of wave amplitude on pressure

Analysis was also carried out for the impact of different wave amplitudes on the pressure applied to the column as shown in Figure 3. It is obvious that there is a linear relationship between amplitude and

Description	Value
Volume	29897.5 m³
Ixx	1.05 109 kg.m²
Iyy	1.05 109 kg.m²
Izz	1.98 108 kg.m²
VCB	17.50 m
Center of Gravity From Bottom	18.54 m
Total upward thrust	161885.79 kN
Draft	35 m
Free Board	30 m
Dia of the Column	24.2 m
Design Wave Height for Normal Sea Environment	2 m
Design Wave Height for Normal Sea Environment	7 m

Table 1: Detail of column.

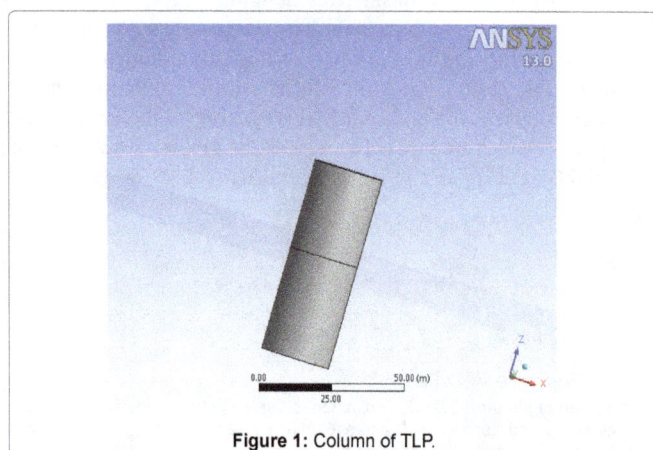

Figure 1: Column of TLP.

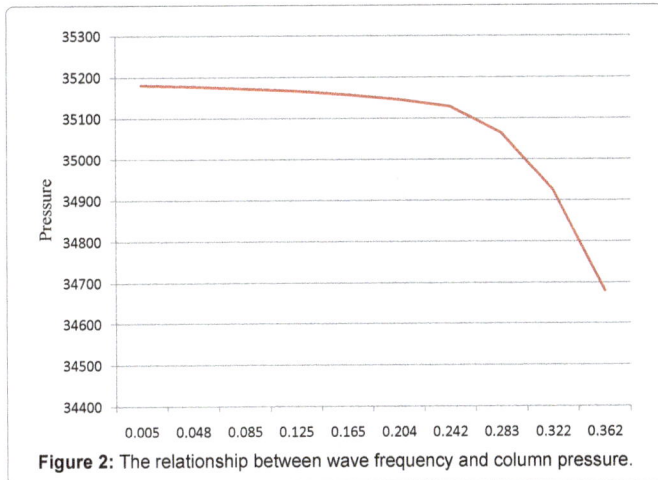

Figure 2: The relationship between wave frequency and column pressure.

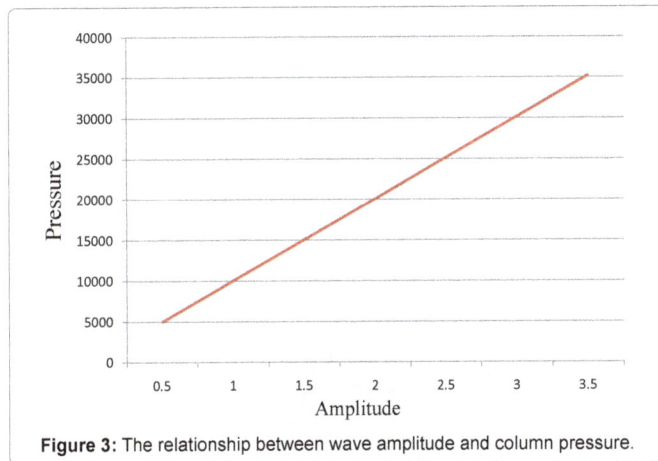

Figure 3: The relationship between wave amplitude and column pressure.

pressure of the waves. As the value of amplitude increases the pressure of the wave on offshore oil rig/TLP column also increases.

The increase in pressure is attributed to the height of the wave. As the wave height increases, the energy or disturbing potential of the wave is also increased. Wave is not a motion of water rather it is the transfer of disturbance from one particle to another particle. Thus, the wave with higher amplitude possesses the higher energy which creates greater pressure on the column as compared to that with lower amplitude.

Analytical verification of pressure mapping

As discussed above a load case file is to be generated/developed to transfer the pressures generated by waves on the column from one software i.e. ANSYS AQWA Lin to another software i.e. ANSYS APDL Mechanical. A load case file was developed in this study along with the summary/entries given as follows:

System Data Area 200000

JOB NEW LINE

PROJECT ANSY

EXTENSION dat

END Model. asas

END AQWAID Analysis stat

FELM

FEPG ANSY

END LOAD CASE 0 7 5 2 0

END STOP

The pressure mapping output file is given in Figure 4.

Analytical verification of hydrostatic pressure mapping is performed using the following expression Journee and Massie [8]:

$$P_{max} = \rho\, g\, h + \rho\, g\, \zeta^{(-k*z)} \qquad (1)$$

where,

P_{max}=Maximum value of pressure at any node/element.

ρ=Density of Sea Water

g=Gravitational Acceleration

h/z=Depth/Length of Column in Sea Water

ζ=Amplitude

k=Wave Number

Now by putting these values in the above equation we get:

= 351933.75 Pa

The analytically calculated pressure value is compared with that computed with the software, as shown in Figure 4. So it is found that pressure mapping is quite right.

Stresses and buckling analysis

After analysis in ANSYS Mechanical APDL, the maximum value for Von Misses stresses at wave height 2.0 m and 7.0 m were obtained as shown in Figures 5a and 5b) which are 291 MPa and 312 MPa, respectively.

When a structural member is subjected to an axial compressive load, it may fail by a condition called BUCKLING. Buckling is not so much a failure of the material (as is yielding and fracture), but an instability caused by system geometry American Bureau of Shipping [8].

The results for the buckling analysis are shown in Figure 6. Buckling analysis shows the results for five modes of extraction which are ranging from 1.23 to 1.59. The value of frequency obtained by analysis is 1.23 as shown in Fig. 6 at serial number 1, which represents a safe design. If the value of frequency becomes less than one then it will be an unsafe design or buckling may lead the designed column towards collapse/failure. The value of frequency greater than one is the description of safe

```
SFE,164980,     2,PRES,0,  3.5177E+05,  3.5177E+05,  3.5177E+05,  3.5177E+05
SFE,164981,     2,PRES,0,  3.5177E+05,  3.5177E+05,  3.5177E+05,  3.5177E+05
SFE,164982,     2,PRES,0,  3.5177E+05,  3.5177E+05,  3.5177E+05,  3.5177E+05
SFE,164983,     2,PRES,0,  3.5177E+05,  3.5177E+05,  3.5177E+05,  3.5177E+05
SFE,164984,     2,PRES,0,  3.5177E+05,  3.5177E+05,  3.5177E+05,  3.5177E+05
SFE,164985,     2,PRES,0,  3.5177E+05,  3.5177E+05,  3.5177E+05,  3.5177E+05
SFE,164986,     2,PRES,0,  3.5177E+05,  3.5177E+05,  3.5177E+05,  3.5177E+05
SFE,164987,     2,PRES,0,  3.5177E+05,  3.5177E+05,  3.5177E+05,  3.5177E+05
SFE,164988,     2,PRES,0,  3.5177E+05,  3.5177E+05,  3.5177E+05,  3.5177E+05
SFE,164989,     2,PRES,0,  3.5177E+05,  3.5177E+05,  3.5177E+05,  3.5177E+05
SFE,164990,     2,PRES,0,  3.5177E+05,  3.5177E+05,  3.5177E+05,  3.5177E+05
SFE,164991,     2,PRES,0,  3.5177E+05,  3.5177E+05,  3.5177E+05,  3.5177E+05
SFE,164992,     2,PRES,0,  3.5177E+05,  3.5177E+05,  3.5177E+05,  3.5177E+05
SFE,164993,     2,PRES,0,  3.5177E+05,  3.5177E+05,  3.5177E+05,  3.5177E+05
SFE,164994,     2,PRES,0,  3.5177E+05,  3.5177E+05,  3.5177E+05,  3.5177E+05
SFE,164995,     2,PRES,0,  3.5177E+05,  3.5177E+05,  3.5177E+05,  3.5177E+05
SFE,164996,     2,PRES,0,  3.5177E+05,  3.5177E+05,  3.5177E+05,  3.5177E+05
SFE,164997,     2,PRES,0,  3.5177E+05,  3.5177E+05,  3.5177E+05,  3.5177E+05
ACEL, -2.4072E-01, -3.9264E-06,  9.8050E+00
CGLOC,  2.8480E-08,  1.1772E-07,  0.0000E+00
DCGOMG,  3.5330E-08,  2.8424E-03,  3.5057E-09
SOLVE
```

Figure 4: Pressure values at different points/nodes.

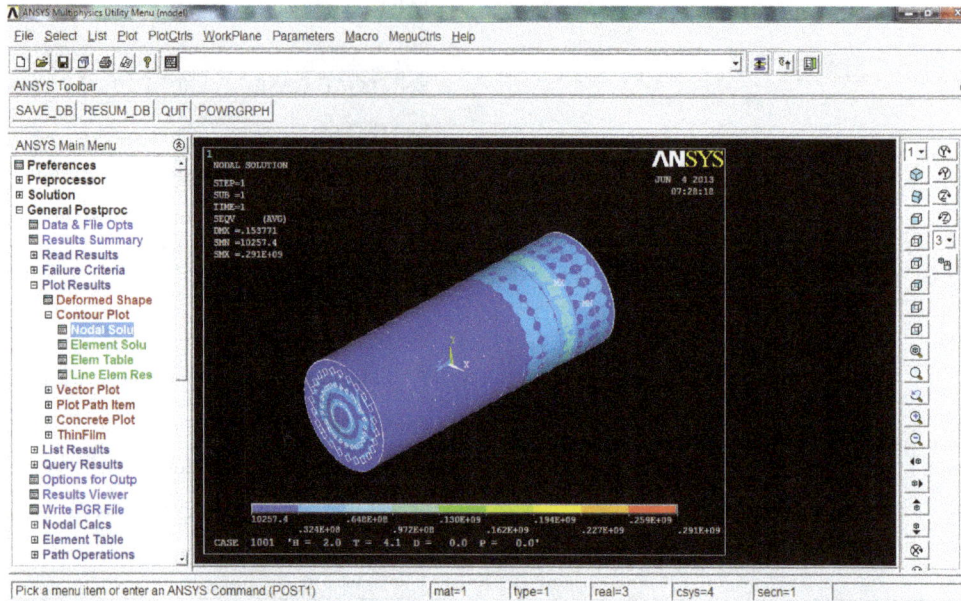

Figure 5a: Von Misses stresses at wave height 2.0 m.

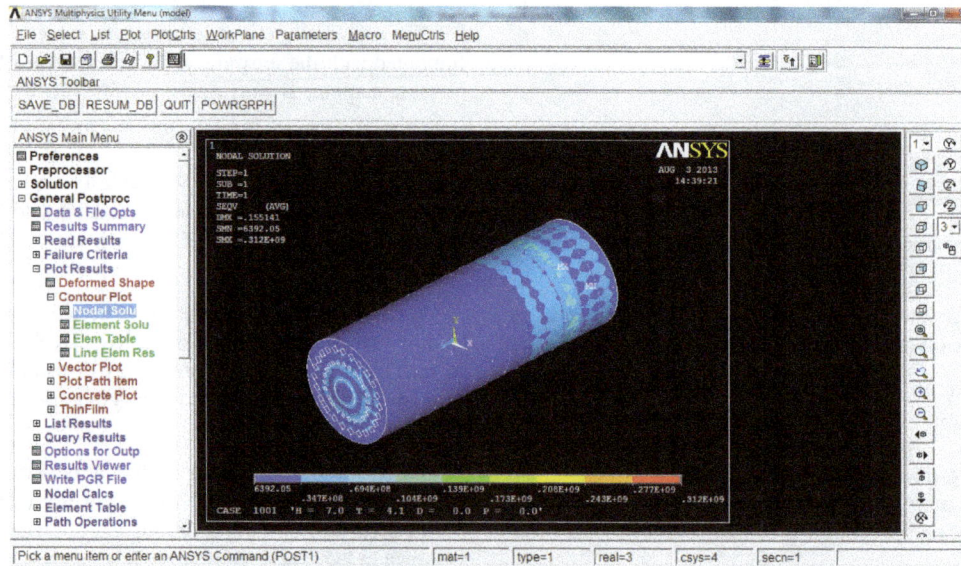

Figure 5b: Von Misses stresses at wave height 7.0 m.

design regarding buckling (here frequency represents the critical load). This value is also greater than the set standards of ASME. According to ASME the value of frequency or critical load should be 1.2 or more.

The designed offshore column must fulfill the following condition to show its safe design.

R>Q

Where,

R=Column's Strength

Q=Applied load on the Column

Above mentioned equation could be given in terms of the total factor of safety "γ", as follows Shama [9]

$$\gamma = R/Q > 1.0 \qquad (2)$$

As discussed earlier that the highest wave encountered in Arabian/Pakistani Sea was of height 16-20 feet but the current model/column was analyzed for worst condition in which waves may have the height up to 23 feet.

After Analysis it was found that the column of offshore oil rig/Tension Leg Platform is quiet safe with respect to the Arabian Sea's working environment and conditions [10,11].

The cyclones which encountered at Pakistani Coastal Area were kept in mind while designing the offshore column. Generally, a Von Misses criterion is applicable to check or verify the structural safety or failure. After the evaluation of all kinds of results, the factor of safety

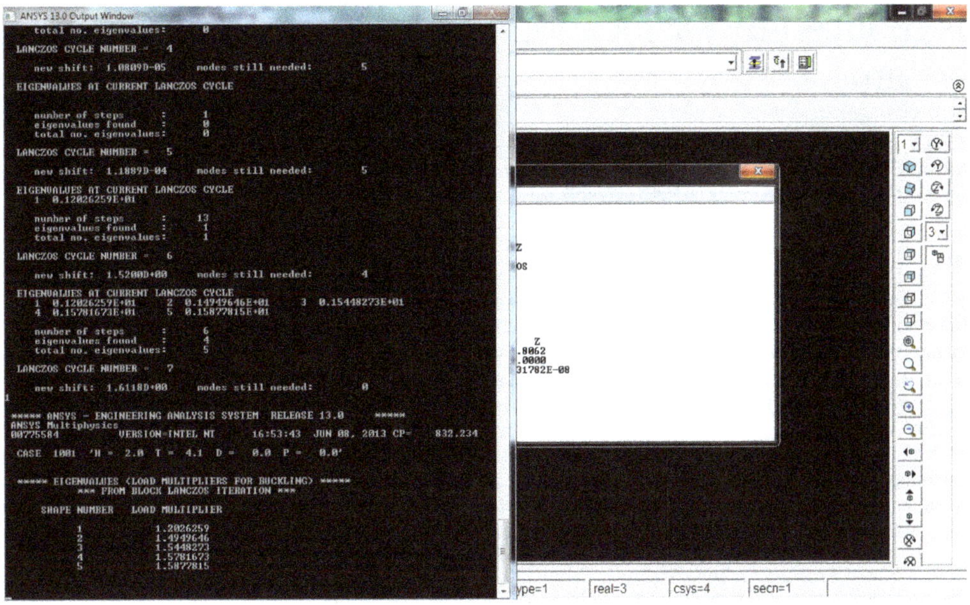

Figure 6: Solution of buckling analysis (output window).

obtained in normal and worst Sea environment/conditions are as under:

Applied Stresses under Von Misses Criteria at Wave Height 2.0 m =291 M Pa

Applied Stresses under Von Misses Criteria at Wave Height 7.0 m =312 M Pa

Material Ultimate Strength=552 M Pa

Factor of Safety under Von Misses Stress=1.90 Approx. (at Wave Height 2.0 m)

Factor of Safety under Von Misses Stress=1.77 Approx. (at Wave Height 7.0 m)

Buckling Frequency/Critical Load=1.23-1.59

Safety factor is of prime importance in the designs where the loss of human lives is expected or involved. In present case there is a danger of wastage of human lives. This is the reason for the factor of safety to be taken more than 1.5, which is however 1.77 in this study.

Conclusion

Current study is focused on the design and analysis of a TLP column that needs to be designed with some extra supports to reduce the effect of stress concentration and to sustain against the unusual sudden impacts of slamming, wave and current loading. The results show that the pressure on the structure decreases as the wave frequency is increased, while increases with the increase in wave amplitude. It was found that the designed column is safe with respect to ultimate stress and buckling phenomenon even in the true and worst hurricane conditions. From the history of Arabian Sea, it is revealed that the cyclones encountered at 18-20 ft wave height, however the designed column was analyzed on the basis of a wave height up to 23 ft. After analysis, it was found that the structure is safe with a factor of 1.90 and 1.77 for normal and worst conditions, respectively. Moreover, buckling modes of extraction were in the limit of 1.23 and 1.6 which are also safe with the view point of the ASME standards.

Recommendations

The designed column is safe for practical use and may be implemented in Pakistani/Arabian Sea in the Makran Coastal Area where Oil and Gas reserves are frequently available.

References

1. Mallory (2010) Deep water Gulf of Mexico oil reserves and production.

2. Arvid N, Moan T (2005) Probabilistic design of offshore structures. Chakrabarti Hand Book of offshore Engineering, Elsevier Oxford: 191-272.

3. Chakrabarti S (2005) Hand book of offshore engineering Elsevier USA.

4. Nuzhat K (2011) Marine resources in Pakistan: A tentative inventory. Pakistan Business Review 12: 837-838.

5. Hussain MB (2011) Pakistan weather portal.

6. Dexter SC (1972) Hand book of oceanographic engineering materials Metals and Alloys: 62-63.

7. ANSYS Inc. (2010) AQWA wave user Manual Hydrodynamic Load Transfer: 1-90.

8. American Bureau of Shipping (2005) Buckling and ultimate strength assessment for offshore structures: Commentary Guide: 1-151.

9. Shama MA (2009) Basic concept of the factor of safety in marine structures 4: 307-314

10. Ansys (2015) Realize your product promise-Explore Engineering Simulation.

11. Von Mises Stress-Von Mises-Hencky criterion for ductile failure.

8

Stress Fields Induced by Dislocation Loops in Isotropic CuNb Film-Substrate System

Wu W[1]*, Qian G[2] and Cui X[3]

[1]State Key Laboratory of Automotive Safety and Energy, Tsinghua University, Beijing 100084, PR China
[2]Paul Scherrer Institute, Nuclear Energy and Safety Department, Laboratory for Nuclear Materials, 5232 Villigen PSI, Switzerland
[3]Sustainable Energy Systems Group, Lawrence Berkeley National Laboratory, 1 Cyclotron Road MS 90R2002, Berkeley CA 94720, USA

Abstract

Based on linear superposition rules and fast discrete Fourier transformation, a semi-analytical solution is developed for calculating the elastic fields induced by dislocation loops in an isotropic thin film-substrate system. The elastic field problem of thin film-substrate system is decomposed into two sub-problems: bulk stress due to a dislocation loop in an infinite space, and correction stress induced by free surface and interface of the film-substrate system. Correction elastic field is linearly superimposed onto bulk elastic field to produce continuous displacement and traction stress across the interface plane of the perfectly-bounded film-substrate system. Firstly, calculation examples of dislocation loops in Cu-Nb film-substrate system are performed to demonstrate the calculation efficiency of the developed semi-analytical approach. Then, elastic fields of dislocation loops within Cu film and Nb substrate of the Cu-Nb film-substrate system are analyzed. Finally, effects of film thickness, loop positions are investigated, and it is found that the elastic fields of dislocation loop are influenced remarkably by these two factors.

Keywords: Elasticity; CuNb; Dislocation loop; Film-substrate; Isotropy

Introduction

Film-substrate structures and systems composed of thin film of finite thickness and substrate of infinite thickness are widely used in micro-chips, smart electronics, micro-sensors and manipulators, protective coatings, etc. During the industrial fabrication process and service life-time of film-substrate structures and systems, dislocation clusters will be generated and will result in the microstructure evolution of the film-substrate system. The performance, reliability and integrity of thin film-substrates are closely related to the material properties of the film and substrate, in particular, the crystal orientation (or texture for polycrystalline material), the dislocation density and distribution. Study of the collective dynamic behaviors of dislocations embedded in thin film-substrate system are of great interest to researchers and engineers, which are of critical importance for understanding the microstructure evolution, plastic deformation process of thin film-substrate systems.

Solutions to the elasticity field induced by dislocations within film-substrate system are important, because the elasticity solution provides a direct means of determining the Peach-Koehler force acting on dislocations, which is of direct relevance in understanding the microstructure evolution and mechanical behaviors of these film-substrate systems. In order to simulate the dynamic evolution process of dislocation and dislocation loops, accurate method are needed to calculate the stresses due to dislocations and dislocation loops in a heterogeneous thin film-substrate system. Making use of mirror dislocation concept and potential theory, Head [1,2] analyzed the changes in the stress field of a straight screw/edge dislocation caused by differences in the shear modulus on either side of bimetallic medium. Moreover, making use of image dislocation concepts, the equilibrium positions of a group of dislocations within bimaterial and film-substrate system are studied. It was found that when dislocations are located on the side with lower modulus and are situated on the slip plane which is perpendicular to the boundary of bimaterial system, dislocations are forced towards the boundary by an applied stress but repelled from the boundary by the change in modulus [3]. Making use of the solution of an edge dislocation in half-plane and that of the reversed traction force

prescribed on the interface plane, Weeks et al. [4] presented an exact analysis for the elastic field and the Peach-Kohler force due to an edge dislocation within the substrate medium of film-substrate system. By using Fourier exponential transform, Lee and Dundurs [5] investigated the elastic field of an edge dislocation situated in the surface layer of the film-substrate system, and the behavior of edge dislocation is discussed through analyzing the Peach-Kochler force acting on these dislocations. By extending the classical mirror dislocation solutions for the interface plane between two semi-infinite elastically isotropic media, the image problem for a screw dislocation in the thin films of multilayered film-substrate system has been solved analytically [6], and the equilibrium position of a single dislocation in thin films was determined as a function of stress in the films. Kelly et al. [7] studied the stress field induced by an edge dislocation of arbitrary orientation in both the surface layer and substrate medium of the layer-substrate system, and the solution can be employed for deriving the solution of crack problems, together with the Peach-Koehler force. Savage [8] studied the solution for the displacement field induced by an edge dislocation in a layered half-space, and four elementary solutions were considered: the dislocation is either in the half-space or the layer, and the Burgers vector is either parallel or perpendicular to the interface plane. It was found that the surface displacement field produced by the edge dislocation in the layered half-space is very similar to that produced by an edge dislocation at a different depth in a uniform half-space. Moreover, a low-modulus (high-modulus) layer causes the equivalent dislocation in the half-space to appear shallower (deeper) than the actual dislocation in the layered half-space. Wu and Weatherly

---

*Corresponding author: Wu W, State Key Laboratory of Automotive Safety and Energy, Tsinghua University, Beijing 100084, PR China
E-mail:wenwang_wu@mail.tsinghua.edu.cn

[9] studied the equilibrium position of misfit dislocations in epitaxial grown systems where the thickness of the epitaxial film is several orders of magnitude smaller than the thickness of the substrate. When the film is elastically stiffer than the substrate, the core of the dislocation is predicted to lie at some distance from the interface in the softer substrate. On the other hand, when the film is softer than the substrate, the core of the dislocation is always predicted to lie close to the interface [9]. The continuous dislocation image dislocation method is used to derive elastic solutions for an edge dislocation in an anisotropic film-substrate system, and it was found that elastic anisotropy and material mismatch play an important role in determining the image forces and the stress components due to an edge dislocation located in film-substrate systems [10]. The stress and displacement fields due to an edge dislocation in a linearly elastic isotropic film-substrate were studied based on image dislocations methods, and it was shown that the film thickness and dislocation position have a significant influence on the image force acting on the dislocation, and the film thickness variation due to an accumulation of dislocations may degrade the performance of optical films [11]. Recently, anisotropic elastic stress fields caused by a dislocation in Ge_xSi_{1-x} epitaxial layer on Si substrate are investigated by Wang [12], and effects of layer thickness, the dislocation position and the crystallographic orientation on the stresses of anisotropic film-substrate system were investigated extensively. It was revealed that the layer thickness and dislocation position strongly affects the stresses, while the crystallographic orientation play a very weak role in determining the elastic stress fields [12]. However, theses analytical solutions are mainly limited to straight dislocation lines parallel to the interface planes, which is invalid for curved dislocations and dislocation loops within film-substrate systems. Making use of potential theory and complete elliptic integrals, the elastic fields of a circular planar dislocation loop in an isotropic two-phase material, and the interaction forces between dislocation loop and the second phase were studied analytically, it was found that Peach-Koehler forces acting locally on the loop always tend to distort it [13,14]. Besides these analytical solutions, Weinberger et al. [15] and Wu et al. [16] developed semi-analytical solutions to calculate the image stress of approaching dislocations and dislocation loops within half space and free standing thin film. It is found that image force has a much stronger effect on 1/2 (111) dislocation loop in anisotropic (111) thin Fe foil, compared to the Voigt equivalent isotropy simulation results. Moreover, for complex dislocation configurations and boundary conditions, the image stress due to free surfaces and interfaces can now, owing to computing power, be computed by finite element method [16].

The objective of this paper is to develop an efficient and accurate method to calculate the stress due to dislocation loops in an isotropic heterogeneous thin film-substrate system, which is composed of a thin film of finite thickness and a substrate of infinite thickness. Based on linear superposition rules and fast discrete Fourier transformation, the problem is decomposed into two sub-problems: the stresses due to a dislocation loop in an infinite space and the correction stresses induced by the upper free surface of film and the film-substrate interface. Firstly, calculation examples of dislocation loops in Cu-Nb film-substrate system are performed to verify the semi-analytical approach; Then, elastic fields of dislocation loops within Cu film and Nb substrate of the Cu-Nb film-substrate system are analyzed; Finally, effects of film thickness, dislocation loop positions on the elastic field of is investigated, and it is found that film thickness, dislocation loop positions has a remarkable impact on the elastic fields of dislocation loops within film-substrate system.

Stress Fields Induced by Dislocation Loop in Perfect Bonding Film-Substrate System

Statement of problem

As shown in Figure 1, perfect bonding isotropic film-substrate system is decomposed into a thin film (**A**) of finite thickness 2h and a substrate (B) of infinite thickness. The elastic properties of the film and substrate medium are assumed to be $(\lambda_f, \mu_f, \nu_f)$ and $(\lambda_s, \mu_s, \nu_s)$, respectively. Two sets of local Cartesian coordinates are employed for describing the film-substrate system: local Cartesian coordinate (x^+, y^+, z^+) for the upper thin film A, and local Cartesian coordinate (x^-, y^-, z^-) for the lower substrate half space B, where (x^+, y^+) and (x^-, y^-) are parallel to the interface plane and simplified as (x, y). The origin of the local Cartesian coordinate for the upper thin film A is located in the middle of the thin film, and the origin of the local Cartesian coordinate for the lower substrate B is located on the interface plane. Accordingly, $-h \leq z^+ \leq h$ is valid for the upper thin film A, and $z^- \leq 0$ is valid for the lower substrate B. Dislocation loop L_1 is located in the thin film medium A, and dislocation loop L_2 is located in the substrate medium B.

The free surface and interface of the perfect bonding film-substrate system should satisfy the following requirements:

(a) Free traction stress should be satisfied on the top free surface of the film-substrate system.

$$\sigma_{i3}^f \Big|_{z^+=+h} = 0 \tag{1}$$

(b) Elastic displacement and traction stress should be continuous across the interface planes of the film-substrate system.

$$\sigma_{i3}^f \Big|_{z^+=-h} = \sigma_{i3}^s \Big|_{z^-=0} \tag{2}$$

and

$$u_i^f \Big|_{z^+=-h} = u_i^s \Big|_{z^-=0} \tag{3}$$

Where 'f' and 's' stand for the film and substrate, and superscripts '+' and '−' distinguish the local Cartesian coordinates of the film and substrate, respectively.

Solution for Perfect Bonding Film-Substrate System

In this subsection, a semi-analytical solution is proposed to solve the elastic fields due to dislocation loops within thin film or substrate of perfect bonding film-substrate system. Elastic field of perfect bonding film-substrate system can be decomposed into two sub-problems: bulk stress due to a dislocation loop in an infinite space, and correction stress induced by the free surface and interface of the film-substrate system.

As shown in Figure 2a, perfect bonding film-substrate system is

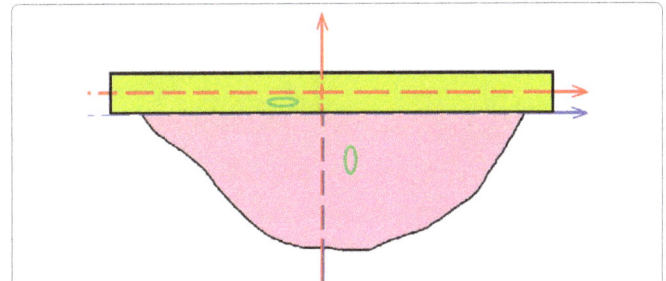

Figure 1: Cartesian coordinates of a film-substrate system: (a) coordinate (x^+, y^+, z^+) for the thin film; (b) coordinate (x^-, y^-, z^-) for the lower half space.

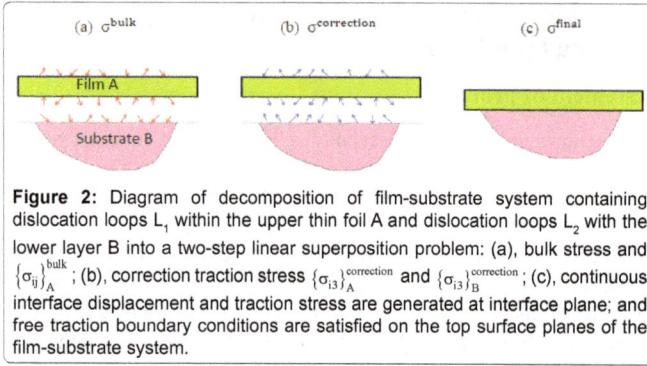

Figure 2: Diagram of decomposition of film-substrate system containing dislocation loops L_1 within the upper thin foil A and dislocation loops L_2 with the lower layer B into a two-step linear superposition problem: (a), bulk stress and $\{\sigma_{ij}\}_A^{bulk}$; (b), correction traction stress $\{\sigma_{i3}\}_A^{correction}$ and $\{\sigma_{i3}\}_B^{correction}$; (c), continuous interface displacement and traction stress are generated at interface plane; and free traction boundary conditions are satisfied on the top surface planes of the film-substrate system.

decomposed into thin film **A** containing dislocation loops L_1 and substrate B containing dislocation loop L_2. Bulk traction elastic field $\{\sigma_{i3}\}_A^{\infty}$ and $\{\sigma_{i3}\}_B^{\infty}$ on the top free surface and interface plane can be calculated out.

As shown in Figure 2b, correction traction stress $\{\sigma_{i3}\}_A^{correction}$ and $\{\sigma_{i3}\}_B^{correction}$ are linearly superimposed onto the bulk elastic fields.

As shown in Figure 2c, after linear superposition of bulk stress and correction stress, displacement and traction stress continuity across the interface plane are satisfied.

Elastic Fields of Dislocation Loops within Infinite Isotropic Medium

Bulk displacement of dislocation loops can be calculated with Volterra's formula, and written as surface integration over dislocation area [17,18]:

$$u_m^{\infty}(x) = b_i \int_S C_{ijkl} G_{km,l}(x-x')dS_j \qquad (4)$$

Where S is the dislocation surface, a cap with its boundary formed by dislocation loop perimeter. C_{ijkl} are the elastic constants, and $G_{km,l}$ is the first order derivative of Green's function.

For an isotropic material, following relation exists:

$$b_i C_{ijkl} G_{km,l}(R) = \frac{1}{8\pi\mu}\left[\begin{array}{c}\mu b_m R_{,ppj} + \mu\left(b_i R_{,ppl}\delta_{jm} - b_j R_{,ppm}\right) \\ +2\left(\frac{\lambda+\mu}{\lambda+2\mu}\right)\mu\left(b_j R_{,ppm} - b_i R_{,mij}\right)\end{array}\right] \qquad (5)$$

Where R = x–x' is the vector from dislocation position x' to the calculated position x, λ, is Lamé's first parameter of isotropic material, and μ is shear modulus of isotropic material, δ_{ij} is the Kronecker delta operator.

Bulk stress induced by dislocation loops in a homogenous medium can be calculated with Mura's formula [18]. Firstly, bulk displacement gradient can be written as:

$$u_{i,j}^{\infty}(x) = \int \varepsilon_{jnh} C_{pqmn} G_{ip,q}(x-x')b_m' dx_h' \qquad (6)$$

where the integration is performed along dislocation loop perimeter.

For an isotropic material, it appears that:

$$C_{pqmn} G_{ip,q}(x-x') = \frac{-1}{8\pi(1-v)}\left[\begin{array}{c}(1-2v)\dfrac{\delta_{ni}(x_m-x_m')+\delta_{im}(x_n-x_n')+\delta_{mn}(x_i-x_i')}{R^3} \\ +3\dfrac{(x_m-x_m')(x_n-x_n')(x_i-x_i')}{R^5}\end{array}\right] \qquad (7)$$

Where R = |X-X'| is the distance from the dislocation position x' to the calculated position x in the space.

Considering the displacement-strain differential relation, bulk stress of dislocation segment can be generated with isotropic Hooke's law:

$$\sigma_{ij}^{\infty} = \lambda.\varepsilon_{kk}.\delta_{ij} + 2\mu\varepsilon_{ij} = C_{ijkl}.\varepsilon_{ij} \qquad (8)$$

Then, bulk stress induced by a circular dislocation loop can be produced through linear integration along dislocation loop perimeter for a round.

Correction Stress of Lower Substrate Medium

In the absence of body forces, the stress equilibrium equation of the lower substrate medium B can be written in terms of the displacement $u_i^-(x)$ as:

$$\mu^s u_{i,jj}^- + \left(\lambda^s + \mu^s\right)u_{j,ji}^- = 0 \qquad (9)$$

Where $\lambda^s = 2\ \mu^s v^s/(1-2v^s)$, μ^s and v^s are the shear modulus and Poisson's ratio for the lower substrate medium B.

Similar to the solutions for the image stress of half space developed by Weinberger et. al [15], an arbitrary correction elastic field written in the form of Fourier series with unknown Fourier coefficients is employed for solving the correction elastic field of the lower substrate medium B. The following correction displacement solution to Eq. (9) is written as sum over different Fourier modes:

$$\begin{cases} u^- = \displaystyle\sum_{k_x}\sum_{k_y}\left(+k_x z^- K_1^- - k_y K_2^- + ik_x K_3^-\right)\cdot e^{+k_z z^-}\cdot e^{+\left(ik_x x + ik_y y\right)} \\ v^- = \displaystyle\sum_{k_x}\sum_{k_y}\left(+k_y z^- K_1^- + k_x K_2^- + ik_y K_3^-\right)\cdot e^{+k_z z^-}\cdot e^{+\left(ik_x x + ik_y y\right)} \\ w^- = \displaystyle\sum_{k_x}\sum_{k_y}\left(\left(-ik_x z^- + i\dfrac{\ddot{e}_s+3\grave{\imath}_s}{\ddot{e}_s+\grave{\imath}_s}\right)K_1^- + k_z K_3^-\right)\cdot e^{+k_z z^-}\cdot e^{+\left(ik_x x+ik_y y\right)} \end{cases} \qquad (10)$$

where $k_z = \sqrt{k_x^2+k_y^2}$ and $\left(K_1^-, K_2^-, K_3^-\right)$ are complex constants. The solution is periodic in the x and y directions and exponential in the z^- direction.

Due to the completeness of the Fourier series, Fourier coefficient components for certain (k_x, k_y) mode can be written as:

$$\begin{cases} \hat{u}^- = \left(+k_x z^- K_1^- - k_y K_2^- + ik_x K_3^-\right)\cdot e^{+k_z z^-} \\ \hat{v}^- = \left(+k_y z^- K_1^- + k_x K_2^- + ik_y K_3^-\right)\cdot e^{+k_z z^-} \\ \hat{w}^- = \left(\left(-ik_x z^- + i\dfrac{\ddot{e}_s+3\grave{\imath}_s}{\ddot{e}_s+\grave{\imath}_s}\right)K_1^- + k_z K_3^-\right)\cdot e^{+k_z z^-} \end{cases} \qquad (11)$$

Thus, the correction displacement field is written as:

$$u_i^-\left(x,y,z^-\right) = \sum_{k_x^-}\sum_{k_y^-}\hat{u}_i^-\left(k_x, k_y, z^-\right).e^{+\left(ik_x x+ik_y y\right)} \qquad (12)$$

The correction displacement components $\left(\hat{u}^-, \hat{v}^-, \hat{w}^-\right)$ on the free surface plane $z^- = 0$ for certain (k_x, k_y) mode can be written as:

$$\begin{Bmatrix} \hat{u}^- \\ \hat{v}^- \\ \hat{w}^- \end{Bmatrix} = \begin{bmatrix} N^- \end{bmatrix} \cdot \begin{Bmatrix} K_1^- \\ K_2^- \\ K_3^- \end{Bmatrix} \tag{13}$$

where the details of $[N^-]$ are shown in Appendix (A. 1).

Following the displacement field solution in Eq. (10), it is straightforward to obtain the strain field $\hat{\varepsilon}_{ij}^-$ through differentiation rule and the stress field $\hat{\sigma}_{ij}^-$ by using Hooke's law. The traction stress field can also be written in the form of Fourier series:

$$\sigma_{i3}^- \left(x, y, z^- \right) = \sum_{k_x} \sum_{k_y} \hat{\sigma}_{i3}^- \left(k_x, k_y, z^- \right) \cdot e^{+\left(ik_x x + ik_y y \right)} \tag{14}$$

The correction traction stress components $\left(\hat{\sigma}_{13}^-, \hat{\sigma}_{23}^-, \hat{\sigma}_{33}^- \right)$ on the free surface plane $z^- = 0$ for certain $\left(k_x, k_y \right)$ mode can be written as:

$$\begin{Bmatrix} \hat{\sigma}_{13}^- \\ \hat{\sigma}_{23}^- \\ \hat{\sigma}_{33}^- \end{Bmatrix} = \begin{bmatrix} M^- \end{bmatrix} \cdot \begin{Bmatrix} K_1^- \\ K_2^- \\ K_3^- \end{Bmatrix} \tag{15}$$

and the details of $\begin{bmatrix} M^- \end{bmatrix}$ are shown in Appendix (A. 2).

The numerical solutions of Eqs. (10)-(15) for the substrate medium are considered in the x and y directions with periodic lengths L_x and L_y. The wave number is set to be $k_x = 2\pi n_x/L_x$ and $k_y = 2\pi n_y/L_y$, where $n_x = n_y = 0, \pm1, 2, \pm3...$

Correction Stress of upper Thin Film

Similar to the efficient semi-analytical image stress solution derived by Weinberger et al. [15] of isotropic thin foils, an arbitrary correction elastic field written in the form of Fourier series with unknown Fourier coefficients is employed for solving the correction elastic field of the film medium A.

In the absence of body forces, the stress equilibrium of an isotropic linear medium composed of upper thin film medium **A** can be written in terms of the displacement $u_i^+ (x)$ as:

$$\mu^f u_{i,jj}^+ + \left(\lambda^f + \mu^f \right) u_{j,ji}^+ = 0 \tag{16}$$

where $\lambda^f = 2\mu^f v^f / \left(1 - 2v^f \right)$, μ^f and v^f are the shear modulus and Poisson's ratio for the upper thin film medium A. The thin film is assumed to have a thickness of 2h in the z^+ direction, and $-h \, z^+ \leq h$.

The correction elastic field can be written as sum over Fourier series, and the solution is periodic in the x and y directions, and hyperbolic in the z^+ direction.

$$\begin{cases} u^+ = \sum_{k_x} \sum_{k_y} \left[\begin{array}{l} \left(k_x z^+ A - k_y F + ik_x G \right) sinh\left(k_z z^+ \right) \\ + \left(k_x z^+ E + k_y B + ik_x C \right) cosh\left(k_z z^+ \right) \end{array} \right] \cdot e^{\left(ik_x x + ik_y y \right)} \\[4mm] v^+ = \sum_{k_x} \sum_{k_y} \left[\begin{array}{l} \left(k_y z^+ A + k_x F + ik_y G \right) sinh\left(k_z z^+ \right) \\ + \left(k_y z^+ E - k_x B + ik_y C \right) cosh\left(k_z z^+ \right) \end{array} \right] \cdot e^{\left(ik_x x + ik_y y \right)} \\[4mm] w^+ = \sum_{k_x} \sum_{k_y} \left[\begin{array}{l} \left(-ik_z z^+ A + i\dfrac{\breve{e}_f + 3i}{\breve{e}_f + i}_f E + k_z G \right) cosh\left(k_z z^+ \right) \\ + \left(-ik_z z^+ E + i\dfrac{\breve{e}_f + 3i}{\breve{e}_f + i}_f A + k_z C \right) sinh\left(k_z z^+ \right) \end{array} \right] \cdot e^{\left(ik_x x + ik_y y \right)} \end{cases} \tag{17}$$

where $k_z = \sqrt{k_x^2 + k_y^2}$, and the unknown terms (A,B,C) and (E,F,G) are

complex constants.

Due to the mathematical completeness of Fourier series, the Fourier coefficient components for each Fourier (k_x, k_y) mode is:

$$\begin{cases} \hat{u}^+ = \left[\begin{array}{l} \left(k_x z^+ A - k_y F + ik_x G \right) sinh\left(k_z z^+ \right) \\ + \left(k_x z^+ E + k_y B + ik_x C \right) cosh\left(k_z z^+ \right) \end{array} \right] \\[4mm] \hat{v}^+ = \left[\begin{array}{l} \left(k_y z^+ A + k_x F + ik_y G \right) sinh\left(k_z z^+ \right) \\ + \left(k_y z^+ E - k_x B + ik_y C \right) cosh\left(k_z z^+ \right) \end{array} \right] \\[4mm] \hat{w}^+ = \left[\begin{array}{l} \left(-ik_z z^+ A + i\dfrac{\breve{e}_f + 3i}{\breve{e}_f + i}_f E + k_z G \right) cosh\left(k_z z^+ \right) \\ + \left(-ik_z z^+ E + i\dfrac{\breve{e}_f + 3i}{\breve{e}_f + i}_f A + k_z C \right) sinh\left(k_z z^+ \right) \end{array} \right] \end{cases} \tag{18}$$

Alternatively, the correction displacement field in Eq. (17) can be written as:

$$u_i^+ \left(x, y, z^+ \right) = \sum_{k_x} \sum_{k_y} \left[\hat{u}_i^+ \left(k_x, k_y, z^+ \right) \right] \cdot \exp\left(ik_x x + ik_y y \right) \tag{19}$$

and the correction traction stress can be obtained from isotropic Hooke's law:

$$\sigma_{i3}^+ \left(x, y, z^+ \right) = \sum_{k_x} \sum_{k_y} \left[\hat{\sigma}_{i3}^+ \left(k_x, k_y, z^+ \right) \right] \cdot \exp\left(ik_x x + ik_y y \right) \tag{20}$$

Then, Eq. (20) was submitted into the equilibrium Eq. (16), correction displacement fields on the surface planes $z^+ = \pm h$ can be combined together, and rewritten into two sets of equations on unknown coefficients (A,B,C) and (E,F,G), which correspond to the symmetrical and the asymmetrical parts, respectively.

The symmetrical correction displacement is:

$$u^S = \frac{1}{2} \begin{bmatrix} \hat{u}^+\big|_{z^+=+h} + \hat{u}^+\big|_{z^+=-h} \\ \hat{v}^+\big|_{z^+=+h} + \hat{v}^+\big|_{z^+=-h} \\ \hat{w}^+\big|_{z^+=+h} - \hat{w}^+\big|_{z^+=-h} \end{bmatrix} = \begin{bmatrix} N^S \end{bmatrix} \cdot \begin{Bmatrix} A \\ B \\ C \end{Bmatrix} \cdot \exp\left(ik_x x + ik_y y \right) \tag{21}$$

and the asymmetrical correction displacement part is:

$$u^A = \frac{1}{2} \begin{bmatrix} \hat{u}^+\big|_{z^+=+h} - \hat{u}^+\big|_{z^+=-h} \\ \hat{v}^+\big|_{z^+=+h} - \hat{v}^+\big|_{z^+=-h} \\ \hat{w}^+\big|_{z^+=+h} + \hat{w}^+\big|_{z^+=-h} \end{bmatrix} = \begin{bmatrix} N^A \end{bmatrix} \cdot \begin{Bmatrix} E \\ F \\ G \end{Bmatrix} \cdot \exp\left(ik_x x + ik_y y \right) \tag{22}$$

The symmetrical correction traction stress is:

$$T^S = \frac{1}{2} \begin{bmatrix} \sigma_{13}^+\big|_{z^+=+h} - \sigma_{13}^-\big|_{z^-=-h} \\ \sigma_{23}^+\big|_{z^+=+h} - \sigma_{23}^-\big|_{z^-=-h} \\ \sigma_{33}^+\big|_{z^+=+h} + \sigma_{33}^-\big|_{z^-=-h} \end{bmatrix} = \begin{bmatrix} M^S \end{bmatrix} \cdot \begin{Bmatrix} A \\ B \\ C \end{Bmatrix} \cdot \exp\left(ik_x x + ik_y y \right) \tag{23}$$

and the asymmetrical correction traction stress is:

$$T^A = \frac{1}{2} \begin{bmatrix} \sigma_{13}^+\big|_{z^+=+h} + \sigma_{13}^-\big|_{z^-=-h} \\ \sigma_{23}^+\big|_{z^+=+h} + \sigma_{23}^-\big|_{z^-=-h} \\ \sigma_{33}^+\big|_{z^+=+h} - \sigma_{33}^-\big|_{z^-=-h} \end{bmatrix} = \begin{bmatrix} M^A \end{bmatrix} \cdot \begin{Bmatrix} E \\ F \\ G \end{Bmatrix} \cdot \exp\left(ik_x x + ik_y y \right) \tag{24}$$

The explicit expressions for the correction displacement and traction stress matrices [M^S], [M^A], [N^S] and [N^A] are given in Appendix (A. 3) - (A. 6).

The calculation procedures of Eqs. (17)-(24) for isotropic thin film are considered in the x and y directions with periodic lengths L_x and L_y. The wave number is set to be $k_x = 2\pi n_x/L_x$ and $k_y = 2\pi n_y/L_y$, where $n_x = n_y = 0, \pm1, \pm2, \pm3\ldots$

Elastic Field of Perfect Bonding Film-Substrate System

Considering the displacement and traction stress continuity requirements in Eqs. (1)-(3) for the perfect bonding film-substrate system, following relation stands for each (k_x, k_y) Fourier mode.

Free traction stress on the free surface plane of the perfect bonding film-substrate system should be satisfied:

$$\{\hat{\sigma}_{i3}\}_{z^+=+h}^{\infty} + \{\hat{\sigma}_{i3}\}_{z^+=+h}^{\text{correction}} = 0 \tag{25}$$

Interface traction stress and displacement continuity across the interface plane of the perfect bonding film-substrate system should be satisfied:

$$\{\hat{\sigma}_{i3}\}_{z^+=-h}^{\infty} + \{\hat{\sigma}_{i3}\}_{z^+=-h}^{\text{correction}} = \{\hat{\sigma}_{i3}\}_{z^-=0}^{\infty} + \{\hat{\sigma}_{i3}\}_{z^-=0}^{\text{correction}} \Big| \tag{26}$$

and

$$\{\hat{u}_{i}\}_{z^+=+h}^{\infty} + \{\hat{u}_{i}\}_{z^+=+h}^{\text{correction}} = \{\hat{u}_{i}\}_{z^-=0}^{\infty} + \{\hat{u}_{i}\}_{z^-=0}^{\text{correction}} \tag{27}$$

After submitting the bulk elastic field and correction elastic field into Eqs. (25)-(27), the following relations stand for each (k_x, k_y) Fourier mode.

(a) Free traction boundary condition should be satisfied on the top free surface plane of the thin film A for each (k_x, k_y) Fourier mode.

$$[M^S]\begin{Bmatrix} A \\ B \\ C \end{Bmatrix} + [M^A]\begin{Bmatrix} E \\ F \\ G \end{Bmatrix} + \begin{bmatrix} \{\hat{\sigma}_{13}\}_{z^+=h}^{\infty} \\ \{\hat{\sigma}_{23}\}_{z^+=h}^{\infty} \\ \{\hat{\sigma}_{33}\}_{z^+=h}^{\infty} \end{bmatrix} = 0 \tag{28}$$

(b) Interface traction stress should be continuous for each (k_x, k_y) Fourier mode.

$$[K_{ij}] * \begin{bmatrix} -[M^S]\begin{Bmatrix} A \\ B \\ C \end{Bmatrix} + [M^A]\begin{Bmatrix} E \\ F \\ G \end{Bmatrix} + \begin{bmatrix} \{\hat{\sigma}_{13}\}_{z^+=-h}^{\infty} \\ \{\hat{\sigma}_{23}\}_{z^+=-h}^{\infty} \\ \{\hat{\sigma}_{33}\}_{z^+=-h}^{\infty} \end{bmatrix} \end{bmatrix} = [M^-]\begin{Bmatrix} K_1^- \\ K_2^- \\ K_3^- \end{Bmatrix} + \begin{bmatrix} \{\hat{\sigma}_{13}\}_{z^-=0}^{\infty} \\ \{\hat{\sigma}_{23}\}_{z^-=0}^{\infty} \\ \{\hat{\sigma}_{33}\}_{z^-=0}^{\infty} \end{bmatrix} \tag{29}$$

in which, $[K_{ij}] = \text{diag}\{1, 1, -1\}$ is a 3*3 diagonal matrix.

(c) Interface displacement should be continuous for each (k_x, k_y) Fourier mode.

$$[K_{ij}] * \begin{bmatrix} [N^S]\begin{Bmatrix} A \\ B \\ C \end{Bmatrix} - [N^A]\begin{Bmatrix} E \\ F \\ G \end{Bmatrix} + \begin{bmatrix} \{\hat{u}_{1}\}_{z^+=-h}^{\infty} \\ \{\hat{u}_{2}\}_{z^+=-h}^{\infty} \\ \{\hat{u}_{3}\}_{z^+=-h}^{\infty} \end{bmatrix} \end{bmatrix} = [N^-]\begin{Bmatrix} K_1^- \\ K_2^- \\ K_3^- \end{Bmatrix} + \begin{bmatrix} \{\hat{u}_{1}\}_{z^-=0}^{\infty} \\ \{\hat{u}_{2}\}_{z^-=0}^{\infty} \\ \{\hat{u}_{3}\}_{z^-=0}^{\infty} \end{bmatrix} \tag{30}$$

In summary, Eqs. (28)-(30) can be written together as:

$$\begin{bmatrix} [M^S] & [M^A] & 0 \\ [K_{ij}]*[N^S] & -[K_{ij}]*[N^A] & -[N^-] \\ -[K_{ij}]*[M^S] & [K_{ij}]*[M^A] & -[M^-] \end{bmatrix} \blacklozenge \begin{Bmatrix} A \\ B \\ C \\ E \\ F \\ G \\ K_1^- \\ K_2^- \\ K_3^- \end{Bmatrix} = \begin{Bmatrix} -\{\hat{\sigma}_{13}\}_{z^+=h}^{\infty} \\ -\{\hat{\sigma}_{23}\}_{z^+=h}^{\infty} \\ -\{\hat{\sigma}_{33}\}_{z^+=h}^{\infty} \\ -\{\hat{u}_{1}\}_{z^+=-h}^{\infty} + \{\hat{u}_{1}\}_{z^-=0}^{\infty} \\ -\{\hat{u}_{2}\}_{z^+=-h}^{\infty} + \{\hat{u}_{2}\}_{z^-=0}^{\infty} \\ -\{\hat{u}_{3}\}_{z^+=-h}^{\infty} + \{\hat{u}_{3}\}_{z^-=0}^{\infty} \\ -\{\hat{\sigma}_{13}\}_{z^+=-h}^{\infty} + \{\hat{\sigma}_{13}\}_{z^-=0}^{\infty} \\ -\{\hat{\sigma}_{23}\}_{z^+=-h}^{\infty} + \{\hat{\sigma}_{23}\}_{z^-=0}^{\infty} \\ -\{\hat{\sigma}_{33}\}_{z^+=-h}^{\infty} + \{\hat{\sigma}_{33}\}_{z^-=0}^{\infty} \end{Bmatrix} \tag{31}$$

Then, unknown coefficient (A,B,C,E,F,G) and (K_1^-, K_2^-, K_3^-) of correction displacement can be solved from Eq. (31), and the correction elastic field of the film medium A and substrate medium B can be generated.

The total elastic field is the sum of the two contribution parts: bulk elastic field $\{\hat{\sigma}_{ij}\}_A^{\text{bulk}}$ and $\{\hat{\sigma}_{ij}\}_B^{\text{bulk}}$; and correction elastic fields $\{\hat{\sigma}_{ij}\}_A^{\text{correction}}$ and $\{\hat{\sigma}_{ij}\}_B^{\text{correction}}$, respectively.

Total displacement within the upper film A of the film-substrate system is:

$$\{\hat{u}_{i}\}_A^{\text{final}} = \{\hat{u}_{i}\}_A^{\text{bulk}} + \{\hat{u}_{i}\}_A^{\text{correction}} \tag{32}$$

Total stress within the upper film A of the film-substrate system is:

$$\{\hat{\sigma}_{ij}\}_A^{\text{final}} = \{\hat{\sigma}_{ij}\}_A^{\text{bulk}} + \{\hat{\sigma}_{ij}\}_A^{\text{correction}} \tag{33}$$

Total displacement within the substrate B of the film-substrate system is:

$$\{\hat{u}_{i}\}_B^{\text{final}} = \{\hat{u}_{i}\}_B^{\text{bulk}} + \{\hat{u}_{i}\}_B^{\text{correction}} \tag{34}$$

Total stress within the substrate B of the film-substrate system is:

$$\{\hat{\sigma}_{ij}\}_B^{\text{final}} = \{\hat{\sigma}_{ij}\}_B^{\text{bulk}} + \{\hat{\sigma}_{ij}\}_B^{\text{correction}} \tag{35}$$

Calculation Examples

In this section, the above semi-analytical approach is employed for analyzing the elastic fields induced by dislocation loop within perfect bonding isotropic Cu-Nb film-substrate system. The local Cartesian coordinate (x^+, y^+, z^+) is along $(11\bar{2})$, $(\bar{1}10)$ and (111) in upper Cu thin foil; and the local Cartesian coordinate (x^-, y^-, z^-) is along $(1\bar{1}2)$, $(1\bar{1}\bar{1})$ and (110) in lower Nb substrate medium. The origin of the upper Cartesian coordinate is in the middle of the Cu thin foil, and the origin of the lower Nb half space is on the interface plane of the film-substrate system. In all the calculation examples below, dislocation loop is segmented into 40 straight dislocation segments along the circular perimeter for a round, and bulk displacement and stress fields are obtained through integrating the dislocation segments around the dislocation loop perimeter, based on Volterra's [17], Devincre's [18] and Mura's integration formulas [19], respectively. The elastic modulus of Cu and Nb is shown in Table 1, and the isotropic equivalent shear modulus and Poisson's ratio are treated with Voigt isotropic model [20,21]. In the following simulation examples, the periodic length on the interface plane is $L_x = L_y = 200$ nm, the meshing density is $n_x = n_y = 200$, and the Fourier wave number range is: $-30 \leq k_x \leq 30$ and $-30 \leq k_y \leq 30$.

Elastic Field due to a Dislocation Loop in the Substrate Medium

In this section, elastic displacement and traction stress due to a dislocation loop in the lower Nb substrate medium of the perfect bonding Cu-Nb film-substrate system are studied, and the thickness of the upper Cu film is assumed to be 40 nm. The circular $1/2a\langle110\rangle(111)$

Material	C_{11} (GPa)	C_{12} (GPa)	C_{14} (GPa)	Voigt shear modulus (GPa)	Voigt Poisson's ratio (-)	Lattice constant (nm)
Cu	168.4	121.4	75.4	54.64	0.3241	a=0.36149
Nb	240.2	125.6	28.2	39.84	0.3875	a=0.33004

Table 1: Elastic parameters of pure Cu and Nb [20,21].

dislocation loop with radius r = 5 nm is located at a distance d = 10 nm below the interface plane in lower Nb medium, and its habit plane is inclined to the interface plane. Figures 3a, 3b, 3e and 3f are the side view of the elastic field plotted in the σ_{xz} plane (y = 0), and Figures 3c and 3d are the side view of the elastic field plotted in the oyz plane (x = 0) for the perfect bonding Cu-Nb film-substrate system. It can be seen that the final interface in plane and out of plane displacement field, traction stress field across the interface plane are identical, and thus continuous displacement and traction stress is generated.

As shown in Figure 4, correction elastic field are superimposed onto the bulk elastic field, thus generating the final interface elastic field, and the contributions from bulk and correction elastic fields are compared with each other. As shown in Figures 4a-4f, interface displacement profile of u and w, and interface traction stress σ_{xz} and σ_{zz} are plotted along x direction (y = 0) on the interface plane of the perfect bonding Cu-Nb film-substrate system; as shown in Figures 4c and 4d, interface displacement profile of v and interface traction stress σ_{yz} are plotted along y direction (x = 0) on the interface plane of the perfect bonding Cu-Nb film-substrate system. It can be seen from Figures 4b, 4d and

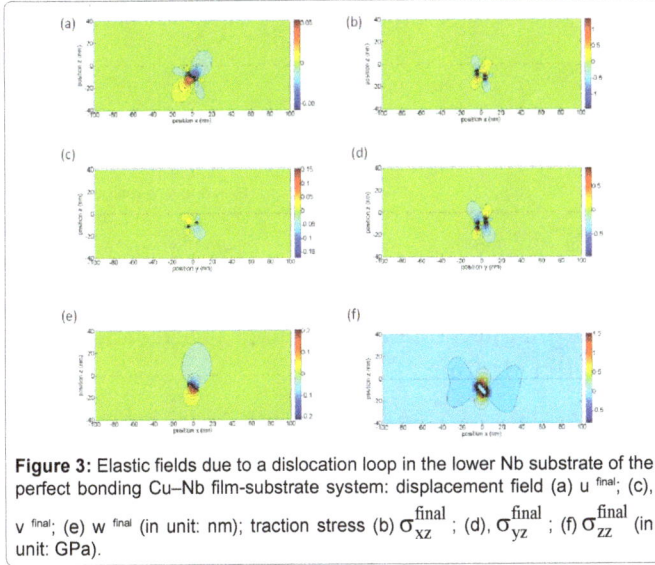

Figure 5: Elastic fields due to a dislocation loop in the upper Cu film of the perfect bonding Cu–Nb film-substrate system: displacement field (a) u ^{final}; (c) v ^{final}; (e) w ^{final} (in unit: nm); traction stress (b) σ_{xz}^{final}; (d) σ_{yz}^{final}; (f) σ_{zz}^{final} (in unit: GPa).

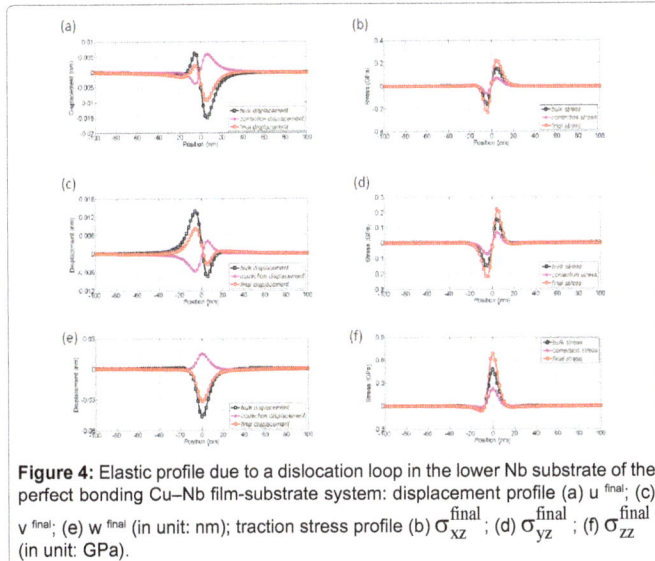

Figure 3: Elastic fields due to a dislocation loop in the lower Nb substrate of the perfect bonding Cu–Nb film-substrate system: displacement field (a) u ^{final}; (c), v ^{final}; (e) w ^{final} (in unit: nm); traction stress (b) σ_{xz}^{final}; (d), σ_{yz}^{final}; (f) σ_{zz}^{final} (in unit: GPa).

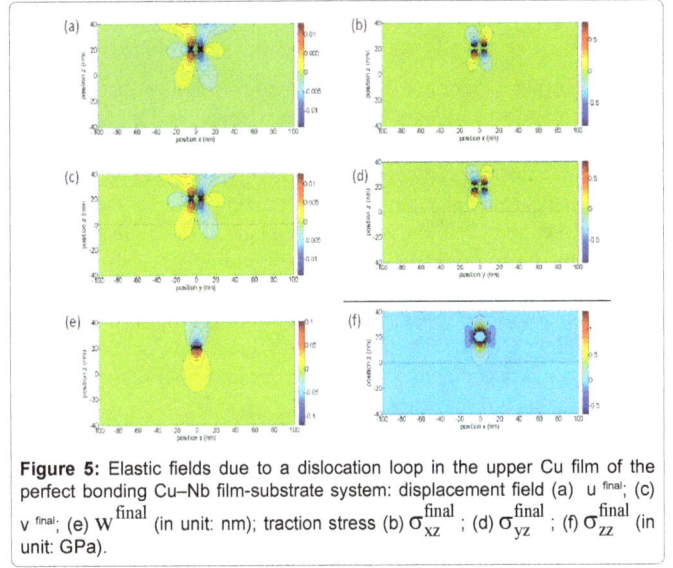

Figure 6: Elastic profile due to a dislocation loop in the upper Cu film of the perfect bonding Cu–Nb film-substrate system: displacement profile (a) u ^{final}; (c) v ^{final}; (e) w ^{final} (in unit: nm); traction stress profile (b) σ_{xz}^{final}; (d) σ_{yz}^{final}; (f) σ_{zz}^{final} (in unit: GPa).

4f that the amplitudes of bulk traction stress are slightly strengthened on the interface plane, as the Voigt shear modulus of upper layer Cu is larger than lower layer Nb, and the traction stress due to a dislocation loop in the lower substrate Nb is slightly strengthened by the upper Cu film at the perfect bonding interface plane.

Elastic Field due to a Dislocation Loop in the Film Medium

In this section, elastic displacement and traction stress due to a dislocation loop in the upper Cu film of the perfect bonding Cu-Nb film-substrate system are studied, and the thickness of the upper Cu film is assumed to be 40 nm. The circular 1/3a(111)(111) dislocation loop with radius r = 5 nm is located in the middle of upper Cu thin film, and its habit plane is parallel to the interface plane. Figures 5a, 5b, 5e and 5f are the side view of the elastic field plotted in the oxz plane (y = 0), and Figures 5c and 5d are the side view of the elastic field plotted in the oyz plane (x = 0) for the perfect bonding Cu-Nb film-substrate system. It can be seen that the final interface in plane and out of plane displacement field, traction stress field across the interface plane are

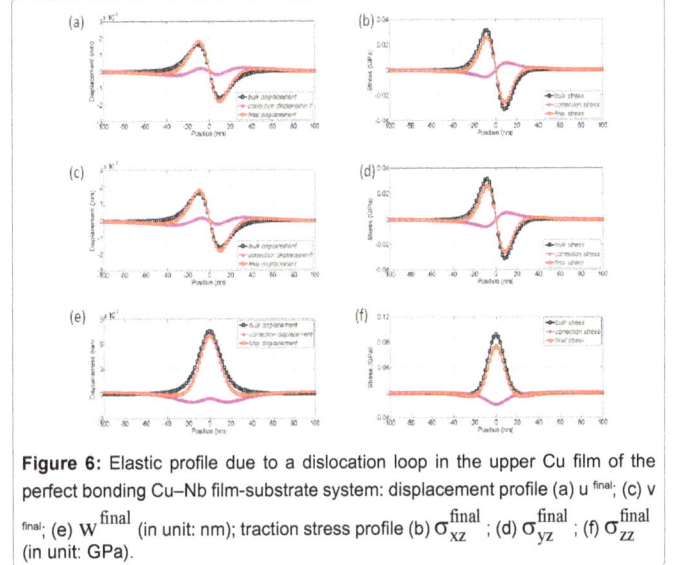

Figure 4: Elastic profile due to a dislocation loop in the lower Nb substrate of the perfect bonding Cu–Nb film-substrate system: displacement profile (a) u ^{final}; (c) v ^{final}; (e) w ^{final} (in unit: nm); traction stress profile (b) σ_{xz}^{final}; (d) σ_{yz}^{final}; (f) σ_{zz}^{final} (in unit: GPa).

identical, and thus continuous displacement and traction stress is generated.

As shown in Figures 6a, 6b, 6e and 6f, interface displacement profile of u and w, and interface traction stress σ_{xz} and σ_{zz} are plotted along x direction (y = 0) on the interface plane of the perfect bonding Cu-Nb film-substrate system; as shown in Figures 6c and 6d, interface displacement profile of v and interface traction stress σ_{yz} are plotted along y direction (x = 0) on the interface plane of the perfect bonding Cu-Nb film-substrate system. It can be seen from Figures 6b, 6d and 6f that the amplitudes of bulk traction stress are slightly weakened on the interface plane, as the Voigt shear modulus of upper layer Cu is larger than lower layer Nb, and the traction stress due to a dislocation loop in the upper Cu film is slightly weakened by the lower Nb substrate at the perfect bonding interface plane.

Film Thickness Effect on Elastic Field of a Dislocation Loop in the Film Medium

In this subsection, effects of film thickness on the interface displacement and interface traction stress amplitudes are investigated. Besides Cu film thickness, the physical and material parameters of dislocation loop and film-substrate system are identical to the simulated Cu-Nb film-substrate system example in subsection 3.2. The thin Cu film thickness is assumed to be: t = 20, 30, 40, 50 and 60 nm, and the dislocation loop is situated in the middle of the upper thin Cu film. The simulation results are shown in Figure 7, and side view of the traction stress σ_{xz} and σ_{zz} are plotted in the oxz plane (y = 0) for the perfect bonding Cu-Nb film-substrate system. It can be concluded from Figure 7 that: with the decrease of film thickness, bulk elastic field at interface plane are changed more remarkable by correction stress.

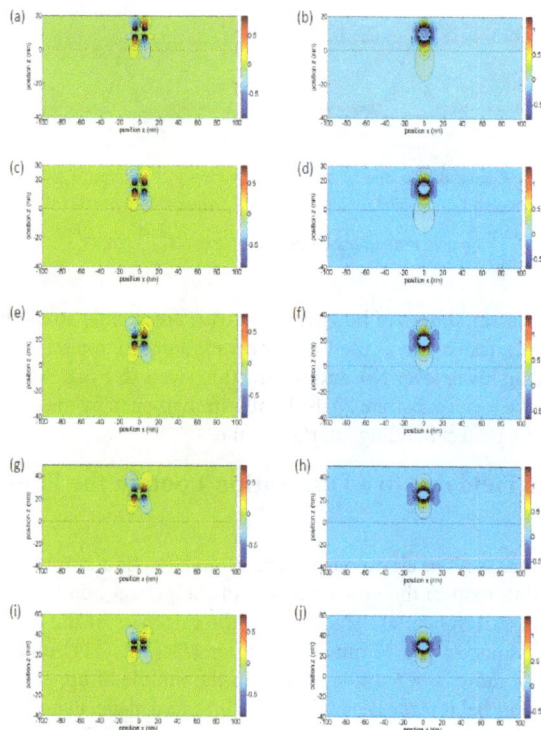

Figure 7: Film thickness effect on elastic fields due to a dislocation loop in the upper Cu film of perfect bonding Cu–Nb film-substrate system. Traction stress field σ_{xz}^{final} and σ_{zz}^{final} : (a)–(b), t=20 nm; (c)–(d), t=30 nm; (e)–(f), t=40 nm; (g)–(h) t=50 nm; (i)–(j), t=60 nm (in unit: GPa).

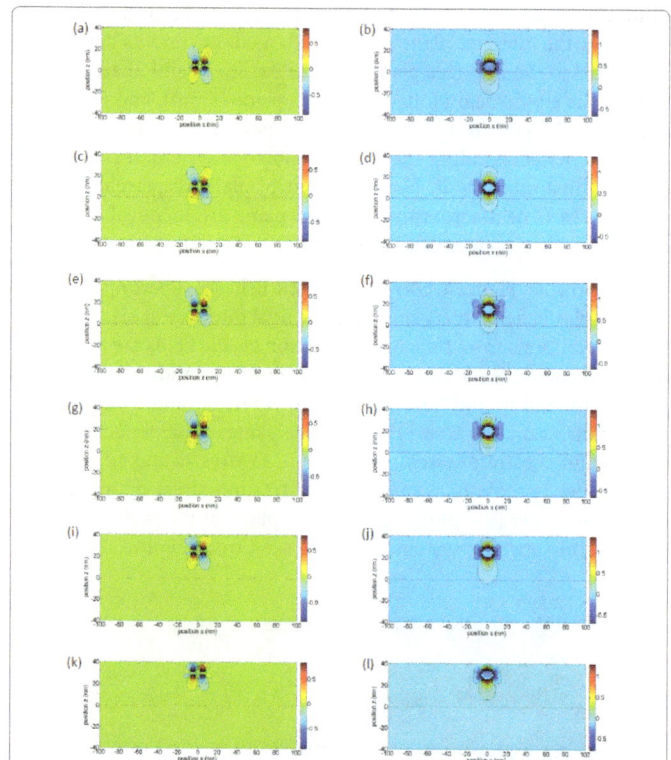

Figure 8: Dislocation loop depth effect on elastic fields due to a dislocation loop in the upper Cu film of perfect bonding Cu–Nb film-substrate system. Traction stress field σ_{xz}^{final} and σ_{zz}^{final} : (a)–(b) d=5 nm; (c)–(d) d=10 nm; (e)–(f) d=15 nm; (g)–(h) d=20 nm; (i)–(j) d=25 nm; (k)–(l) d=30 nm; (in unit: GPa).

Loop Depth Effect on Elastic Field of a Dislocation Loop in the Film Medium

In this subsection, effects of dislocation loop depth within the upper Cu thin film of the Cu-Nb film-substrate system on the elastic field are studied. Besides loop depth, physical parameters of the dislocation loop and perfect bonding film-substrate system are identical to the simulated Cu-Nb film-substrate system example in subsection 3.2. The distance from dislocation loop center to the interface plane is assumed to be: d = 5, 10, 15, 20, 25 and 30 nm, and the film thickness is assumed to be t = 40 nm. The simulation results are shown in Figure 8, and side view of the traction stress σ_{xz} and σ_{zz} are plotted in the oxz plane (y = 0) for the perfect bonding Cu-Nb film-substrate system. It can be concluded from Figure 8 that the interface and free surface can influence the bulk elastic field drastically. With the increase of the distance from dislocation loop center to the interface plane, bulk elastic field at interface plane are changed more remarkable by correction stress.

Conclusion

A semi-analytical calculation approach based on 2D discrete FFT is developed for studying the elastic field due to dislocation loop in perfect bonding thin film-substrate system. Final elastic field is calculated as the linear superposition of bulk stress and the correction stress.

Reliability of the semi-analytical solution is verified by studying the elastic field of dislocation loop within perfect bonding Cu-Nb film-substrate system. Effects of film thickness and loop depth within thin film on the elastic field are analyzed, demonstrating that these two factors have a significant impact on the elastic fields of dislocation loops in the thin film.

Acknowledgement

The research project was Supported by the China Postdoctoral Science Foundation under Grant No. 2015M80091.

References

1. Head AK (1953) The interaction of dislocations. Philosophical Magazine 44: 92-94.

2. Head AK (1953) Edge dislocations in inhomogeneous media. Proc Physical Society 66: 793-801.

3. Head AK (1960) The interaction of dislocations with boundaries and surface films. Australian J of Physics 13: 278-283.

4. Weeks R, Dundurs J, Stippes M (1968) Exact analysis of an edge dislocation near a surface layer. Int J Impact Eng 6: 365-372.

5. Lee MS, Dundurs J (1973) Edge dislocation in a surface layer. Int J Eng Sci 11: 87-94.

6. Öveçoğlu ML, Doerner MF, Nix WD (1987) Elastic interactions of screw dislocations in thin films on substrates. Acta Metallurgica 35: 2947-2957.

7. Kelly PA, O'Connor JJ, Hills DA (1995) The stress field due to dislocation in layered media. J Physics D: Applied Physics 28: 530-534.

8. Savage JC (1998) Displacement field for an edge dislocation in a layered half-space. J Geophysical Research: Solid Earth 103: 2439-2446.

9. Wu X, Weatherly GC (2003) Equilibrium position of misfit dislocations in thin epitaxial films. Semiconductor Sci and Tech 18: 307-311.

10. Wu MS, Wang HY (2007) Solutions for edge dislocation in anisotropic film substrate system by the image method. Mathematics and Mechanics of Solids 12: 183-212.

11. Zhou K, Wu MS (2010) Elastic fields due to an edge dislocation in an isotropic film-substrate by the image method. Acta Mechanica 211: 271-292.

12. Wang HY, Yu Y, Yan SP (2014) Elastic stress fields caused by a dislocation in Ge_xSi_{1-x}/Si film-substrate system. Science China Physics, Mechanics & Astronomy 57: 1078-1089.

13. Salamon NJ, Dundurs J (1971) Elastic fields of a dislocation loop in a two-phase material. J Elasticity 1: 153-163.

14. Salamon NJ, Dundurs J (1977) A circular glide dislocation loop in a two-phase material. J Physics C: Solid State Physics 10: 497-507.

15. Weinberger CR, Aubry S, Lee SW, Nix WD, Cai W (2009) Modelling dislocations in a free-standing thin film. Modelling and Simulation in Mater Sci and Eng 17: 075007.

16. Wu W, Schäublin R, Chen J (2012) General dislocation image stress of anisotropic cubic thin film. J Applied Physics 112: 093522.

17. Volterra, V (1907) Sur l'équilibre des corps élastiques multiplement connexes. Gauthier-Villars Paris 24: 401-517.

18. Devincre B (1995) Three dimensional stress field expressions for straight dislocation segments. Solid State Communications 93: 875-878.

19. Mura T (1987) Micromechanics of Defects in Solids. (2ndedn). Martinus Nijhoff Dordrecht.

20. Bower AF (2012) Applied Mechanics of Solids. CRC Press London.

21. Voigt W (1889) Ueber die Beziehung zwischen den beiden Elasticitätsconstanten isotroper Körper. Annals of Physics 274: 573-587.

9

A Computational Fluid Dynamics Study of the Swirl Generation Analysis in Four-stroke Direct Injection Engine

Chun Xu C and Muk Choa H*

Division of Mechanical and Automotive Engineering, Kongju National University 275, Budae-dong, Cheonan-si, Chungcheongnam-Do 331-717, South Korea

Abstract

In this study, the Computational Fluid Dynamics (CFD) simulation to investigate the effect of two piston crowns to the field inside the combustion chamber of a four-stroke direct injection automotive engine when at the motoring condition. The field flow into the chamber for air-fuel mixing to obtain the better swirl ratio, engine combustion, performance and efficiency when the swirl is appeared. So this analysis is important on this study of the effect of the piston shapes to the fluid flow and the air into the cylinder, The behaviors of fluid flow occurred inside combustion chamber is represented by two parameters which influences the air streams to the cylinder during intake stroke and improves the swirl of the air-fuel to produce better air-fuel mixing during the compression stroke. To investigate the effect of air-fuel swirl when change the intake valves conditions and the piston crown shapes. The numerical simulation analysis can be showed the swirl ratio when the air-fuel into the combustion chamber that the condition of the intake valves and the shapes of the piston crown in the combustion chamber by use the Computational Fluid Dynamics (CFD) mode.

Keywords: Computational Fluid Dynamics (CFD); Piston crown shape; Swirl ratio; Fluid flow; Air-fuel mixture; Intake valve

Introduction

The in-cylinder flows of Internal Combustion Engine (ICE) that the flow structure generated by intake flows is related closely to the design and performance of the ICEs. One of the most important factors for stabilizing the ignition process and fast propagation of flame is the production of high turbulence intensity. In general, the swirls and tumble flows which two types of vortices are utilized in order to generate and preserve the turbulence flows efficiency. The two types of the vortices are organized rotations in the horizontal and vertical plane of the engine cylinder respectively. They improved the engine performance by the mixing of fuel and induced air [1]. It is very important for the development of the ICE with high compression ratio to realize higher turbulence intensity and lean burn combustion.

On the other hand, we can add the shrouds to the intake valves to produce airflow swirl. Consider the two sets of intake valves' structures that the first set of conventional valves and the second set of valves which can be set up back shrouds to prevent airflow on the backside of the valves into the combustion chamber. In this case, we can change the ports with different diameters to investigate the air flow and swirl motion. In addition, one of the ports is blocked to determine the effect of the swirl generation as the one intake port. So the results with the single port and a shrouded valve show the higher swirl generation to improve the efficiency of an engine through increasing the combustion rate of the air-fuel mixture.

In the DI engine, two experimental simulations can be achieved to increase the air-fuel mixture: one way is designing the combustion chamber structure for increasing the air-fuel mixture. Another way is designing the intake systems by blocking one of the ports or using the shrouded valve to impact a swirling motion into the combustion chamber. The swirl ratio and the fluid motion are important effect on air-fuel mixing, combustion, heat transfer and exhaust emissions.

CFD Modeling of ICE Procession

The experimental simulation works in cylinder flows have been conducted to measure velocity fields by using the hot wire anemometry or lased Doppler velocimetry (LDV) [1]. It is a really hard task to perform it because the measurements of in-cylinder flows in the reciprocating engine are characterized by highly complex three-dimensionality turbulence and unsteadiness [1]. Therefore, the numerical approach could be changed, use the CFD model to develop for the in-cylinder flow predictions. The computer codes are utilized to solve the Navier-Stokes equations to produce detailed descriptions of the mean velocity and the turbulence velocity fields. While showing ICE case, the CFD model should solve the specific problems about the air-fuel mixture into the combustion chamber turbulence flow, high Reynolds number, compressible flow and the air-fuel mixture through the geometry model.

Navier-Stokes Equations

Fluid dynamics deals with the motion of liquids and gases. Which when studied macroscopically, appear to be continuous in structure. All the variables are considered to be continuous functions of the spatial coordinates and time. The Navier-Stokes equations are a set of nonlinear partial differential equations that describe the flow of fluids. The Navier-equations for irrotational flow are shown below.

$$\frac{\partial \rho}{\partial t} + \nabla \cdot (\rho u) = 0 \quad \text{Continuity Equation} \tag{1}$$

$$\frac{\partial u}{\partial t} + (u \cdot \nabla) u = -\frac{1}{\rho} \nabla p + F + \frac{\mu}{\rho} \nabla^2 u$$

Equations of Motion (2)

$$\rho \left(\frac{\partial}{\partial x} + u \cdot \nabla \varepsilon \right) - \nabla \cdot (K_H \nabla T) + p \nabla \cdot u = 0$$

***Corresponding author:** Muk Choa H, Division of Mechanical and Automotive Engineering, Kongju National University 275, Budae-dong, Cheonan-si, Chungcheongnam-Do 331-717, South Korea
E-mail:hmcho@kongju.ac.kr

Conservation of Energy   (3)

Where u=velocity vector field, ε= thermodynamic internal energy, p = pressure, T= temperature, ρ= density, μ= viscosity, K_H = heat conduction coefficient, F= external force per unit mass.

$$\nabla = \frac{\partial}{\partial x}\,i + \frac{\partial}{\partial y}\,j + \frac{\partial}{\partial z}\,k \qquad (4)$$

$$\nabla^2 = \frac{\partial^2}{\partial x^2} + \frac{\partial^2}{\partial y^2} + \frac{\partial^2}{\partial z^2} \qquad (5)$$

In this study, the CFD code of the capability of moving mesh and boundary algorithms to investigate the effect of the piston crown shape to the fluid flow field that use the valves and piston movement capability. The first to find the better piston crown shape to measure when the same composition of air structure for fuel mixing preparation that occurred inside engine cylinder. Therefore, this study is that the fluid flow with two different combustion chambers influence in air-flow mixing preparation and generation of turbulence for great impact on engine performance. During intake and compression stroke, numerical calculation is performed to obtain the optimum parameters for engine, which influences to accomplish the better air and fuel mixing for the rapid combustion process for studying the effect of the combustion chamber shape to the fluid flow field.

The numerical computation was using the analysis of intake and compression stroke for the swirl generation of air-fuel mixture into the chamber with two different piston crowns, with its boundary conditions. This engine speed considered in this work is 2000 rpm with the fixed valve timing and lift. The two parameters for the fluid flow characteristics acquired from the simulation will be taken into account to verify the same composition of air structure for mixture preparation so that the better air-fuel mixture when compression stroke combustion process can be achieved [1]. In particular, in the experiment process, the difference test observed for the two different piston bowl shapes can be drawn out with the parameters conformation of fluid flow characteristics during intake and compression stroke [2]. From the simulation experiment, we can observe the piston crown shape impact the air-fuel swirl combustion processes to improvement of the piston crown shape for obtaining the better engine performance and exhaust emissions. The CFD simulations use the moving mesh-boundary algorithm that to use the different mesh and boundary geometries for different crank angle in each step of engine cycle for getting different air-fuel swirl. In order to perform the better CFD simulations for an internal combustion process, the analysis and calculation should be carried out by using the unsteady calculation, moving meshes and boundaries, high compressible Reynolds number, high fluid dynamics characteristics, momentum, heat and mass transfer and the air-fuel mixture through complex geometries model and chemical-thermal dependent as well [1]. In addition, the calculation of the flow field in different combustion chambers was compared with Laser Doppler Velocimetry measurements (LDV).

The CFD code developed to simulate three-dimensional curvilinear domains using the finite volume method in a collocated grid. It employs the standard k-ε turbulence model in the vector notion with some slight modifications to introduce the compressibility of a fluid. It solves the follows transfer equations for describing mass, momentum and energy [3].

$$\frac{\partial(\rho\phi)}{\partial t} + \nabla\cdot(\rho\phi\,\vec{u}) = \nabla\cdot(\Gamma_\phi(\phi)) + S_\phi \qquad (6)$$

Where, $\vec{u}$ is the velocity vector with components u, v and w, ρ the

Generalized flow field variable (ϕ)	Equation	Γ_ϕ
I	Continuity	0
u	u-Momentum	μ_{eff}
v	v-Momentum	μ_{eff}
w	w-Momentum	μ_{eff}
h	Enthalpy	μ_{eff}/μ_h
k	Turbulent kinetic energy	μ_{eff}/μ_k
ε	Dissipation of turbulent kinetic energy	$\mu_{eff}/\mu_\varepsilon$

Table 1: Variables representing the generalized variable ϕ in Equation (6) [3].

Variable	Γ_ϕ	S_ϕ
k	μ_{eff}/σ_k	$G_K - \rho\varepsilon$
ε	$\mu_{eff}/\sigma_\varepsilon$	$C_1 G_K \dfrac{\varepsilon}{k} - C_2\rho\dfrac{\varepsilon^2}{k} + C_3\rho\varepsilon(\nabla\cdot\vec{u})$

Table 2: Source terms of the turbulence variables [3].

density, ϕ the generalized transport property and S_ϕ the source term. In Table 1, all the variables corresponding to the general transport property are shown [3].

For the turbulence model, specifically the source terms of the turbulence kinetic energy and its dissipation can be seen in Table 2, where G_k is the turbulent kinetic energy production rate [3].

The turbulent viscosity is calculated from the following Eq. (7), and the effective viscosity used in the calculations is the sum of the turbulent and laminar viscosity. Also, all the constants used in the implementation of the k-ε turbulence model and the enthalpy equation are shown in Table 3 [3].

$$\mu_t = \frac{\rho k^2 C_\mu}{\varepsilon}\;,\; \mu_{eff} = \mu_1 + \mu_2 \qquad (7)$$

Engine Geometry Analysis

The engine model studied is a typical single-cylinder of a CNG-DI engine with two intake and exhaust valves as shown Figure 1 and with two pistons shapes structure shown Figure 2. From above mentioned, two piston crowns were considered to investigate the behavior of the swirl and tumble flows occurred inside cylinder in order to obtain the better piston shape for the combustion process in the engine. Piston shapes shown in Figure 2, two piston shapes are typical of the real engine geometry model that affect the air-fuel mixture swirl for obtaining the higher swirl ratio as well as the optimum combustion process in the DI engine. Piston A has a bowl at the center of its piston crown while piston B has the deeper bowl volume than piston A, and piston B bowl is different from piston A that the bowl is not at the center of its piston crown [1].

The compare of the bowls on piston crown surface when air-fuel flows into the combustion chamber to generate swirl, which according to the piston shapes to form the air-fuel swirl of influence to obtain the better piston shape for combustion process in the engine, and operation the piston boundary conditions use the CFD model analysis during the intake and compression process.

During the compression stroke, the swirl generated by the annular jet flows during the previous intake stroke decay quite quickly and remains seem to be distributed homogeneously along the engine cylinder. And the air-fuel with quite quickly velocity to make air-fuel mixture full mix whether or not depending on the piston crown shape,

Constant	c_1	c_2	c_3	c_μ	c_h	c_k	c_ε
Value	1.44	1.92	-0.373	0.09	1.0	1.0	1.3

Table 3: Constants used in the k-ε turbulence model and the enthalpy equation [3].

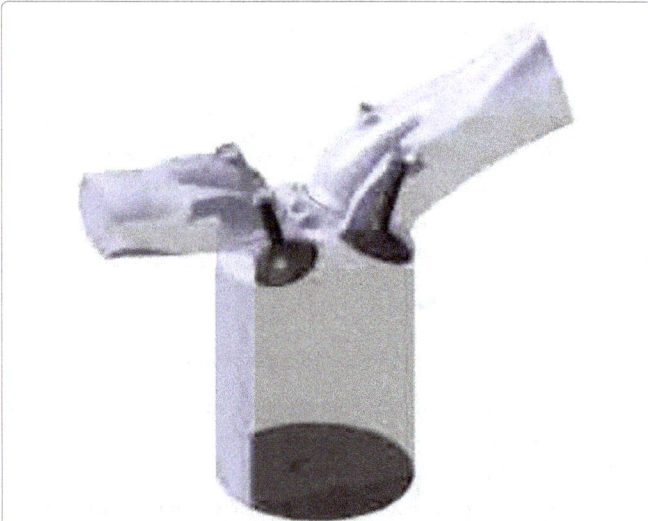

Figure 1: A schematic view of typical engine model with piston crown A [1].

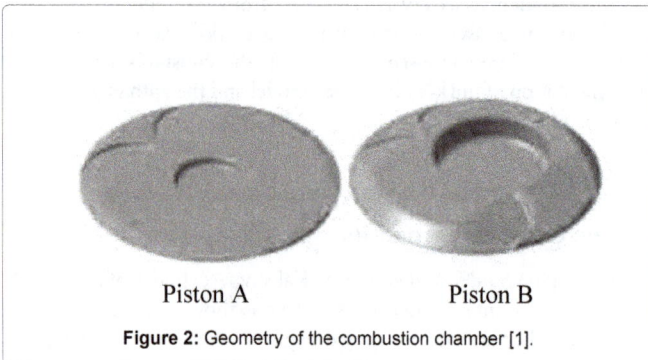

Piston A   Piston B

Figure 2: Geometry of the combustion chamber [1].

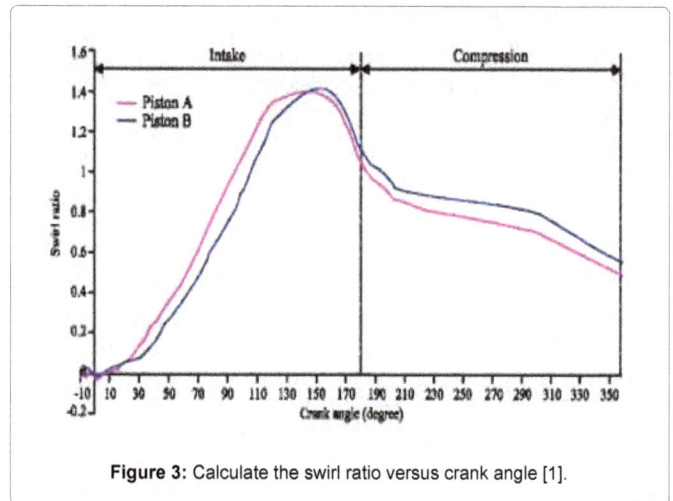

Figure 3: Calculate the swirl ratio versus crank angle [1].

when the combustion chamber shape of an ICE is affect to the velocity field close to the piston top surface during the latter of the compression stroke when the air-fuel injection start and the spark plug ready to burn the air-fuel mixture start end of the compression process. In the compression process, combustion chamber with the different of the piston crown bowl structure that for the purpose of comparing swirl ratio in the combustion chamber.

The experiment analysis obtains variations of swirl ratio for two pistons are calculated and display in Figure 3. As can be seen from the graph, the swirl is generated before the start of the intake stroke in the cylinder. Because air-fuel before into the cylinder start to become mixture and inside the cylinder due to the piston crown shape air-fuel with quite quickly velocity generate swirl by annular jet to full mix, as shown in Figure 3. The maximum is achieved near around 140° after TDC, where the piston reaches the maximum speed and the valve opening is in maximum distance. And after that, the air-fuel velocity into the cylinder decreases and swirl will fail slowly during the intake stroke. In the first part of the compression stroke, swirl velocity due to the friction at the cylinder wall the trend continues to fail slowly. Therefore, follow this graph the swirl is developed as the flow accelerates in preserving its angular momentum within the smaller diameter

piston bowl when approaching TDC [3]. During the compression stroke, swirl is necessary needed to generate a complex flow field for the purpose of enhancing air-fuel mixture during fuel injection process. From the analysis, the piston A due to smaller diameter crown shape is able to generate higher swirl ratio than piston B with small range differences during intake and compression strokes. But because of the fraction within the cylinder wall and its intake air flows influence by combustion chamber head shape.

Swirl Ratio Analysis Use and the CFD Code of KIVA-3V

This case is conducted to investigate the effect of the intake system design on swirl ratio use the computer code KIVA-3V in a DI engine [4]. The KIVA-3V code is used to study the effect of the one or two ports and intake position with shrouded valves for the swirl ratio.

The engine studied in this case is a single-cylinder direct-injection Diesel engine with typical two direct intake ports whose outlet is tangential to the wall of the cylinder, as shown in Figure 4 [4].

The simulations were performed through the intake and compression strokes. And the mesh had about 150,000 cells [4]. Pressure inflow boundaries were imposed at the open ends of the intake runners. For this cases, when shrouded valves were used on the backside of the valves prevent the air-fuel into the cylinder through backside position of the valve is shown Figure 5. Conventional engine, air can flow around all sides with un-shrouded valve through intake port into the cylinder in all directions. In order to improve the air-fuel swirl ratio in the cylinder that in the intake and compression strokes processes, therefore a shrouded valve is added to the backside of the valve, air flow will only flow in one direction of the valve into the cylinder is shown in Figure 6 [4].

The values of air-fuel swirl ratio for the different crank angles were calculated and plotted in Figure 7 [4] shown that this case with one port and a shrouded valve obtained the highest air-fuel swirl ratio. The case with one port intake the air-fuel into the cylinder is stronger than two or more ports intake through the pipe. Similarly the shrouded valve in the backside of the valve prevents the flow through the backside of the valve into cylinder, and makes the flow through the front side into cylinder strongly for enhancing the swirl motion. So the experiment simulation results shown this case with one port or a shrouded valve can generate higher air-fuel swirl ratio than two or more ports or un-shrouded valve. Therefore, with intake and compression strokes, the shrouded valve can

Figure 4: Schematic of engine and computational mesh [2,4].

Figure 5: conventional valve with a shroud.

(a)

(b)

Figure 6: Show the flow through (a) shroud valve and (b) un-shroud valve [4].

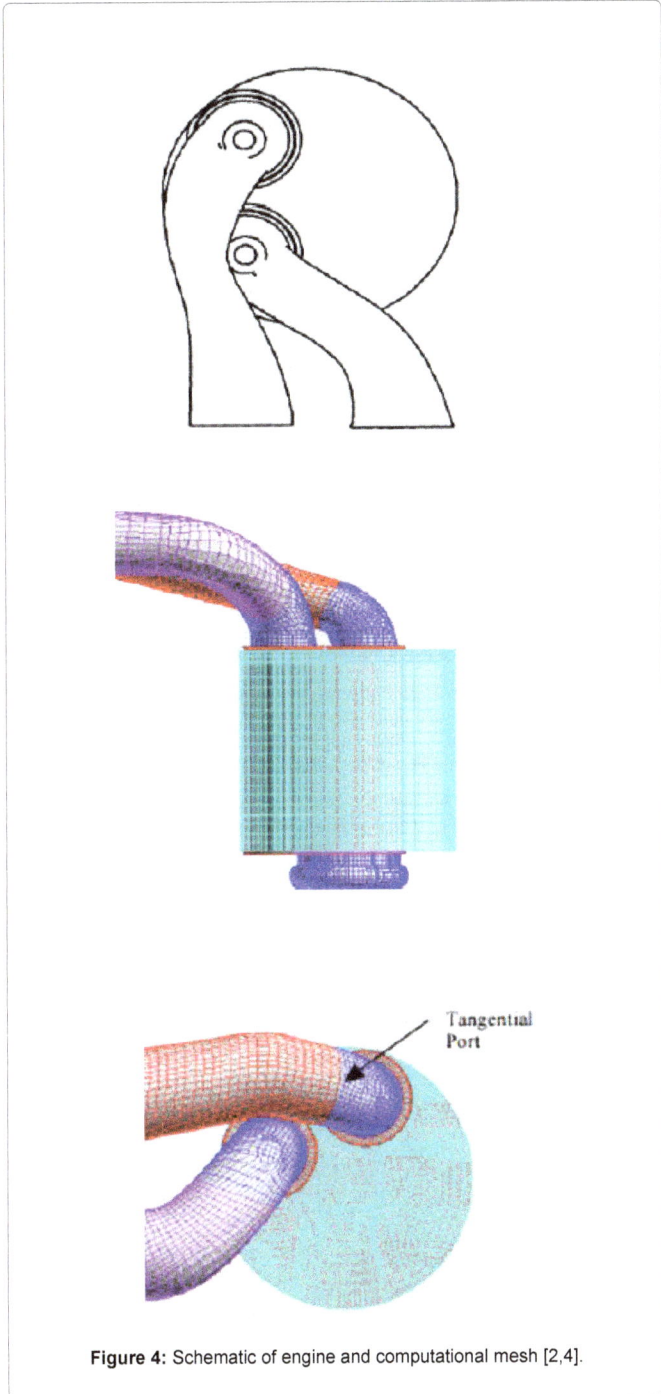

improve and enhance the swirl ratio when the air-fuel through the valve into the cylinder chamber in the CNG-DI engine [4].

For this case, the effect of one or two ports on swirl ratio can be considered. This method is similarity to added shroud valve on the backside of valve that block one port to determine if a single port would generate a higher swirl ratio than two ports. Then similarly this air-fuel swirl ratio can be shown in Figure 7 [4].

Other Methods for Enhancing Swirl Ratio

From above methods, other methods can enhance swirl ratio for intake

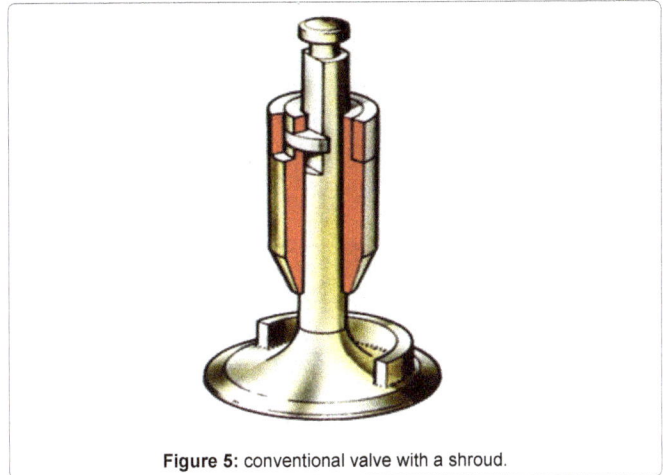

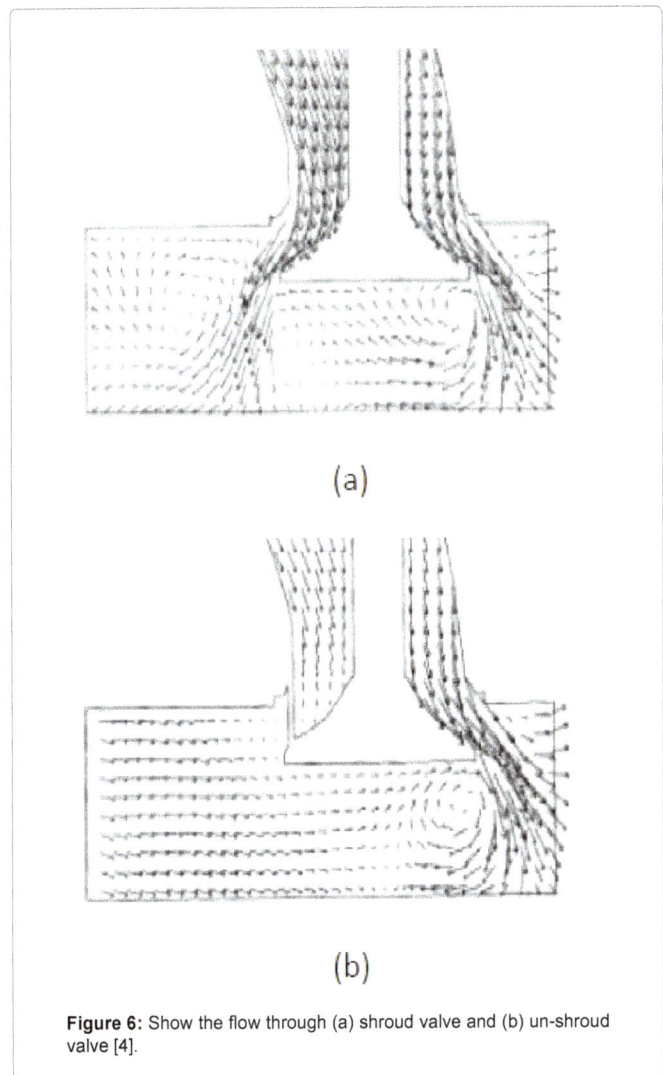

and compression strokes in the engine. From above mentioned, we can change port number to enhance the swirl ratio. The effect of port diameter on swirl ratio can be considered [4], reduce the port diameter for enhancing the flow velocity and air-fuel mixture into the combustion chamber. The narrow port obtains the better swirl ratio than width port, because as the

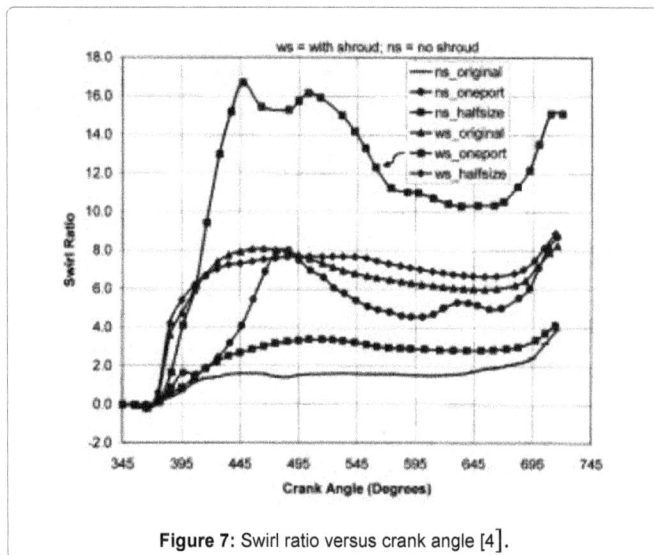

Figure 7: Swirl ratio versus crank angle [4].

port is narrow, the flow pressure is smaller than combustion chamber, so air-fuel is inhaled rapidly due to the pressure difference. Other methods involve change the injection timing [5-8] and variable valve timing and dual-fuel inside the cylinder [7] and late second injection variations [9-11] can affect the air-fuel mixture.

Conclusions

The Computational fluid dynamic investigation is carried out to study the effect of the piston crown structure to the fluid flow field in inside the combustion chamber, and investigation the intake system and compression strokes to enhance swirl ratio to solve CFD code.

(1) This paper is carried out for two piston crown shapes to evaluate the effect of different combustion chamber shape to the fluid flow field for the air-fuel mixture before fuel injection begins, and to obtain the better piston crown shape for swirl ratio for the combustion chamber in the DI engine.

(2) For the intake system design can be considered in this case,

and investigated through the CFD code of KIVA-3V model. Increase the burn rate of the air-fuel mixture for the efficiency of the engine that change the port number and add shroud valve for considering on the swirl ratio. The results can be derived that use shrouded in the cylinder and get the higher swirl ratios than with un-shrouded valves.

References

1. Kurniawan WH, Abdullah S, Shamsudeen A (2007) A Computational fluid dynamics study of cold–flow analysis for mixture preparation in a motored four-stroke direct injection engine. J Applied Sciences 7: 2710-2724.

2. Payri F, Benajes J, Margot X, Gil A (2004) CFD modeling of the in-cylinder flow in direct-injection diesel engines. Computers & Fluids 33: 995-1021.

3. Rakopoulos CD, Kosmadakis GM, Pariotis EG (2010) Investigation of piston bowl geometry and speed effects in a motored HSDI diesel engine using a CFD against a quasi-dimensional model. Energy Conversion and Management 51: 470-484.

4. Ramadan B (2003) A study of swirl generation in DI engines using KIVA-3V [J] Kettering University.

5. Jayadhankara B, Ganesan V (2010) Effect of fuel injection timing and intake pressure on the performance of a DI diesel engine - A parametric study using CFD. Energy Conversion and Management 51: 1835-1848.

6. kailas PP, Hemant WK, Vijayenra Maharu P, Anil HK (2014) Study of combustion in DI diesel engine for different compression ratios using experimental and CFD approach. Int J Research in Engineering and Technology (IJRET) 3: 2321-7308.

7. Hussain SM, Sudheer Prem Kumar B, Vijaya Kumar Reddy K (2012) CFD analysis of combustion and emissions to study the effect of compression ratio and biogas substitution in a diesel engine with experimental verification. Int J Engineering Science and Technology (IJEST) 4: 0975-5462.

8. Vijayendra Maharu P, Agrawal A (2014) Optimization of time step and CFD study of combustion in DI diesel engine. IJRET Int J Research in Engineering and Technology 3: 2319-1163.

9. Coskun G, Soyhan HS, Demir U, Turkcan A, Ozsezen AN et al. (2014) Influences of second injection variations on combustion and emissions of an HCCI-DI engine: Experiments and CFD modelling. Fuel 136: 287-294.

10. Holkar R, Sule-Patil YN, Pise SM, Godase YA, Jagadale VS (2015) Numerical simulation of steady flow through engine intake system using CFD. IOSR J Mechanical and Civil Engineering (ISOR-JMCE) 12: 30-45.

11. Abdul Gafoor CP, Gupta R (2015) Numerical investigation of piston bowl geometry and swirl ratio on emission from diesel engines. Energy Conversion and Management 101: 541-551.

10

An Integrated Modeling Approach for Management of Process-Induced Properties in Friction Stir Welding Processes

El-Gizawy A. Sherif [1, 2], Chitti Babu S[2], and Bogis Haitham[1]

[1]Center of Excellence for Industrial Design and Manufacturing Research (CEIDM) Mechanical Engineering, King Abdulaziz University, Jeddah, Saudi Arabia
[2]Industrial and Technological Development Center, Mechanical and Aerospace Engineering, University of Missouri-Columbia Columbia, Missouri 65211, USA

Abstract

Numerical and physical modeling techniques are used to predict process behavior in friction stir welding (FSW) high strength aluminum alloys. The numerical approach uses a non-linear finite element method to characterize thermal and deformation behavior along the welded structure during FSW. Coupled temperature-displacement analysis is applied in order to determine temperature, displacement, and mechanical responses simultaneously. The physical modeling approach uses the response surface methodology (RSM) to evaluate the effects of the process controlling parameters on the properties of the welded joints. The results obtained, offer insights into the effects of the major process parameters in establishing successful FSW joints that satisfy further processing requirements and product service conditions.

Keywords: Friction stir welding; Non-linear finite element analysis; Process-induced properties; Response surface methodology

Introduction

Friction stir welding (FSW) is a solid state welding process where localized deformation at the joint interface establishes the bond between the base metals Figure 1. In this process, a rotating tool generates heat and deformation at the joint interface [1-4]. The interface temperature never exceeds the melting point of the base metals (maximum 90% of melting temperature) [5]. Therefore FSW does not involve liquid phase transformation. This makes the process superior to all other welding processes that result in unfavorable microstructures and properties associated with solidification mechanisms in fusion welding. FSW has many other significant advantages including controlled properties and microstructure, improved material utilization (light weight structures), improved energy utilization (only 2.5% of energy needed for fusion welding), and reduced harmful effects on environment [2-4]. Research in FSW has focused on developing experimental, analytical, and numerical models in order to characterize the different zones in FSW [5-11]. They include the heat affected zone (HAZ), the thermo-mechanically affected zone (TMAZ), and the base metal. The properties of FSW joints are related to the properties and microstructure of different zones in the joint. FSW cycle consists of four stages: plunge stage (the rotating tool is plunged vertically into the joint);dwell stage (tool is held in the plunging position while still rotating); welding stage (rotating tool travel along the joint at constant velocity); and pulling tool out stage (tool is pulled out of the joint leaving behind an exit hole). The parameters that influence the performance of FSW are displayed in Figure 1. They include tool rotational speed, travel speed of the tool, plunge force, plunge depth, and tool design [2-13]. These parameters affect the thermo-mechanical and metallurgical changes established during FSW which in turn are related to the evolved properties, microstructure, and process-induced damage in the course of the welding process.

Research has been carried out to develop the numerical models to simulate friction stir welding. Chao and Qi [14] used a constant heat flux input from tool shoulder/work piece to increase the temperature and trial and error method was used to adjust the heat input. Frigaard, Grong and Midling [15] developed a model for friction stir welding. Heat input was assumed to be the frictional heat and co-efficient of friction and other conditions were adjusted to keep the temperature below the melting point. Zahedul, Khandkar, and Khan [16] modeled the friction stir welding with a moving heat source. The above researches account only for the heat generated by friction between tool shoulder and sheet surface. They did not account for heat generated at the interface between tool pin and sheet materials as well as the heat produced by plastic deformation. Other investigators have attempted to include all components of heat source in their models. Askari used

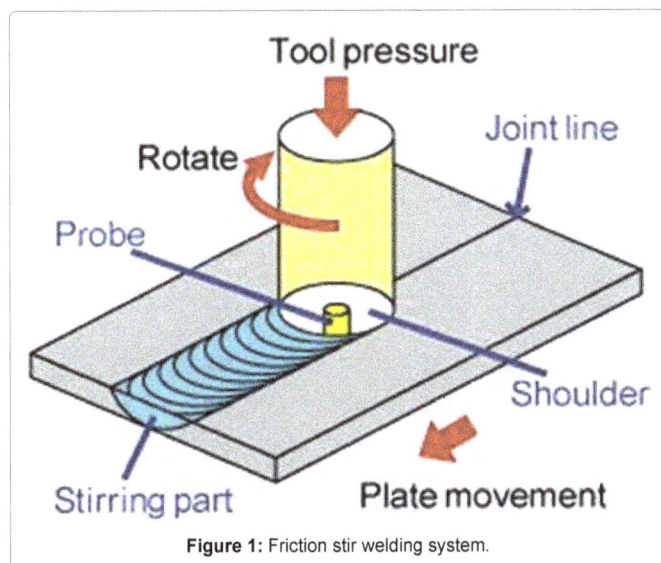

Figure 1: Friction stir welding system.

*Corresponding author: El-Gizawy A. Sherif, Center of Excellence for Industrial Design and Manufacturing Research (CEIDM) Mechanical Engineering, King Abdulaziz University, Jeddah, Saudi Arabia, E-mail: sherifelg@yahoo.com

CTH hydro-code based on finite volume to model the flow of the material on the assumption that the material sticks to the tool surface [9]. Residual stress analysis was investigated using 3D thermal and thermo-mechanical numerical simulations [17]. This model used symmetry about the weld line, however this won't be accurate since friction stir welding is not symmetrical due the presence of advancing and retreating sides that are not identical.

The present work aims at development of a numerical and physical models to predict the process behavior in Friction Stir Welding. The numerical model is used for process design of friction stir welding operations using a non linear finite element techniques to characterize thermal and deformation behavior during friction stir welding high strength aluminum alloys. An empirical model is developed using design of experiments to relate the process parameters to the properties of the weld. Finally, case studies were conducted to validate the numerical model.

Numerical Modelling

Coupled temperature-displacement analysis

Coupled Temperature-Displacement analysis is used to model friction stir welding using ABAQUS-Explicit [18]. In this analysis, temperature and mechanical responses are determined simultaneously. Heat generated during friction stir welding is produced by friction between tool's shoulder and probe with sheet material and plastic deformation energy. Heat loss from sheet is due to convection from exterior surfaces of sheet and conduction from bottom surface of sheet to the backing plate.

Finite element method

Friction stir welding processes are inherently nonlinear because of the large strains, high temperature, and plastic behavior of the material in the welding zone. Complex nature of the interface friction conditions between the material and tooling surfaces adds to the difficulty of modeling the process. The standard "implicit" finite element formulation is a true quasi-static solution. However, in applications such as FSW, the standard method would yield very large linear matrix equation which must be solved for each load step [19]. The irregular nature of the interface friction will add to the complexity of the solution and make convergence extremely difficult and time consuming for each load step. The explicit method on the other hand [16] is basically a dynamic solution procedure through the application of explicit time integration to the discrete equation of motion. Total computation time in this method can be reduced by scaling up velocity and mass. Scaling velocity beyond certain limit might introduce non-realistic dynamic effects that can result in inaccuracy of the solution. Limiting punch speed in the simulation to less than 1% of the wave speed in the workpiece material would not significantly affect the solution accuracy [19]. In the present case of Aluminum Alloys with wave speeds of 5600 m/s, a welding speed up to 50 m/s will not affect the accuracy of the solution. Interface friction can be simulated in much simpler way than in the standard method. Furthermore, there is no convergence problems associated with the explicit approach. In the present work, the explicit method in ABAQUS is used to model the behavior of friction stir welding processes.

Three types or surfaces used in finite element analysis are Lagrangian, Eulerian and Sliding. Displacement in normal and tangential directions will follow the material in Lagrangian Surface. In Eulerian surface, the material is allowed to flow through the mesh. The mesh is fixed in all direction for eulerian surface. For sliding surface, mesh will follow the material in normal direction, while in tangential direction the mesh is fixed. Thus, top and bottom surfaces are modeled as sliding surface and other surfaces are modeled as Eulerian surface.

In Friction Stir Welding, there is considerable amount of material flow around the tool pin and contact forces. This will lead severe element distortion and ultimately lead to premature termination of the analysis. In order to overcome this mesh distortion during large plastic deformation, Arbitrary Lagrangian-Eulerian (ALE) technique has been used. Adaptive meshing with re-meshing has been employed to reduce large element distortion in modeling friction stir welding. Re-meshing improves the quality of the mesh as the analysis sweeps the mesh for every preset increment and the results are remapped to the improved mesh.

Governing Equations for Mechanical Analysis

ABAQUS/Explicit solves for a state of dynamic equilibrium at the start of current time increment t:

$$M^{NM}\ddot{u}^N\big|_t = \left(P^M - I^M\right)\big|_t \tag{1}$$

where M^{NM} is the mass matrix, u^N is the acceleration vector, P^M is the external force or applied load vector, and I^M is the internal force vector (the 'internal force' created by stresses in the elements). In the explicit procedure, a diagonal mass matrix is used for efficiency. Thus the nodal acceleration can be easily obtained:

$$\ddot{u}^N\big|_t = \left[M^{NM}\right]^{-1}.\left(P^M - I^M\right)\big|_t \tag{2}$$

The central difference integration rule is used to update the velocities and displacements:

$$\dot{u}^N\bigg|_{t+\frac{\Delta t}{2}} = \dot{u}^N\bigg|_{t-\frac{\Delta t}{2}} + \left(\frac{\Delta t\big|_{t+\Delta t} + \Delta t\big|_t}{2}\right)\ddot{u}^N\bigg|_t$$

$$u^N\big|_{t+\Delta t} = u^N\big|_t + \Delta t\big|_{t+\Delta t}\,\dot{u}^N\bigg|_{t+\frac{\Delta t}{2}} \tag{3}$$

No iterations are required in the equation solver to update the accelerations, velocities, and displacements; so it is computationally economical for big model size like the cases addressed in this research. The stability of the solution depends on the time increment size, which is approximated as the smallest transit time of a dilatational wave across any of the element in the mesh.

$$\Delta t \approx \frac{L_{min}}{c_d} \tag{4}$$

In which $Lmin$ is the smallest element dimension in the mesh and cd is the dilatational wave speed.

Governing Equations for Thermal Analysis

In ABAQUS/Explicit, the heat transfer equations are integrated using the explicit forward-difference time integration rule.

$$\theta^N_{(i+1)} = \theta^N_{(i)} + \Delta t_{(i+1)}\dot{\theta}^N_{(i)} \tag{5}$$

where θ^N is the temperature at node N and the subscript i refers to the increment number in an explicit dynamic step. The forward-difference integration is explicit in the sense that no equations need to be solved when a lumped capacitance matrix is used. The current temperatures are obtained using known values of $\dot{\theta}^N_{(i)}$ from the previous increment.

In order to simulate friction stir welding process accurately, heat generation by both friction and plastic deformation are modeled rather than adding heat flux to the tool.

$$qA = f\eta Pfr \text{ and } qB = (1-f)\eta Pfr \qquad (6)$$

Where qA is the heat flux into the sheet and qB is the heat flux into the tool.

f is the percent of heat flux that flows into the sheet (0.0 to 1.0)

η is the factor of converting mechanical to thermal energy Pfr is the frictional energy dissipation.

Plastic straining gives rise to a heat flux per unit volume.

$$r^{pl} = \eta\sigma : \overset{.pl}{\varepsilon} \qquad (7)$$

Where r^{pl} is the heat flux that is added into the thermal energy balance, η is the factor for percent of heat converted, σ is the flow stress of the material, and ε^{pl} is the rate of plastic straining.

The Johnson-Cook Strain Rate dependent flow stress is used as the constitutive equation to describe the flow stress behavior of the material during processing.

$$\bar{\sigma} = \left[A + B\left(\bar{\varepsilon}^{pl}\right)^n\right]\left[1 + C\ln\left(\frac{\bar{\varepsilon}^{pl}}{\dot{\varepsilon}_0}\right)\right]\left(1 - \hat{\theta}^m\right) \qquad (8)$$

Where $\bar{\sigma}$ is the effective flow stress, ε^{pl} is the effective plastic strain is $\dot{\varepsilon}_0$ the normalizing strain rate, A, B, C, n, m are the material constants (Table 1).

$$\hat{\theta} = \begin{cases} 0 & for & \theta \leq \theta_{transition} \\ (\theta - \theta_{trension}) / (\theta_{melt} - \theta_{transition}) & for & \theta_{transition} \leq \theta \leq \theta_{melt} \\ 1 & for & \theta \geq \theta_{melt} \end{cases} \qquad (9)$$

Where θ is the current temperature, $\theta melt$ is the melting temperature and $\theta transition$ is the transition temperature below which there is no temperature dependence on flow stress.

Heat loss due to conduction and convection is considered in this research. Heat loss is modeled from bottom of the plate to the backing plate. Conductive heat loss is given by,

$$q = k(\theta_A - \theta_B) \qquad (10)$$

Where q is the heat flux, K is the conductivity, θ^A and θ^B are the temperatures at point A and B on the surface.

Heat loss due to convection is considered from all exterior surfaces of the sheet. Heat flux due to convection is given by:

$$q - h(\theta - \theta^0) \qquad (11)$$

Where q is the heat flux, h is the film co-efficient, θ is the temperature at the surface θ^0 is the sink temperature.

The developed model deals with characterization of the responses of the material to the mechanical and thermal loading environment generated by friction stir welding. This model provides guidance for selecting the appropriate process conditions that result in desirable properties of the joint.

Experimental Investigation

Experimental setup

Friction Stir Welds were produced using an in-house built FSW system based on an available milling machine. Two Al 2024-T3 sheets of dimensions 50 mm × 50 mm × 2 mm were rigidly clamped on a titanium backing plate which is fixed to a steel base. The two sheets are then butt welded using the rotating FSW tool with a shoulder diameter of 18 mm. The tool's probe has diameter of 6 mm and a height of 1.9 mm. The base metal properties are given in Table 1.

Statistical design of experiments

A set of experiments was designed using the response surface methodology (RSM) [20]. RSM methodology is a collection of statistical and mathematical techniques effective for modelling and optimization of manufacturing process designs. Central composite designs in RSM, are vastly used for fitting second-order response surface because of both their statistical properties and the practical attraction of their expanded coverage around a centre point [20]. In the present research, RSM is used to investigate the effects of the control process parameters that include rotational speed and feed rate (welding speed) on the important quality characteristics of the joint. The experimental matrix used in the present investigation is presented in Table 2.

In the present study, a second-order response surface model (equation 12) is used to formulate a least square relationship between the input parameters and the output response measures.

$$Z = \beta_0 + \beta_1 X + \beta_2 Y + \beta_{11} X_2 + \beta_{22} Y_2 + \beta_{12} XY \qquad (12)$$

Where Z are the observed responses (formability of the joint measured by reduction of area % at fracture), as a function of the main influences of factors X (rotational speed, RPM) and Y (feed rate, IPM), their interaction (XY), and their quadratic components (X², Y²). β_0, β_1,…etc. are estimated regression coefficients.

Evaluation of Properties of FSW joints

Mechanical properties of the friction stir welded joints were evaluated using standard tensile test procedures. The gage dimensions of the specimens were 25.4mm (1") long and 19mm (3/4") wide and with the weld zone running across the gage length.

The test speed was kept constant at 1 cm /min for the duration of the test.

Results and Discussions

Prediction of process behaviour using numerical models

In order to evaluate the effectiveness of the developed numerical model, two case studies for friction stir welding of Al 2024-T3 were conducted. One case was chosen to represent sound joint and another one represents the condition of bad joint. The process parameters for both cases were selected based on preliminary experimental evaluation conducted by the authors on the investigated material. Table 3 presents

Material Properties	Al 2024-T3
Density (Kg/cc)	2780
Modulus of Elasticity (GPA)	73
Poisson's Ratio	0.33
Ultimate Tensile Strength (MPa)	483
Yield Stress, A (MPa)	369
Strain Factor, B (MPa)	684
Strain Exponent, n	0.73
Temperature Exponent, m	1.7
Strain Rate Factor, C	0.0083
θmelt (°C)	502
θtransition (°C)	25
Thermal Conductivity (W/m-K)	121
Specific Heat Capacity (J/Kg °C)	875

Table 1: Material properties of Al 2024-T3.

Exp. #	1	2	3	4	5	6	7	8	9(c)	10(c)	11(c)	12(c)	13(c)
Rotational Speed(RPM)	840	840	1300	1300	675	1300	1045	1045	1045	1045	1045	1045	1045
Feed Rate(IPM)	4.625	7.625	4.625	7.625	5.75	5.75	3.625	7.625	5.75	5.75	5.75	5.75	5.75

Table 2: Experimental matrix using RSM.

Process Parameters	Case 1	Case 2
Welding Velocity (mm/sec)	2.43	2.43
Rotational Velocity of Tool (rpm)	1045	445
Co-efficient of Friction	0.3	0.3
Effective Plunge Depth (mm)	2	1.9

Table 3: Process parameters for case studies.

the levels of the control parameters used for good weld (Case 1), and bad weld (Case 2). Other friction stir welding process parameters were kept constant. They include: Plunge time (5 seconds), Dwell time (15 seconds), and Pull time (5 seconds).

Numerical modeling results using the conditions listed for Case 1 are displayed in Figures 2-5. Figure 2 displays temperature distribution fields along the weld line at the end of welding process. Temperature distribution across the width of entire blank is show in Figure 3. The temperature in area adjacent to the tool is the highest (538°C). The evolved temperature field is extended to cover the entire blank because of the active heat transfer by convection due to the flow of heated material and conduction to the neighboring zones. The lowest temperature in the field during FSW is 45°C. The close spacing between isotherms in the neighborhood of tool/work-piece interface is anindication that heat is generated at that spot and then dissipated by convection and conduction to other zones in the panel. Figure 4 displays the progress of the heat source as the tool travels along the weld line. Equivalent plastic strain along the joint-line is shown in Figure 5. Plastic strains represent degree of deformation as result of the stirring action during the process. The extent of the zone with high level of plastic strain (highly deformed region) along the weld line is an indication of the size of the weld nugget and the adjacent thermo-mechanical affected zone (TMAZ). One can conclude that simulation results presented in Figure 2 through Figure 5 can be used to evaluate alternative process designs in order to explore the optimum one.

In the second case study, conditions not favorable for a successful weld are investigated. All process parameters are kept the same except using reduced rotational velocity and smaller plunge depth. These conditions do not allow the establishment of sticking friction between the tool shoulder and the surface of the base metal. Hence no sufficient heat is generated and the hot flow of material needed to fill the joint gab is ceased. A groove is generated along the joint line as a result of removing instead of depositing material behind the FSW tool. This is evident in Figure 6. The evolved temperature fields for this case are also displayed in Figure 6. Maximum temperature along the joint never exceeded 110°C which is not sufficient to establish softening by dynamic re-crystallization in the material Figure 7.

Experimental characterization of FSW process behaviour

Case studies simulated using the numerical models in section 4.1 were verified experimentally through running FSW of Al 2024 T3 blanks according to the process conditions described in Table 3. Figure 8 shows an established FSW joint using parameters of case 1. Conducting FSW according to conditions described in case 2 resulted in poor joint. Because of insufficient temperature rise along the joint, FSW tool created a groove instead of depositing material along the weld

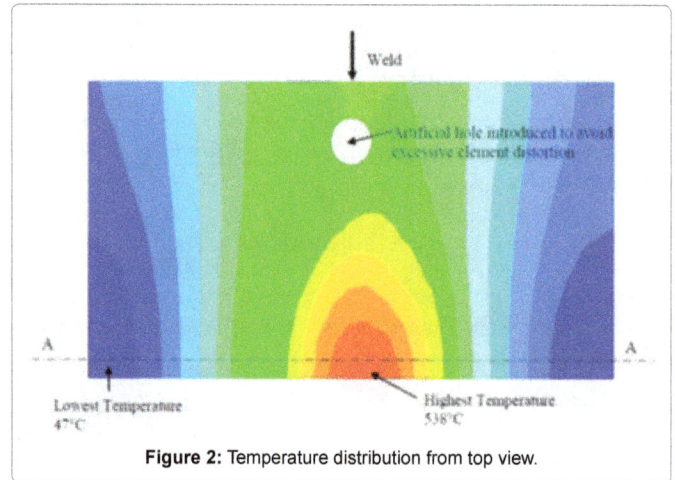

Figure 2: Temperature distribution from top view.

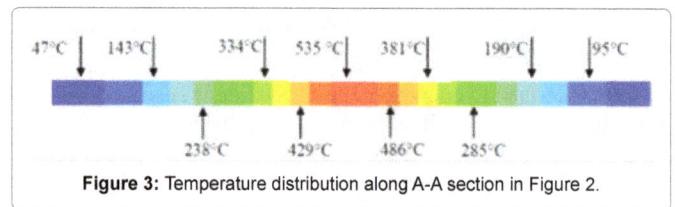

Figure 3: Temperature distribution along A-A section in Figure 2.

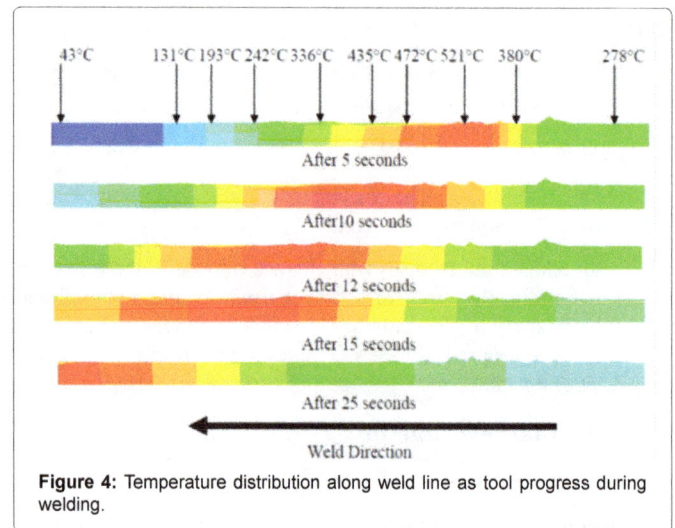

Figure 4: Temperature distribution along weld line as tool progress during welding.

line see Figure 9.

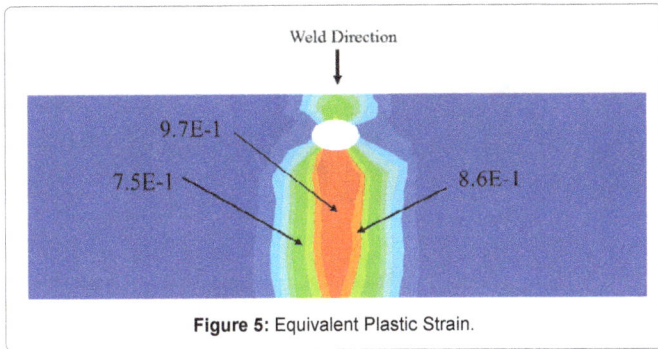

Figure 5: Equivalent Plastic Strain.

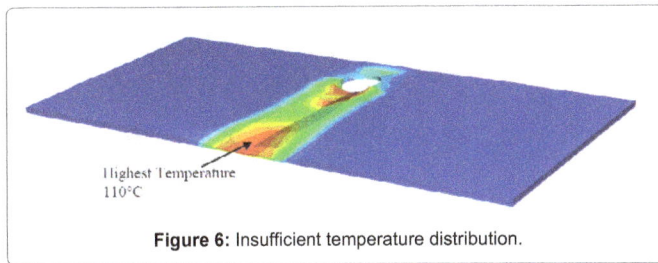

Figure 6: Insufficient temperature distribution.

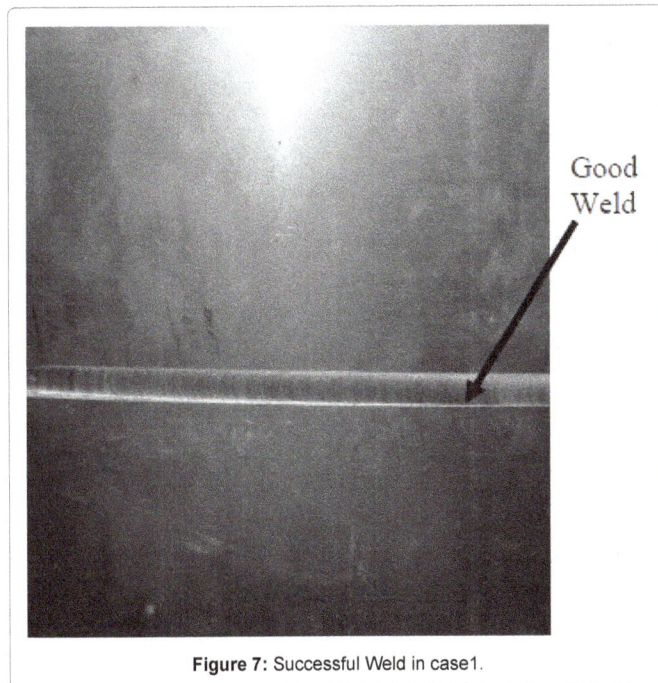

Figure 8: Unsuccessful Weld with no metal deposition in case 2.

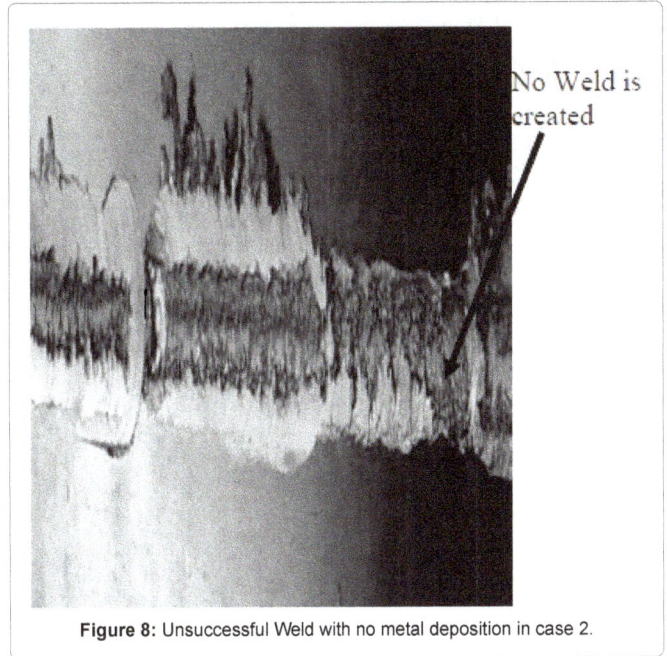

Figure 7: Successful Weld in case1.

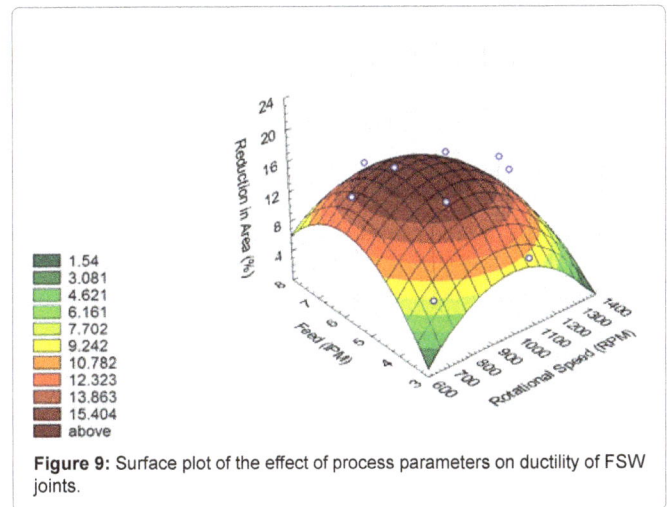

Figure 9: Surface plot of the effect of process parameters on ductility of FSW joints.

These results compare very well with the prediction of the numerical model. Temperature indicating paints have been used check the establishment of particular temperature field during FSW process. The measured temperatures were very close to the predicted ones using the developed numerical models.

Friction stir welding process-induced properties

The design of experiments defined in section 3.2, were used to collect data on the effect of FSW process parameters on evolved strength and ductility of the welded joints. Data in Table 4 represents strength and ductility measures of joints established using different combinations of rotational speed and feed rate (welding velocity). From the results shown in Table 4, condition in experiment #5 (rotational speed of

1045rpm and welding speed of 2.43 mm/sec) created the weld with the most desirable properties. It has the best combination of strength and ductility (high toughness). All the results were also fitted to three dimensional surfaces relating the effect of the independent variables (rotational speed and feed) on one of the quality characteristics of the process. A typical presentation of response surface for ductility is displayed in Figure 10. In this plot, ductility of the joint as measured by the reduction of area% was selected as the major response. The process contour map for joint ductility extracted from the surface plot is shown in Figure 10. Both figures indicate that rotational speed in the range of 800 to 1100 rpm and welding speed of 4.5-6.5 inch/min (114-165 mm/min) would yield optimum joint ductility. Similar surface plots and contour maps for the ultimate strength and yield strength were also generated from the results.

Conclusions

1. A numerical model uses a non-linear finite element method is developed to characterize thermal and deformation behavior along the

#	Rot. Vel.(RPM)	Weld Vel.		% Elong	% Red.In Area	YS		UTS	
		(mm/sec)	(IPM)			(MPa)	(PSI)	(MPa)	(PSI)
1	1300	0.9	2.125	12.37	14.58	211.52	30678.44	324.28	47032.92
2	1300	1.53	3.625	16	12.13	319.36	46319.34	437.66	63477.33
3	1300	2.43	5.75	10.67	5.95	313.17	45421.55	407.01	59031.92
4	1300	3.23	7.625	6.49	9.23	301.38	43711.55	429.48	62290.92
5	1045	2.43	5.75	18.18	17.08	309.75	44925.52	449.86	65246.79
6	1045	1.53	3.625	11.4	7.09	330.66	47958.27	445.23	64575.27
7	1045	3.23	7.625	8.49	8.85	314.57	45624.6	424.62	61586.04
8	840	1.96	4.625	12.73	11.17	303.69	44046.59	418.08	60637.49
9	840	2.43	5.75	11.18	10.03	310.56	45043	424.73	61601.99

Table 4: Mechanical properties from design of experiments.

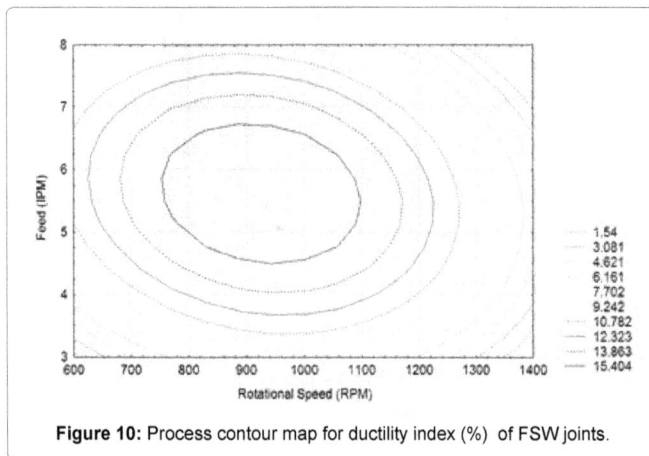

Figure 10: Process contour map for ductility index (%) of FSW joints.

weld line during friction stir welding process.

2. Coupled Temperature-Displacement analysis is used in the FEM model in order to allow for simultaneous determination of temperature, displacement, and mechanical responses.

3. Experimental verification of the proposed numerical models for friction stir welding was conducted using two case studies. The experimental observations confirm the predictions of the models.

4. The results obtained using the design of experiments and surface response methodology offer insights into the effects of the major process parameters in establishing successful FSW joints with optimum strength and ductility that satisfy further processing requirements and product service conditions.

5. The results generated from the present investigation were used for constructing process maps for FSW of Al 2024-T3. These maps are effective tools that can be used by industry as road maps in selecting process designs that satisfy both quality requirements and productivity constraints.

Acknowledgements

This work was funded by the Deanship of Scientific Research (DSR), King Abdulaziz University, under grant number (135-590- D1435). The authors, therefore, acknowledge the technical and financial support of King Abdulaziz University. The authors wish also to acknowledge the experimental support of Industrial Technology Development Centre at University of Missouri. Our appreciations are extended to Mr. Edward Gerding, and Mr. Dick Lederich of the Boeing Company for their technical support, valuable discussions and encouragements.

References

1. Thomas WM, Nicholas ED, Need ham JC, Murch MG, Templesmith P, et al. (1991) Friction Stir Welding.

2. Mishra RS (2003) Friction stir processing technologies. Advan Mater & Processes 43-46.

3. Mishra RS, Ma ZY (2005) Friction stir welding and processing. Mater Sci Eng 50: 1-78.

4. Nandan R, DebRoy T, Bhadeshia HKDH (2008) Recent advances in friction-stir welding - Process weldment structure and properties. Progr in Mater Sci 53: 980-1023.

5. Tang W, Guo X, McClure JC, Murr LE, Nunes A (1998) Heat input and temperature distribution in friction stir welding. J Mater Processing Manufacturing Sci 7: 163-172.

6. Heurtier P, Jones MJ, Desrayaud C, Driver JH, Montheillet F, et al. (2006) Mechanical and thermal modeling of friction stir welding. J Mater Processing Technol 171: 348-357.

7. Palm F, Hennebohle U, Erofeev V, Earpuchin E, Zaitzev O (2004) Improved verification of FSW-process modeling relating to the origin of material plasticity. P Fifth Int Symposium of Friction Stir Welding TWI Ltd Metz France.

8. Ulysee P (2002) Three dimensional modeling of the friction stir welding process. Int J Mach. Tools Manufacture 42: 1549-1557.

9. Askari A, Silling S, London B, Mahoney M (2001) Modeling and analysis of friction stir welding processing. In: Jata KV (ed.) Friction Stir Welding and Processing TMS Warrendale, PA : 43-54.

10. Padmanaban R, Ratna V, Balusamy V (2014) Numerical simulation of temperature distribution and material flow during friction stir welding of dissimilar aluminum alloys. P Eng 97: 854-863.

11. Kesharwania RK, Pandab SK, Palc SK (2014) Multi Objective Optimization of Friction Stir Welding Parameters for Joining of Two Dissimilar Thin Aluminum Sheets. P Mater Sci 6: 178-187.

12. Kimapong K, Wanabe T (2004) Friction stir welding of aluminum alloy to steel. Welding J 83: 277-282.

13. Hunt F, Badarinarayan H, Okamoto K (2006) Design of experiments for friction stir stitch welding of aluminum alloy 6022-T4. SAE Int: 173-178.

14. Chao YJ, Qi X (1998) Heat Transfer and Thermo-Mechanical analysis of friction stir joining of AA 6061-T6 plates. J Mater Processing Manufacture Sci: 215-233.

15. Frigaard F, Grong F, Midling OT (2001) A process model for friction stir welding of age hardening aluminum alloys. Metallurgical Mater Transactions A 32: 1189-1200.

16. Zahedul M, Khandkar H, Khan J (2001) Thermal modeling of overlap friction stir welding for Al alloys. J MaterProcessing Manufacture Sci 10: 91-106.

17. Zhu XK, Chao YJ (2004) Numerical simulation of transient temperature and residual stresses in friction stir welding of 304L stainless steel. J Mater Processing Technol 146: 263-272.

18. ABAQUS version 6.9 Internet Manual.

19. El-Gizawy AS, Chitti Babu S, Yeh T (2004) An integrated virtual model for characterization and management of process-induced damage in sheet forming processes. P 4th CIRP Int. Seminar on Intelligent Computation in Mfg Eng Naples Italy: 321-326.

20. Mayers RH, Montgomery DC, Anderson Kook CM (2009) Response surface methodology-process and product optimization using design of experiments. 3rd edn John Wiley & Sons Inc Hoboken New Jersey.

11

Influence of Sm_2O_3 Ion Concentration on Structural and Thermal Modification of TeO_2-Na_2O Glasses

Mawlud SQ*[1,2], Ameen MM[1], Md. Sahar R[2] and Ahmed KF[1,2]

[1]Department of Physics, College of Education, University of Salahaddin, Erbil, Kurdistan Region, Iraq
[2]Advanced Optical Material Research Group, Department of Physics, Faculty of Science, University of Technology Malaysia, Skudai and Johor, Malaysia

Abstract

The effect of Sm^{+3} ions concentration doped TeO_2-Na_2O glasses on structural and thermal parameters have been discussed. Glass samples with molar composition $(80-x)$ TeO_2-$20Na_2O$-xSm_2O_3 glasses (x=0, 0.3, 0.6, 1, 1.2, 1.5) are prepared by melt quenching technique. Crystallization temperature (T_c), melting temperature (T_m) and glass transition temperature (T_g) are measured by using differential thermal analysis (DTA), it is found that the stability factor (ΔT) increases from (58.5-97.8) °C with the increasing of Sm_2O_3. The amorphous phase nature of the glass samples are observed by using X-ray diffraction (XRD), scanning electron microscopy (SEM) and energy dispersive x-ray (EDX) spectrometer are applied to study the structural properties of the glass samples. Values of density (ρ), molar volume (V_M), and ionic packing density (V_i) were calculated for each of the glass compositions. The effect of the Sm_2O_3 on the glass structure have been investigated by using FTIR and Raman spectroscopies, the FTIR spectra are characterized by a band of 637 cm^{-1} for the telluride glass, high frequency peak at 668 cm^{-1} presented by Raman spectra which indicates that these glasses network are basically consists of TeO_4 and TeO_3/TeO_{3+1} structural units. The spectra of Raman shows the presence of Sm-O bond, Na-O bond, Te-O-Te bridging configurations, vibrations of Te-O-Te bonds and stretching modes of non-bonding oxygen found on the TeO_3/TeO_{3+1} structural unit.

Keywords: Telluride glass; Rare Earth; Glass transition; FTIR; Raman spectra

Introduction

Under normal quenching conditions tellurium dioxide (TeO_2) does not have the ability of forming a glassy network without the assistant of the secondary material component which is called a network modifier. This specific feature makes TeO_2 a conditional glass former [1,2]. Recent studies indicate that heavy metal oxide (HMO) glasses have been found to be more affirmative convenient glassy materials especially for photonic applications with acceptable low phonon energies [3,4], their wide advantages of tellurite glasses makes them to be represented as a more attractive structure among the HMO glasses such as low temperature of melting, good corrosion resistance, excellent chemical durability, low glass transition temperature T_g, better thermal stability, high thermal expansion coefficient, low phonon energy, wide optical transmission region (especially in the infrared region) (0.36 μm–6.3 μm), high refractive index and required rare earth concentration in the matrices [5]. Basically, the structure of TeO_2 glasses contains of three dimensional network of TeO_4 trigonal bipyramids (tpb) units having two each of oxygen at two equatorial and axial sites with one other equatorial site being occupied by a lone pair of electrons, once the modifier is introduced to the structure, the three dimensional network will break down with the conversion of TeO_4 units into TeO_{3+1} and TeO_3 units [6,7]. In addition to the glass former conditionality and fast quenching requirements properties of tellurite glasses, they are also counted as a good host for the formation of rare earth doped glasses. The reason behind this characteristic is the Te-O bonds are weak and can easily be broken by heavy metal and the rare earth atoms can enter in the glass networks [8]. In the literature there are relatively few reports on structural of glasses doped with Sm_3O ions [9,10]. Because of that reason the present work is to investigate the influence of Sm^{3+} ion on the structural and thermal properties of TeO_2-Na_2O glasses, the thermal modification of the glass system were analyzed by DTA technique and the structure of the Sm_2O_3 doped glasses has been studied through Raman and FTIR Spectra.

Experimental Procedure

TeO_2-Na_2O-Sm_2O_3 glass samples having compositions $(80-x)$ TeO_2-$20Na_2O$-xSm_2O_3 with (x=0,0.3,0.6,1,1.2,1.5) are prepared using melt quenching technique, glasses code of the sample and their molar ratio of the compositions are listed in Table 1. High purity 99.9% of raw materials from Sigma Aldrich with an appropriate amounts of chemicals for 15 gm batch were weighed by using electronic balance Precisa 205A SCS model with accuracy ± 0.001 gm, the chemical compositions were mixed well by using a milling machine. Platinum crucible of about 30 ml capacity containing the batch was used and preheated at 250°C for about 20 min in order to reduce the batch blanket coverage on the top of the glass and enlarges the free non coverage glass melt surface [11], then the batch is melted at 900ºC for 40 min by a controlled electric furnace. The melts are poured on to a stainless steel mold and annealed at 250°C for 3 hours, then the sample cooled down until its temperature reaches the room temperature, sample powder of the glass prepared and used for characterizing each of T_g, T_c and T_m by Pyris Dymond TG/DTA technique at a heating rate 50°C to 1000°C at 10°C/min. To analysis the amorphous state of the samples an advanced powder XRD Bruker D8 were used, the CuKα radiation specification was (1.54A°) at 40 kV, 100 mA and scanning angel 2θ ranges between 10°-80°. A Perkin–Elmer FTIR double beam spectrometer over the range 400-4000 cm^{-1} are used for studying the transmission measurements, the spectra resolution of 4 cm^{-1} at room temperature for samples made by

*Corresponding author:** Mawlud SQ, Advanced Optical Material Research Group, Department of Physics, Faculty of Science, University of Technology Malaysia, Skudai and Johor, Malaysia, E-mail: samanqadir@gmail.com

using KBr pellets were investigated. Raman measurement is performed using a confocal Horbia Jobin Yvon (Model HR800 UV) with Argon ion laser (excitation wavelength 514.55 nm) operates at 20 mW in the range of 100-900 cm⁻¹. Relatively fine glass powders are used for Raman and FTIR measurements.

Traditional Archimedes method is used for determining the density of the glass sample after measuring the weight of the sample by using a sensitive analytical balance Precisa XT220A, toluene (ρ_t=0.8669 g cm⁻³) is used as an immersed liquid. The density ρ is calculated using the expression [12]:

$$\rho = \left[\left(\frac{W_a}{W_a - W_t}\right) * \rho_t + \rho_a\right] \tag{1}$$

Where W_a and W_t are weight of the glass sample in air and in toluene respectively, ρ_a is the density of air and ρ_t is the density of the toluene. The molar volume (V_M) of the glasses was calculated from density values according to [13]:

$$V_M = \frac{M}{\rho} \tag{2}$$

where M is the molar weight and ρ is the density of the glass. The ionic packing density (V_t) is calculated using Makishima and Mackenzie approach [13,14],

$$V_t = \left(\frac{1}{V_M}\right) * \sum (V_i * x_i) \tag{3}$$

Where x_i is the mole fraction (mol%) and V_i is packing density parameter (m³/mol). For an oxide glass of the form M_xO_y, the value of V_i yields [13],

$$V_i = \left(\frac{4\pi N_A}{3}\right)\left[Xr_M^3 + Yr_o^3\right] \tag{4}$$

where N_A is Avogadro's number (mol⁻¹), r_M and r_o are the Shannon's ionic radius of metal and oxygen, respectively. Table 1 represents the glass sample code and their molar ratio effect on each of the density, molar volume and ionic packing density. The effect of Sm_2O_3 % on both of the density and molar volume for each composition of the glass samples is shown in Figure 1, from Figure 1 it can be seen that the density of TNS glass system increases from 4.903 g cm⁻³ for x=0 mol% to 5.019 g cm⁻³ for x=1.2 mol% of Sm^{3+} ion. Higher molecular mass of Sm_2O_3 (348.74 g mol⁻¹) and those for TeO_2 (159.60 g mol⁻¹) might be responsible for increase its density [15]. The decrease in density at x=1.5 mol% might be caused by Sm^{3+} ions which take a part in the structure of glass and make the density to be decreased [16]. A molar volume of the TNS glass system which is plotted against Sm_2O_3 % in the Figure 1 reveals that the molar volume increased from TNS1 to TNS3 glasses as the Sm^{3+} content increased from 0 mol% to 0.6 mol%, while the molar volume of glass TNS4 and TNS5 decreased with increasing of Sm^{3+} content, this effect can be explained as the Sm^{3+} mole friction content more increased the structure of glasses will get more compact due to increase in packing density of oxygen [17] and this indicates that the structure of TNS glasses changed, this behavior of molar volume is in consistent with the increasing behavior of rigidity and compactness of glass samples [18] (Table 1).

Results and Discussion

XRD and SEM-EDX spectral analysis

In order to check the non-crystalline nature of the glass samples, the XRD measurement is performed for all samples and the result can be seen in Figure 2. From Figure 2 it can be seen that the results did not reveal any kinds of sharp peaks, therefore, proving the amorphous

Sample Code	TeO₂ %	Na₂O %	Sm₂O₃ %	M g mol⁻¹	Vt	VM cm³ mol⁻¹	ρ g cm⁻³
TNS1	80	20	0	140.07	0.4012	28.571	4.903
TNS2	79.7	20	0.3	140.64	0.4017	28.623	4.914
TNS3	79.4	20	0.6	141.21	0.4019	28.700	4.920
TNS4	79	20	1	141.97	0.4051	28.593	4.965
TNS5	78.8	20	1.2	142.34	0.4092	28.359	5.019
TNS6	78.5	20	1.5	142.91	0.4025	28.919	4.942

Table 1: Glasses sample code and calculated density, molar volume and ionic packing density.

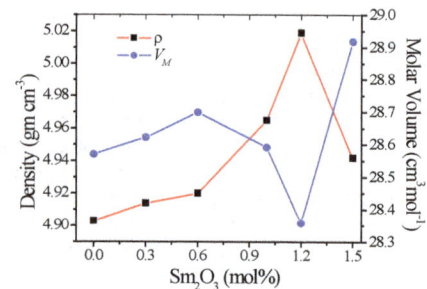

Figure 1: Density and molar volume of the glasses (80-x) TeO_2-20Na_2O-$x$$Sm_2O_3$ as a function of Sm_2O_3 mol%.

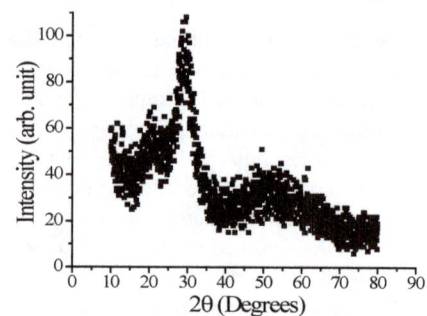

Figure 2: Typical XRD pattern of (80-x) TeO_2-20Na_2O-$x$$Sm_2O_3$ glasses (x=0, 1.5) glasses.

nature structure of the present glass.

SEM-EDX investigations were performed on the (80-x) TeO_2-20Na_2O-$x$$Sm_2O_3$ with (x=1) % mole fraction glasses in order to identify changes in morphology and chemical composition. No crystals were detected by SEM in any of the investigated glasses. The SEM micrograph in Figure 3a for the glass containing 1% mole fraction of Sm_2O_3 showing homogeneous glassy phase is typical for these glasses. The EDX analysis are shown in Figure 3b on several different spots on glass sample gives almost identical spectra confirming homogeneous character of the sample. The chemical composition calculated from EDX spectra is in good agreement with nominal composition of the glass. Similar results were obtained for all investigated TeO_2-Na_2O-Sm_2O_3 glasses.

DTA spectral analysis

The DTA thermograms for (80-x) TeO_2-20Na_2O-$x$$Sm_2O_3$ glasses sample are shown in Figure 4. Endothermic peaks due to the glass transition and melting point and exothermic peak due to the crystallization are clearly observed. The T_g value of 80 TeO_2 - 20 Na_2O

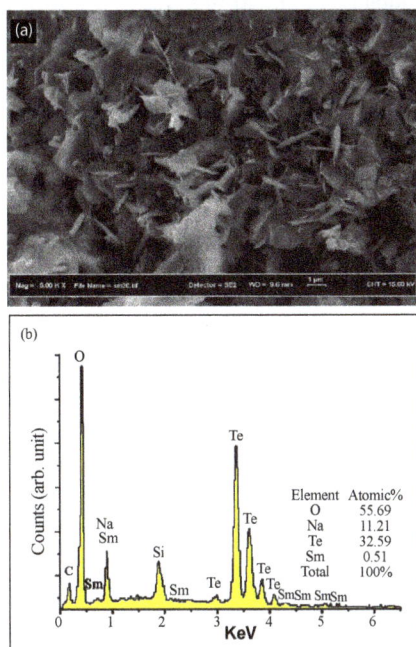

Figure 3: (a) SEM micrograph and (b) EDX spectra of TNS4 sample.

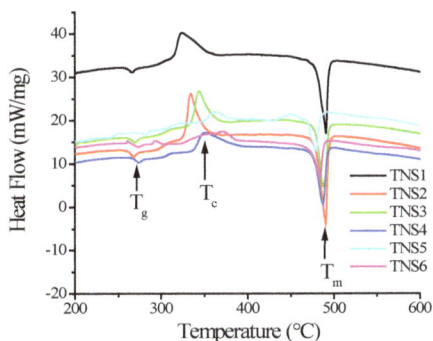

Figure 4: DTA traces of the (80-x) TeO$_2$-20Na$_2$O-xSm$_2$O$_3$ glasses at a heating rate is 10°C/min.

have been determined to be 265.8°C and by adding 0.3 mol% Sm$_2$O$_3$ into the composition it increased T_g to 267.6°C. Obtained results indicate that by increasing the amount of mol% Sm$_2$O$_3$, the T_g of the samples also increases, the small increase of T_g in these glasses shows that the structure is strongly and progressively modified.

The thermal stabilities ΔT of the TNS reference glass and Sm^{+3}: TNS glass has been evaluated from their T_g, T_c and T_m values, the results are listed out in Table 2. ΔT is generally used as a rough measure of glass thermal stability and it is desirable for a glass host to have ΔT as large as possible [19]. The higher the ΔT is, the stronger the inhibition of nucleation and crystallizations will occur [20]. The ΔT of present glasses are calculated to be in the range of (58.5-97.8)°C, it is noted that the ΔT increases with increasing % Sm$_2$O$_3$ to the composition, results indicate that the present glass possesses good thermal stability and anti-crystallization ability. Hruby's parameter also calculated by using eq. (5), the greater values of the Hruby's parameter indicate higher glass forming tendency, the values of H in our glasses increased with the addition of the Sm$_2$O$_3$. Figure 5 shows the effect of increasing % Sm$_2$O$_3$ on ΔT, T_c, T_g, T_m and Hruby's parameter (Table 2).

$$H = \frac{T_c - T_g}{T_m - T_c} \qquad (5)$$

FTIR spectral study

In order to understand the structural units of glasses studied, FTIR transmission spectra of TNS glasses and pure KBr have been recorded in the range 400-4000 cm^{-1} as shown in Figure 6. From Figure 6 it can be seen that there are seven important absorption peaks appear in the FTIR transmission spectra. The identical peaks assignments are then listed in Table 3. For giving more detail about the FTIR spectra, the FTIR spectra are divided into three regions: 400-900 cm^{-1}, 1000-2500 cm^{-1} and 2500-4000 cm^{-1} as shown in Fig. (6a, 6b and 6c) respectively. The spectra in Figure 6a present three bands at optical visible range 759.6, 637.5 and 471.8 cm^{-1} and are assigned to stretching vibrations of TeO$_3$ or TeO$_{3+1}$, TeO$_4$ structural units and bending vibrations of Te-O-Te or O-Te-O linkages with Na-O respectively. The change in sodium tellurite structure can be predicted as the shift in the position is attributed to the change in the bonding length predicting [21,22]. The shifts in the stretching vibration of TeO$_4$ and TeO$_3$ towards 637.5 and 759.6 cm^{-1} are observed by increasing the Sm^{3+} ion concentration up to 1 mol%. This shift in the transmission bands is attributed to the deformation of TeO$_4$ group into TeO$_3$ through TeO$_{3+1}$ intermediate coordinate formation [23,24]. The ratio of 759.6/637.5 cm^{-1} represents the relative concentrations of the TeO$_3$ and TeO$_4$ structural units, which is absolutely dependent on the glass composition.

According to the electronegativity theory, the covalency of the bond will become stronger with the decrease of the difference of electronegativity between cation and anion ions. From the periodic table, since the values of electronegativity for Te, Na, Sm and O elements are 2.1, 0.9, 1.17 and 3.5, respectively, the covalency of

Sample Name	% Sm$_2$O$_3$	T_g °C	T_c °C	T_m °C	$\Delta T = T_c - T_g$ °C	$H = \frac{T_c - T_g}{T_m - T_c}$
TNS1	0	265.8	323.6	489.5	57.8	0.349
TNS2	0.3	267.6	334.1	490.2	66.5	0.426
TNS3	0.6	269.1	343.5	486.9	74.3	0.518
TNS4	1	273.8	350.4	486.4	76.6	0.563
TNS5	1.2	281.7	362.2	478.6	80.6	0.692
TNS6	1.5	273.1	370.6	483.0	97.5	0.868

Table 2: Thermal parameters determined from the DTA traces of (80-x) TeO$_2$-20Na$_2$O-xSm$_2$O$_3$.

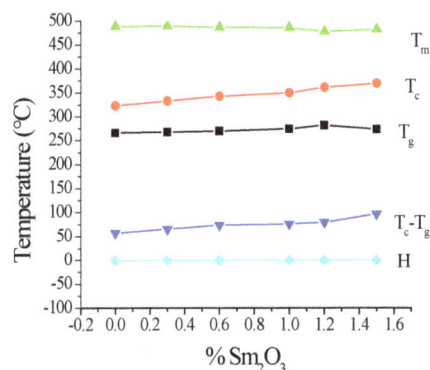

Figure 5: Effect of Sm$_2$O$_3$ % on each of the on ΔT, T_c, T_g, T_m and Hruby's parameter.

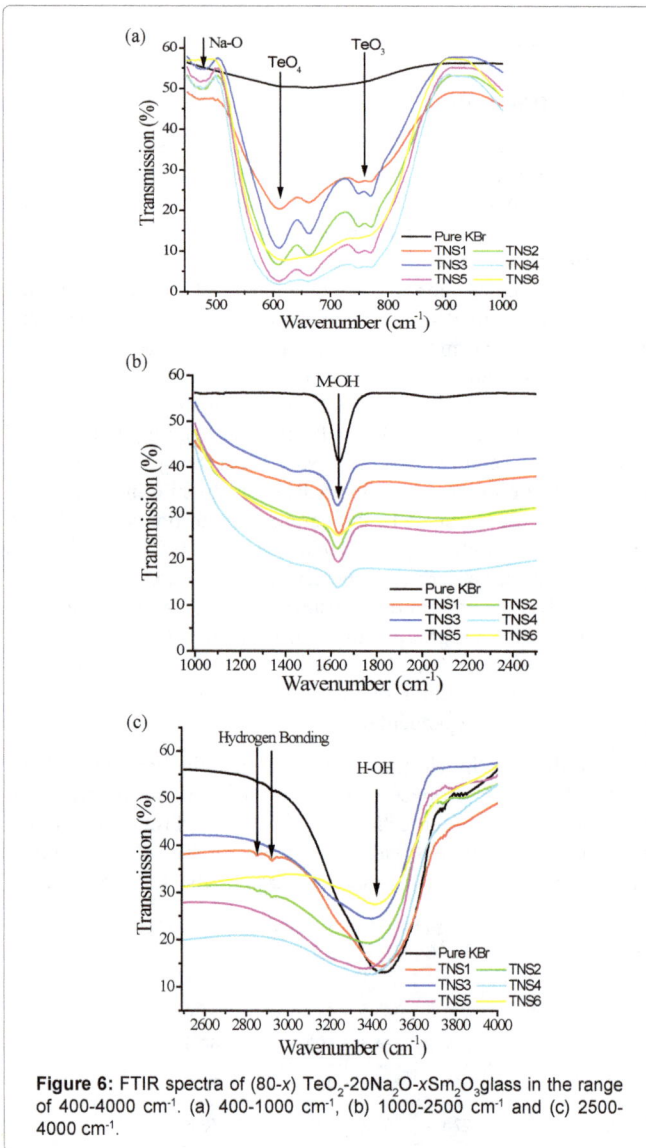

Figure 6: FTIR spectra of (80-*x*) TeO$_2$-20Na$_2$O-*x*Sm$_2$O$_3$ glass in the range of 400-4000 cm^{-1}. (a) 400-1000 cm^{-1}, (b) 1000-2500 cm^{-1} and (c) 2500-4000 cm^{-1}.

Sample Code	Sm$_2$O$_3$ (mol%)	Na-O	TeO$_4$	TeO$_3$	M-OH	Hydrogen Bonding	Hydrogen Bonding	OH-Group
Pure KBr	-	-	-	-	1637.53	2855.22	2933.75	3447.77
TNS1	0	471.80	637.51	759.61	1633.99	2851.32	2923.83	3443.73
TNS2	0.3	472.19	637.21	760.20	1627.15	2853.64	2923.07	3389.21
TNS3	0.6	473.30	635.9	765.61	1629.76	2855.22	2922.24	3395.65
TNS4	1	478.36	637.31	759.35	1630.26	2768.96	2927.44	3400.04
TNS5	1.2	475.12	619.19	759.95	1623.00	2848.98	2915.79	3382.01
TNS6	1.5	476.65	636.42	759.63	1632.26	2848.92	2921.24	3370.68

Table 3: Peaks position (in cm^{-1}) in the FTIR spectra of (80-*x*) TeO$_2$-20Na$_2$O-*x*Sm$_2$O$_3$ glass.

Te-O are stronger than Na-O and Sm-O, respectively. As a result, the higher affinity of the tellurium ions to attract oxygen atoms yields the apparition of TeO$_4$ structural units [25]. These three bands do not exist in the pure KBr spectra.

Figure 6b shows the hydroxyl-metal (M-OH) stretching vibrations bond which is observed at 1637.53 cm^{-1}. This band is shifted to 1632.26 cm^{-1} with increasing amount of Sm^{3+} concentration. This shift might

be due to the addition of rare earth in the host matrix of glasses that slightly increase the IR transmission and shift it to longer wavelength [26]. M-OH stretching vibration peaks also appeared for pure KBr sample spectra. The IR transmission bands occur around 2900 and 3400 cm^{-1} belong to stretching vibration of the hydroxyl group and hydrogen bond [27]. This is observed clearly in Figure 6c, the band ranges 2768.96-2927.44 cm^{-1} and 3400.04-3447.77 cm^{-1} which correspond to hydroxyl group and hydrogen bond, respectively (Table 3).

Raman spectra

One of the most well-known types of vibrational spectroscopy is Raman spectroscopy. It is considered as a finger print about structural information of the glass. The Raman spectra of TNS glass series in the frequency range of 100-900 cm^{-1} is shown in Figure 7, and the deconvoluted of the Raman spectra for sample TNS2 is presented in Figure 8. The observed spectrum is then fitted to Gaussian peaks and to get four distinct solid line peaks named by (A,B,C and D). The sum of these picks are represented by a dotted line which is well coincide with the obtained solid line from the Raman spectra. These peaks are centered at 294.43 (A), 471.09 (B), 668.33 (C) and 760.52 (D) cm^{-1}. The corresponding absorption peak shifts are listed in Table 4. It can be observed that the shift is dominant by all band regions around 279.4-306.63 cm^{-1}, 465.11-473.45 cm^{-1}, 668.33-684.12 cm^{-1} and 759.06-773.3 cm^{-1}. The Raman band in the region around 279.4-306.63 cm^{-1} can be assigned to both Sm^{3+}-O and TeO$_3$ tp which indicate that the presence of rare earth ions might significantly change the Te-O networking structure in glasses [27,28]. The Raman shift corresponding to the band in a range of 465.11-473.45 cm^{-1} is due to Te-O-Te linkages vibration. Meanwhile, the Raman peaks shift in the range of 668.33-684.12 cm^{-1}

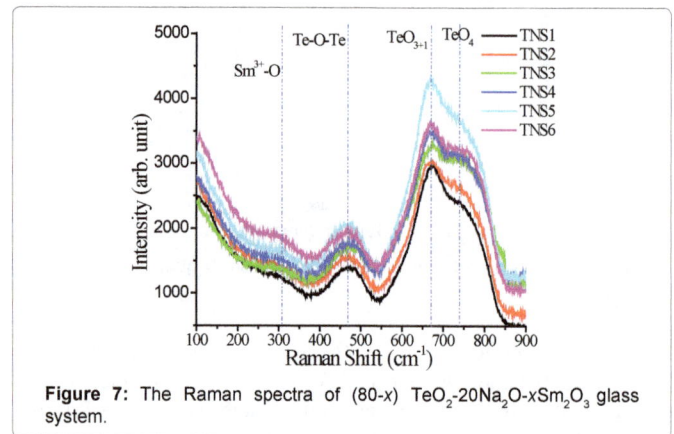

Figure 7: The Raman spectra of (80-*x*) TeO$_2$-20Na$_2$O-*x*Sm$_2$O$_3$ glass system.

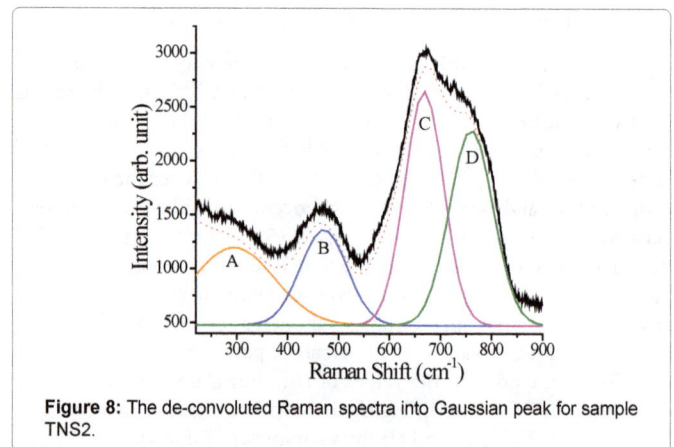

Figure 8: The de-convoluted Raman spectra into Gaussian peak for sample TNS2.

Sample Code	% Sm_2O_3	Raman Shifts/ cm^{-1}			
		A	B	C	D
TNS1	0	284.34	473.17	671.8	759.06
TNS2	0.3	294.43	471.09	668.33	760.52
TNS3	0.6	289.74	473.45	671.01	767.81
TNS4	1	303.549	467.33	682.23	770.37
TNS5	1.2	306.63	465.11	684.12	773.3
TNS6	1.5	276.94	467.54	677.61	773.02

Table 4: Raman shift peaks position in cm^{-1} for (80-x) TeO_2-20Na_2O-xSm_2O_3.

and 759.06-773.3 cm^{-1} are corresponding to TeO_3 bp unit and TeO_4 tbp unit respectively. It can also be observed that increasing the Sm^{3+} ions from 0 to 1.5 mol % and adding into the TNS glass will turn the glass structure to transform slightly by the perturbation of TeO_4 tbp unit into TeO_3 tp unit through TeO_{3+1} intermediate coordination, this has been satisfied by the Raman band shift at the certain region [29,30]. Moreover, it is noted that the intensity of Raman peaks around 668.33-684.12 cm^{-1} and peaks around 759.06-773.3 cm^{-1} are increased by the increasing of the concentration of Sm^{3+} ions as shown in Figure 7. The change in intensity might be due to the creation of TeO_3 tp unit by perturbation of TeO_4. It is asserted that the addition of Sm^{3+} ions in TNS glasses creates more number of TeO_3 tp units (Table 4).

Conclusion

In this study, glass formation range of (80-x) TeO_2-20Na_2O-xSm_2O_3 system is prepared by melt quenching method. The XRD results how all samples are amorphous in nature. The physical parameter such as glass density, molar volume and ionic packing density were found to be (4.903-5.019) g cm^{-3}, (28.359-28.919) $cm^{-3}mol^{-1}$ and (0.4012-0.4092) respectively. Thermal and structural properties are investigated by DTA, FTIR and Raman spectroscopies. The thermal characteristics reveal that the glass transition temperature and stability factor increases with the increasing of Sm_2O_3 mol% content. This is due to the increase in bond number per unit volume which is an indication of change in packing density in the structure. The vibrational spectra of the glass system suggested that the glass network consists of TeO_4, TeO_3/TeO_{3+1}, Na_2O and Sm_2O_3 units. Raman spectra for the (80-x) TeO_2-20Na_2O-xSm_2O_3 glasses indicated that the TeO_4 tbps convert to TeO_3 tps with the increasing of Sm_2O_3 % mol fraction. The Raman shifts due to the structural units TeO_4 and TeO_3 have almost equal to the intensities in these glasses and overlapped to each other.

Acknowledgments

The authors gratefully acknowledge the financial   support from Ministry of Higher Education, RMC, UTM and University of Salahaddin/ Ministry of Higher Education/KRG through the research grant (vote 4B182) is highly appreciated.

References

1. El-Mallawany RAH (2002) Tellurite glasses - Handbook of physical properties and data. Boca Raton CRC Press.

2. Çelikbilek M, Ersundu AE, Solak N, Aydın S (2011) Investigation on thermal and microstructural characterization of the TeO2-WO3 system. J Alloys and Compounds 509: 5646-5654.

3. Sun H, Duan Z, Zhou G, Yu C, Liao M, et al. (2006) Structural and up-conversion luminescence properties in Tm3+/Yb3+-co-doped heavy metal oxide-halide glasses. Spectrochim Acta A Mol Biomol Spectrosc 63: 149-153.

4. Pan Z, Morgan SH, Dyer K, Ueda A, Liu H (1996) (1996) Host-dependent optical transitions of Er3+ ions in lead–germanate and lead-tellurium-germanate glasses. J Appl Phys 79: 8906

5. Xu S, Fang D, Jiang Z, Zhang J (2005) New synthesis of excellent visible-light TiO"2"-"xN"x photo catalyst using a very simple method. J Solid State Chem 178: 1817

6. John Kieffer, Jacqueline AJ, Nickolayev O, Bass JD (2006) Structures and visco-elastic properties of potassium tellurite: glass versus melt. J Phys Condens Matter 18: 903-914.

7. Rada S, Culea E, Rada M (2011) Materials chemistry and physics the experimental and theoretical investigations on the structure of the gadolinium–lead–tellurate glasses. Mat Chem Physic 128: 464-469.

8. Murali A, Sreekanth Chakradhar RP, Lakshmana Rao J (2005) EPR studies of Gd3O ions in lithium tetraboro–tellurite and lithium lead tetraboro-tellurite glasses Physica B 364: 142-149.

9. Bolunduta L, Culeaa E, Borodib G, Stefanc R, Munteanua C, et al. (2015) Influence of Sm3O:Ag codoping on structural and spectroscopic properties of lead tellurite glass ceramics. Ceramic Int 41: 2931-2939.

10. Yusoff NM, Sahar MR, Ghoshal (2014) Sm3O: Ag NPs assisted modification in absorption features of magnesium tellurite glass. J Mol Struct 1079: 167-172.

11. Xia H, Nie Q, Zhang J, Wang J (2003) Spectroscopic studies of TeO2–ZnO–Er2O3 glass system. Mater Lett.

12. Siti Amlah M, Azmi, Sahar MR, Ghoshal SK, Arifin R (2015) Modification of structural and physical properties of samarium doped zinc phosphate glasses due to the inclusion of nickel oxide nanoparticles. J Non-Crystalline Solids 411: 53-58.

13. Yusoff NM, Sahar MR (2015) Effect of silver nanoparticles incorporated with samarium-doped magnesium tellurite glasses Physica B 456: 191-196.

14. Ahmmad MA, Samee A, Edukondalu S, Rahman M (2012) Physical and optical properties of zinc arsenic tellurite glasses. Results Phys 2: 175.

15. Jaba N, Mermet A, Duval E, Champagnon B (2005) Raman spectroscopy studies of Er3+ doped zinc tellurite glasses. J Non-Cryst Solids 351: 833-837

16. Plotnichenko VG, Sokolov VO, Koltashev VV, Dianov EM (2005) Raman band intensities of tellurite glasses. Optics Lettrs 30:1156-1158.

17. Aye NN (2011) Study of xP2O5-(1-x-y)V2O5-yCuO semi-conducting glass system. Univ Res Jour 4: 39.

18. Sae-hoon K, Yoko T (1995) Nonlinear optical properties of TeO2- based glasses: MOx-TeO2 (M= Sc, Ti, V, Nb, Mo, Ta, and W) binary glasses. J Am Ceram Soc 78:1061-1065.

19. Baki SO, Tan LS, Kan CS, Kamari HM, Noor ASM et al. (2013) Structural and optical properties of Er3+-Yb3+codoped multi-composition TeO2-ZnO-PbO-TiO2-Na2O glass. J Non-cryst Solids 362: 156-161.

20. Wang M, Yi L, Zhang L, Wang G, Hu L et al. (2009) 2-μm fluorescence and Raman spectra in high and low Al (PO3)3 content fluorophosphate glasses doped with Er–Tm–Ho. Chin Opt Lett 7:1035-1037.

21. Ozen G, Demirata B, Ovecoglu ML, Genc A (2001) Thermal and optical properties of Tm3+ doped tellurite glasses. J Spectrochimica Acta Part A 57: 273-280.

22. Saritha D, Markandeya Y, Salagram M, Vithal M, Singh AK (2008) Effect of Bi2O3 on physical optical and structural studies of ZnO–Bi2O3–B2O3 glasses. J Non-Cryst Solids 354: 5573-5579.

23. Sahar MR, Jehbu AK, Karim MM (1997) TeO –ZnO–ZnCl glasses for IR transmission. J Non-Cryste Solids 213 & 214: 164-167.

24. Ahmed MM, Holland D (1985) Oxychloride glasses for infra-red transmission. Mater Sci Forum 5: 175-184.

25. Rada S, Pascuta P, Rada M, Culea E (2011) Effects of Samarium (III) oxide content on structural investigations of the samarium–vanadate–tellurate glasses and glass ceramics. J Non Crystalline Solids 357: 3405-3409

26. Upender G, Chandra Mouli V, Sathe T, Vasant G (2009) EPR Raman infrared and optical absorption studies of Cu2+ ions in 60TeO2–(40-x)WO3–xPbO glasses. Indian J Pure and Applied Physics 50: 399-406.

27. Hager IZ, El-Mallawany R (2010) Preparation and structural studies in the (70-x)TeO2–20WO3–10Li2O–xLn2O3 glasses. J Mater Sci 45: 897.

28. Nazabal V, Todoroki S, Nukui A, Matsumoto T, Suehara S (2003) Oxyfluoride tellurite glasses doped by erbium: thermal analysis, structural organization and spectral properties. J Non-Cryst. Solids 325: 85-102.

29. Suresh Kumar K, Pavani A, Babu M, Kumar N, Giri SB (2010) Fluorescence characteristics of Dy 3+ ions in calcium fluoroborate glasses. J Lmin 130: 1916.

30. Som T, Karmakar B (2010) Structure and properties of low phonon antimony glasses and nano-glass ceramics in K2o-B2o-B2O3-Sb2O3 system. J Non-Cryst Solids 356: 987.

12

CFD Analysis of a Gasoline Engine Exhaust Pipe

Pengyun Xu*, Haiyong Jiang and Xiaoshun Zhao

Mechanical and Electronical Engineering College, Agriculture University of Hebei, Baoding 071001, P.R. China

Abstract

The exhaust pipe is an important part of gasoline engine. Its structure and performance have a direct impact on the engine power, economy and emissions, and it is one of the key technologies of multi valve engine development. In or-der to test the theoretical design of a 1.5 L gasoline engine exhaust pipe, Solidworks Flow simulation was used to analyze the exhaust pipe. Pressure and velocity of the position near the three-way catalytic converter and the oxygen sensor were selective analyzed. CFD Simulation results show that the internal flow is laminar flow state, and the sensor position is reasonable. The design is reasonable, and can achieve the design goal.

Keywords: Exhaust pipe; Catalytic converter; Carrier component; Fluid uniformity; CFD analysis

Introduction

The exhaust pipe is an important part of gasoline engine. It connects all parts of the automobile exhaust system, and prevent the leakage of waste gas. Its structure and performance have a direct impact on the engine power, economy and emissions, and it is one of the key technologies of multi valve engine development.

Automobile exhaust pipe is a space curved surface geometry, it has a certain difficult to manufacture, and the exhaust pipe is worked in a bad condition, and some problems will occur in the process of production and use. The traditional design method of the exhaust pipe is in steady flow test stand experiments, to obtain or test shape parameter, this method takes long time and cost. However, using computer fluid analysis technology (CFD), it is convenient and intuitive to analyze the three-dimensional model of the exhaust pipe, the analysis process is visual, and easy to adjust the parameters, the analysis results are intuitive, determine whether the structure meets the design requirements quickly. In this paper, SolidWorks Flow Simulation is used to simulate the structure of a certain type of engine. Pressure and velocity of the position near the three way catalytic convert-er and the oxygen sensor were selective analyzed. The design goals are verified by simulation analysis (CFD) results.

In order to reduce the concentration of CO, NOx and CxHy in the exhaust gas, there are (mostly) two catalytic converters installed in the exhaust pipe system. The crucial quantity to control the efficiency of a catalytic converter is the temperature in the catalytic converter. Due to this reason, one is interested in how to ensure a sufficient high temperature in the catalytic converters in a short time after the engine start. A special method of heating after the engine start is the combustion of unburnt gas in the catalytic converters. Modern cars can control the ratio of oxygen and fuel in the combustion chamber of the engine. By choosing a ratio with more fuel and less oxygen some unburnt fuel gets to the catalytic converters and can be used there for an exothermic reaction.

In order to test the theoretical design of a 1.5 L gasoline engine exhaust pipe, carrying out a natural experiment in respect gas flow hydrodynamics. To estimate the spectrum of possible technological innovations in the existing struc-ture expedient, first carry out numerical simulations using fluid dynamics software packages solidworks flow simula-tion. This paper presents the numerical simulation of classical structure of exhaust pipe; the results are compared with the calculation for the case of a conical nozzle on the exhaust pipe [1-5].

Material and Methods

The calculation was carried out in gas dynamics software package Solidworks Cosmos FloWorks, which uses a finite volume method, the movement of the fluid is modeled by the Navier Stokes equations, the Reynolds averaged. Their closures are used for the transport equation and its kinetic energy dissipation within the k-ε turbulence model.

During the calculation was the condition of the grid and the iterative convergence, which were determined by the self-similarity of the final result of the number of cells and iterations.

Calculation of the motion of particles in the flow was conducted under the following assumptions:

- Uniform air and standard atmospheric pressure;

- The particles have a spherical shape;

- The drag coefficient of the particle is calculated by Henderson's equation;

A series of pertinent indoor and outdoor experiments are carried out to know the engineering properties of exhaust pipe, and provide theoretical references for engineering quality evaluation and improvement scheme selection.

3-D model analysis of exhaust pipe

Using SolidWorks software to establish the 3D model of the exhaust pipe, as shown in Figure 1. The model parameters are shown in Table 1.

In order to control the automobile harmful gases such as NOx, HC and CO, the three-way catalytic converter must be installed in the exhaust pipe. The harmful gases can be converted into harmless carbon dioxide, water and nitrogen by oxidizing and reducing gases. The carrier component of the three-way catalytic converter is a porous ceramic material, which is installed in the specific position of the

*Corresponding author: Pengyun Xu, Mechanical and Electronical Engineering College, Agriculture University of Hebei, Baoding 071001, PR China
E-mail: pengyun99@qq.com

exhaust pipe, and is the most important equipment in the automobile ex-haust system. The three-way catalytic converter and oxygen sensor are generally installed in the exhaust manifold (natural gas engine) or after the turbocharger (turbocharged gasoline engine). The vector parameters and grid partition are shown in Tables 2 and 3. Grid partitioning result is shown in Figure 2.

Boundary conditions setting

Fluid computational domain is shown in Figure 3. Boundary conditions are shown in Table 4. The velocity in the inlet section was set as a fully developed turbulent flow in a pipe, its mean value of 0.13 kg/s; in the outlet section was set at ambient pressure 101325 Pa; on the walls of all the components of the velocity zeroed (condition "sticking") [6-9].

Additional information:

(a) fixed wall boundary

Adiabatic slip free, fixed temperature wall 293 K, the boundary layer is treated by turbulent wall law.

(b) import and export boundary

Use the parameters in steady flow test, the values shown in Table 4.

Analysis target

The inhomogeneous flow of the front end face of the catalytic converter can produce the phenomenon of vortex flow and air separation, which cause the temperature distribution no uniform, also cause the carrier component damage, and then affect the engine's

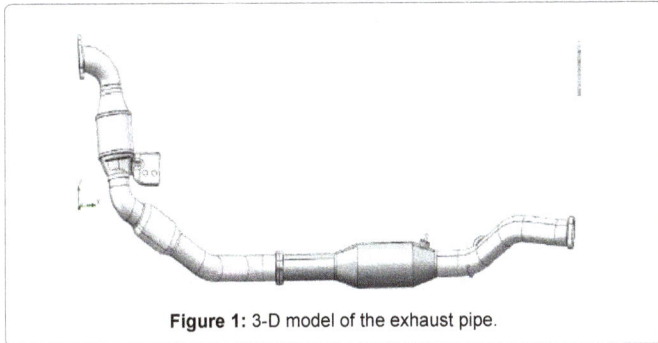

Figure 1: 3-D model of the exhaust pipe.

Grid type grid	Grid number
Tetrahedral mesh and prismatic boundary layer grid	692162

Table 1: Shape parameter (mm).

Name	Cell (mil)	Thickness (mm)	Coating (g/ft³)	Diameter (mm)	Length (mm)
Pre-catalytic converter	600	4	130	101.6	90
Main catalytic converter	400	6	130	101.6	152.4

Table 2: The carrier component parameters.

Name	Data
Pre-catalytic converter inlet pipe diameter	62
Pre-catalytic converter length	90
Main catalytic converter inlet pipe diameter	62
Main catalytic converter length	152.4
Total length	2150

Table 3: Grid partition.

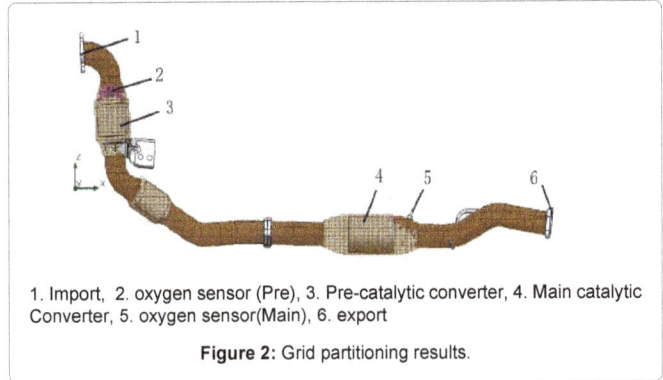

1. Import, 2. oxygen sensor (Pre), 3. Pre-catalytic converter, 4. Main catalytic Converter, 5. oxygen sensor(Main), 6. export

Figure 2: Grid partitioning results.

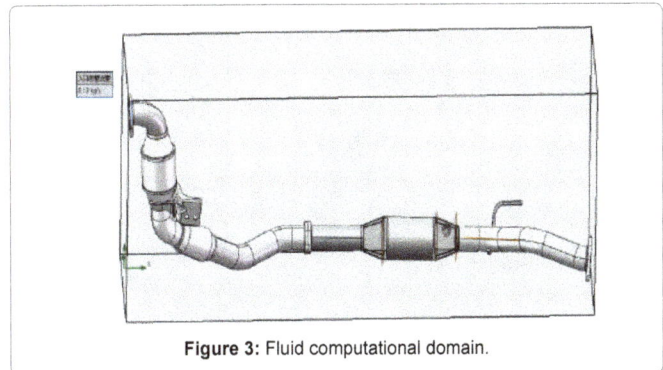

Figure 3: Fluid computational domain.

Parameter	Setting
Import	Flow 0.13 kg/s temperature 85°C
Export	Relative pressure 30 KPa
Wall	No slip
Pre-catalytic converter	Porosity 0.81
Main catalytic converter	Porosity 0.75

Table 4: Import and export boundary parameters.

work. Therefore, it is necessary to analyze the flow uniformity index γ, the range is between 0 ~ 1, and the 1 means completely uniform. When γ is more than 0.9, the flow uniformity of the cross section is better.

If the position of the oxygen sensor in the exhaust pipe is not suitable, the oxygen sensor cannot measure the oxygen concentration accurately [10-15]. It will affect the air fuel ratio of the ECU calibration, and directly affect the engine's power per-formance and emission performance. So it is necessary to use CFD to analyze the flow field around the exhaust pipe.

The CFD analysis process is shown in Figure 4.

Results and Discussion

The computed result of all design variables were analysed and discussed in detail to identify the optimum performance of exhaust pipe. The computed result obtained at different design variables; effects of one design variable on other variables were assessed. Effects of flow rate, pressure, uniformity index were discussed in detail [16].

Flow field analysis

The whole pressure field shows that the fluid pressure decreases along the axis of the tube, and the pressure gradi-ent is obvious in the position of the expanding port and the shrink port. The result is shown as Figure 5.

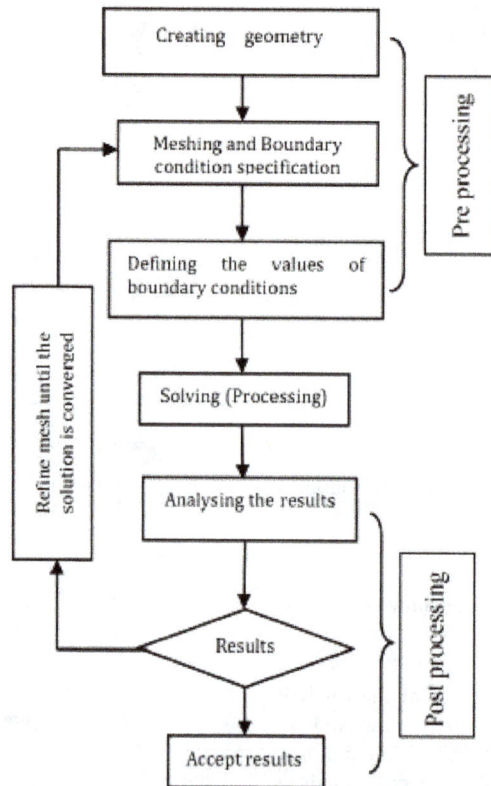

Figure 4: CFD analysis process.

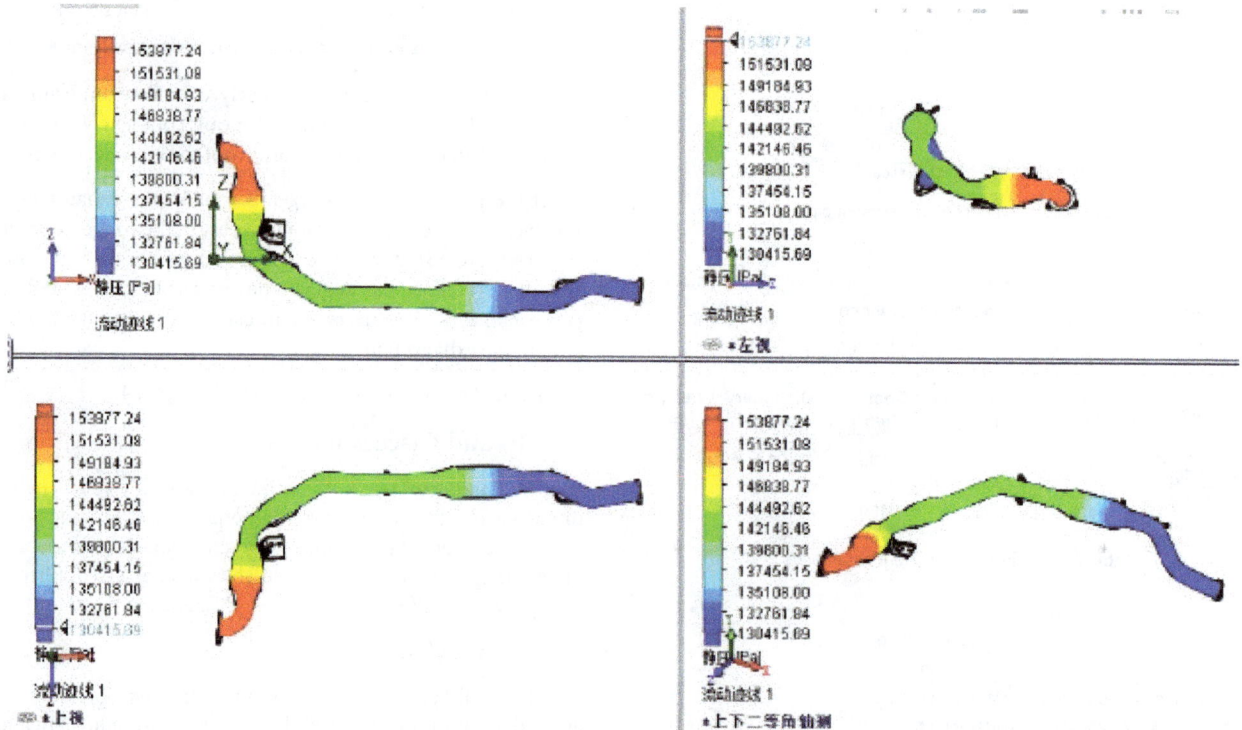

Figure 5: Pressure field analysis.

Pressure field distribution

As shown in Figure 6, near the inlet elbow outside wall, the radial pressure of pre-catalytic converter carrier component is relatively large. Near axis center, the axial pressure of main- catalytic converter carrier component is relatively large [17].

Pressure drop analysis

The main pressure drop detection position of the air flow in the pipe is shown in Figure 7, and the pressure loss value is shown in Table 5.

Velocity field analysis

The Figure 8 shows that when the inlet flow is 0.13 kg/s, the highest flow rate can reach 104.8 m/s, and the velocity muta-tion mainly occurred in the import (export) conical surface. Fluid through the catalytic converter, Reynolds number Re<2000, is laminar flow state.

As is shown in Figure 9, the maximum speed of the pre-catalytic converter end face is 34.92 m/s, the maximum speed of the main catalytic converter end face is 48.91 m/s, which is less than 100 m/s, that is in accordance with the design requirements.

Oxygen sensor position CFD analysis .

The front oxygen sensor is located in the main flow area, which is in accordance with the design requirements; The rear oxygen sensor is located in the main flow area, which is in accordance with the design requirements. The analysis result is shown as Figure 10.

Fluid uniformity analysis

In general, the calculation of the fluid uniformity coefficient γ is only for the end face of catalytic converter. When γ <0.9, it is necessary to optimize the import.

After calculation, the fluid uniformity coefficient γ of the main catalytic converter end face is 0.97, and the precatalytic converter is 0.696. The results are shown in Figure 11.

Velocity Index is a criterion for judging the radial force of the carrier component, as shown in Figure 12, the calculation method is shown in formula (1~3). Under normal circumstances, when the velocity index ε ≤0.7, that is in accordance with design requirements, but when the fluid uniformity γ ≥ 0.94, without considering the influence of Velocity Index [18].

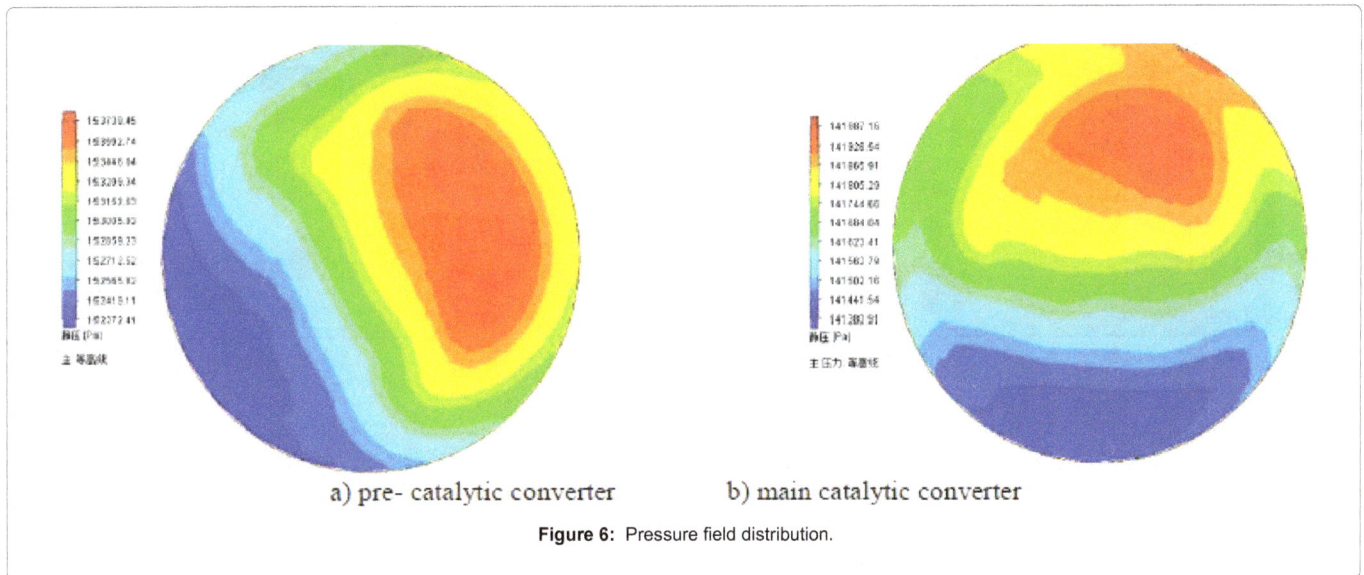

a) pre- catalytic converter   b) main catalytic converter

Figure 6: Pressure field distribution.

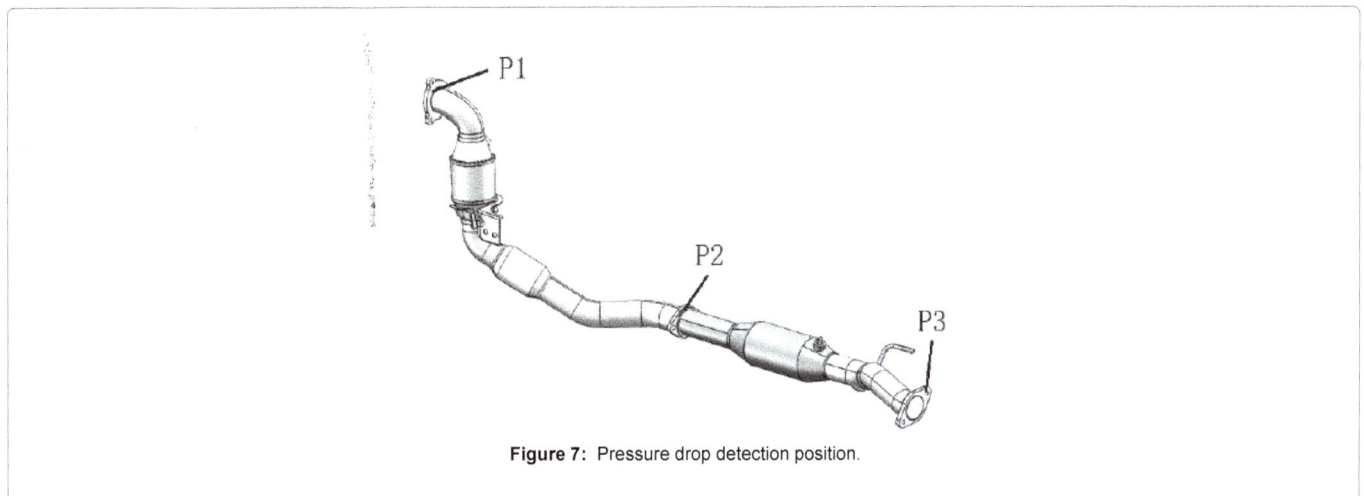

Figure 7: Pressure drop detection position.

Position	Pressure drop data (KPa)
P1	12.57
P2	9.09
P3	21.65

Table 5: Pressure drop.

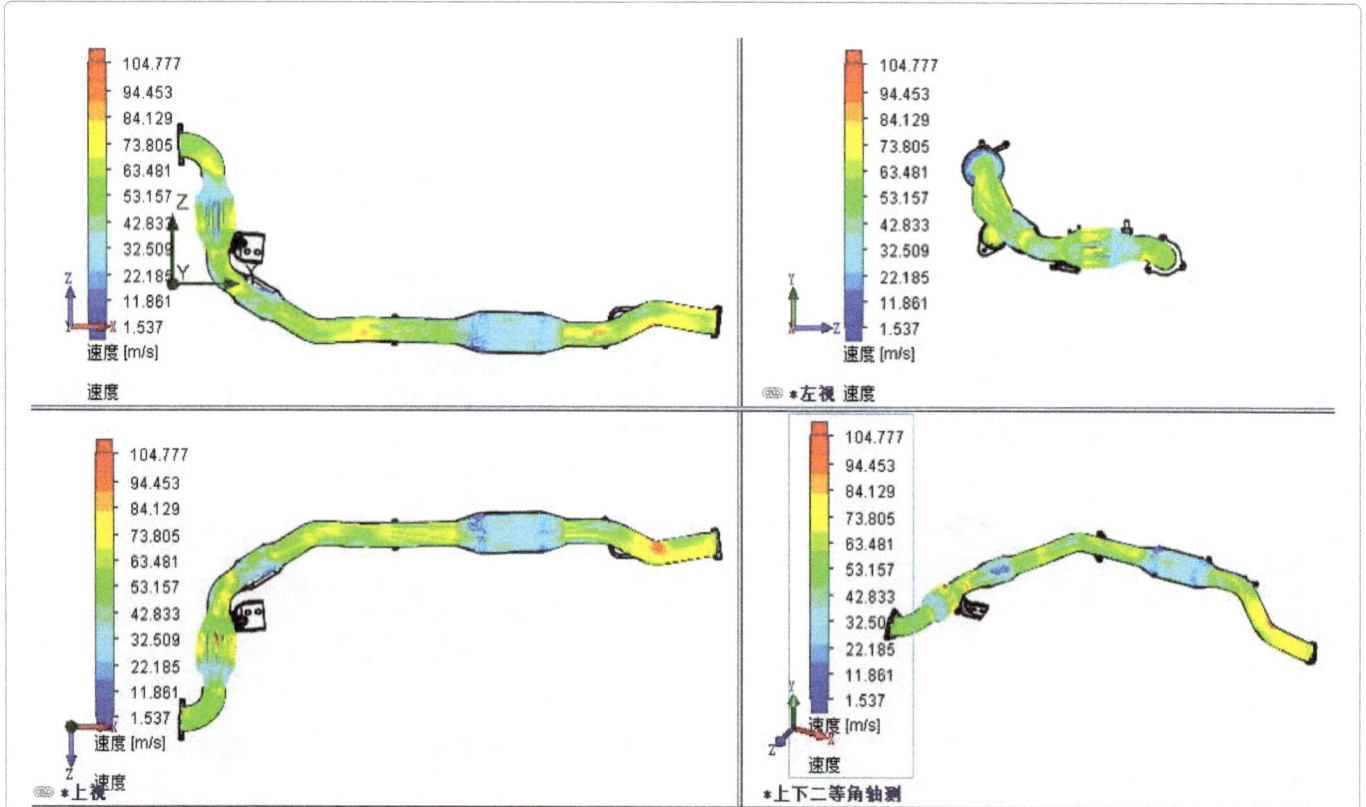

Figure 8: Velocity field analysis.

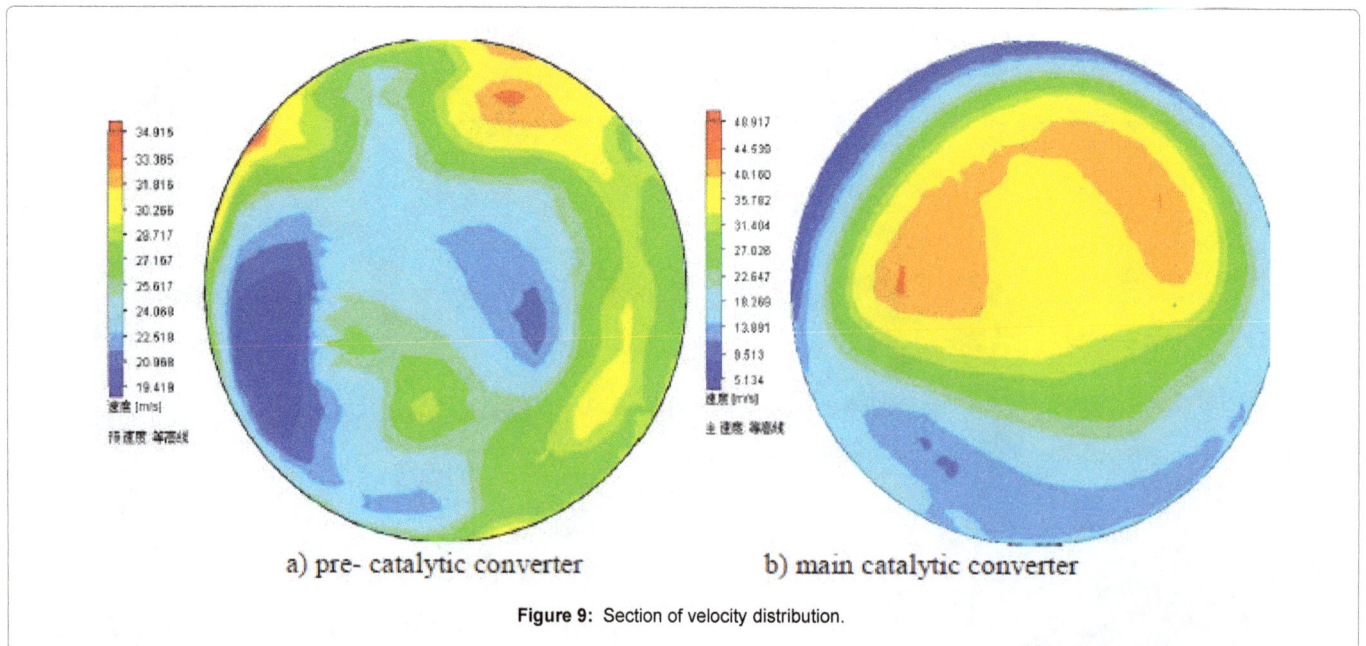

a) pre- catalytic converter   b) main catalytic converter

Figure 9: Section of velocity distribution.

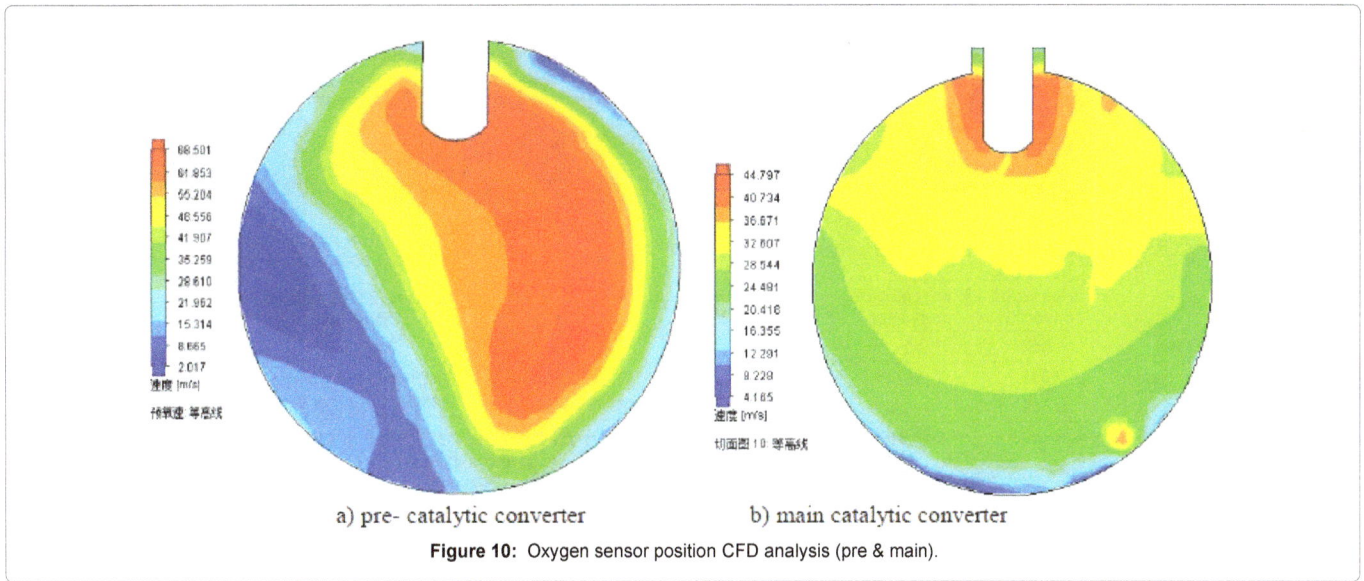

a) pre- catalytic converter   b) main catalytic converter

Figure 10: Oxygen sensor position CFD analysis (pre & main).

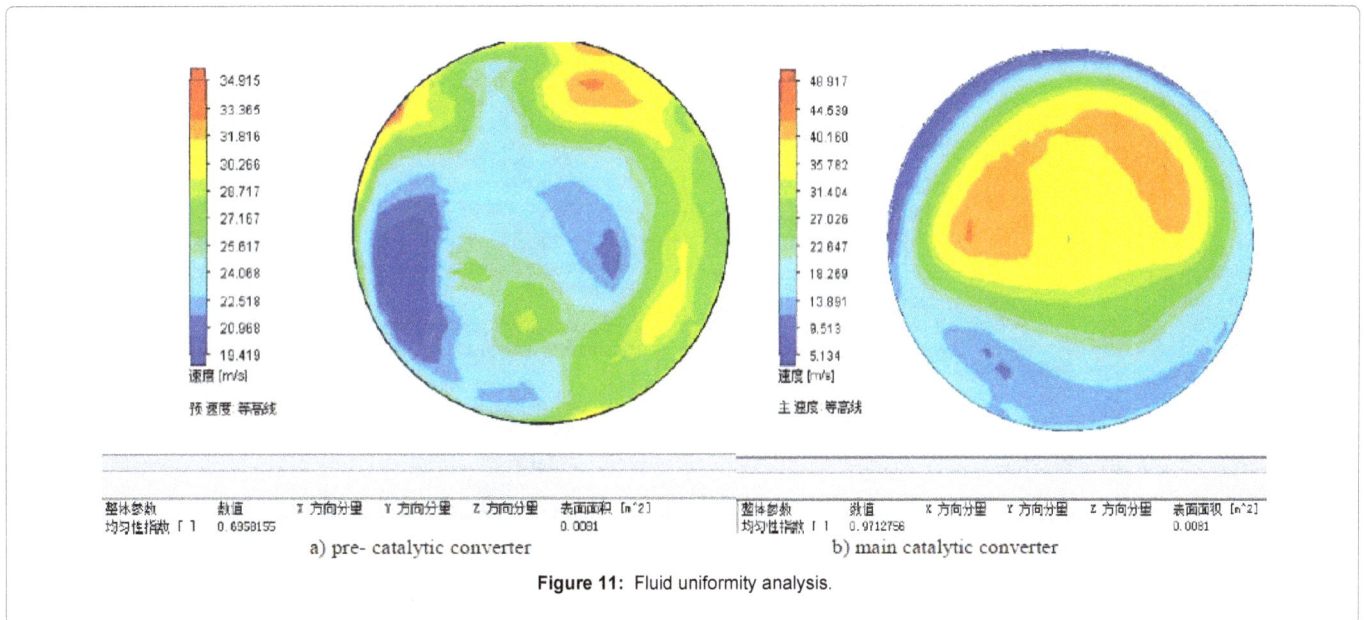

a) pre- catalytic converter   b) main catalytic converter

Figure 11: Fluid uniformity analysis.

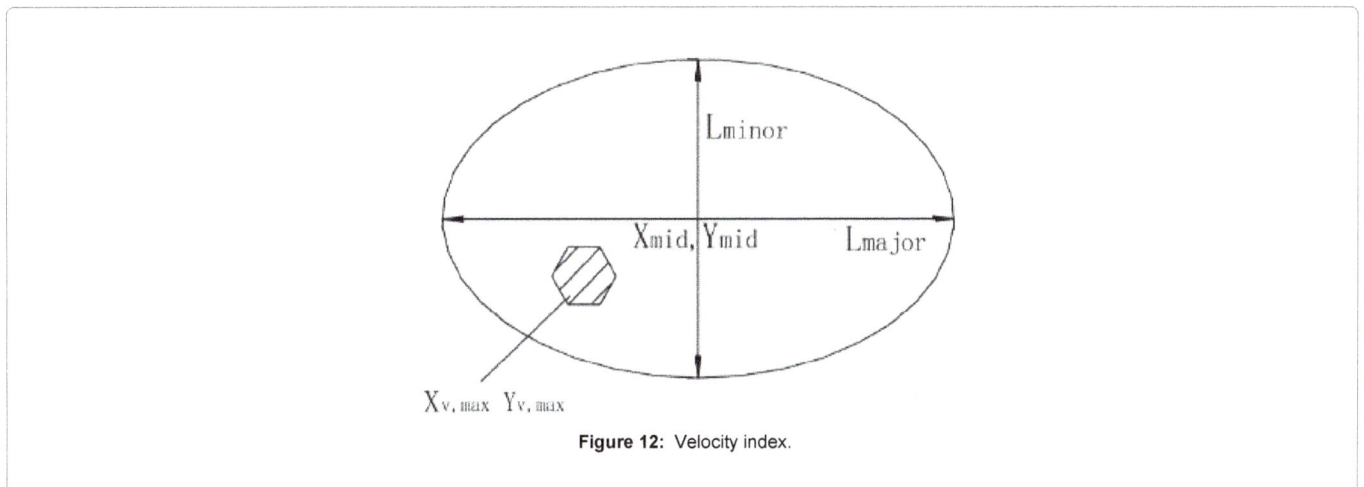

Figure 12: Velocity index.

Parameter	Pre-catalytic converter	Main catalytic converter
Xv, max	408.358	137.256
Xv, mid	392.497	120.36
Yv, max	-109.333	209.776
Yv, mid	-134.719	171.64
L	101.6	101.6
ε	0.589	0.821

Table 6: Velocity index calculation results.

Parameter	Pre-catalytic converter	Main catalytic converter	Total
Pressure drop (KPa)	12.57	9.09	21.65
Fluid uniformity coeffi-cient γ	0.696	0.97	
velocity index ε	0.589	0.821	
The maximum speed	34.92	48.91	

Table 7: Results of comprehensive analysis.

$$\varepsilon_x = \frac{2(x_{v,\max} - x_{mid})}{L_{major}} \tag{1}$$

$$\varepsilon_y = \frac{2(y_{v,\max} - y_{mid})}{L_{minor}} \tag{2}$$

$$\varepsilon = \sqrt{\varepsilon_x^2 + \varepsilon_y^2} \tag{3}$$

Velocity Index calculation results are shown in Table 6.

The velocity index ε of the precatalytic converter is 0.589, which is less than 0.7, means that is in accordance with design requirements. Even if the main catalytic converter velocity index ε (0.821) is more than 0.7, but also in accordance with design requirements, because the main catalytic converter fluid uniformity coefficient γ =0.97, is more than 0.94.

The comprehensive analysis of the exhaust pipe is shown in Table 7.

Conclusion

Take integrated analysis of the above test results, and the following conclusions can be drawn:

(1) The exhaust pipe pressure loss of the catalytic converters is 12.57 kPa, 9.09 kPa, 21.65 kPa, which is in accordance with the design requirements.

(2) The fluid uniformity coefficient of the inlet end face of the catalytic converter is satisfied.

(3) The velocity index of the main and the pre catalytic converter meets the design requirements.

(4) The maximum flow velocity of the pre and main catalytic converter is less than 100 m/s, which meets the design requirements.

(5) The flow velocity of the oxygen sensor is higher and the oxygen sensor place is more reasonable.

Acknowledgement

The authors thank the Science research project of Hebei Province, Youth Science and Technology Fund of agriculture university of Hebei, and Baoding science and technology research and development plan for support.

References

1. Tao J, Maji L (2012) CFD automatic analysis process of the engine ports. Journal of Wuhan University of Technology 34: 310-312.

2. Zhi W, Ronghua H (2002) Research based on CAD/CAM/ CFD for engine port development. Chinese Internal Combustion Engine Engineering 23: 26-29.

3. Kharkov N, Vatin N, Strelets K (2014) Gas dynamics in a counter-flow cyclone with conical nozzles on the exhaust pipe. Applied Mechanics & Materials: 635-637.

4. Gasser I, Rybicki M (2013) Modelling and simulation of gas dynamics in an exhaust pipe. Applied Mathematical Modelling 37: 2747-2764.

5. Usmanova RR, Zaikov GE, Stoyanov OV, Klodzinska E (2013) Research of the mechanism of shock-interial deposition of dits-persed particles from gas flow. Herald of Kazan Technological University 16: 203-207.

6. Usmanova RR, Zaikov GE (2014) The new equipment for modernization of system of clearing of flue gases Herald of Kazan Technological University 17: 246-251.

7. Usmanova RR, Zaikov GE (2014) Experimental researches and calculation of boundary concentration of an irrigating liquid. Herald of Kazan Technological University 17: 183-187.

8. Usmanova RR, Zaikov GE, Zaikov VG (2008) Calculation of dust separation efficiency of new dising dynamic gas washer. Journal of the Balkan Tribological Association 14: 247-251.

9. Panov AK, Usmanova RR, Zaikov VG, Zaikov GE (2007) Complex aero hydrodynamic research and the effectiveness or arresting dispersed particles for barbotage rotation. Journal of Applied Polymer Science 104: 2088-2091.

10. Vatin NI, Chechevickin VN, Chechevickin AV (2011) About sorp-tion-catalytic air cleaning in premises for people habitation in megapolises. Magazine of Civil Engineering 1: 24-27.

11. Wangwenhai W, Cho HM (2014) A Study on the Fluid Dynamic of Catalytic Converter in Exhaust Pipe", Journal of Energy Engi-neering 23: 114-118.

12. Kumar S, Bergada JM (2013)The effect of piston grooves performance in an axial piston pumps via CFD analysis. International Journal of Mechanical Sciences 66: 168-179.

13. Omidbeygi F, Hashemabadi SH (2013) Exact solution and CFD simulation of magnetorheological fluid purely tangential flow within an eccentric annulus. International Journal of Mechanical Sciences 75: 26-33.

14. Li J, Uttarwar RG, Huang Y (2013) CFD-based modeling and design for energy-efficient VOC emission reduction in surface coat-ing systems. Clean Technologies and Environmental Policy 15: 1023-1032.

15. Chica Arrieta EL, Florez SA, Sierra NI (2013) Application of CFD to the design of the runner of a propeller turbine for small hydroelectric power plants. Rev Fac Ing Univ Antioquia: 181-192.

16. Mitiku Y, Ramayya V, Shunki G (2015) Turbine Driven Pump CFD Modelling and Simulation a Centrifugal Pump Optimization forIrrigation. International Journal of Engineering and Technical Research (IJETR) 3: 154-159.

17. Ramos HM, Simao M, Borga A (2012) CFD and experimental study in the optimization of an energy converter for low heads. CSCanada. Energy Science and Technology 4: 1029-1035.

18. Nawaz H, Yuan YS (2013) Thermal Comfort Analysis of a Ship Air-Conditioning System Using Solidworks Flow Simulation. Advanced Materials Research 4: 883-888.

13

Mechanical Behavior of Corroded Protruding Rebars From Unfinished Concrete Structures

Drakakaki Arg[1], Diamantogiannis G[1], Apostolopoulos Ch[1]* and Apostolopoulos Alk[2]

[1]University of Patras, Panepistimioupoli Patron 265 04, Greece
[2]University of Ioannina, Epirus, Greece

Abstract

In the current study, the effects of chloride-induced corrosion on B500c semi-embedded steel bars, both immersed in a salt-spray chamber and on protruding areas of existing structures, are evaluated in terms of mass loss and mechanical characteristics. Comparison of corrosion damage rates between bare and semi-embedded specimens at the early stages of the corrosion test, indicates that the bare steel bars present quite high mass losses, which however, over exposure time reach approximately similar rates to the ones presented by the semi-embedded steel bars. As far as mechanical characteristics are concerned, the mass loss of semi-embedded steel rebars, at the protruding site, is related to the strength and ductility properties drop. At the same time, while the corrosion exposure time is increasing in bare samples, a continuous-almost linear and proportional to the mass loss-reduction of the strength properties and the uniform elongation is observed. Finally, a worth referring point is the appearance of two Ludder areas on stress-strain diagrams of semi-embedded steel bars, a point that confirms the different corrosion mechanism of protruding areas in comparison to the bare or the wholly embedded ones.

Keywords: Mechanical behavior; Semi-embedded steel bars; Rebars

Introduction

As it is widely known, the protruding rebar consist steel bars emerging from the structural components (columns and beams) intended for future use and more specifically for further extension of the frames of reinforced concrete structures. In certain Mediterranean countries, such as Greece or Turkey, it is quite often for the protruding rebar to be left without any protection. This fact, however, often raises speculations concerning the structural integrity of steel, mainly due to the corrosive environmental action.

According to Figure 1, region A consists the rebar part which is emerging from the concrete and since it is bare, inevitably it is exposed to the corroding environment of the atmosphere. The regulatory framework concerning the rebars of region A, is conducted by ISO 12944 [1]. According to ISO 12944 [1], the corrosiveness is directly related to the environmental exposure category, the type of polluting compounds such as SO_4^{2-}, NO_2, Cl^- and other gas phase compounds like H_2S, CO_2, etc. An important parameter on the mechanism of atmospheric corrosion also consists the percentage of relative humidity on the steel surface. In terms of regulatory framework, region B which consists the protruding part of the steel reinforcement bars, is conducted according to EN 206-1 [2] with the reference that concrete thickness, which is coating the steel, is defined zero. The contact of free concrete surface with the atmospheric pollution, as well as the given

penetration of the water into the concrete, makes region B particularly complex. This complexity is directly related to the interaction between chemical and electrochemical corrosion or just with corrosion using atmosphere and polluting deposits as electrolytes or even concrete. Also, a quite important percentage of steel corrosion (region B), may happens due to microbiological induced corrosion (MIC) [3,4], derived from the animal (exudates of microorganisms) and the natural environment (aureobasidium pullulans-cladosporium cladosporioides -alternaria alternata etc).

In existing structures, at steel bar protruding sites, can be quite often noticed an increased corrosion rate and as a result a significant decrease of sectional area, which comes in contrast with low decrement of sectional areas which appeared in bare rebars.

This is the reason why the seismic vulnerability of existing reinforced concrete (RC) buildings in coastal areas should be determined with reliability, even in an approximate manner, using the extent of reinforcement corrosion in the structural elements).

Browsing in the literature, on topics such as the mechanical behavior of steel bars due to corrosion, several studies came of concerning both bare and embedded steel bars. There comes a conclusion, that corrosion causes a reduction in mass of the rebar reinforcement and a change in the mechanical characteristics of the steel (strength and ductility properties) [5,6].

Additionally, Apostolopoulos et al. [7,8] has proved through extended studies that the corrosive action localized on the constructions,

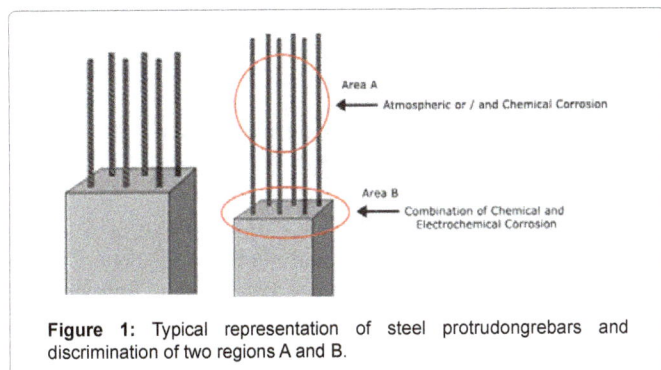

Figure 1: Typical representation of steel protrudongrebars and discrimination of two regions A and B.

*Corresponding author: Apostolopoulos Ch, University of Patras, Panepistimioupoli Patron 265 04, Greece, E-mail: charrisa@mech.upatras.gr

is time dependent and is associated with the type of concrete and reinforcement used and the intensity of environmental exposure to penetrating chlorides.

The initiation and rate of increase of reinforcement corrosion and the parameters that influence corrosion have been described using physicochemical models that take into account the material composition and the geometric characteristics of the RC elements [9].

Several investigations have been published in the literature on the evaluation of the chloride corrosion damage, as well as, the mechanical behavior of steel reinforcing bars.

However, the extent and the consequences of the induced corrosion damage on the protruding steel bars have not been studied to a similar range [10].

The induced damage on the protruding part is associated with a micro crack network development in concrete and thus should be expected to influence the rate of chloride ingress and more aggressive agents as well. The protruding area (concrete- steel) is considered to be a critical region for a future development of a frame structure, however, a little attention is paid to structural preparation and much more is paid on the valuation of the mechanical behavior of exposed steel bars.

To approach the mechanical behavior valuation of protruding found in existing structures, primarily loss measurements concerning the sectional areas of protruding steel bars were demanded. Semi-embedded bars, after a process of tensile tests, were compared to reference samples and they were experimentally simulated using laboratory corrosion.

Measurements in Existing Structures

There took place local measurements of sectional area loss which corresponds to mass loss of protruding steel bars of existing structures, which were categorized to exposure levels XS_2 and XS_3 according to EN 206 standard [2]. Particularly, these are the pedestals of a bridge located an Ionian Street, in place Gavrolimnis, as shown in Figure 2, (near the famous Rio- Antirrio bridge). This structure as well as many others located to the network of Highway of Greece has remained unfinished, because of economic hardship, for the last two years.

Measurements were performed in numerous protruding steel bars exposed to environmental conditions. Exposure rules concerning the bare parts of protruding steel bars are conducted according to ISO 12944 [1] and they enlist the test to the C5-M category, which is characterized by high or very high corrosion rates, which means about 50 μm/200 μm/per year.

On the contrary, the examination of the embedded part of the protruding bars is taking place according to European Standard EN206-1 [2] and is enlisted to XS_2 and XS_3 categories (Table 1).

Figure 2: View of existing pedestals and of a typical protruding.

Classification of region A according to ISO 12944 standard	Classification of region B according to EN 206-1 standard
C2	XA1
C3	XC3,XC4,XA2
C4	XD1,XA2,XC3,XC4
C5-1	XC3,XC4,XA2,XA3
C5-M	XC3, XC4, XS2, XS3, XA2, XA3

Table 1: Environmental classification of protruding steel bars according to specific standards.

Steel bar 1	Steel bar 2	Steel bar 3	Steel bar 4	Steel bar 5	Steel bar 6	Steel bar 7	Steel bar 8	Steel bar 9	Steel bar 10
17.40	17.55	17.67	17.41	17.12	17.29	17.34	17.2	17.24	17.63

Table 2: Results of diameter measurements (mm) on bare part, region A.

To calculate the remaining sectional area corresponding to mass loss on specific heights of the bar, two diameter measurements were taken in two perpendicular directions. The final diameter was calculated as the average of these measurements. Before this procedure, a local cleaning was performed according to ASTM G1 standard [11].

Table 2 presents the results of the average diameter, referred to the free (bare) region A, of 10 protruding steel bars with nominal diameter Φ18. Each measurement consist the average of 15 partial measurements.

Coastal areas, such as in the case of the previously referred pedestals, the constructions environment is characterized by high humidity and salt, which in combination with relatively high temperatures are favoring the chloride ions penetration in concrete. The free chloride ions in concrete consist a fixed percentage of total chloride ions (bound to concrete and free). The ratio of free Cl^- to the total of the individual Cl^-, varies in a range between 0.28 and 0.92 according to each situation. In cases where super-saturation of chemically bound Cl^- is taking place, each addition of new Cl^- increases the percentage of free Cl^- in the solution, situation which is related to the corruption of steel constructions.

Experimental Procedure

Induced corrosion

For further testing of mechanical behavior and for extensive examination of the phenomena, an experimental simulation was performed on the existing protruding steel bars. This was the reason for the examination of a dual-phase, high strength steel, appearing ductility B500C, with the configuration of 10 rebar of 12 mm in diameter, which would consist tensile testing samples and would directly get exposed to a chloride rich environment as well. The same procedure was followed for 18 semi-embedded rebar in prismatic shape concrete (32 mm × 32 mm) too. For each bar, were recorded before testing its total length and mass. Prior to tensile testing, the specimens were separated into three different groups (a,b,c) and were inserted in a laboratory salt-spray exposure chamber, in accordance to the ASTM B117-94 [12] specification (directly exposed to the corrosive medium), for a period of 30, 60 and 90 days respectively, subsistent to 8 cycles wet/dry per day.

The ASTM B117 [12] specification covers every aspect of the apparatus configuration, procedure and conditions required to create and maintain a salt spray (fog) testing environment. The selection of such a procedure for corroding the specimens, relies on the fact that the salt spray environment lies qualitatively closer to the natural coastal (rich in chlorides) conditions than any other accelerated laboratory corrosion test. In principle, the testing apparatus consists of a closed chamber in which a salted solution atomized by means of a nozzle, produces a corrosive environment of dense saline fog. In this particular study a special apparatus, model SF 450 (made by Cand W. Specialist Equipment Ltd.) was used. The salt solution was prepared by dissolving 5 parts by mass of sodium chloride (NaCl) into 95 parts of distilled water (pH range 6.5-7.2). The temperature inside the salt spray chamber was maintained at 35°C (+1.1–1.7)°C.

As far as the semi-embedded in concrete steel bars are concerned, prismatic wooden tubes were used as concrete molds, with dimensions 32 × 32 × 500 (mm). The steel bars were placed in the molds prior to concrete casting and were held in position using specific grips. A CEM IV (according to EN-197 [13]) cement type and crushed sand (and other fine aggregate) were used in the cement mix giving a water/cement (W/C) and aggregate/cement (A/C) ratios of 0.66 and 3.8, respectively.

The tubes were filled with concrete and vibrated for 20 s on a vibration table. After 24 h the molds were stripped and the concrete/steel specimens were washed with fresh water. Following this initial curing period, 18 specimens were placed in the laboratory salt spray exposure chamber, divided into three groups. The groups of specimens remained into the chamber for 30, 60 and 90 days respectively. In Figure 3 is presented a group of specimens after the corrosion process. It should be noted that since the primary aim of this particular study is to evaluate (and correlate) the nature of the corrosion damage on bare and on semi-embedded steel bars, time of exposure is not a factor of comparison. Given the fact that the corrosion rate in concrete is much slower than when the bars are directly exposed to the corrosive medium, corresponding conclusions were expected. By the end of the exposure time the surrounding concrete was crushed and the corroded steel bars were removed for tensile testing. The next step was tensile testing which took place at different time intervals. At each testing date specimens were removed from the salt spray chamber, washed with clean running water to remove any salt deposits from their surfaces and air dried. The corrosion products were removed from the surface of the specimen by means of a brittle brush, according to ASTM G1 specification [11]. The specimens were then weighed and the mass loss due to corrosion exposure was calculated as:

$$Xp = \frac{m_o - m_c}{m_o} * 100\%$$

where m_0 is the mass of un corroded specimens and m_c the reduced mass of the corroded specimen.

The tensile tests were performed according to the ISO/FDIS 15630-

Figure 3: View of corroded rebars.

-50 mm (from the surface)	0 mm (on the surface)	+50 mm (from the surface)
17.55	16.25	16.72
17.48	16.34	16.79
17.42	16.43	17.05
17.33	16.47	16.88
17.26	16.31	16.94

Table 3: Results of diameter measurements (mm) on the protruding area, region B.

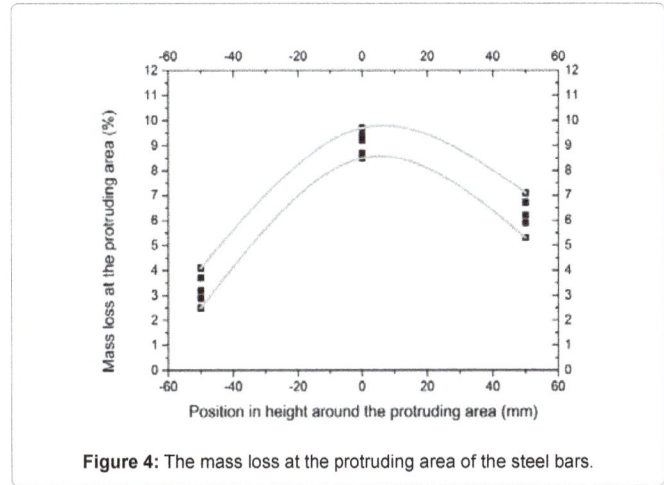

Figure 4: The mass loss at the protruding area of the steel bars.

1 [14] specification, using a servo-hydraulic MTS 250KN machine with a constant elongation rate of 2 mm/min. The mechanical properties, yield strength R_p, ultimate strength R_m, and uniform elongation A_{gt}, were determined. It should be noted that A_{gt} was measured according to the manual method described in the relevant standard (on a gauge length of 100 mm, at a distance of 50 mm away from the fracture).

According to ELOT 1421-1 [15] the chemical composition of B500c steel in maximum by weight permissible values is C=0.24, S=0.055, P=0.055, N=0.014, and Cu=0.85 and the equivalent carbon content Ceq was determined to be 0.52. The material yield stress is ≥ 500 (MPa), the ratio 1.15 ≤ Rm/Ry ≤ 1.35 and elongation at maximum load ≥ 7.5%.

Results and Discussion

Mass loss

As shown in Table 3 the mean diameter of the protruding bars of the pedestals, after a period of about 2 years of exposure to the natural environment (exposure class XS$_2$, XS$_3$ according to EN 206 [2]) is proved to be impaired in three discrete positions in height of the bars. In the free (bare) sections of the bars (region A), the percentage drop is 2.5-4.1%, at the protruding area (region B) the percentage drop is 8.5-9.7% and in depth of 50 mm from the protruding area the percentage drop is about 5.2-7.1%.

Based on the results of the average diameter measurements (Table 3) in reference to their position in height around the protruding area of the bars, the sectional loss/mass loss was calculated and is depicted in Figure 4. These measurements showed a significant cross sectional reduction of the protruding steel bars in the short duration of two years, which directly raises structural integrity issues concerning constructions of high importance, just like the bridge's pedestals.

To control the degradation of semi –embedded steel bars after the laboratory corrosion, mass loss was measured for both groups of rebar (group A-bare, group B-semi embedded). After cleaning, according to ASTM G1 specification [11] and weighting, the percentage mass loss was calculated. Table 3 presents the results of mass loss for the two rebar groups (bare and embedded).

As far as the damage extension that the rebar were sustained is concerned, attention should be paid to the following: The embedded part of the rebar remained almost unaffected in all exposure times, in contrast to the bare part where the damage recorded was visible and obvious, since it functioned as a sacrificial anode (Figure 5). Due to this fact, the calculation of rebar mass loss of group B was performed with reference to the length of the exposed part which suffered (failed) corrosion damage.

Comparing the corrosion damage of the bare and the semi-embedded specimens, it is obvious that at the early stages of group's A corrosion test, have been recorded quite high mass losses, which over exposure time (for 90 days) have balanced by the corresponding group B. A reference to these data is made on Table 3 and Figures 6 and 7.

Mechanical properties

Corrosive effects of chloride ions on reinforcement rebar, lead to local reduction of the cross sectional area and as a result to steel bar's mechanical properties reduction. Figure 8 is related to these conclusions, where are given indicative diagrams for a reference specimen and for the corroded specimens 1a, 4b and 5c of group B. From the results of Table 3, a notable percentage drop was recorded in strength properties (Rp, Rm), since it is almost two times the respective mass loss. As a result, yield strength (Rp) drop is 6.22%, 13% and 21.7% for mass loss about 3.43%, 6.29% and 11.08% respectively. On the contrary, on bare specimen cannot confirmed the same behavior as the percentage drop of yield strength (Rp) is 9.76%,1.63% and 19.26% which are corresponding to an overrun of about 1.50% , 6.70 and 9.47% mass loss respectively. Mass loss for group B samples around 6.29%,

Figure 5: View of the protruding area of a specimen (left),view of the bare area of a specimen (right).

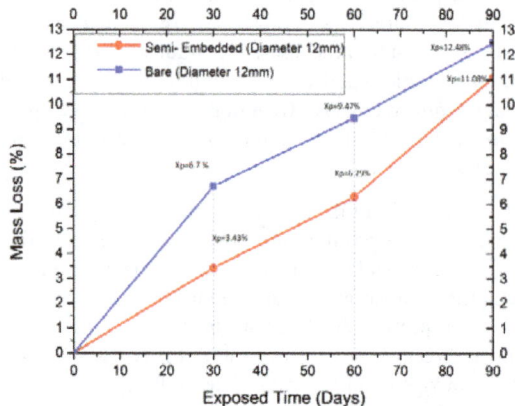

Figure 6: The mass loss rate of the bare and the semi-embedded specimens.

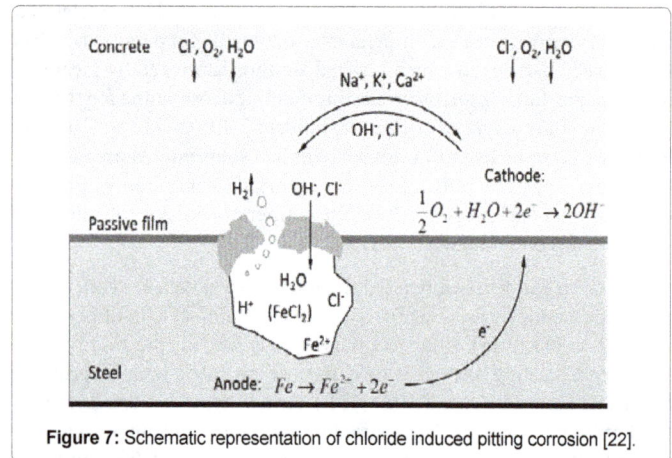

Figure 7: Schematic representation of chloride induced pitting corrosion [22].

gave values of yield strength (Rp) under the threshold of 500 MPa which is defined by EC_2 for class c steels. The fact that mechanical behavior of group B was worse than that of group A, can be partly explained by the local noticeable reduction of the sectional area, at protruding height. In fact, the findings concerning corrosion damage of naturally corroded steel bridge pedestals, were confirmed and are shown in Figure 4. The surface cracking due to corrosion of RC structures has been investigated by other researchers [16-21]. When the production of rust begins, it gradually builds pressure around the reinforcing steel. In this case, at the emergence height of protruding steel bars, this buildup of pressure eventually cracks the concrete around the steel, and the crack or cracks propagate with further increase of pressure. The consequence of this fact is the surface propagation of cracks and the spalling or breaking off initiation of the adjacent concrete. This is the point where free chloride ions penetration begins. At the same time begins an interaction with concrete's $Ca(OH)_2$,which not only results in $CaCl_2$ production (which is more aggressive than NaCl), but also these compounds are in constant touch with the steel bar surface. Such a phenomenon, has as a result a high local steel bar damage, which is disproportionate to the bare part. Also, the presence of elemental sulfur (coming from atmospheric deposits) in a wet sour line can cause very severe localized corrosion of the steel in the sulfur particles come in contact with the steel. Although there is a number of papers in the literature discussing this phenomenon, at these areas can get formed a variety of FeS, FeS_2 and MnS corrosion products.

The local macroscopic damage of the bar is always accompanied by pitting corrosion. The pitting can be displayed under various mechanisms. The mechanism of pitting corrosion of embedded steel bars is shown graphically in Figure 7 [22].

The acute local reduction of the cross sectional area of the naturally corroded protruding steel bars (Table 3) in comparison with the corresponding laboratory tests on protruding areas, consist a major problem and requires further investigation. Nevertheless, it is known that in cases where is obvious the appearance of bacteria or algae and fungi, is implied the presence of Microbiological Induced Corrosion (MIC). Generally, the degree of microbial colonization on the metallic surfaces is disparate, this aids in generating corrosion potentials as large as that generated between incompatible metals [3,4]. But given the strong local vegetation and generally the microclimate due to standing water in the area of bridge pedestals, the worst steel bar mass loss may be attributed to the coexistence with Microbiological Induced Corrosion (MIC).

In Figure 8 is shown the uniform elongation percentage drop, which reflects the mean value between the two distinct parts of the semi-embedded specimens, since only one of the two parts is corroded. However, the uniform elongation measurement results in each distinct part of the whole steel bar depict two completely different mechanical behaviors.

A worthy of attention point in this Figure 9, is the behavior of the embedded steel bars which are exposed to salt spray environment until 60 days, which present two Ludder regions, something that implies high deformations. . It is thus a single material with two yield points. This fact introduces disfavor issues concerning the response of structural elements made of reinforced concrete. Moreover, the appearance of two Ludders regions in the [σ-ε] diagrams of the steel protruding introduces important issues for different regulations concerning structures design. In this sense, the steel is calculated (obtained) with a maximum deformation of 0.2% in its initial state of yield point and respectively the concrete with a maximum deformation of 0.35% before the fracture. Given that a reliable estimate of any construction of reinforced concrete is subject to meeting these constraints, a question of credibility is raised at least for the "batch" of steel tested in this study.

On Table 4, is presented the mean value of the maximum plastic elongation that was measured on the bare part of the semi-embedded steel bars in reference to the corrosion level. Thus, the corresponding measurement taken from a specific part, with recorded corrosion

Rebar type B500c	Days in salt spray chamber	Mass Loss (%)	Ultimate Strength Rm (MPa)	Yield Strength Rp (MPa)	Uniform Elongation Ag (%)	Energy Density U (MPa)
Bare	0	0	654.13	561.43	9.36	58.63
Semi-embedded	0	0	654	562	9.3	54.84
Bare	30	6.69	595.62	506.61	7.14	38.65
Semi-embedded	30	3.43	622	527	6.36	37.91
Bare	60	9.47	572.78	490.53	6.08	34.17
Semi-embedded	60	6.29	580	489	4.67	27.08
Bare	90	12.48	530.99	453.29	4.85	25,17
Semi-embedded	90	11.08	520	440	2.78	15.65

Table 4: Average mechanical properties for bare and semi-embedded steel bars in corrosive environment for different exposure times.

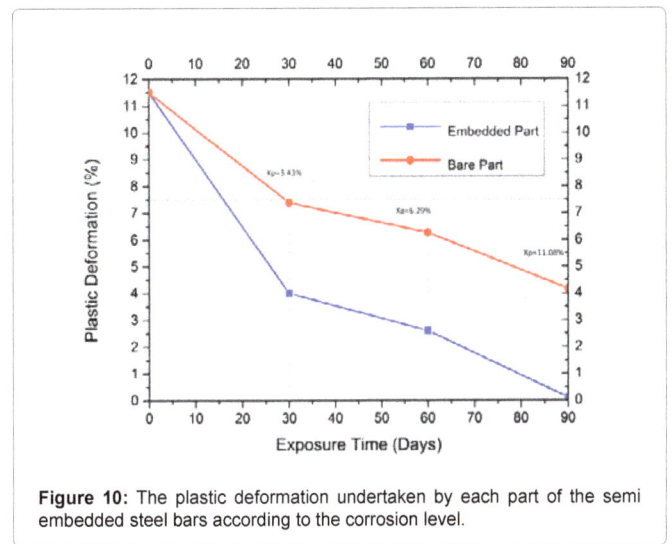

Figure 10: The plastic deformation undertaken by each part of the semi embedded steel bars according to the corrosion level.

about 3.43%, noted maximum plastic elongation 7.38%, while the non-corroded part noted 3.99% maximum plastic elongation. Respectively, for mass loss 6.29%, the max plastic elongation which was recorded was 6.27% and for the non-corroded part 2.60%. At the same time, for mass loss about 11.08%, the maximum plastic elongation which was taken was 4.17% while the non-corroded part had not received any plastic deformation (Figure 10). For mass loss 3.43%, 6.29%, 11.08%, the percentage drop of maximum plastic elongation was found to be 21.23%, 33.16% and 55.54% respectively.

In any case, the values of uniform elongation are found to be quite low in comparison with the threshold set by EC2 for high ductility steels class c, which is defined about 7.50%.

From both Table 4 and Figures 6-8 it is obvious that the corrosive agent, for both groups A and B, is time dependent and has a direct impact on the steel mechanical properties.

Tensile testing of these samples gave a series of strength and ductility properties. However, the observed local discontinuities in the stress-strain curves of the corroded semi- embedded steel bars can be attached to any local detachments in the interface of martensite and transition zone. Related to this point, is the Apostolopoulos-Diamantogiannis JAME [23] manuscript which refers to the same dual-phase steel B500c. Also, the existence of internal micro voids, which were formed near the fracture surfaces, in combination with surface pits (because of chloride ions) may be responsible for their fast

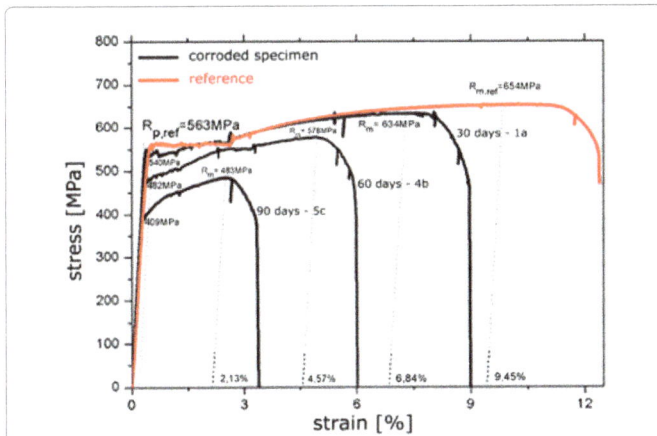

Figure 8: Tensile graphs for semi-embedded steel bars B500c.

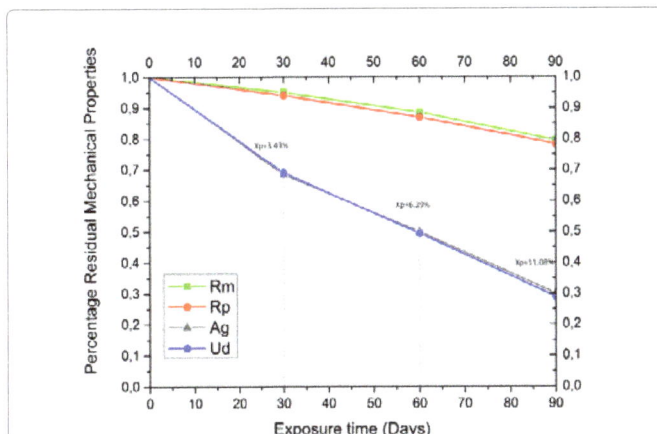

Figure 9: Percentage residual mechanical properties in reference to the percentage mass loss for semi-embedded steel bars.

coalescence and the specimens' failure at low deformation values.

In Figure 9 is depicted the dramatic drop of semi-embedded steel bar ductility properties in reference to the percentage mass loss. In all cases, with the increase of the corrosion exposure time, a continuous almost linear decrease in the yield and tensile strength and the uniform elongation was observed. The mass loss of corroded reinforcing steel bars is directly related with their mechanical strength properties. Additionally, the fact that the corrosion level had a great impact on the degradation rate of mechanical properties was expected since the pits and notches, which were generated due to corrosion, are especially high and determinative for the creation of stress concentration points (Figure 10).

Although it is widely known that chloride induced reinforcement corrosion is one type of highly localized (pitting) corrosion and corrosion experiments of steel embedded in concrete often aim at determining the chloride threshold values (C_{th}), in this case of protruding steel bars the damage mechanism in emergence area is obviously different. Moreover, the protruding corrosion damage focuses on a small area of the bar, which should arise reflection and concern as these areas may have unpredictable and unfavorable behavior in case of strong seismic events.

This investigation was carried out to determine the influences of the reinforcement corrosion level on the mechanical behavior of steels and in particular the effect of corrosion on existing protruding steel bars. Investigation results may help field engineers in making decisions concerning the use of corroded bars in construction. Accordingly, in order for the conclusions of the present study to get used for the seismic analysis of RC structures suffering from corrosion of reinforcement we need to understand how corrosion of reinforcement changes the constitutive behavior of reinforcing steel under monotonic and cyclic loading.

Conclusions

The following conclusions can be drawn from the present study:

- Mass loss measurements were taken from protruding area of B500c dual phase rebar after natural and laboratory accelerated corrosion due to chloride ions.

- Combination of ex situ characterization and in situ locally probing is very powerful for the study of localized corrosion of the reinforcement steel.

- The mechanical behavior (tensile tests) study of both laboratories corroded bare and semi-embedded steel bars was useful for a further understanding of the localized corrosion behavior of protruding rebar.

- Protruding steel bars with an average mass loss at about 4% of a reference bar showed Yield strength and Uniform Elongation below the limits that are set by the EC2 for high ductility steels.

As it was proved by the comparison of the mass loss results of the bare and the semi-embedded specimens, the corrosion mechanism differs for the two categories. The confirmation is given by tensile tests diagrams [σ-ε] which depict the appearance of two Ludders areas on semi-embedded steel bars, something that needs further investigation.

References

1. ISO 12944: International standard for paints and varnishes. Corrosion protection of steel structures by protective paint systems.

2. EN 206-1: Concrete, Part 1: Specification, performance production and conformity

3. Little B, Wagner P, Mansfeld M (1991) Microbiologically influenced corrosion of metals and alloys. Int Mater Rev 36: 253-272.

4. Videla HA, Herrera LK (2005) Microbiologically influenced corrosion: looking to the future. Int Microbiol 8: 169-180.

5. Almusallam AA (2001) Effect of degree of corrosion on the properties of reinforcing steel bars. Construc. Build Mater 15: 361-8.

6. Apostolopoulos Ch Alk, Apostolopoulos (2009) The influence of corrosion and Cross-Section Diameter on the Mechanical Properties of B500c steel, Journal of Materials Engineering and Performance 18: 190-195.

7. Apostolopoulos CH, Kappatos V (2013) Tensile properties of corroded embedded Steel Bars B500c in concrete. International Journal of Structural Integrity 4: 275-294.

8. Kappatos V, Apostolopoulos CH (2013) Tensile Mechanical properties of Reinforced Steel Bars B500c in coastal structures. Journal in Materials Evaluation 71.

9. Papadakis VG, Fardis MN, Vayenas CG (1996) Physicochemical processes and mathematical modeling of concrete chlorination, Chemical Engineering Science 51: 505-513.

10. Chung- Ho H (2014) Effects of Rust and Scale of Reinforcing Bars on the Bond Performance of Reinforcement Concrete, Journal of Materials in Civil Engineering 26: 576-584.

11. ASTM Standard G (2011) Standard practice for preparing, cleaning and evaluating corrosion test specimens ASTM International.

12. ASTM Standard B 117: Standard practice for operating salt (fog) testing apparatus.

13. EN 197 (2000) European Standard, Cement, Composition, specifications and conformity criteria for common cements.

14. UNI EN ISO 15630-1:2010, Steel for the reinforcement and pre-stressing of concrete. Test methods. Part 1: Reinforced bars, wire rod and wire. Eur. Com Stand.

15. ELOT 1421-1: Hellenic Standard (Draft), Steel for the reinforcing of concrete. Weldable Reinforcing steel

16. Vidal T, Castel A, Francois R (2004) Analyzing crack width to predict corrosion in reinforced concrete. Cement and Concrete Research 34: 165-174.

17. Arya C, Ofori-Darko FK (1996) Influence of crack frequency on reinforcement corrosion in concrete. Cement and concrete composites 26: 345-353.

18. Maaddawy ET, Soudki K, Topper T (2005) Long-term Performance of Corrosion- Damaged Reinforced Concrete Beams. ACI Structural Journal 102: 649.

19. Mohammed TU (2001) Effect of Crack Width and Bar Types on Corrosion of Steel in Concrete. Journal of Materials in Civil Engineering 13: 194-201.

20. Francois R, Arliguie G (1999) Effect of micro cracking and cracking on the development 72 of corrosion in reinforced concrete members. Magazine of Concrete Research 51: 143-150.

21. Schiebl P, Raupach M (1997) Laboratory studies and calculations on the influence of crack width on chloride-induced corrosion of steel in concrete. ACI Materials Journal 94: 56-62.

22. Nelson S (2013) Chloride induced Corrosion of Reinforcement Steel in Concrete. Threshold values and Ion Distributions at the Concrete-Steel Interface, Thesis for the Degree Of Doctor of Philosophy, Department of Civil and Environmental Engineering, Chalmers University of Technology, Gothenburg, Sweden.

23. Apostolopoulos CH, Diamantogiannis G (2012) Structural Integrity Problems in Dual-Phase High Ductility Steel Bar. Journal of Applied Mechanical Engineering 1(5).

14

Manageable Reactor Pressure Vessel Materials Control Surveillance Programme-Flexible and Adaptable to Innovations

Krasikov E*

National Research Centre, Kurchatov Institute, 123182, Moscow, Russia

Abstract

As a main barrier against radioactivity outlet reactor pressure vessel (RPV) is a key component in terms of safety and extended light water reactor (LWR) life. The surveillance programme (SP) calls upon to predict ahead RPV materials characteristics conservatively to guarantee RPV structural integrity without any compromise. General vice of existing SPs is an impossibility of SP changing and development during reactor operation (30, 60 and even more years). Up to day, approach based on initial hard nomenclature of surveillance specimens installed in capsules. Therefore, practically it is impossible to change anything in SP during RPV service life. Anachronistic principle of ahead of time, for some decades of years in advance fabrication and installation into reactor vessel the sets of surveillance specimens (SS) contradicts to request of RPV innovative monitoring technologies development during long-term operation.

Besides there is a deficiency of SP portliness relative to conditions of the RPV irradiation during operation. Most important is the discrepancy of the actual thermal condition of RPV wall from SSs irradiation temperature. This fact carries in the element of non-conservatism into the system of control. Ideally, surveillance metal has to be irradiated in contact with coolant. Metal placement in perforated capsules that is immediately in running water provides the minimum irradiation temperature and therefore guarantees the most conservative data on RPV metal mechanical properties getting. Clearly, that at this case there is no need in temperature monitors. Moreover, today there is no hard confidence in SS capsules integrity during RPV operation. In the event of capsule depressurization SSs damage occurs. At the same time in reality it is impossible to exclude environmentally assisted cracking of the primary circuit stainless steel components during 60 and more years of operation. Surveillance metal contacting with water in perforated capsules emulate RPV metal-water corrosion reaction appearance as a result of possible cladding cracking and hydrogen (as a corrosion product) - metal interaction. Therefore for materials susceptible to hydrogen embrittlement, the degree of SP conservatism grows.

We suggest to improve LWR SPs by means of passage from existing «hard» SPs to «flexible» manageable SPs (MSP) that would give the possibility of SP adaptation to requirements of time and to strengthen technical and scientific potential of investigators and researchers in the future. So, we believe that is no sense to leave present-day level of knowledge and technology in congeal state to next generation of researchers. Thus for new LWRs with the service life of 60 and more years we propose pass on from the SSs of routine nomenclature to MSP i.e. sets of archive materials coupons placed in non-hermetic containers and cooled directly by running water. It gives a perspective in case of need put into practice an innovative MSP taking into account the state-of-the-art safety standards, technical progress, present day level of science and technology. In support of the above-mentioned MSP conception 5 year duration prototype version of the MSP is under execution at operating commercial LWR.

Keywords: RPV materials; Manageable surveillance programme; Innovations

Introduction

Modern nuclear power engineering is based on LWR type plant reactors. As a main barrier against radioactivity outlet reactor pressure vessel (RPV) is a key component in terms of safety and LWR plant life extension when needed. The surveillance programme (SP) calls upon to predict ahead RPV materials characteristics conservatively to guarantee RPV structural integrity without any compromise. General vice of existing SPs is an impossibility of SP changing and development during reactor operation (30, 60 and even more years). Up to day, approach based on initial hard nomenclature of surveillance specimens installed in capsules. Therefore, practically it is impossible to change anything in SP during RPV service life. Anachronistic principle of ahead of time, for some decades of years in advance fabrication and installation into reactor vessel the sets of surveillance specimens (SS) without taking into account quantitative and qualitative changes of norms; state of the present-day science, testing methods and technique contradict to request of RPV operational monitoring technologies innovative development during long-term LWR operation.

LWRs surveillance programme improvement actuality

It is necessary to recognize that there is a deficiency of routine SP adequacy to real conditions of the RPV operation. The most important item is the discrepancy of the actual thermal condition of RPV wall from SSs irradiation temperature. At any case because of γ-heating, SSs irradiation temperature exceeds the real RPV temperature. This fact carries in the element of non-conservatism into the system of control. Moreover, because of specimen-to-specimen clearance temperature gradients through the SS exist. Ideally, surveillance metal has to be irradiated in contact with coolant. Archive metal blocks placement

---

*Corresponding author: Krasikov E, National Research Centre, Kurchatov Institute, 123182, Moscow, Russia, E-mail: ekrasikov@mail.ru

immediately in running water (in perforated capsules) would provide the minimum irradiation temperature and therefore would guarantee the most conservative data on mechanical properties getting. Clearly, that at this case there is no need in temperature monitors.

The second reason is that inasmuch as there is no hard confidence in SS capsules integrity during RPV operation (capsules depressurization can take place) the idea made sense to put archive metal billet in coolant beforehand. To solve the problem of metal corrosion archive metal billets (instead finished specimens) for surveillance irradiation are proposed. It means that test specimens have to be machined after irradiation and immediately before testing.

In reality, it is impossible to exclude environmentally assisted cracking of the primary circuit stainless steel components during, for instance, 60 years of operation. Surveillance metal contacting with water in perforated capsules emulate base metal-water corrosion reaction appearance as a result of possible RPV clad cracking and hydrogen (as a corrosion product)-RPV metal interaction. By this means for materials susceptible to hydrogen embrittlement the degree of SP conservatism grows.

Evaluation of the SSs testing long-term practice and experience allows proposing the new conception of RPV metal control by means of passage from existing «hard» SPs to «flexible» adaptable, «open» SPs. This approach would give the possibility of SP adaptation to requirements of time and to strengthen technical and scientific potential of investigators and researchers in the future.

Thus for new LWRs with the service life of 60 and more years we propose pass on from SPs, that are based on SSs of routine nomenclature to manageable SP (MSP), which will be based on sets of archive material billets placed in non-hermetic containers inside the RPV and will cooled directly by running water.

It clears the way to a perspective in case of need put into practice an innovative MSP of anyone content and complexity, taking into account state-of-the-art of the safety standards, technical progress, level of science and technology. Consequently, we believe that is no sense to leave present-day level of knowledge and technology in congeal state to next generation of researchers.

Prerequisites for going to manageable surveillance programmes

Routine SPs are characterized by high laboriousness (Figure 1) [1] because call for precious rigging, containers pressurizing and tightening, necessity of the SSs temperature control, in case of depressurizing or temperature exceeding the SSs may be lost. In fact, we propose going to adaptable, «open» SPs that in potentiality allow the actualization and specialization of SPs and SSs types. These manageable SP will be based on sets of archive material billets placed inside the RPV closely to wall and will cooled directly by primary circuit water. It clears the way to a perspective in case of need put into practice an innovative MSP of anyone content and complexity, taking into account state-of-the-art of the safety standards, technical progress, level of science and technology. Certainly MSPs development and application have to be based on disposable similar experience understanding and utilization. Let remember it.

It is known [2,3] that for the first generation of the Russian PWRs (WWERs) instead of the cancelled SPs just RPV (100% surveillance material) serve as billet for thin plates cutting and test specimens manufacturing as needed. As a matter, this practice is the first prerequisite of the proposed SP technology.

The second prerequisite is a worldwide experience on the through wall probes (trepans) of the ex-service RPVs using for actual metal properties examination [4-12].

The third prerequisite is our own long-term practice in SSs testing and experience in decommissioned LWR pressure vessel material properties study [13]. Recently for the first time in the history of the RPV materials study set of the 1T-CT type specimens for fracture mechanics tests was produced from 140 mm in diameter EPR RPV trepan. Figure 2 shows the steps of 1T-CT manufacturing and testing. Encouraging results are obtained and analyzed now.

In a certain sense, proposed MSP procedure (technology) is the closest analogy to trepans investigation with the exception surveillance billets (SB) in advance should be placed inside RPV and ready for examination in case of need without extra complex RPV cutting. SBs placement inside the RPV as close as possible to RPV wall should be the best decision in SP performance from all points of view. In the upshot, one can say that the scientific and technological prerequisites to LWRs surveillance programme improvement by means of going to manageable SPs (MSP) exist.

In support of the idea, experimental elaboration of the MSP prototype version is under development. Placed in the stainless steel perforated capsules Figure 3 RPV Cr-Ni-Mo steel (base and weld

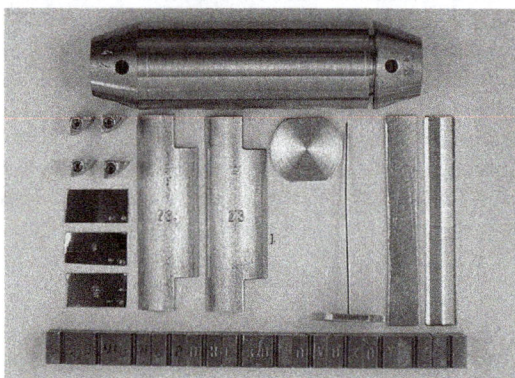

Figure 1: Set of modern SS capsule internals [1].

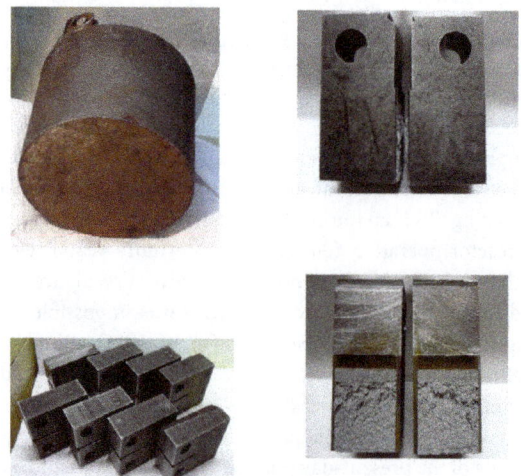

Figure 2: Steps of 1T-CT type specimens from 140 mm in diameter EPR RPV trepanmanufacturing (left side) and testing.

metal, Table 1) billets of the cylindrical shape are under irradiation in WWER-440/213 SS channels immediately in running water. Sketch of the full size and sub size Charpy specimens manufacturing from irradiated billets by means of electro discharge machining (EDM) is depicted in Figure 4.

New experimental results in the routine form of the transition temperature shift (TTS)-fast (E>0,5 MeV) neutron fluence (FNF) dependence are represented in Figure 5 (crosses and diamond) against a background of the disposable data [14]. It is seen that new data are in a good agreement with «old» data that were received during numerous experiments in commercial and test reactors earlier.

As an example of the MSP potentialities, experiment on so-called «wet» annealing effectiveness of the reactor vessel was conducted. Pre-irradiated in WWER-440/213 SS channels immediately in running water up to 9×10^{19} cm^{-2} at 270°C base metal Table 1 billets of the cylindrical shape were additionally irradiated in test reactor IR-8 at 330°C and neutron flux level of 3×10^{11} sm^{-2}s^{-1} during 87 hours. Figure 6

Figure 6: Experimental device for RPV steel billets irradiation. (1 and 2 – RPV metal billets, 3 - capsule for neutron monitors, 4 – heaters).

Figure 3: Set of the perforated capsules with cylindrical billets inside.

Material	C	Si	Mn	P	Cu	Cr	Mo	Ni
Base	0,14	0,34	0,59	0,009	0,08	2,00	0,90	1,15
Weld	0,04	0,45	0,76	0,005	0,02	1,46	0,65	1,26

Table 1: RPV materials chemical composition [wt. %].

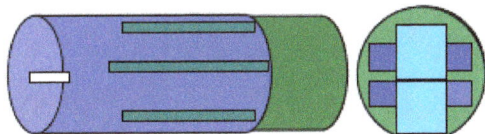

Figure 4: Sketch of the specimens by EDM manufacturing.

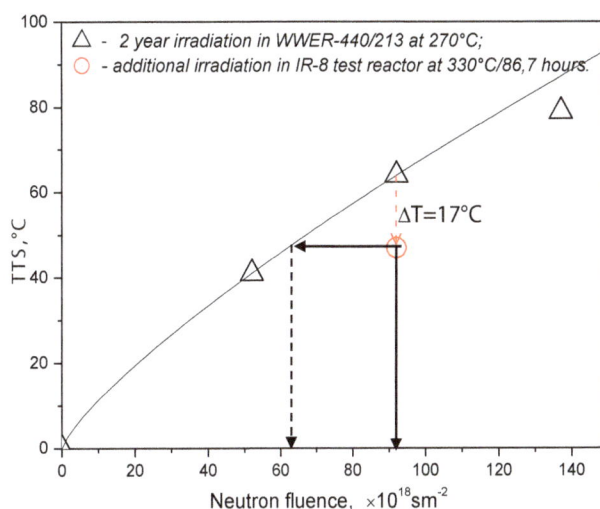

Figure 7: Results of the experiment on potential effectiveness of the RPV «wet» annealing.

shows the experimental billets (pos.1, 2), arrangement and irradiation device. One can see and understand that simple forms of the billets and device components allow providing the possibility of operative and inexpensive irradiation process. As it is seen from Figure 7, where experimental results are demonstrated, 17°C recovery of the TTS take place. This value is equivalent to 1,5-fold neutron fluence reduction and therefore «wet» annealing technology has evident practical benefit.

Conclusion

Development of the new SP conception based on substitution of the surveillance specimens irradiation in sealed capsules by the surveillance billets irradiation in perforated containers with following test specimens manufacturing allows:

1. To strengthen the contribution of surveillance investigations to improve the safety and performance of LWRs;

2. To increase the level of LWR type safety on account of more adequate conditions of the surveillance metal irradiation;

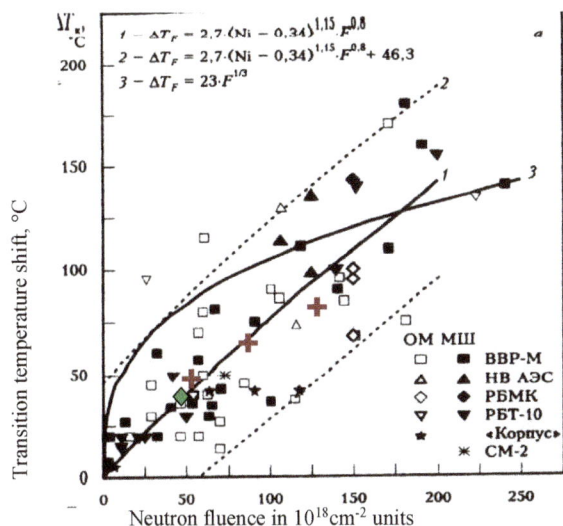

Figure 5: Comparison of the measured Charpybase ( ) and weld ( ) metal TTS - fast (E≥ 0.5 MeV), Neutroninfluence dependence with disposable results from irradiation experiments produced in power and test reactors.

3. To improve the informativeness owing to carrying over the specimens of actual nomenclature manufacturing process immediately to moment of testing from initial stage of RPV producing;

4. To decrease the laboriousness and specific quantity of rigging metal for surveillance metal irradiation and to reduce the quantity of radioactive wastes;

5. To release funds and resources, to reduce the cost of the joint RPV metal surveillance programme execution;

6. To make better LWR's competitiveness.

References

1. Kupka L (2003) Irradiation embrittlement monitoring programmes in Slovak Republic. Topical information meeting in prediction of irradiation damage effects on reactor components. Brussels 1: 433-441.

2. Ya I (1998) Shtrombach assessment of irradiation response of WWER-440 welds using samples taken from Novovoronezh unit 3 and 4 reactor pressure vessels. Nucl.Eng.Des.185 309-317.

3. Ya I (2000) Shtrombach properties of WWER-440 type reactor pressure vessel steels cut out from operated units. Nucl.Eng.Des.195 137-142.

4. Kussmaul K (1989) Assurance of the pressure vessel integrity with respect to irradiation embrittlement: activities in the federal republic of Germany. ASTM STP 10: 3-26.

5. Suzuki M (1994) Investigation on irradiation embrittlement of reactor pressure vessel steel using decommissioned technology for lifetime management of nuclear power plants specialists meeting organized by the IAEA Tokyo 14-17 Proceedings.

6. Iskander SK, Nanstad RK (1997) JPDR Vessel steel examinations. Heavy-section steel irradiation program. NUREG/CR-5591 ORNL/TM-11568.

7. Fabry A, Van Walle E, Gerard R (1994) Enhancing the surveillance of LWR steel components. Technology for lifetime management of nuclear power plants specialists meeting organized by the IAEA Tokyo Proceedings.

8. Godinn R, Kudriavtsev B, Chernykh L (1994) Manned journey to the center of a reactor. Nuclear Engineering International 39:18-20.

9. Curry A, Clyton R (1997) Remote through-wall sampling of the trawsfynydd reactor pressure vessel: An overview. Nuclear energy 36: 59-64.

10. Bischer PJE (1999) Microstructural examination of irradiated reactor pressure vessel welds samples. IAEA Meeting on Irradiation Embrittlement and Mitigation Madrid Spain Proceedings 77: 356-373.

11. Stoller E, Nanstad RK (2002) A proposal for sampling the Songs-1 reactor pressure vessel ORNL/NRC/LTR-02/12.

12. Brillaud C (2001) Vessel investigation programme of chooze A PWR reactor after shutdown. Effects of Radiation on Materials: 20th Intern. Symposium ASTM STP 1405.

13. Amaev A (2010) Decommissioned LWR pressure vessel material properties study. International symposium contribution of materials investigations to improve the safety and performance of LWRs-Fontevraud 7. 26-30 September Avignon France. Paper A041-T01.

14. Morozov AM (2002) In Reactor materials under irradiation behavior and structural strength CRISM Prometey publication Saint Petersburg 200-211.

15

Application Framework for Aero-based Design Optimization of Passenger Cars using NURBS

Ghani AO, Agelin-chaab M* and Barari A

Department of Automotive, Mechanical and Manufacturing Engineering, University of Ontario Institute of Technology, Oshawa, Canada

Abstract

This paper presents a new application framework for aerodynamics-based shape optimization of passenger cars. The rear geometry of a passenger car is the focus of this study due to its significant influence on the aerodynamic characteristics of vehicle. The rear body of a generic car model (the Ahmed body) was represented by Non-Uniform Rational B-Spline (NURBS) curve and NURBS parameters were employed for geometric parameterization. These geometric parameters were systematically modified to alter the geometry using the model developed through a design of experiments process. Computational Fluid Dynamics (CFD) simulations were performed on these geometries to obtain drag coefficients. A polynomial response surface model of drag coefficient was then constructed using linear regression to relate design parameters to the drag coefficient. This response surface model was then used as a starting point for the optimization process. The proposed framework was implemented on a generic notch back car model and the optimized geometric parameters for minimum drag were obtained.

Keywords: CFD; Design optimization; Design of experiments; Response surface modelling; Non-Uniform Rational B-Spline (NURBS)

Introduction

In recent years, the improvement of fuel efficiency has become a major factor in passenger car development due to increasing population, global decline in fossil fuel reserves, rising fuel prices and the damaging effects of global warming. The aerodynamic drag of passenger cars is responsible for a large part of a vehicle's fuel consumption and can contribute to as much as 50% of the total vehicle fuel consumption at highway speeds [1]. Reducing the aerodynamic drag offers an inexpensive solution to improve fuel efficiency and therefore shape optimization for low drag has become an essential part of the overall vehicle design process [2]. Although wind tunnels can provide most realistic data when the test condition are close to actual road condition, the large number of design variables and geometric configurations involved at the conceptual stage of vehicle design make wind tunnel experiments very expensive and time consuming. The availability of high performance computers and relatively accurate turbulence models has led to increased use of CFD in the development of passenger vehicles.

An important aspect of shape optimization through CFD is the parameterization of the model geometry. A common method of parameterization for automotive bodies is the use of geometric parameters such as back light angle (α), boat tail angle (β) and diffuser angle (γ) as shown in Figure 1 [3-5]. Another method is

shape modification by displacing particular edges (R_1, R_2 & R_3) on the body in the desired direction as shown in Figure 2 [6]. These parameterization techniques can be implemented in all modern parametric Computer Aided Design (CAD) systems. The drawback of using these parameterization techniques is that only simple shapes with small changes in geometry can be studied. In the present work, the generic notch back model was parameterized using parameters of NURBS curves. The advantage of using NURBS is that it provides a single mathematical formulation to represent a variety of shapes including free form curves and surfaces [7].

The process usually employed for aerodynamic shape design can be either direct or indirect shape optimization [6]. In the direct shape optimization approach, the process starts with random combinations of design parameters. An optimization algorithm is used which

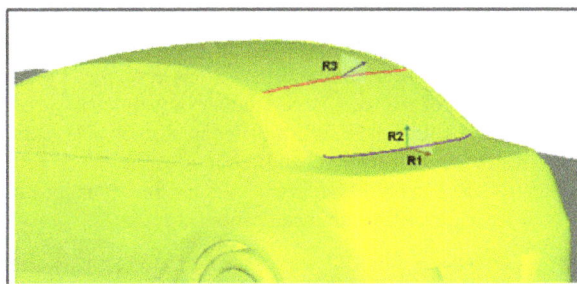

Figure 2: Parametric geometry using edge displacement [6].

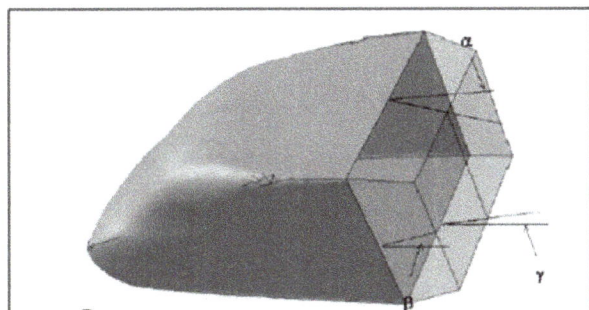

Figure 1: Parametric geometry using simple geometric parameters [5].

*Corresponding author: Agelin-Chaab M, Department of Automotive, Mechanical and Manufacturing Engineering, Oshawa, Canada
E-mail: martin.agelin-chaab@uoit.ca

requires CFD simulation at each iteration to find parameters in the design space for minimum drag [4]. This approach requires a large number of CFD simulations and takes significant amount of time to complete the optimization process. On the other hand, in the indirect approach, a design of experiments method is used to obtain geometries from combinations of design parameters and response surface function is built which describes the aerodynamic behaviour of the entire design space. In this study, the indirect shape optimization technique was employed. Linear regression was used to obtain a response surface models that relate the aerodynamic drag coefficient to the NURBS parameters. This response surface model was then used for shape optimization.

Background

Mathematical modelling of geometry

The NURBS curves and surfaces have been used extensively in the aerospace industry for parameterizing complex surfaces of wings and fuselages. The NURBS is also the industry standard tool for representing curves and surfaces in CAD, Computer Aided Manufacturing (CAM), and Computer Graphics. Moreover, NURBS is also used for representing curves and surfaces in Initial Graphics Exchange Specification (IGES) which is one of the standard formats to exchange design information between CAD and CAM systems.

Samareh [8] proposed a free form deformation technique for aerodynamic shape optimization using the NURBS. The optimization was performed on a fuselage of an air plane using the aerodynamic drag coefficient as the objective function. The NURBS parameters changed in their study were the NURBS

control points. The knot vector and the weights of the control points were kept constant. Lepine [9] and Bentamy [10] also performed shape optimization of air foil using NURBS. The design variables in their studies were control points and weights. Lepine [9] showed that a large number of complex airfoil shapes could be represented using only 13 control points. It was shown that NURBS minimizes the number of design variables and provides smooth profiles. Thus, the main advantage of using NURBS is that free form geometrical shapes can be generated with very few design variables. However, the drawback of using NURBS control points as design variables is the difficulty in changing the relative position of control points, which only allows for the control points to be changed in a small range [11]. In the present study, only the weights of the control points were used for geometric parameterization and it was observed that by careful placement of control points, a large number of geometric variations can be generated. Although NURBS are used in modern CAD software tools for creating free form curves and surfaces, to the best of author's knowledge, NURBS have never been used for aerodynamics-based automotive body geometric parameterization.

Mathematically, a NURBS curve $C(u)$ of degree p is defined Samareh:

$$C(u) = \frac{\sum_{i=0}^{n} N_{i,p}(u) w_i P_i}{\sum_{i=0}^{n} N_{i,p}(u) w_i} \tag{1}$$

where $N_{i,p}(u)$ is the B-spline basis function given by:

$$N_{i,0}(u) = \begin{cases} 1 & if\ u_i \leq u \leq u_{i+1} \\ 0 & otherwise \end{cases} \tag{2}$$

$$N_{i,p}(u) = \frac{u - u_i}{u_{i+p} - u_i} N_{i,p-1}(u) + \frac{u_{i+p+1} - u}{u_{i+p+1} - u_{i+1}} N_{i+1,p-1}(u) \tag{3}$$

Equation 3 can result in $0/0$; which is defined to be zero. The breakpoints of the B-spline are defined by knots and the sequence of knots called a knot vector. There are two fundamental types of knot vectors: clamped and unclamped, which can be either uniform or non-uniform. In uniform knot vector, the individual knots are evenly spaced, whereas non-uniform knot vector may have unequally spaced or multiple internal knots. In clamped knot vector, the knot at the ends has a multiplicity equal to $p+1$, which is of the form:

$$U = \left[\underbrace{a,.....a}_{p+1}\ u_p + 1,.....................u_{m-p+1}, \underbrace{b,.....b}_{p+1} \right] \tag{4}$$

The knot vector U consists of $m+1$ elements where m is calculated from:

$$m = n + p + 1 \tag{5}$$

In Equation 4, the first (a) and last (b) elements are repeated $p+1$ times and are usually set equal to 0 and 1, respectively.

In Equation 1, when the weights of all control points are equal to 1, the resulting curve is a B-spline. The weight of the control point defines how much that control point "attracts" the curve towards itself relative to other control points. Figure 3 shows the effect of changing the weight ($h3$) of the control point ($B3$). It can be seen that by modifying the weight of just one control point several different curves can be obtained. This feature of NURBS curves was exploited in this study to generate free form curves that represent the rear geometry of the passenger car. Moreover, NURBS weights were used as design parameters to obtain parametric geometry which was used for shape optimization.

In the present study, a degree 3 NURBS curve with uniform spacing between knots was used. Thus the knot vector for 10 control points using Equations 4 and 5 is:

$U = [0\ 0\ 0\ 0\ 0.1429\ 0.2857\ 0.4286$

$0.5715\ 0.7143\ 0.8571\ 1\ 1\ 1\ 1]$

Response surface modeling

Response surface methodology (RSM) is a set of mathematical and statistical techniques used to develop adequate functional relationship between an objective function $y(x)$ and the control or design variables x_1, x_2, x_k [12]. A response surface is a smooth analytical function which is often approximated by lower order polynomials. Mathematically, the approximation can be expressed as:

$$y(x) = f(x) + e, \tag{6}$$

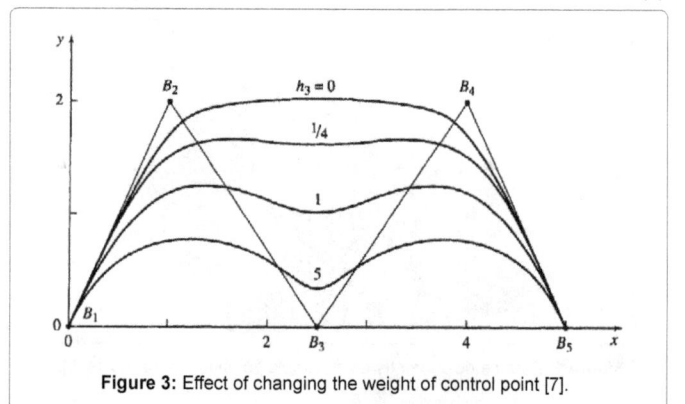

Figure 3: Effect of changing the weight of control point [7].

where $y(x)$ is the unknown function, $f(x)$ is the polynomial function of x, and e is the random error. The two most common models used for RSM are 1st degree and 2nd degree polynomials. The 2nd degree polynomial is used in this study since it has been used successfully for similar problems [13-15]. The model can be expressed as:

$$y(x) = \beta_0 + \sum_{i=1}^{d} \beta_i x_i + \sum \sum_{i<j} \beta_{ij} x_i x_j + \sum_{i=1}^{d} \beta_{ii} x_i^2 + e \qquad (7)$$

where d is the number of design variables and β are the unknown coefficients. In matrix notation, polynomial response surface can be expressed as:

$$Y(X) = X^T b \qquad (8)$$

where b is the matrix of unknown coefficients:

$$b = \left(X^T X\right)^{-1} X^T Y \qquad (9)$$

Note that $\beta 0$, β_i, β_{ij} are the unknown coefficients determined from the least-square regression which minimizes the sum of the squares of the deviations (SS_E) of predicted values and the actual values obtained from experiments and $x_i x_j$ are the design variables.

$$SS_E = \sum_{i=1}^{t} \varepsilon^2 = \varepsilon^T \varepsilon = (Y - Xb)^T (Y - Xb) \qquad (10)$$

To obtain the polynomial response surface model, a series of experiments need to be performed in which the response variable Y is measured for different combinations of control variables.

Design of Experiments

The combinations of control variables are obtained by a systematic procedure called design of experiments. The design of experiments is a concept that uses a set of selected experiments to draw information about the general behavior of the studied object against a set of factors which affect the response [12]. It helps to keep the number of performed experiments as low as possible and to obtain most of information with this set of experiments. For this study, a D-optimal array is used which enables more efficient construction of a quadratic response surface model [16].

Methodology

As stated earlier, the drag characteristics of car strongly depend on the rear geometry. Therefore, only the rear geometry of Ahmed body [17] was represented with NURBS and parameterized. In addition, the weights of control points of NURBS (from here on referred to as NURBS parameters) curve determine how much a control point attracts the curve. The geometry was parameterized with only the weights of control points. This dramatically reduced the number of design variables and the complexity of parameterization. Since the effect of the control point is local, only the part of the curve in the vicinity of the control point was modified. The accuracy of the fitting model can be assessed by various criteria. The most commonly used criteria are R^2 and its adjusted form R^2a which also accounts for the number of experiments and degree of freedom.

$$R^2 = 1 - \frac{SS_E}{SS_T} \qquad (11)$$

$$R^2_a = 1 - \left(\frac{t-1}{t-r}\right)\left(1 - R^2\right) \qquad (12)$$

where t is the number of experiments and r is the number of regression coefficients. SS_E and SS_T are given by

$$SS_E = \sum_{i=1}^{t} (y_i - \hat{y}_i)^2 \qquad (13)$$

$$SS_T = \sum_{i=1}^{t} y_i^2 - \frac{1}{t}\left(\sum_{i=1}^{t} y_i\right)^2 \qquad (14)$$

where $\hat{y}$ is the predicted response. The values of R^2 and R^2a are between 0 and 1 and the values closer to 1 signify good fit. Another relevant quantity that measures the accuracy of the fit is the Root Mean Squared Error (RMSE).

$$RMSE = \left[\sum_{i=1}^{t} \frac{(y_i - \hat{y}_i)^2}{t}\right]^{1/2} \qquad (15)$$

The obtained response surface models must be validated for robustness and accuracy. A common technique is to perform validation experiments over the entire design space and compute the RMSE for validation experiments. The model can then be improved globally by updating the original experiment

data with the test experiments which have high prediction errors. However, this does not guarantee that the region with optimal design is improved. Another approach is to add the data points with the predicted optimal design to the original experiments and update the model. Unfortunately, it is not obvious which method is preferable since it is very much model and problem dependent [15-18]. The flow chart in Figure 4 outlines the procedure developed by this study for response surface model generation and improvement.

Parameterization of Geometry

The geometric parameterization was done using NURBS. The rear body of Ahmed model was represented using 10 control points (P1 to P10). Figure 5 shows the control polygon of the NURBS curve. The number and positions of control points were chosen such that a large variety of shapes could be obtained by changing the NURBS parameters, without the need to change the position of any control point. The weights of the end points W1 and W10 were 1 in all cases since this ensured that the curve passed through the end points. Figure 6 shows a sample notch back geometry which was obtained by setting appropriate values of NURBS parameters also shown in the figure.

Numerical Modelling

In order to simplify the problem and reduce computational resources, a two-dimensional computational domain was considered. The rear geometry created using NURBS was imported to ANSYS® Design modeler and attached to the front end of the Ahmed body. The total length (L) and height (H) of the simplified two-dimensional car model were 1.044 m and 0.288 m respectively. To ensure that the domain was sufficiently large for this simulation, the common practical guidelines for automotive external aerodynamics were followed [19]. The domain inlet was 3 model lengths upstream of the model and outlet was 5 model lengths downstream. Thus the total domain was 9 model lengths long. The domain far field was 3 model lengths above the model and the total domain height was 3.3 model lengths.

Computational Domain

ANSYS® meshing software in ANSYS® workbench package was

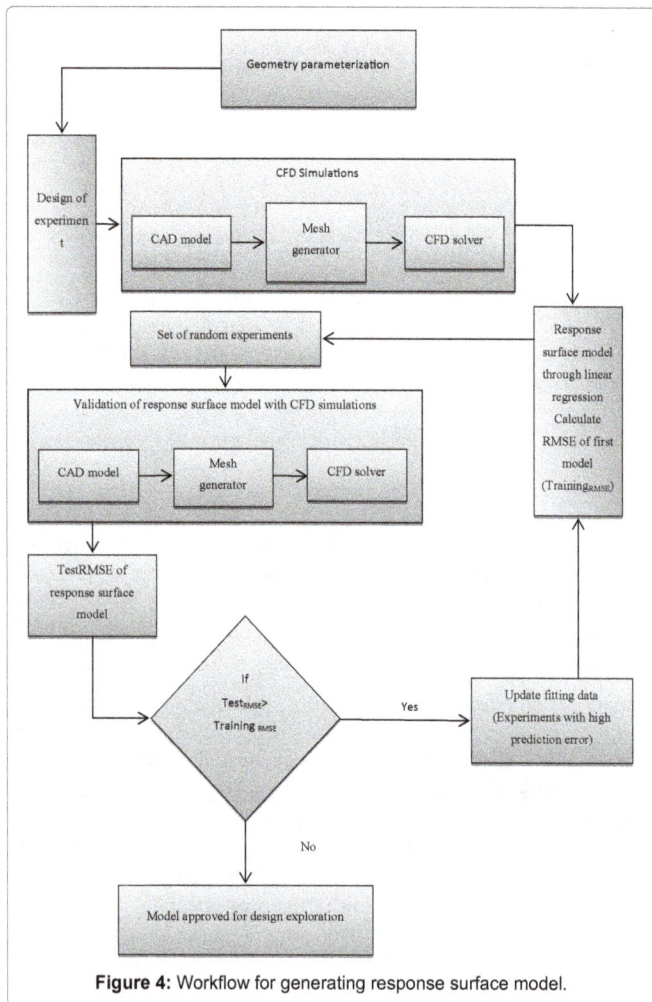

Figure 4: Workflow for generating response surface model.

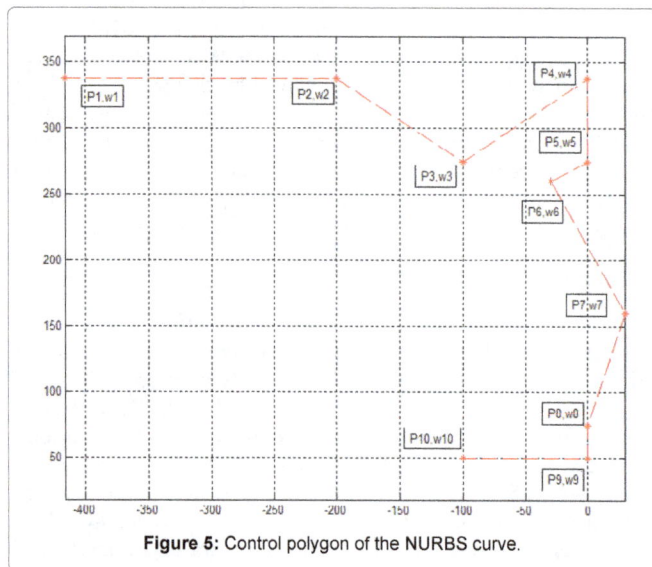

Figure 5: Control polygon of the NURBS curve.

The size of the cells near the model should be adjusted such that the wall functions remain valid and the boundary layer should be adequately resolved. Thus a dense mesh is required close to the model and a coarse mesh can be used away from the model. This strategy to divide the computational domain in coarse and fine regions drastically reduced the total cell count and computational time. Figure 7 shows the details of the computational domain.

Mathematical Modelling

In this study, the flow was considered to be two-dimensional,

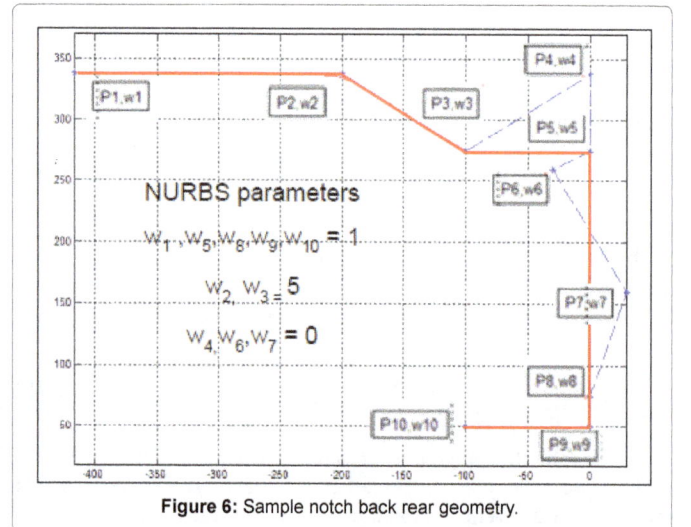

Figure 6: Sample notch back rear geometry.

Figure 7: Numerical domain with boundary layer and grid refinement zones. (a) Boundary layer grid, (b) grid refinement region, and (c) computational mesh.

used for meshing [20]. Unstructured, non-orthogonal grid with quadrilateral elements was used to create the computational domain. The accuracy of the computational results and the required time directly depend on the number of cells in the computational domain.

incompressible, steady, and turbulent. The fluid was Newtonian with constant density, ρ and dynamic viscosity, μ. The Reynolds Averaged Navier-Stokes (RANS) equations for continuity and momentum conservation can be written as:

$$\frac{\partial u_i}{\partial x_i} = 0 \tag{16}$$

$$\frac{\partial}{\partial x_j}\left(\rho u_i u_j\right) = -\frac{\partial p_m}{\partial x_i} + \frac{\partial}{\partial x_j}\mu\left(\frac{\partial u_i}{\partial x_j} - \overline{\rho u_i' u_j'}\right) \tag{17}$$

where $\rho u_i' u_j'$ is the Reynolds stress, p_m is the mean flow pressure and u_i is the mean flow velocity in the x_i direction.

As the goal of this study was to obtain the trends of aerodynamic drag coefficients and to verify the developed framework efficiently, a computationally inexpensive turbulence model was selected. The k-ε model employed in this study is a two equation model based on transport equations for turbulent kinetic energy k and its dissipation rate ε. The k-ε model solves the following two equations of turbulent kinetic energy k and its dissipation rate ε:

$$\frac{\partial}{x_j}\left(\rho u_j k\right) = \frac{\partial}{\partial x_j}\left(\left(\mu + \frac{\mu_t}{\sigma_k}\right)\frac{\partial k}{\partial x_j}\right) + P_k - \rho\varepsilon \tag{18}$$

$$\frac{\partial}{\partial x_j}\left(\rho u_j \varepsilon\right) = \frac{\partial}{\partial x_j}\left(\left(\mu + \frac{\mu_t}{\sigma_\varepsilon}\right)\frac{\partial \varepsilon}{\partial x_j}\right) + \frac{\varepsilon}{k}\left(C_{\varepsilon 1} - \rho C_{\varepsilon 2}\varepsilon\right) \tag{19}$$

where P_k is the production term given by:

$$P_k = \mu_t\left(\frac{\partial u_i}{\partial x_j} + \frac{\partial u_j}{\partial x_i}\right)\frac{\partial u_i}{\partial x_j} \tag{20}$$

and the turbulent viscosity is related to turbulent kinetic energy and dissipation rate by:

$$\mu_t = \rho C_\mu \frac{k^2}{\varepsilon} \tag{21}$$

The k-ε model constants are

$C_{\varepsilon 1} = 1.44$, $C_{\varepsilon 2} = 1.92$, $C_\mu = 0.09$, $\sigma_k = 1.0$ and $\sigma_\varepsilon = 1.3$ [20].

Boundary Conditions

The inlet velocity must be specified to obtain the desired value of Reynolds number. A velocity of 40 m/s was specified normal to the domain inlet. The Reynolds number based on model length was 2.8 million. In addition, the inlet turbulence intensity was set to 1% and viscosity ratio $\frac{\mu_t}{\mu}$ was set to 3 since these values are appropriate for external flow analysis [19]. The turbulence intensity is the ratio of the root-mean-square of the velocity fluctuations to the mean flow velocity. When turbulence intensity and viscosity ratio are specified, FLUENT® solver calculates the dissipation rate using the relation:

$$\varepsilon = \rho C_\mu \frac{k^2}{\mu}\left(\frac{\mu_t}{\mu}\right)^{-1} \tag{22}$$

The outflow condition was specified with zero pressure at the domain outlet. To avoid the effect of shear layers of domain far field on flow field around the model, free-slip wall was specified on domain far field. For car model walls and domain ground, a no-slip boundary condition was imposed.

Numerical Simulations

FLUENT® uses finite volume method to discretize the flow equations. The pressure based coupled solver available in FLUENT® was used which solves the momentum and pressure based continuity equations in a coupled manner. This method accelerates the convergence of the solution [21]. For better accuracy of solution, second order discretization was used for discretizing the equations of momentum, kinetic energy and its dissipation rate. The convergence of the solution was based on the drag coefficient as well as the residuals of kinetic energy and dissipation rate. The solution was considered to be converged when there was no change in drag coefficient for at least 100 iterations and the residuals of kinetic energy and dissipation rate were less than 1×10^{-6}. Moreover, it was also ensured that the dimensionless wall coordinate, y^+ remained in the desired range ($30 < y^+ < 300$) on the car model walls. The average value of y^+ for the simulations was approximately 150.

To ensure that the numerical results were independent of the mesh density, mesh independence tests were performed using coarse, medium and fine grids. These were used to estimate the drag coefficient. The change in drag coefficient from the coarse to medium grid was 0.3% and from medium to fine grid was 0.7%. Based on these results a medium grid with approximately 50,000 elements was selected for present study.

Shape Optimization of Notch Back

The proposed framework was employed to optimize the aerodynamic shape of a simple notch back model for minimum drag coefficient. The NURBS parameters of top edge of rear window, bottom edge of rear window, boot lid, the base bulge and the diffuser shown in Figure 8 were used for shape optimization and response surface modelling.

To obtain a pure quadratic model with five variables, a total of eleven regression coefficients were needed to be calculated which required at least eleven data points.

Moreover, to acquire enough data for the entire design space, each of the five NURBS parameters was studied at the four levels shown in Table 1. Thus a D-optimal array with 16 runs was used to design the CFD experiments. The drag coefficients obtained from the CFD simulations and the corresponding parameter combinations were then used for linear regression to construct the response surface model.

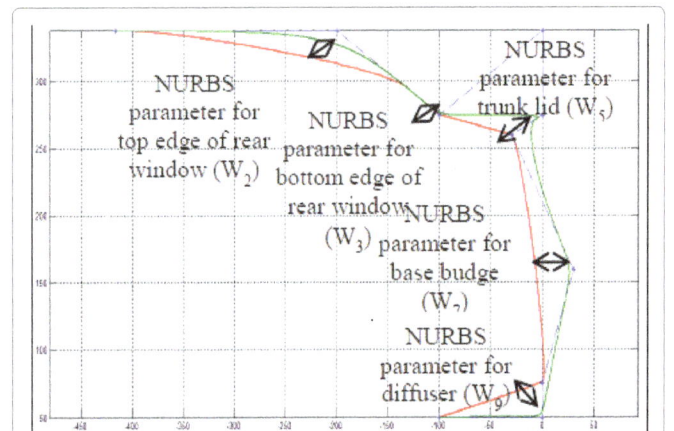

Figure 8: Geometry parameterization for notch back design with five NURBS parameters.

Parameters	Level 1	Level 2	Level 3	Level 4
Rear window top edge W_2	0.1	0.5	1	5
Rear window bottom edge W_3	0.1	0.5	1	5
Trunk lid W_5	0	0.2	0.6	1
Base bulge W_7	0.1	0.5	1	5
Diffuser W_9	0	0.5	2	10

Table 1: Parameter levels for notch back design.

	R^2	R^2-ADJ	$TRAINING_{RMSE}$	$TEST_{RMSE}$
Model 1	0.924	0.735	0.0112	0.0098
Model 2	0.834	0.715	0.0107	0.0067

Table 2: Statistics of Model 1 & 2.

NURBS parameters for minimum C_d					C_d
W_2	W_3	W_5	W_7	W_9	
0.623	2.875	0.0	0.1	0.0	0.186

Table 3: NURBS parameters for minimum drag coefficient.

The parameter levels were chosen considering the changes in the geometry they provided and the sensitivity of the curve. The set of initial 16 CFD simulation experiments performed with different combinations of geometry parameters are shown in Table 1 in the appendix.

A quadratic response model with intercept, linear and quadratic terms of the form of Equation 23 was used to obtain Model 1 for the notch back geometry. Table 2 in the appendix shows the coefficients of Model 1.

$$C_d = I + W_2 + W_3 + W_5 + W_7 + W_9 + W_2^2 + W_3^2 + W_5^2 + W_7^2 + W_9^2 \qquad (23)$$

The statistics of Model 1 are shown in Table 3 in the appendix. The obtained response surface model was then validated with additional experiments which were designed to test the entire range of parameters and the RMSE was calculated dynamically. A model updating scheme discussed in Figure 4 was applied to reduce the RMSE and the CFD experiments with absolute prediction error greater than 0.01 were added to the fitting data to obtain Model 2 with coefficients shown in Table 4 in the appendix. The validation experiments were performed until the change in RMSE value was considerably small. Table 2 compares the statistics of Models 1 and 2.

It can be seen that Model 2 is superior since it predicts the drag coefficient more accurately over the entire range of design parameters. Figure 9 compares the test RMSE of the two models. It should be noted that although the Model 2 shows better performance globally, it does not guarantee good performance in the region of optimal design.

Summary of Results

The constrained non-linear optimization was used to minimize the drag coefficient function represented by Model 2. The starting point of the optimization was Model 2 with all parameters set to level 1 as given in Table 1 with variable bounds between level 1 and level 4. Once the values of parameters were obtained by minimizing the drag coefficient function, CFD simulations were performed with the optimal design parameters and Model 2 was updated with the new data. The process was continued until the drag coefficient value predicted from response surface model and the CFD simulations converged. The developed optimization framework is summarized in the flow chart in Figure 10. The optimization process required 18 iterations to achieve the design parameters for minimum drag coefficient. The optimum design

parameters are summarized in Table 3. The optimized rear geometry for minimum drag is shown in Figure 11 and the drag convergence history is shown in Figure 12.

Velocity streamlines and pressure contour around optimized geometry depicted in Figure 13 indicate that the flow separates only at the slanted base and the diffuser geometry at bottom augments the pressure recovery. In case of the high drag geometry obtained during

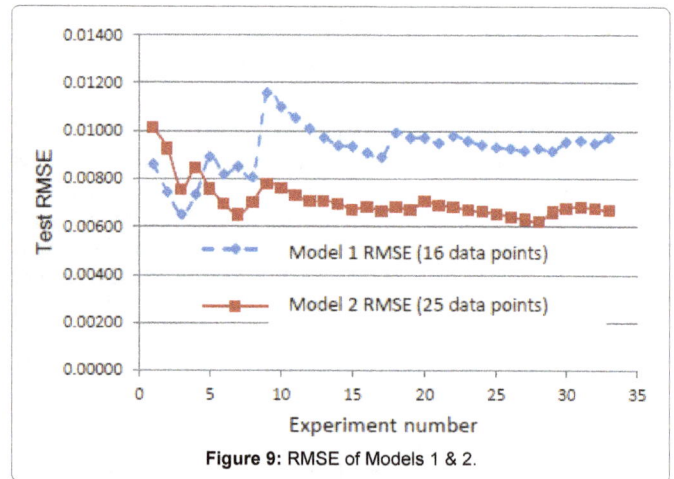

Figure 9: RMSE of Models 1 & 2.

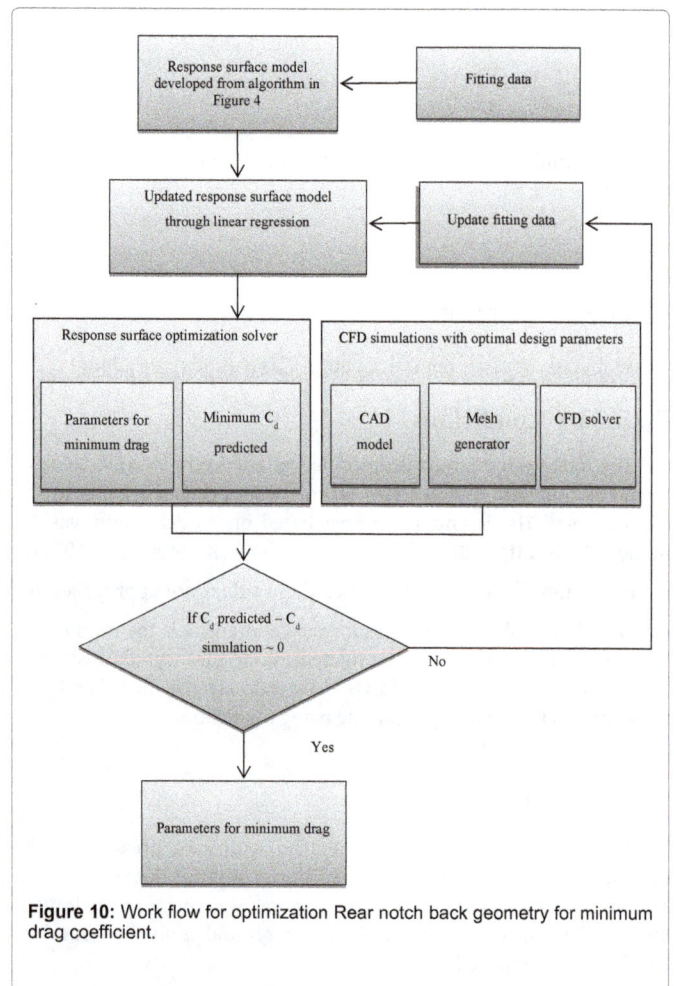

Figure 10: Work flow for optimization Rear notch back geometry for minimum drag coefficient.

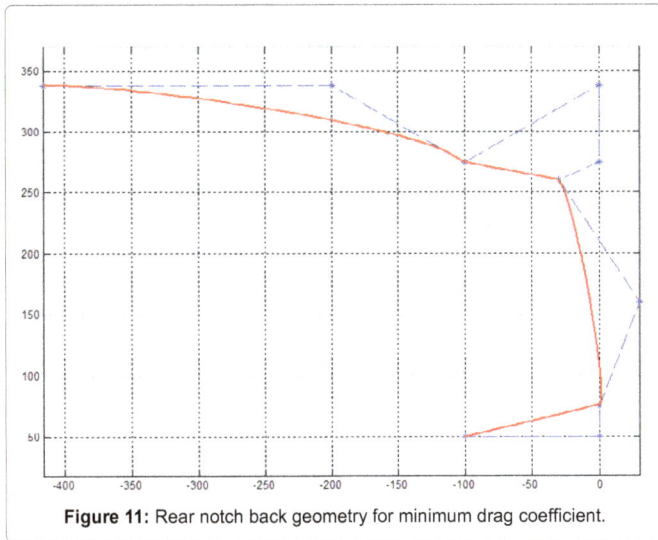

Figure 11: Rear notch back geometry for minimum drag coefficient.

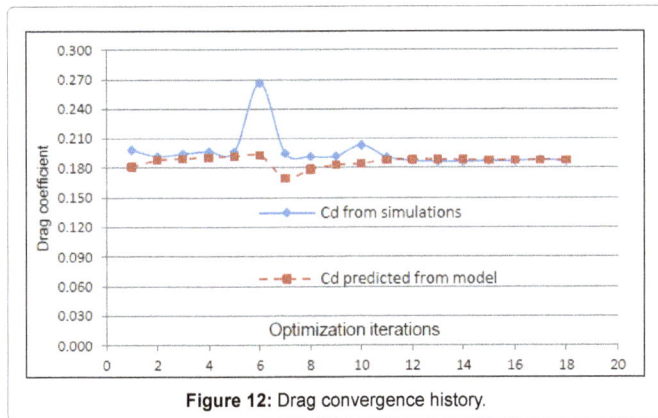

Figure 12: Drag convergence history.

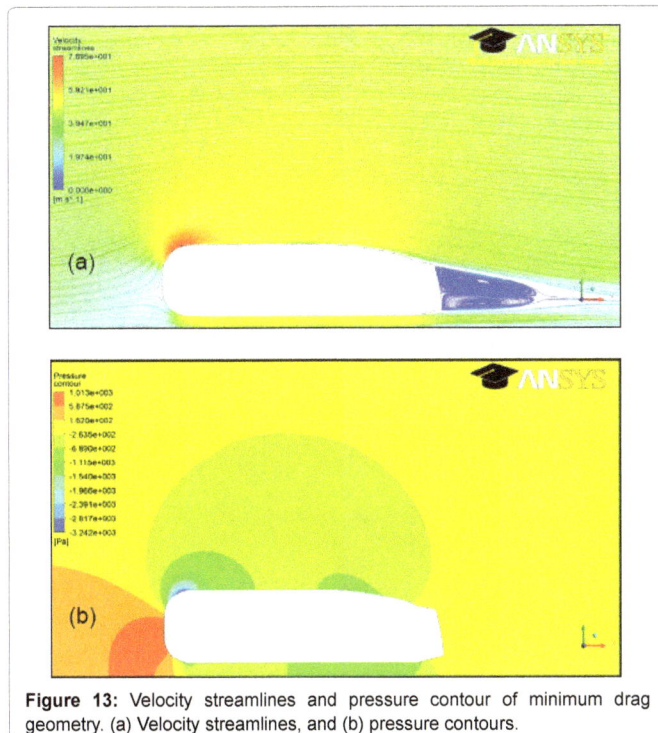

Figure 13: Velocity streamlines and pressure contour of minimum drag geometry. (a) Velocity streamlines, and (b) pressure contours.

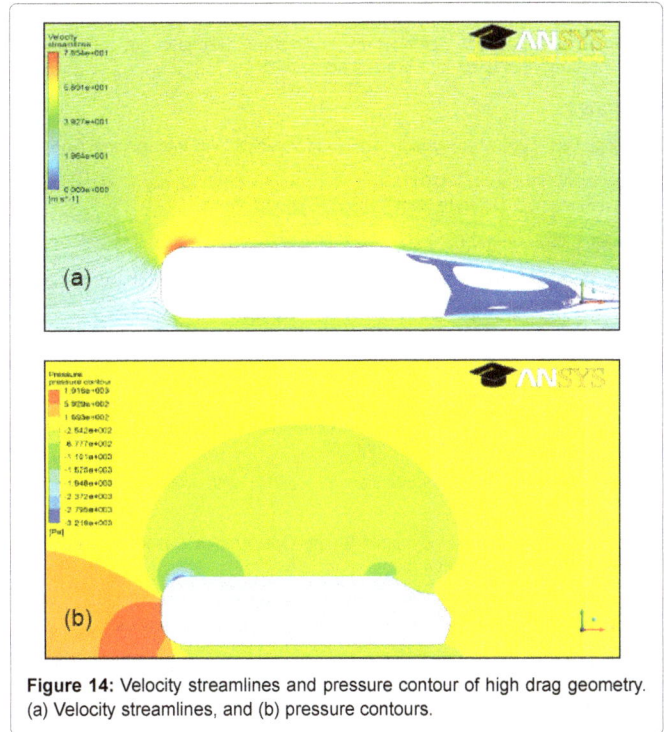

Figure 14: Velocity streamlines and pressure contour of high drag geometry. (a) Velocity streamlines, and (b) pressure contours.

the optimization process, the flow separates at the top edge of the rear window and forms a large recirculation region in the wake as shown in Figure 14. The large recirculation region causes a huge pressure drop in the wake resulting in a higher pressure drag.

Conclusion

This paper presented a new framework for passenger cars rear geometry parameterization and aerodynamics-based shape optimization. The geometric parameterization was implemented using NURBS parameters. The proposed technique greatly simplifies the parameterization process. It also provides the flexibility to generate free form shapes which cannot be obtained using conventional parameterization techniques employed in automotive body design optimization. The developed methodology uses the proposed parameterization technique to construct response surface models with NURBS parameters as the design variables. It should be noted that the framework was developed using 2D case in order to validate and demonstrate the method. For the results to be practical and applicable, the method should to be applied to 3D cases.

The framework was employed for the aerodynamics-based shape optimization of a simplified notch back model. The response surface method efficiently directs the optimization process since it correctly predicts the parameters of minimum drag as confirmed by the CFD simulations. The proposed framework was implemented successfully using NURBS parameters for a car's rear geometry aerodynamic shape optimization.

The response surface model of aerodynamic drag was constructed for five design variables and the optimization process required only 18 iterations to obtain the geometric parameters of minimum drag.

Acknowledgement

Financial support of this work by the Natural Sciences and Engineering Research Council of Canada is grateful acknowledged.

References

1. Hucho WH (1998) Aerodynamics of Road Vehicles SAE International SAE.

2. Mayer W, Wickern G (2011) The New Audi A6/A7 Family - Aerodynamic Development of Different Body Types on One Platform.

3. SAE Int J Passenger Cars – Mech. Syst 4: 197-206.

4. Han T, Hammond DC, Sagi CJ (1992) Optimization of Bluff Body for Minimum Drag in Ground Proximity .AIAA Journal 30: 882-889.

5. Muyl F, Dumas L, Herbert V (2004) Hybrid Method for Aerodynamic Shape Optimization in Automotive Industry. Computers and Fluids 33: 849-858.

6. Peddiraju P, Papadopoulous A, Singh R (2009) CAE Framework for Aerodynamic Design Development of Automotive Vehicles. Presented at 3rd ANSA & μETA International Conference.

7. Rogers DF (2001) An Introduction to NURBS with Historical Perspective. Morgan Kaufmann Pub.

8. Samareh JA (2004) Aerodynamic Shape Optimization Based on Free-Form Deformation. AIAA Paper 4630 .

9. Lepine J, Trépanier J, Pepin F (2000) Wing Aerodynamic Optimization Using an Optimized NURBS Geometrical Representation. AIAA paper 0669.

10. Bentamy A, Trépanier JY, Guibault F (2002)Wing Shape Optimization Using a Constrained NURBS Surface Geometrical Representation. Paper presented in ICAS Congress.

11. Song W, Keane AJ (2004) A Study of Shape Parameterization Methods for Airfoil Optimization . Presented In 10th AIAA/ISSMO Multidisciplinary Analysis and Optimization Conference: 2031-2038 .

12. Baker CA, Grossman B, Haftka RT, Mason WH, Watson LT (1998) HSCT Configuration Design Space Exploration Using Aerodynamic Response Surface Approximations. AIAA paper 4803 (2004).

13. Krajnović S (2009) Optimization of Aerodynamic Properties of High-Speed Trains with CFD and Response Surface Models. The Aerodynamics of Heavy Vehicles II: Trucks, Buses, and Trains 197-211 .

14. Marjavaara D (2006) CFD Driven Optimization of Hydraulic Turbine Draft Tubes Using Surrogate Models. PhD diss Luleå University of Technology .

15. Alvarez L (2000) Design Optimization Based on Genetic Programming. PhD diss University of Bradford .

16. Sobester A, Leary SJ, Keane AJ (2004) A Parallel Updating Scheme for Approximating and Optimizing High Fidelity Computer Simulations. Structural and Multidisciplinary Optimization 27: 371-383.

17. Ahmed SS, Ramm G, Faltin G (1984) Some Salient Features of the Time-Averaged Ground Vehicle Wake. SAE Technical Paper 840300.

18. Lanfrit M (2005) Best Practice Guidelines for Handling Automotive External Aerodynamics with Fluent. Version 1.2. Fluent Deutschland Gmbh .

19. Keating M (2011) Accelerating CFD Solutions. ANSYS Advantage Magazine.

20. (2010)ANSYS® Academic Research Release 13.0 Help System Workbench User's Guide ANSYS Inc .

21. (2010)ANSYS® Academic Research Release 13.0 Help System Fluent Theory Guide ANSYS Inc .

16

Emissions Reduction of Regulated and Unregulated Hydrocarbon, Gases in Gasoline Bi-mode SI/HCCI Engine by TWC Converter

Hasan AO[1]* and Abu-jrai A[2]

[1]*Department of Mechanical Engineering, Al-Hussein Bin Talal University, Maan, P.O. Box-20, Jordan*
[2]*Department of Environmental Engineering, Al-Hussein Bin Talal University, Maan, P.O. Box-20, Jordan*

Abstract

A specific case of HCCI is gasoline fuelled HCCI. It is attractive due to the simplicity of implementing such a technology into existing SI engines as well as the existing fuelling infrastructure. Lean and highly diluted homogeneous charge compression ignition HCCI engines offer great potential in improving vehicle fuel economy and contribute in reducing CO_2 emissions. Gasoline is a complicated mixture of many different hydrocarbons which results in rather poor auto-ignition properties. Hydrocarbons and CO emissions from HCCI engines can be higher than those from spark ignition (SI) engines, especially at low engine load when the EGR rate or the residual gas required to control NOx emission are elevated. Toxic chemicals emitted by SI engines, Carbonyl compounds and poly aromatic hydrocarbons PAH generated by V6 (SI/HCCI) gasoline engine especially in HCCI mode. A qualitative and quantitative analysis of hydrocarbon compounds, alkenes. Alkanes, aromatics and aldehydes was analysed before and after catalyst, alkanes, alkenes and aromatic were conducted using Gas Chromatography-Mass Spectrometry (GC-MS) apparatuses. Aldehydes were conducted using High Performance liquid chromatography (HPLC) on reversed phase. HPLC system, although, bi-functional after treatment, the device will be required to control the regulated and unregulated hydrocarbon, CO, and NOx emissions under lean and stoichiometric (oxygen free) engine operating conditions. This paper describes studies on the regulated and unregulated hydrocarbons, NOx, and CO emissions coming out of HCC/SI gasoline engine. Comparative study of catalyst performance will be analysed under HCCI stoichiometric and SI operation under different engine loads, analysis indicate that, the HC and CO emissions reduction over the prototype catalyst was in the range of 90-95% while the maximum NO_x emissions reduction under lean engine operating conditions was in the range of 35-55%. The catalytic converter showed an excellent efficiency of eliminating unregulated hydrocarbons (alkenes, alkanes, and aromatics) and aldehydes compounds; achieved reduction efficiency was up to 92%.

Keywords: Emissions; Combustion; IC engines; Catalysts; HCCI

Introduction

Homogeneous Charge Compression Ignition (HCCI) engine is an alternative piston-engine combustion process which can deliver high engine efficiency comparable with those of compression ignition (CI) engines. It has been proposed that the engine technology employing HCCI will have dual mode combustion systems, where SI or CI combustion is in use together with HCCI. One of the future directions for HCCI engine development is to widen the operating range i.e. load/speed map [1-6]. Even if HCCI operation at high loads is obtained, the NOx reduction benefit is small compared with conventional SI engine with (TWC). Meanwhile, the fuel consumption advantage of gasoline HCCI over SI combustion is reduced at high load due to the reduced level of throttling. For high-load operation, it may require to again switch to traditional SI or CI operation. In HCCI engines the ignition timing is subject to chemical kinetics of the reactants and dependent on the fuel [7-9].

Hydrocarbon and carbon monoxide levels of HCCI engines vary between experimental conditions (e.g. engine technology and combustion modes). Though they are found to be similar or higher than those of SI engines and both CO and HC emissions are increase at light loads. Although, TWC technology has been perfected over the years for use in SI stoichiometric combustion engines, under HCCI operation or lean operation those catalysts are not effective in reducing NOx emissions. In addition switching engine operation to lean, stoichiometric or even rich is making the application of conventional oxidation catalysts challenging especially at low exhaust temperatures associated with HCCI engines [10-12]. Total hydrocarbon covered a wide range of unregulated compounds such as alkynes; alkenes; alkanes; and aromatics. Alkanes (paraffins) molecules contain single hydrogen-

carbon bonds (e.g. ethane) and are referred as 'saturated'. Alkene (olefins) molecules contain double bonds (e.g. ethylene) 'unsaturated', and alkynes (i.e., acetylene), triple bonds unstable and reactive. Aromatic molecules typically display enhanced chemical stability (e.g. benzene, toluene) are building blocks for Poly (cyclic)-Aromatic Hydrocarbons (PAH) and are molecules containing two or more simple aromatic rings. Most of hydrocarbons in fuel are burnet completely during the combustion process. Small amounts of HCs, however, gets partial combustion, this produces different lower molecular HCs (in range of C1-C5) and some oxidized compounds such as aldehydes [13]. Aromatics in a fuel may survive the combustion process and could be more emitted through the exhaust as unburned HCs. At poor combustion, and engine misfiring, large amounts of hydrocarbons are emitted from the combustion chamber. The mass distribution of these species in the exhaust is a function of the engine design, fuel composition, and the engine operation conditions. Another cause of excessive HC emissions is related to combustion chamber deposits. Because these carbon deposits are porous, HC is forced into these pores as the air/fuel mixture is compressed. When combustion takes place,

*Corresponding author: Hasan AO, Department of Mechanical Engineering, Al-Hussein Bin Talal University, Maan, PO Box-20, Jordan
E-mail: Ahmad_almaany@yahoo.com

this fuel does not burn; however, as the piston begins its exhaust stroke, these HCs are released into the exhaust stream [14-17].

Wall quenching occurs as the combustion flame front burns up to the relatively cool walls of the combustion chamber. This cooling extinguishes the flame before all of the fuel is fully burned, leaving a small amount of HC to be pushed out the exhaust valve. This is the major source of HC emissions from four-stroke engines.

This manuscripts presents study on a new catalytic converter design, aiming to control regulated and unregulated hydrocarbons, CO and NOx emissions under stoichiometric, HCCI/SI engine operation and different engine loads. The experiments were conducted on a V6 gasoline engine equipped with the prototype catalyst at the one bank. The effects of engine loads and modes were analysed.

Experimental Procedure

Engine - The experimental work was performed on a V6 HCCI/SI mode gasoline direct injection (GDI) engine image 1. The engine intake and exhaust camshafts were built with variable cam timing (VCT) and cam profile switching (CPS) system. Fuel direct injection pulse width is adjusted by the engine management system to maintain the required value of λ. The engine was coupled to an eddy-current dynamometer for speed and load control, twelve thermocouples type K were installed to monitor the exhaust and inlet temperature. Intake and exhaust cam timing for minimum residuals were chosen to start the engine with spark ignition combustion. When engine block achieved steady condition of temp level in SI mode, the HCCI combustion mode was started.

Catalyst - The prototype three zones monolith catalyst, with optimized order of the zones, was connected to the actual engine exhaust manifold. The ceramic monolith dimensions and hence volume were the same as the production TWC, the engine was originally equipped. The first catalyst zone contains base metal catalyst and was designed to control part of the exhaust hydrocarbons and CO. The second zone again contains base metal catalyst was designed to reduce NOx with HC under lean engine operation while the third zone which contain precious metals (e.g. Pt) was designed to reduce HC, CO and NOx under lean and stoichiometric engine conditions.

Emissions analysis - Horiba MEXA 7100 DEGR (HC, CO, NOx, and CO_2) was connected to measure the exhaust emissions upstream and downstream the catalyst.

Hydrocarbon speciation of C5-C11 compounds was carried out using an on-line GC-MS. An 8000 series GC equipped with direct injector was connected to a Fisons MD 800 mass spectrometer, used as a detector. The gas samples were introduced via a heated line into a six port Valco valve outfitted with a 0.1 ml sample loop. The gas sampling apparatus was kept at a constant temperature of 200°C. A 30 meter long x 0.53 mm i.d. DB-1 capillary column with a 3 μm film thickness was used, this type of column allows for separation of both the polar and non-polar compounds.

Carbonyls: The samples were carried out by passing the engine exhaust gas with flow rate of 1.0 L/min in to 25 ml midget impinger and the exhaust gas bubbled inside the DNPH solution reagent for 20 minutes, during the test the sample must be surrounding by bath of ice to prevent any vaporising of the compound species during the sampling. When the exhaust gas bubbles in the DNPH solution the carbonyl compounds react with the reagent to produce DNPH- carbonyl derivatives. Each sample was kept in the fridge below 4ºC until HPLC analysis takes place. Carbonyl species analysis, were performed by

High Performance liquid chromatography (HPLC) on reversed phase. HPLC system was used Dionex Acclaim 120, separation of the sample components was achieved on a 250 mm x 4.6 mm ID. Column backed with 5 μm C18 Acclaim 120. UV detector was used and the wavelength was set to 365 nm. a mobile phase solvent gradient of acetonitrile and deionised water was (10:90 v/v) used as an eluent, the flow rate of the eluent was adjusted to 1.0 ml/min and the mixing ratio was changed linearly to reach 75:25 v/v acetonitrile to deionised water at 60 min, and then kept constant to the end of the run, the analysing time for the a single measurement was 70 minutes (Figure 1).

Results and Discussion

Hydrocarbon and CO emissions are influenced by the in-cylinder conditions (i.e., temperature) and the homogeneity of the fuel with air. Increasing engine load from 3-4 bar improves HC and CO oxidation which leads to significant reduction of HC and CO emissions, on the other hand the opposite trend seen in the case of NOx emission when the engine load was increased. In case of HC this was presumably because crevice loading of HC with lower load has a bigger effect on HC emissions than that with a higher load. The burned gas temperature at low load seems to be too low for the crevice hydrocarbons to be oxidized during the expansion stroke, comparing to the higher gas temperature. The lower gas temperature is not favourable for HC oxidation in the exhaust ports. Furthermore, the low combustion gas temperature presented at low load can result in incompletion of the combustion process and cause higher HC [18]. Similarly when load is decreased CO emissions increased. This rapid increase resulted from the quenching of the CO oxidation process in the lower load as the gas temperature dropped. Dec [19] predicted that for low loads, incomplete bulk-gas reactions should play a significant and perhaps dominant role in CO emissions and this contributes to HC emissions. NOx reduction has been a result of lower combustion temperature in the homogeneous combustion associated with dilution and an increased heat capacity as an exhaust gas is used to dilute the cylinder mixture. Under most engine operation the prototype catalyst covered a wide range of engine conditions, HCCI stoichiometric and SI mode, this led to approval consistency of HC and CO conversions above 90% over the catalyst independently from in-cylinder conditions (Figure 2).

Figure 3 shows that the HCCI combustion with residual gas trapping produced much less CO emissions than SI engine mode which was confirmed by other studies using the residual gas trapping method [20]. The reduction in CO emission is likely caused by the recycling of burned gases and their subsequent conversion in to CO_2 in the

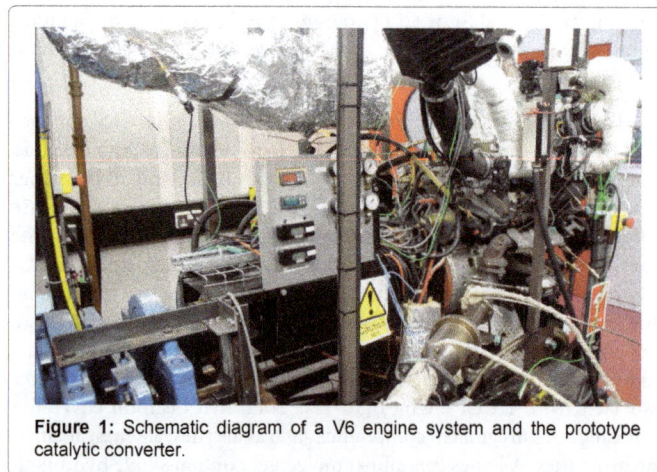

Figure 1: Schematic diagram of a V6 engine system and the prototype catalytic converter.

next cycle. It should be noted that HCCI combustion in the 4-stroke gasoline engine had been always associated with higher CO emissions than the SI combustion until the residual gas trapping method was employed [21].

Minimal catalytic conversion of NOx emissions up to 55% efficiency occurred at HCCI mode this is correlated with CO and HC emissions. When CO and HC emissions are relatively low (and exhaust temperature is low) the CO oxidation deteriorates and allows higher efficiency of NOx reduction (more CO available for reduction reaction). Despite the catalyst lower NOx conversion efficiencies under HCCI mode, NOx emissions after the catalyst were kept at lower values compared to SI (Figure 4). Further discussion is in (Figures 5-7).

Alkenes

Unsaturated hydrocarbon compounds (olefins) presented in the engine tails in a great concentration, with HCCI mode at engine load of 3 bar, alkenes showed a high concentration just over 600 ppm and was reduced about 50% when shifting to higher load of 4 bar. Changing engine mode from HCCI to SI engine mode did not influence much the alkene concentration, the proto type convertor showed high conversion efficiency at high load 4 bar for both engine modes. Catalyst efficiency reached over 90%.

Alkanes

Paraffin saturated hydrocarbons with open – chain compound has high concentration in HCCI mode with engine load of 3 bar, shifting

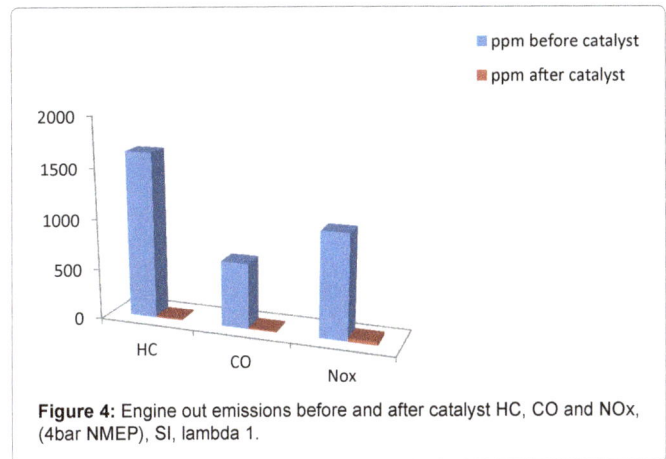

Figure 2: Engine out emissions before and after catalyst HC, CO, and NOx (3 bar NMEP), HCCI, lambda 1.

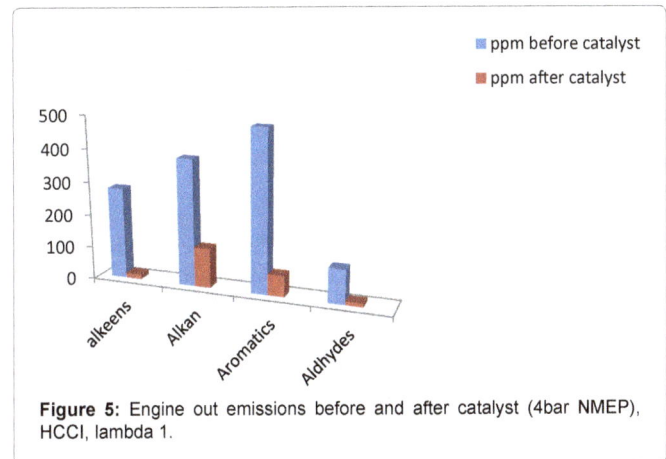

Figure 3: Engine out emissions before and after catalyst of HC, CO, and NOx (4bar NMEP), HCC, lambda 1.

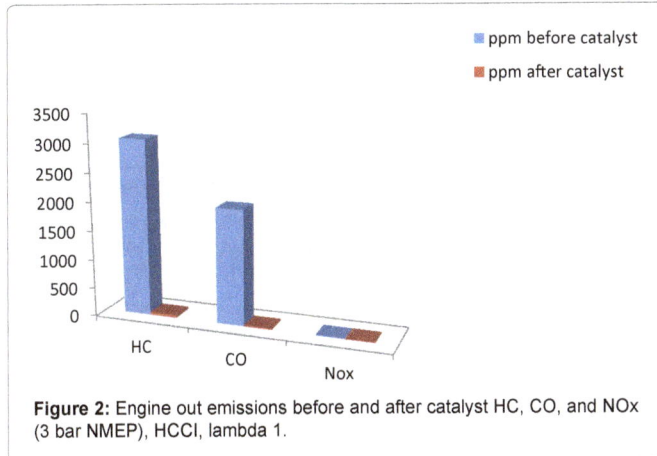

Figure 4: Engine out emissions before and after catalyst HC, CO and NOx, (4bar NMEP), SI, lambda 1.

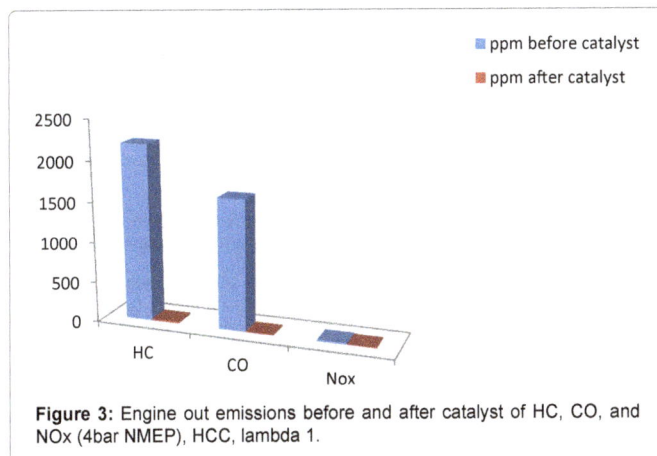

Figure 5: Engine out emissions before and after catalyst (4bar NMEP), HCCI, lambda 1.

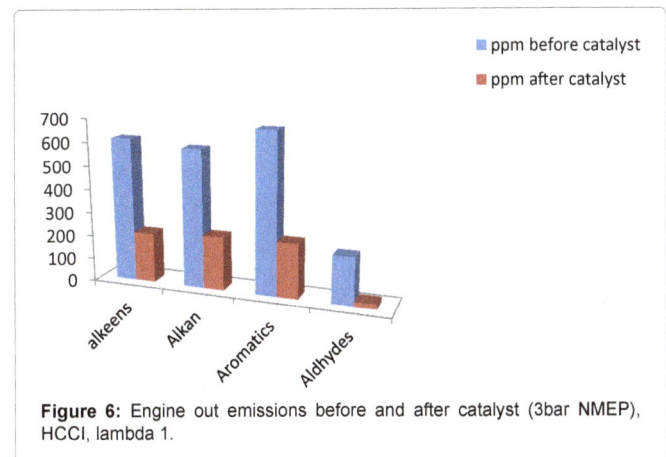

Figure 6: Engine out emissions before and after catalyst (3bar NMEP), HCCI, lambda 1.

to higher engine load 4 bar has reduced alkane concentration down to 30%, on the other hand alkanes presented with less concentration in SI engine mode with the same load, the best conversion efficiency found in SI mode, followed by HCCI mode with engine load of 3 bar.

Aromatics

The chemical compound of aromatics contain a conjugated planer ring stable ring tends to form easily, and once it is formed tends to be difficult to brake in chemical reactions, this compounds was presented in a largest concentration among other chemical compounds especially

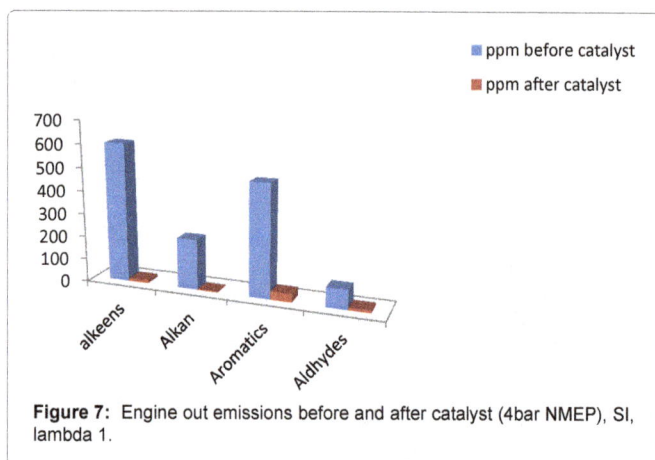

Figure 7: Engine out emissions before and after catalyst (4bar NMEP), SI, lambda 1.

in HCCI engine mode at load of 3 bar, when shifting to higher load of 4 bar concentration was reduced 20%, in SI engine mode aromatics concentration were similar to HCCI mode at the same engine load of 4 bar, the best conversion efficiency presented with SI engine mode, and HCCI mode at the same engine conditions. Conversion efficiency reached 92%.

Aldehydes

Aldehydes are hydrocarbons with additional embedded oxygen atoms. These O-H-C compounds are produced mainly during the combustion of fuels with high oxygen content, (e.g. aldehydes, ketones and alcohols). Aldehydes are formed from incomplete combustion in the cylinder, and by thermal decomposition in the exhaust system, the highest concentration of aldehydes presented in HCCI mode at engine load of 3 bar, and reduced about 55% when engine shifted to higher load of 4 bar, the minimum concentration existed in SI mode at 4 bar, high conversion efficiency presented at all engine modes and different loads, it ranged from 86% - 90%.

Conclusions

(i) Engine-out emissions: Hydrocarbon, CO and NOx are influenced by engine operation and combustion mode (i.e. HCCI or SI). HC and CO emissions were higher in HCCI mode while NOx emission is found to be more in SI. Analysis shows HC, CO in HCCI mode are heavily dependent on engine load.

(ii) Unregulated compounds: in HCCI mode, aromatic concentration were the highest followed by alkanes, alkenes and aldehydes. At lower load all compounds have increased regardless of engine conditions. Alkenes were influenced by changing engine load more than engine modes. Alkane's compound influenced more by engine mode than engine load. Aromatic compound presented with high concentration in all engine modes and operation. Aldehydes species found to be in a low concentration in all engine conditions.

(iii) Post-catalyst emissions: High consistency of HC and CO conversion was approximately 90% over the prototype catalyst independent from in-cylinder conditions which has been designed to combine the catalytic functions required to oxidise CO and hydrocarbons, while simultaneously reducing NOx over the expected range of exhaust-gas temperatures and stoichiometries, during operation of an engine in both HCCI and SI modes. During HCCI higher load, this catalyst shows more efficient conversion of alkenes and aldehydes up to 95%, on the other hand only 70% was achieved in reducing alkanes. Catalyst was more efficient in reducing the chemical

compounds in higher load than lower engine load. Excellent catalyst efficiency was achieved in SI mode, all chemical compounds were eliminated.

References

1. Lü X, Hou Y, Zuz L, Huang Y (2006) Experimental study on the auto-ignition and combustion characteristics in the homogeneous charge compression ignition (HCCI) combustion operation with ethanol/n-heptane blend fuels by port injection. Fuel 85: 2622-2631.

2. Machrafi HS, Cavadias J, Amouroux P (2008) A parametric study on the emissions from an HCCI alternative combustion engine resulting from the auto-ignition of primary reference fuels. Applied Energy 85: 755-764.

3. Su H, Vikhansky A, Mosbach S, Kraft M, Bhave A, et al. (2006) A computational study of an HCCI engine with direct injection during gas exchange. Combustion and Flame 147: 118-132.

4. Wang Z, Shuai SJ, WangG JX, Tian H (2006) A computational study of direct injection gasoline HCCI engine with secondary injection. Fuel 85: 1831-1841.

5. Ying W, Li H, Jie ZZ (2009) Longbao study of HCCI-DI combustion and emissions in a DME engine. Fuel 88: 2255-2261.

6. Zheng ZM, Yao L (2009) Charge stratification to control HCCI: Experiments and CFD modeling with n-heptane as fuel. Fuel 88: 354-365.

7. Abu-Jrai A, Tsolakis A, Megaritis Y (2007) The influence of H_2 and CO on diesel engine combustion characteristics, exhaust gas emissions, and after treatment selective catalytic NOx reduction. International Journal of Hydrogen Energy 32: 3565-3571.

8. Ludykar D, Westerholm J, Almén R (1999) Cold start emissions at +22, -7 and -20°C ambient temperatures from a three-way catalyst (TWC) car: regulated and unregulated exhaust components. The Science of the Total Environment 235: 65-69.

9. Zervas E, Montagne X, Lahaye J (2004) Influence of fuel and air/fuel equivalence ratio on the emission of hydrocarbons from a SI engine. 1. Experimental findings. Fuel 83: 2301-2311.

10. Sartipi S, KhodadadiY AA, Mortazavi S (2008) Pd-doped $LaCoO_3$ regenerative catalyst for automotive emissions control. Applied Catalysis B: Environmental 83: 214-220.

11. Schlatter JC, Taylor KC (1977) Platinum and palladium addition to supported rhodium catalysts for automotive emission control. J Catalysis 49: 42-50.

12. Winkler A, Dimopoulos P, Hauert R, BachM C (2008) Aguirre catalytic activity and aging phenomena of three-way catalysts in a compressed natural gas/gasoline powered passenger car. Applied Catalysis B: Environmental 84: 162-169.

13. Andrews GE, Ahmed FM, Li H (2007) Condensable and gaseous HC emissions and their speciation for a real world SI car test. SAE International.

14. Leclerc B (2008) Detailed chemical kinetic models for the low-temperature combustion of hydrocarbons with application to gasoline and diesel fuel surrogates. Progress in energy and combustion science 34: 440-495.

15. Kaiser EW, Siegl WO, Henig YI, Anderson RW, Trinker FH (1991) Effect of fuel structure on emission from a spark-ignited engine. Environ Sci and Technol 26: 1581-1586.

16. Kaiser EW, Siegl WO, Cotton DF, Anderson RW (1992) Effect of fuel structure on emission from a spark-ignited engine with naphthene and aromatic fuels. Environ Sci and Technol 26: 1581-1586.

17. Kaiser EW, Siegl WO, Cotton DF, Anderson RW (1993) Effect of fuel structure on emission from a spark-ignited engine with olefinic fuels. Environ Sci Technol 27: 1440-1447.

18. Maurya K, Agarwal K (2008) Combustion and emission behaviour of ethanol fuelled homogeneous charge compression ignition (HCCI) engine. SAE.

19. Dec E (2002) A computational study of the effects of low fuel loading and EGR on heat release rates and combustion limits in HCCI engines. SAE.

20. Koopmans L, Denbratt L (2001) A four stroke camless engine operated in homogeneous charge compression ignition mode with a commercial gasoline. SAE.

21. Zhao H (2007) HCCI and CAI engines for the automotive industry. CRC Press Cambridge England.

17

Variation of Driver's Arousal Level when Using ACC and LKA

Suzuki K*

Department of Intelligent Mechanical Systems Engineering, Kagawa University 2217-20, Hayashi-cho, Takamatsu-city, Kagawa 761-0396, Japan

Abstract

In this study, we specifically considered decrease in arousal level stemming from using semi-autonomous driver assistances featuring ACC and LKA functions in operation. 10 young men took part in the experimental study using a driving simulator. For an average of 10 people, the participant rate for those who reached level 4 or above (level 1; Seems not sleepy at all, level 5: Very sleepy) on the sleepiness scale when the system was enabled increased by 34 percent compared with the system-disabled condition. The reaction time to visual stimuli was noticeably delayed and significant differences are acknowledged for a level 4 or above on the sleepiness rating scale. We believe that continuous long term investigation on actual driving (FOT; field operation test) should be carried out to validate the results in this study. We also analysed the relationship between eye closure ratio and the sleepiness rating scale and confirmed that eye closure ratio value of 0.26 corresponded to level 4 on the sleepiness scale where the reaction time delayed remarkably. Drawing on these findings, it is believed that a device can be designed to detect arousal level based on eye closure rate.

Keywords: Driving support; Drowsiness; Facial expression/Arousal level; Reaction time; ACC; LKA; Eye closure ratio

Introduction

LKA (Lane Keeping Assistance) and ACC (Adaptive Cruise Control) have variety of benefits about the lessening of driving workload and the enhancement of the risk-avoidance behaviour. A previous study has reported that the ACC and LKA system remarkably decrease the driver's workload in terms of braking and steering operation [1]. This experimental study reported that test subjects mentioned it is easy to concentrate on driving to avoid collision to a leading vehicle when ACC and LKA are implemented in the vehicle. For this reason, these systems are gradually pervading into the market, and are expected to be even more widespread in the future.

However, previous studies have reported that a lack of understanding of performance requirement and function of these driver assistance systems by drivers can result in inattentive behaviour at the wheel. On the backdrop of their spread, measures coping with these inattentive behaviours should be considered in the design of functionality of such system to enhance user and social acceptance. The case below exemplifies the kind of inattentive behaviour which drivers are exposed to when driving an ACC system-enabled vehicle. In this example, the driver is overestimating the driver assistance functions, which led her/him to rely on the system to perform specific tasks and to be oblivious of the unreliability of the system [2]. Besides, although there is no remarkable difference in arousal level for a vehicle non-equipped with an automatic stop function, it has been reported that such arousal level is more likely to decrease if a stop function system is enabled [3].

Driver assistance featuring ACC or LKA has obviously various merits in lessening driver's workload and supporting risk-averse behavior. However, actual utilization of such driving assistances has risen concerned over increase in inattentive behavior e.g. decrease in arousal level.

In this study, we specifically considered decrease in arousal level stemming from using semi-autonomous driver assistances featuring ACC and LKA functions in operation. We investigated variation of driver's arousal level through experimental studies using a driving simulator. 10 young men took part in the experimental study. The level of driver assistance functionality that is subjected to analysis corresponds to level 2 on the automatic driving level by NHTSA [4], whose quick diffusion is expected to occur in a near future.

The purpose of this research is broadly divided into the following four points.

I. Analysis of driver's arousal level

We analysed how the driver's arousal level is affected by comparing two running conditions: (1) featuring active ACC and LKA driver assistance; and (2) featuring disabled driving assistances. In this study, we estimated arousal level of drivers in terms of the sleepiness level. To estimate the arousal level of the initial timing of the drive, we used the Stanford Sleepiness Scale. For quantifying the fluctuation of arousal level during the use of ACC and LKA, we used the sleepiness objective evaluation values.

II. Correlation analyses of visual stimulation and arousal level

The purpose of this study is to clearly determine two conditions if marked drop in arousal level was observed owing to the utilization of ACC/LKA-enabled driver assistance system e.g.

(1) Arousal level presenting remarkable delay in reaction time; and

(2) Intensity of the driver's reaction time delay to visual stimuli.

III. Analysis of the driver's behavior when the arousal level is noticeably dropping

We made analysis in the feature of driving behavior of a driver whose arousal level markedly changed when the use of driver assistance featuring active ACC and LKA systems were found to drop the driver arousal level.

*Corresponding author: Suzuki K, Department of Intelligent Mechanical Systems Engineering, Kagawa University 2217-20, Hayashi-cho, Takamatsu-city, Kagawa 761-0396, Japan, E-mail: ksuzuki@eng.kagawa-u.ac.jp

IV. Correlation analysis between arousal level and eye blink

We examined the average eye closure rate (the ratio between eye closing and opening during for every 60 seconds) for each different arousal level and clarified the eye closure average rate at which the reaction time to visual stimuli is noticeably delayed.

Experimental Methods

ACC and LKA operating patterns

We made use of the driving simulator constructed by the Kagawa University and created an environment in which ACC and LKA systems were in operation. To help participants to practically get a grasp of the functions limitation as well as the different working patterns of the system, we set up a trial run practice before the experiment. This run practice featured the function of distance auto-regulation to an accelerating and decelerating vehicle ahead.

The ACC operating parameters were set forth as below: The distance to a leading car was set up so that the vehicle could maintain the time head way of 1.5 sec and the maximum deceleration control system was set to 3m/s². The system control was also configured to stop functioning if the participants applied the brakes when the ACC system was in operation. Subsequently, we equipped the ACC system with a forward vehicle collision warning system. A warning alarm was set to be triggered for a Time to Collision value equivalent to 2.5 sec. The operating conditions of the LKA were set forth as explained below. A steering torque was set to minimize the lateral drift between the centre of vehicle and the lane one second after its estimation based on lateral acceleration and lateral velocity of the vehicle. This LKA was set to allow the system to automatically control the cornering motion without any steering correction by the driver up to 2.4 m/s² lateral acceleration at the vehicle center of gravity. The system was also configured to be overridden if the steering torque exerted by the driver exceeds that generated by the LKA system while in use. The LKA system was also equipped with a lane departure warning system. The system was set to prompt a warning alarm when the spare distance between the front tire center and the lane marker before a lane departure was less than 0.2 m. The driver assistance information about the maintenance of the car-to-car distance and the lane departure prevention were presented on the instrument panel, as shown in the Figure 1.

Running course

We designed a two-lane, highway-type of road with a lane width of 3.5 m in the simulator. Details of the shape-related features of the driving courses are listed below:

a) Minimum curvature radius; 700, 3000 and 5000 [m]

b) Longitudinal gradient; 0 [deg]

c) Percentage of straight line segments and curves section; straight segment: curves section = 40%: 60%

We also added straight and curvilinear easement segments based on actual Japanese road construction standard.

Participants

10 young men took part in the experiment. The average age and the standard deviation of ages were 22.3 ± 1.0 years old. The experiment participants took part in the experiment after following the inform consent procedure, which was carried out upon distribution of a paper-based explanatory document sent to them one week prior to the start of the test. Furthermore, the conduct of the experiments took place upon acceptance of the test content by the Kagawa University Ethics Committee, which was granted prior to the experiments.

Control methods of the initial value of arousal level

The Figure 2 shows a flow of the experiment. We set up a nondriving task just after the start of the test. During this task, experiment participants sat down on the seat in a driving simulator and did not drive a car. When we confirmed the experiment participants became low arousal state, we did an arousal-stimulating task for getting awakening effect. After this arousal-stimulating task, we initiated the actual running test with driving and we investigated what kind of differences would be actualized in terms of the decrease of arousal level with and without using the system. Our analysis of the driver's alertness level when using the semi-automatic driver assistance aimed at considering the differences in term of arousal level, which were observed by comparing the system-enabled configuration with the disabled one. Therefore, we unified the following two conditions. The first condition is the level of arousal at the start of the test without driving before the arousal-stimulating task and the level of arousal at actual start of the running test just after the arousal stimulating task. We asked the test participants to evaluate the arousal level at these two timings and proceeded with the experiments when we confirmed that

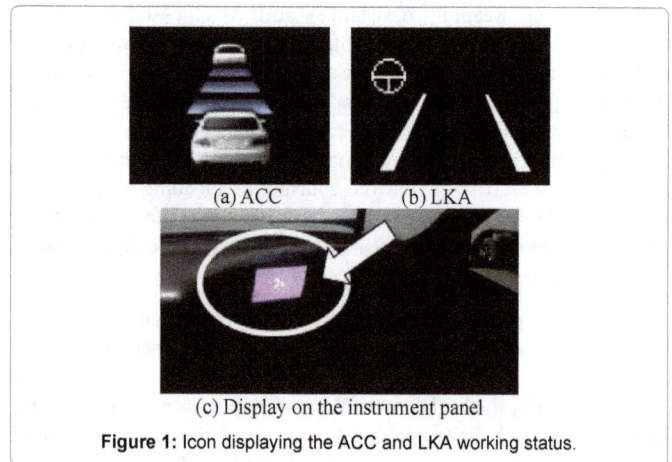

(a) ACC (b) LKA

(c) Display on the instrument panel

Figure 1: Icon displaying the ACC and LKA working status.

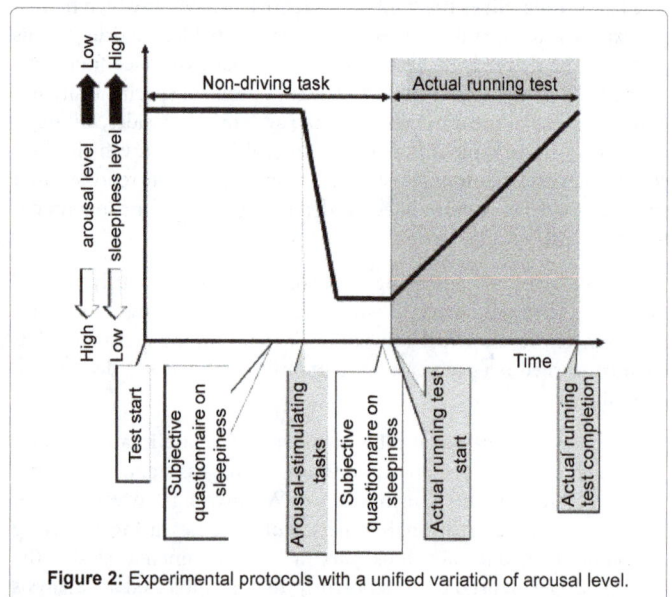

Figure 2: Experimental protocols with a unified variation of arousal level.

the test participants could realize the predetermined arousal level. The reason why we focused on not only at actual start of the running test but also the start of the test without driving is that it is important to realize the same transition of the arousal throughout two-day of the test for each experiment participant, as mentioned below.

To evaluate the arousal level, we conducted a subjective survey on sleepiness by applying the Stanford Sleepiness Scale (SSS) [5] as shown in Table 1. SSS is a sevenfold level-designed method of assessing the subjective feel of sleepiness of the driver. One indicates a normal state of arousal and seven shows the lowest state of vigilance. The test was launched after experiment participants had declared score equivalent to level 5 in the SSS questionnaire. The reason why level 5 of the SSS survey was chosen as a low arousal initial level lies in the fact the lowest level of arousal throughout the course of a whole day never exceeded level 5, according to the findings of a pre-survey in which the evolution of arousal level was monitored every hour over a period of one day.

The second condition is the transition of the arousal level. The conduct of the test was spread out over two days for two reasons:

(1) Time was needed, (2) To avoid ethical issues in terms of minimizing workload for participating in the test. Even if each experiment participant starts the experiments at the exact same time, it doesn't imply that the participant's arousal level transition is identical throughout the test. That is why we cannot succeed in determining whether the activation of the system was of any influences on the arousal level transition when the test was conducted in system disabled condition on the first day and in enabled condition on the second day. The experimental protocol was established to assign test participants with arousal stimulating task (stimulation) to provoke a unified variation of arousal. We could establish such protocol and carry out the test accordingly since we could ensure that the arousal level transition was identical throughout the two-day of test launched upon completion of the arousal-stimulating task and the subsequent alertness recovery. Besides, we managed to lessen the influence exerted by the order in which the experiment participants took part in the test, which was conducted with and without the driver assistance system (ACC+LKA) activated. Participants took strongly refreshing tablet candy (including menthol component) as an arousal-stimulating task. Products sold on the market effecting on level of arousal are composed of caffeine. The awakening effect cannot be quickly obtained from its use since the effect occurs 30 mins after consumption, upon absorption [6]. Thus, we tapped into caffeine-free, highly refreshing candy tablet during this experiment.

Evaluation method of arousal level

Facial expression-based sleepiness scale, which was reported as a useful rating tool by previous studies, was used in this study. This method consists of two examiners looking at the driver's facial expression and assessing the arousal level on a 5-fold level based scale, as shown in Table 2 [7]. For the test, we assessed the arousal level with a six-fold level rating scale, which included an extra level designated as "sleeping state". We added this extra level since drivers could possibly fall completely asleep during the experiment.

Experimental Results

Analysis of driver's arousal level

The Figure 3 shows a measure sample of the transition of the facial expression-based sleepiness evaluation value over a period of

30 minutes while a driver assistance system (ACC+LKA) was in use. In this measurement sample, the experiment participant reached level 5 on the sleepiness scale after a 20-minute drive. We subsequently considered the question of how many experiment participants out of the 10 involved in the experiment showed sign of low level of arousal for the two following drive conditions: (1) driver assistance system (ACC+LKA) enabled, and (2) driver assistance system disabled. The Figure 4 shows the average level of sleepiness over the entire run by each participant.

Only experiment participants in the green frame fell to a lower arousal level when the driver assistance was disabled. 8 experiment participants out of 10 fell to a lower level of arousal at the driver assistance system was in operation.

Afterwards, we focused on the proportion of sleepiness levels and presented the average values of all the ten experiments participants in

Degree of Sleepiness Scale Rating Feeling	Rating
Feeling active, vital, alert, or wide awake	1
Functioning at high levels, but not at peak; able to concentrate	2
Awake, but relaxed; responsive but not fully alert	3
Somewhat foggy, let down	4
Foggy; losing interest in remaining awake; slowed down	5
Sleepy, woozy, fighting sleep; prefer to lie down	6
No longer fighting sleep, sleep onset soon; having dream-like thoughts	7

Table 1: The Stanford sleepiness scale.

Sleepiness Evaluation Values	Behavior Characteristics
Seems not sleepy at all	Quick and frequent gaze shift Regular eye blink Intense activity of the body
A little sleepy	Slow eyes shift Lips ajar
Sleepy	Slow eye blink Mouth in motion Sitting position readjustment and hands facial contact
Pretty sleepy	Seemingly conscious eye blink Needless movement of the whole body i.e. Shoulder shrug Frequent yawning and deep breathing Eye blink and eyes movement slowdown
Very sleepy	Closed eyelids The head is leaning forward The head is leaning backward
Sleeping	Closed eyes more than 90 percent of the time

Table 2: Sleepiness objective evaluation values and behaviour.

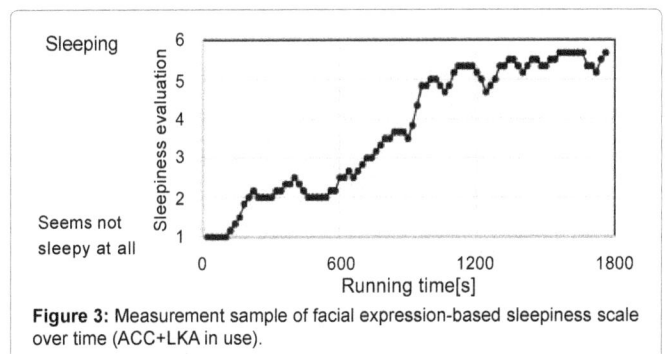

Figure 3: Measurement sample of facial expression-based sleepiness scale over time (ACC+LKA in use).

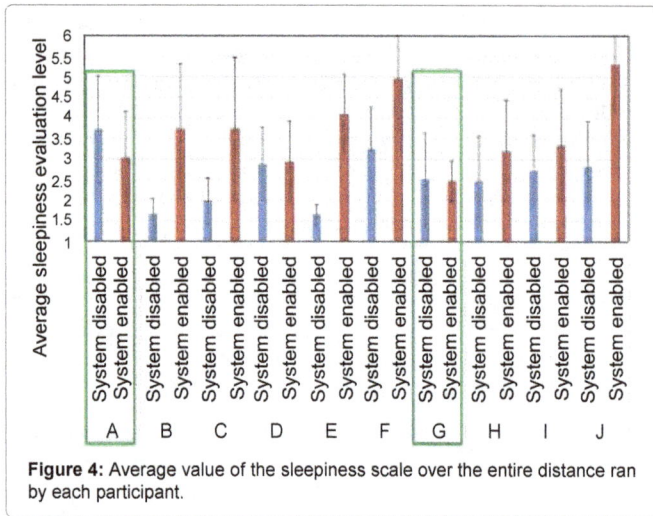

Figure 4: Average value of the sleepiness scale over the entire distance ran by each participant.

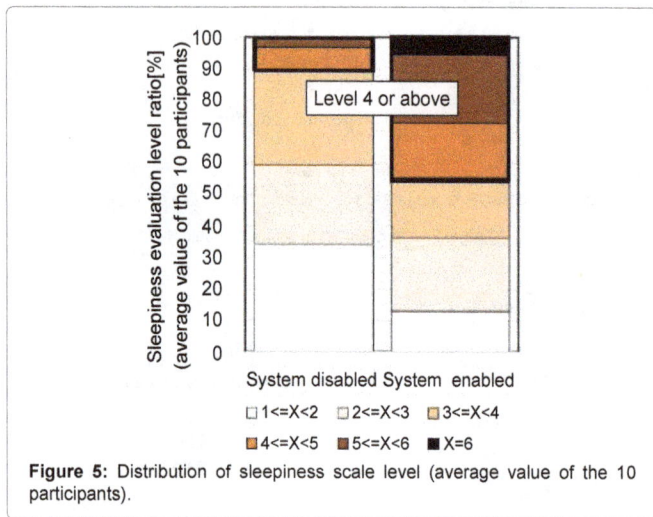

Figure 5: Distribution of sleepiness scale level (average value of the 10 participants).

the Figure 5. When the system is in operation, the ratio characteristically decreased for a sleepiness level lower than 4 and increased for a level higher than 4. When using a driver assistance system, 8 experiment participants out of 10 fell to a lower level of arousal than in absence of the system. The increased proportion of participants who showed at least level 4 of sleepiness (sleepy) is also salient if a driver assistance system is in use. For an average of 10 people, the time frequency of the driving that reached level 4 or above on the sleepiness scale when the system was enabled increased compared with the system-disabled condition.

Correlation analysis of visual stimulation and arousal level

We analysed the relationship between the arousal level and the experiment participants' reaction time to a red lamp, which was placed near the central visual field of the simulator, simulating the brake lights of the vehicle ahead. The reaction time was investigated by the means of a switch installed on the steering wheel. We used the sleepiness rating scale shown in Table 1 above for the estimation of the level of arousal. An average sleepiness level over a period of one minute before the lamp lights was leveraged in this experiment. Since we endeavored to parallel the difference in response time with the difference in arousal level, the reaction times obtained were classified by arousal level. It is known that the probability density function associated with reaction time doesn't

pertain to a normal distribution but a logarithmic normal one. The distribution of the data must be of a normal nature to achieve a test of statistical significance, such as t-test. Therefore, we have carried on analysis with a verification method using t-test upon conversion into logarithmical data of the reaction time data, as reported in previous study [8]. The sleepiness level-specific reaction times are shown in the Figure 6 along with the significance test findings.

According to the Figure 6, the reaction time is noticeably delayed and significant differences are acknowledged for a level 4 or above on the sleepiness rating scale. It was confirmed that the reaction time is further delayed for at least level 4 of arousal on the sleepiness scale in contrast with a perfect state of awakening.

For levels less than 4, the findings suggest that drivers tend to react the same way as if perfectly awaken even if arousal levels drop. It is also suggested that a correlation exists between increase in the ratio of level 4 or above and the delay of reaction time to visual stimuli. This is a result describing the reaction time to visual stimuli and is different from the reaction time for taking over the system when the systems are difficult to support driving.

But, same tendency that reaction time in level 4 or above on the sleepiness scale is prolonged can be estimated. To investigate the relation between sleepiness level and the reaction time for taking over the systems will be an additional study.

The differences in arousal level with and without driver assistance system (ACC+LKA) are referred to in the Figure 5 above.

The ratios within the black boxes in the Figure 5 indicate the level 4 and above share on the sleepiness scale. Our findings suggest that the arousal level has lessened due to the use of driver assistance system.

Analysis of the drivers' behavior when the arousal level noticeably dropping

We analyzed the behavioral changes associated with the fluctuation of the driver's arousal level. We carried out analysis to contribute to the basic data intended to detect low arousal level in keeping with the driving behavior. We also considered the characteristics of the drowsiness-prone driver when a driver assistance system was in use. The horizontal axis of the chart below indicates the increase of level 4 ratio if driving assistance was in use compared to it was not in use. We

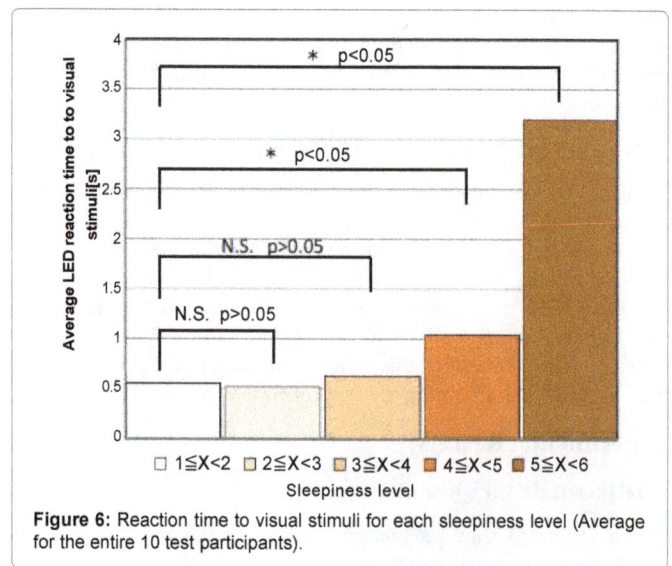

Figure 6: Reaction time to visual stimuli for each sleepiness level (Average for the entire 10 test participants).

brought our attention on level 4 and above because the reaction time was further delayed in such condition. More specifically, we focused on the increased percentage of level 4 or above and expressed on the horizontal axis. We expressed the resulting increase of each variable on the vertical axis when the system was in use. Yet, additional analysis will be necessary in the future to substantiate these findings since the amount of data collected presently was insufficient.

The Figure 7 shows the relationship between the increase percentage of level 4 or above and accelerator/brake readiness rate. Almost no correlation can be seen ($0 < | r | \leqq 0.2$).

The Figure 8 shows the relationship between the increase percentage of level 4 or above and gripping percentage e of the lower half of the steering wheel. Low correlation was confirmed ($0.2 < | r | \leqq 0.4$). We observed that the more the arousal level of the driver dropped down the more likely she/he was to grip the lower half of the handle.

The Figure 9 shows the relationship between the increase percentage of level 4 or above and single-handed steering wheel grip ratio. A correlation was confirmed in this relationship ($0.4 < | r | \leqq 0.7$). We observed that the more the arousal level of the driver dropped down the more likely she/he was to grip the steering wheel of the car single handedly. We asked test participants after the investigation why they tended to show the single-handed grip. During the use of ACC and LKA, test participants felt that they can rely on the driving support devices and showed a tendency to grip a steering wheel with a single hand.

It is believed that this driver behaviour like the rate of singlehanded grip of the steering wheel can be used to detect the situation that the driver becomes low arousal.

Correlation analysis between arousal level and eye blink

There are many possible methods to estimate the arousal level by analysis of the eye blink. The eye closure average rate has been used as an indicator in this study. Eye closure rate is reported as a highly

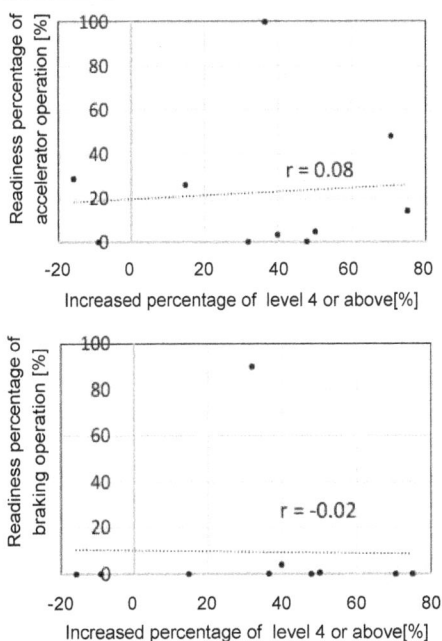

Figure 7: Relationship between the increase of sleepiness level and accelerator/brake readiness rate.

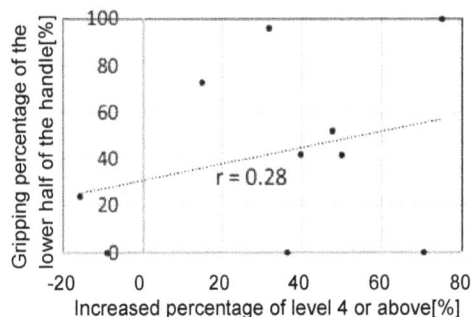

Figure 8: Relationship between the increase of sleepiness level and the gripping rate of the lower half of the handle.

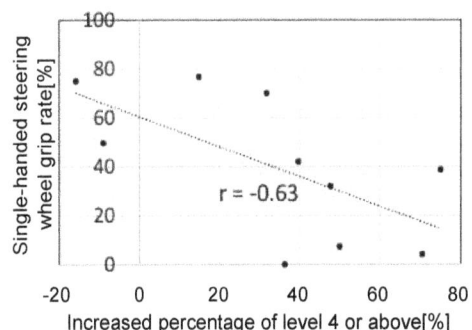

Figure 9: Relationship between the increase of sleepiness level and the rate of single-handed grip of the steering wheel.

accurate arousal level indicator by previous study [9]. The eye closure rate is obtained by the ratio between eye closure and eye opening for a given unit of time.

We analysed the eye closure ratio for the 5 different levels of the facial expression-based sleepiness rating scale. Results are shown in the Figure 10.

The Figure 10 reveals that increase in eye closure rate is commensurate with an increase in facial expression-based sleepiness levels. However, the rate did not increase in a linear function but exponentially. In a previous study by author [9] to evaluate the sleepiness level, we have already clarified that there is a non-linear relationship between the sleepiness level and eye closure rate. We make an exponential approximation to pass through the median values, a red dashed line graph can be obtained, as shown in the figure above. Working on the principle of a relationship between eye closure rate and sleepiness level as shown by the dashed-line chart, we confirmed that eye closure ratio value of 0.26 corresponded to level 4 on the sleepiness scale where the reaction time delayed remarkably. Drawing on these findings, it is believed that HMI device can be designed to prevent the low arousal situation by emitting arousal-stimulating alarm signals before the eye closure rate value exceeds 0.26.

Conclusion

We have listed below a summary of the findings presented above.

Analysis of the arousal level of the driver

We can draw the conclusion that the arousal level is likely to drop down if a driver assistance system (ACC+LKA) is in operation since 8 out of 10 experiment participants saw their arousal decreasing when such a system was enabled.

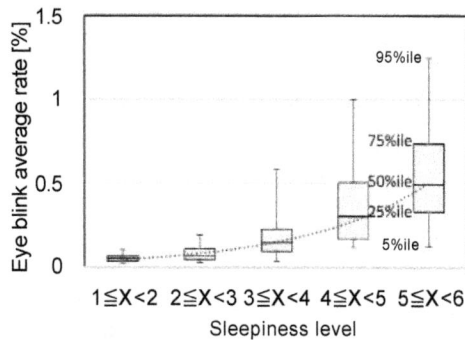

Figure 10: Relationship between sleepiness levels and eye closure rate.

The increased temporal frequency of sleepiness scale level 4 or above (sleepy, very sleepy, and sleeping) is apparent when a driver assistance system is in use.

Correlation analysis of visual stimulation and arousal level

The reaction time is further delayed for a level 4 or above of arousal on the facial expression-based sleepiness scale rather than in a perfect state of awakening. Comparing both conditions e.g. driver assistance system enabled or disabled, the arousal level has lessened due to the use of driver assistance system and that the inattentive driving rate increased in comparison with the system disabled condition. This is a result describing the reaction time to visual stimuli and is different from the reaction time for taking over the system when the systems are difficult to support driving. To investigate the relation between sleepiness level and the reaction time for taking over the systems will be an additional study.

Analysis of the driver's behaviour when the arousal level is noticeably dropping

It was shown the relationship between single-handed steering wheel grip ratio and the arousal level. A correlation was confirmed in this relationship ($0.4 < | \ r \ | \leqq 0.7$). We observed that the more the arousal level of the driver dropped down the more likely she/he was to grip the steering wheel of the car single-handedly.

Correlation analysis between arousal level and eye blink

We analysed the eye closure ratio for the 5 different levels of the facial expression-based sleepiness rating scale. Working on the principle of a relationship between eye closure rate and sleepiness level, we confirmed that eye closure ratio value of 0.26 corresponded to level 4 on the sleepiness scale where the reaction time delayed remarkably. Drawing on these findings, it is believed that a device can be designed to detect arousal level based on eye closure rate. In this study, we have investigated the actualization of low arousal situation of driver in using semi-autonomous driving support systems featuring adaptive cruise control and lane keeping assistance. We have observed that the driver tends to become low arousal situation earlier when using the system than when the system is disabled. This investigation was carried out in a driving simulator with participants of only 10 young test subjects. We think it is difficult to say somewhat that the completely same tendency will be actualized in real driving on public road. We believe that continuous long term investigation on actual driving (FOT; field operation test) should be carried out to validate the results in this study.

References

1. Suzuki K, Oda K, Miichi Y (2015) Driver behavior when installing with ACC and lane keeping assistance System. Transaction of JSAE 46: 145-151.

2. Inagaki T (2010) Human's over trust in and overreliance on driver assistance systems, Technical research report of the Institute of Electronics, Information and Communication Engineers 110: 21-24.

3. Uno H, Uchida N, Noguchi M (2002) Effect of driver support system on arousal level of drive. Proceedings of JSAE 114: 17-20.

4. NHTSA (2013) Preliminary statement of policy concerning automated vehicles.

5. Hoddes E, Zarcone E, Smythe H, Philips R, Dement WC (1973) Quantification of sleepiness: A new approach. Psychophysiology 10: 431-436.

6. Ministry of Education Culture, Sports, Science and Technology (2002) Report of general research on reservation of the comfortable sleep in everyday life.

7. Kitajima H, Numata N, Yamamoto K, Goi Y (1997) Prediction of automobile driver sleepiness, 1st report. Transaction of JSME 63: 93-100.

8. Morita K, Mashiko J, Okada T (1998) Delay in response time of automobile driver due to gazing at an in-vehicle navigation display device. Transaction of the Illuminating Engineering Institute of Japan 82: 121-130

9. Suzuki K, Aoki H, Yamada K, Minakami Y, Kawamura H (2010) Detection of the low wakefulness level of car driver based on the eye blinking behaviour and the effectiveness of awakening alarm. Transaction of JSME 76: 2611- 2620.

18

The Simulation of Non-Newtonian Power-Law Fluid Flow in a Centrifugal Pump Impeller

Xuan-Linh N* and Huan-Xin L

Key Laboratory of Pressurized Systems and Safety, Ministry of Education, East China University of Science and Technology, Shanghai 200237, P.R. China

Abstract

The performances of a centrifugal pump are affected when handling non-Newtonian liquids, requiring a larger power input than same pump handling water. In this paper, the power-law fluid model was implemented to study the non-Newtonian effects on the centrifugal pump impeller when the working fluid is the 7.5% by weight aqueous solutions of carboxy-methycellulose (CMC). The standard k- ε two-equation model is used for the closure of turbulent incompressible flow. The results of non-Newtonian power law fluid are compared with those of water in the turbulence flow regime. It is found that the steady flow characteristics of the non-Newtonian power law fluid and of the water have a plenty of differences. This indicates that the non-Newtonian power law characteristics of the CMC 7.5% might be an important factor in determining the internal flow patterns and the general performances of a centrifugal impeller.

Keywords: Centrifugal Pump Impeller; CFD technique; Non-Newtonian fluid; Numerical simulation

Introduction

The flows of non-Newtonian fluids in centrifugal pumps are very common and important in numerous engineering applications such as water supply and irrigation, flood control, sewage handling and treatment, chemical, biomedical and food processes. Understanding of the characteristics of such non-Newtonian fluid flows is of considerable interests in many areas of science and engineering JJN Kalombo et al. [1]. In recent years, with the development of computation fluid dynamics, a large amount of studies have been carried out and good progresses in understanding those flows have been obtained. EYK. Ng et al. [2] studied the non-Newtonian behavior of blood, using both the Casson and power-law relations for the fluid model with the dynamic viscosity fitted by experimental data. The predicted results are compared with those of Newtonian fluid flow in the mixed-flow "blood" pump. C. Arakawa et al. [3] found that the performance of a slurry pump of transporting solid-liquid two-phase fluid is lower than that of the single-phase water transport. The particles have a great influence on the pressure distribution and pressure changes in the pump impeller. BK Gandhi et al. [4] experimentally investigated a centrifugal slurry pump to explore the applicability of affinity relations when the rotation speed of the pump is changed. B. Y. Zhao et al. [5], LJW Granham et al. [6] and L. Pullum et al. [7] proposed a modification of the hydraulic institute head deration method that is suitable for all homogeneous non-Newtonian fluids. Genguang Zhang et al. in 2008 studied the hydrodynamic characteristics and scaling law of the model pump when aqueous Xanthan gum solution was the working fluid. The pump rotation speed and the solution concentration vary in relatively wide ranges. Special attention was focused on the effects of non-Newtonian fluid properties and pump operating conditions on the pump performances. YC Yassine et al. [8] discussed the effect of sand/water slurry flow on the pump performance. J Crawford et al. [9] demonstrated that it was possible to pump very viscous, high yield stress slurries with limited head and efficiency de-rating using centrifugal pumps, provided that positive suction conditions were maintained at all time.

This paper aims to study the internal flow and performance of a centrifugal impeller when the working fluid is the 7.5% by weight aqueous solutions of carboxy-methycellulose (mentioned as CMC

7.5% for simplicity later). The non-Newtonian constitution of the fluid is described by the power-law model. The predicted results are compared with those for the flow of water, which is a well-known Newtonian fluid, to analyze the non-Newtonian effects of the working fluids on the centrifugal pump.

Impeller and Grids

Centrifugal impeller

The studied impeller is shrouded, representing the rotating part of an industrial multi-stage low-specific speed centrifugal pump as reported by N. Pedersen et al. in 2003 (Figure 1).

Figure 2 shows the three-dimensional (3D) view of the impeller, the flow domain and the grids. The impeller consists of six simple curvature backward swept blades of constant thickness. The axial height is tapered nearly linearly from 13.8 mm at the inlet to 5.8 mm at the outlet. Table 1 summarizes the main dimensions of the impeller.

Grid-independence verification

It is necessary to carry out the independency verification of the grid system before CFD computation. The verified case of the current study is the pump at the design point, i.e., the rotation speed of 725 rpm and the flow-rate of 3.06 l/s. The head is 1.75 m, according to the designation. Seven grid-systems are generated in the computational domain to perform this independency verification. The calculated heads using different meshes are compared in Figure 3. The results shows the mesh which has more than 1,395,952 grid nodes is able to obtain a grid-independent solution, as the calculated head doesn't change apparently when the mesh is further refined. Therefore, the calculated results discussed in the follows are obtained using this mesh.

*Corresponding author: Huan-Xin L, Key Laboratory of Pressurized Systems and Safety, Ministry of Education, East China University of Science and Technology, Shanghai 200237, P.R. China, E-mail: hlai@ecust.edu.cn

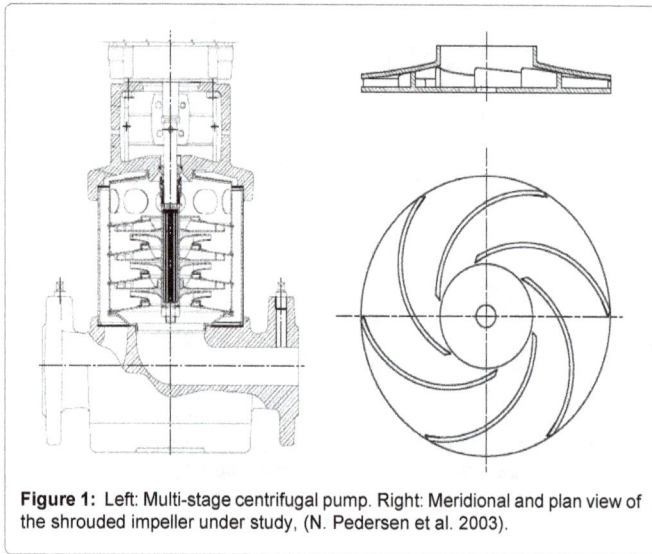

Figure 1: Left: Multi-stage centrifugal pump. Right: Meridional and plan view of the shrouded impeller under study, (N. Pedersen et al. 2003).

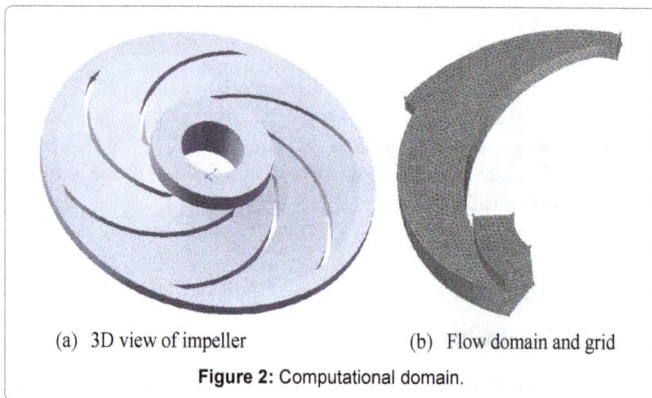

(a) 3D view of impeller   (b) Flow domain and grid

Figure 2: Computational domain.

Geometry			
Inlet diameter	D_1	77	(mm)
Outlet diameter	D_2	190	(mm)
Inlet height	b_1	13.8	(mm)
Outlet height	b_2	5.8	(mm)
Number of blades	Z	6	(-)
Blade thickness	t	3.0	(mm)
Inlet blade angle	β_1	19.7	(deg.)
Outlet blade angle	β_2	18.4	(deg.)
Specific speed	N_s	26.3	(-)
Flow conditions			
Flow rate	Q	3.06	(l/s)
Head	H	1.75	(m)
Rotational speed	n	725	(rpm)

Table 1: Impeller geometry and operating conditions.

Control Equations and Boundary Conditions

The control equations

In the present calculation, only the flow in the rotating impeller is considered. In the rotating reference frame, the flow is regarded as steady and incompressible. The governing equations are as follows:

Continuity:

$$\nabla \cdot \vec{u} = 0 \qquad (1)$$

Momentum equation:

$$\frac{\partial}{\partial x_j}\left(u_i u_j\right) = f_i - \frac{1}{\rho}\frac{\partial p}{\partial x_i} + \mu \frac{\partial^2 u_i}{\partial x_i x_j} \qquad (2)$$

Energy conservation equation:

$$\nabla \cdot \left(\rho \vec{u} T\right) = \nabla \cdot \left(\frac{\lambda}{c_p}\nabla T\right) + \Phi \qquad (3)$$

Where f_i is the mass force acting on the fluid. ρ is the density, λ and c_p are the heat conductivity and the constant-pressure specific heat, respectively. Φ is the viscous dissipation term. The viscosity μ in equation (2) distinguishes the fluids. For a Newtonian fluid, μ is a fluid property and is independent of the flow. But for a non-Newtonian fluid, μ is generally mentioned as the apparent viscosity, whose value is decided by not only the fluid itself but also by the status of fluid flow. In the present study, the apparent viscosity of CMC 7.5% is described by the power-law fluid model, $\mu = K\dot{\gamma}^{n-1}$, where $\dot{\gamma}$ is the shear-rate, K is the consistency index in power law model, and n is power law index. These parameters for CMC 7.5% are summarized in Table 2.

Numerical method and turbulence model

In engineering applications, the control equations (1)~(3) are often further simplified to the steady Reynolds Averaged Navier-Stokes (RANS) equations, in which turbulence model is needed for closure. In the present study, four turbulence models (the standard k-ε, RNG k-ε, realizable k-ε and RSM) are selected respectively and compared. The RANS equations coupled with the turbulence model are numerically solved using a commercial CFD package, the FLUENT. The simple method is employed for calculation, and the iteration residual is set to 10^{-5} for the convergence.

Figure 4 compares the calculated head of the pump. It reveals that compared with the design value of head, the error of 2.3% by the

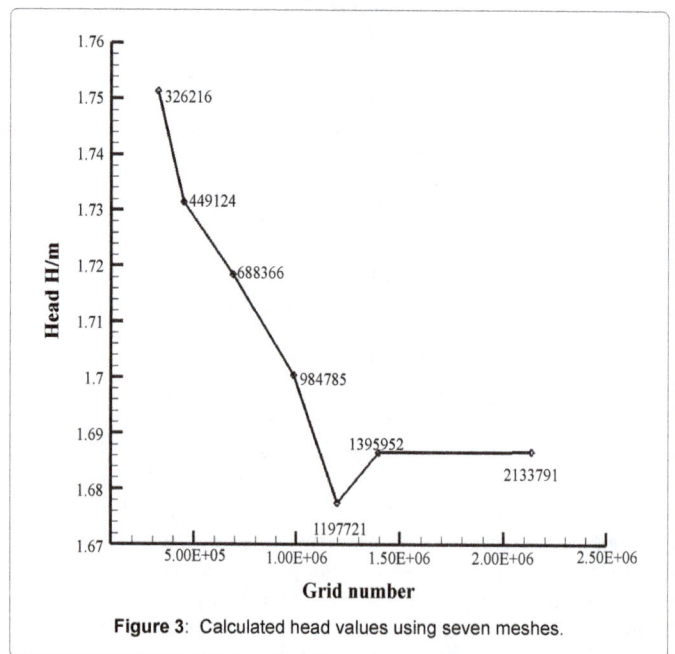

Figure 3: Calculated head values using seven meshes.

K (Pa.sⁿ)	n	μ_{min} (Pa.s)	μ_{max} (Pa.s)	ρ (kg/m³)
0.525	0.533	0.03	0.17	1001

Table 2: Power-law parameters for CMC 7.5% [14].

standard k-ε is the lowest. The Realizable k-ε model is the second, has an error of 4.8%. The RNG k-ε is the third (its error is 5.6%) and the worst is the RSM (the error is 6.6% for this case).

Using the same grids and the convergence criterion, the computational time for the four models is shown in Table 3. The RSM model reaches convergence taking much longer time than the three two-equation turbulence models. Considering the results in Figure 4, the standard k-ε is selected for further analysis.

Boundary conditions

At the inlet of the flow domain shown in Figure 2b, the inflow velocity u_{in} is imposed according to the flow-rate,

$$u_{in} = \frac{4Q}{\pi \times D_{in}^2} \tag{4}$$

Where Q is the flow rate, D_{in} is the impeller inlet diameter. Because turbulence model is employed for the closure of RANS equations, boundary conditions for turbulent kinetic energy k and turbulent dissipation rate ε are also needed. They are specified as follows:

$$\left.\begin{array}{l} k_{in} = 0.005u_{in}^2 \\ \varepsilon_{in} = C_\mu^{\frac{3}{4}} \dfrac{k_{in}^{\frac{3}{2}}}{l} \end{array}\right\} \tag{5}$$

Where l is the turbulence length scale, which is calculated empirically as $l=0.07D_{in}$; $C_\mu =0.09$.

At the outflow boundary, reference pressure is specified. The flow at this bound is assumed to be not returning, so the general transporting quantity φ ($\varphi = u, v, w, k, \varepsilon$) satisfies:

$$\frac{\partial \varphi}{\partial s} = 0 \tag{6}$$

Where s denotes the local flow direction at the outflow bound.

On solid walls of the flow passage of the impeller, no-slip conditions are imposed for the velocity components. Because the turbulence quantities k and ε are for fully turbulent flows, they are not directly integrable to solid walls. Alternatively, standard wall function is applied to solid walls as follows (Versteeg and Malalasekera in 2011 [10]:

$$u^+ = \frac{u_P}{u_\tau} = \frac{1}{\kappa}\ln\left(Ey_P^+\right), \ y_P^+ = \frac{y_P u_\tau}{\nu}, \ k_P = \frac{u_\tau^2}{\sqrt{C_\mu}}, \ \varepsilon_P = \frac{u_\tau^3}{\kappa y_P} \tag{7}$$

Where u_τ is the shear velocity, y is the coordinate normal to a solid wall. Von Karman constant $\kappa = 0.41, E = 9.8$. The subscript P denotes variables pertaining to the first inner node away from a solid wall. With the wall function, turbulence quantities k and ε near solid walls can be calculated according to equation (7). Based on all these boundary conditions, the RANS equations coupled with the turbulence models are well posed and are numerically solvable.

Results and Discussion

Performance curve

Head curves are generated using the results of the various conditions, calculated using CFD analysis. A comparison of head curves for different working fluids is shown in Figure 5. The head of pump is defined as follows:

$$H = \frac{1}{\rho g}\left(P_{out} - P_{in}\right) \tag{8}$$

Where P_{in} and P_{out} are the total pressure at the impeller inlet and outlet, respectively. g is the gravity.

Figure 5 clearly shows that the pump head for handling water is higher than that for handling CMC 7.5%. This can be explained, as the friction loss of the centrifugal pump consumes mechanical energy, so when the viscosity of the fluid is increased, and the head will be reduced under the same flow rate.

Pressure distribution

The pressure in the impeller has been expressed by the total pressure coefficient C_p, defined as:

Turbulence model	CPU time (h)
Standard k-ε	0.67
RNG k-ε	1.67
Realizable k-ε	1.73
RSM	13.3

Table 3: Computation time of different turbulence models.

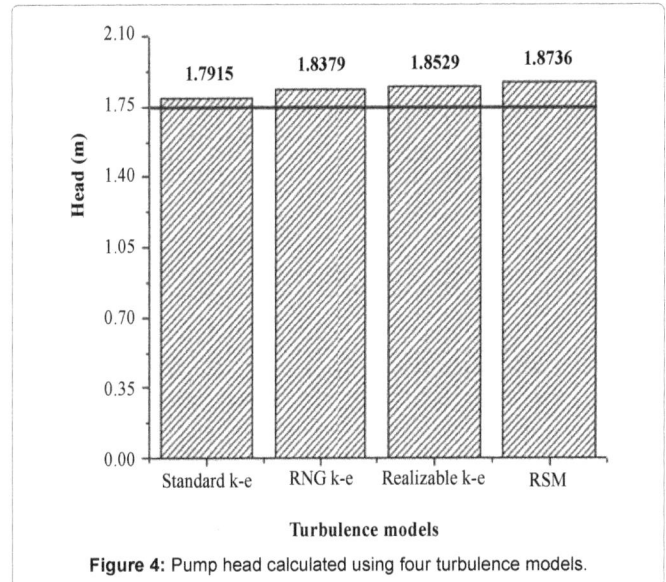

Figure 4: Pump head calculated using four turbulence models.

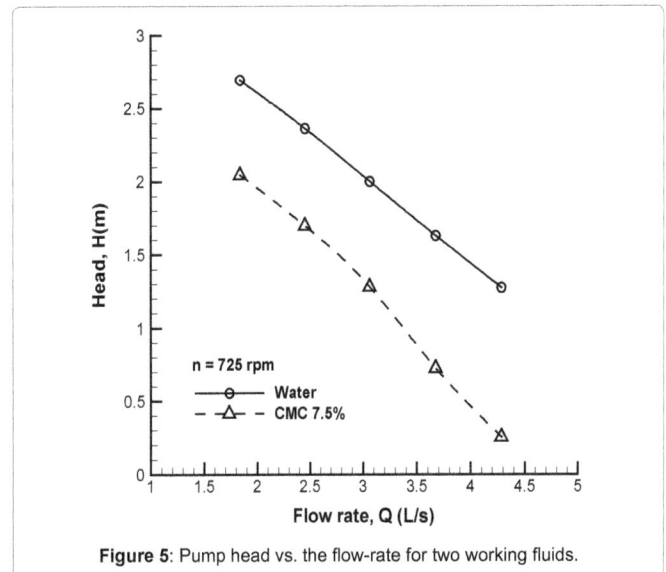

Figure 5: Pump head vs. the flow-rate for two working fluids.

$$C_p = \frac{P_i - P_{t0}}{\frac{1}{2}\rho U_2^2} \tag{9}$$

Where P_i is the local total pressure while P_{t0} is the total pressure at inflow boundary, U_2 is the tangential velocity at the exit of the impeller.

In order to discuss the pressure distribution, 20 observation points are arranged on the blade, the positions of these points are shown in Figure 6. Figure 7 shows the values of C_p on the pressure and suction sides, for CMC 7.5% at the design point. In these results, the pressure difference between the pressure side (PS) and suction side (SS) is quite clear. From the impeller inlet to $r/R_2 = 0.7$, the total pressure coefficient on the PS is larger than that on the SS. But in the range of $r/R_2 > 0.7$, the total pressure on the PS is lower. However, the total pressure on the SS near the trailing edge of the blade decreases, this is possibly due to flow separation and recirculation.

Figure 8 compares of the differences of C_p value between PS and SS, denoted by ΔC_p. It can be observed that ΔC_p for the pump handling CMC 7.5% are lower than that for handling water. This means the work done by the impeller to pump this non-Newtonian fluid is less that when pump water.

Velocity distributions

Shown in Figure 9 are the contours of the simulated relative velocity magnitude. The difference for the water and the CMC 7.5% cases occurs mainly near the SS. The relative velocity near the SS for handling CMC 7.5%, which is shown in Figure 9b, is lower than that for pumping water, which is shown in Figure 9a. This results in that the impeller locally transfers more energy to CMC 7.5% than that to water in the same work condition. This means that the flow slip for pumping CMC is less than that for pumping water.

The causes of relative velocity differences between the Newtonian (water) and non-Newtonian (CMC 7.5%) cases are further analyzed in detail. Figure 10 shows the relative velocity of the water and CMC 7.5% cases between blade-to-blade in middle span, z = 2.9 mm at design condition (n = 725 rpm, Q = 3.06 l/s), at two radial positions r = 0.65 R_2 and 0.9 R_2. Figure 10a shows that the relative velocity near SS is higher than that near blade PS at r = 0.65 R_2. But at r = 0.9 R_2, the relative velocity near SS is almost equal to near PS.

Figure 10b shows the relative velocity of non-Newtonian fluid (CMC 7.5%). The relative velocity near SS is higher than that near PS

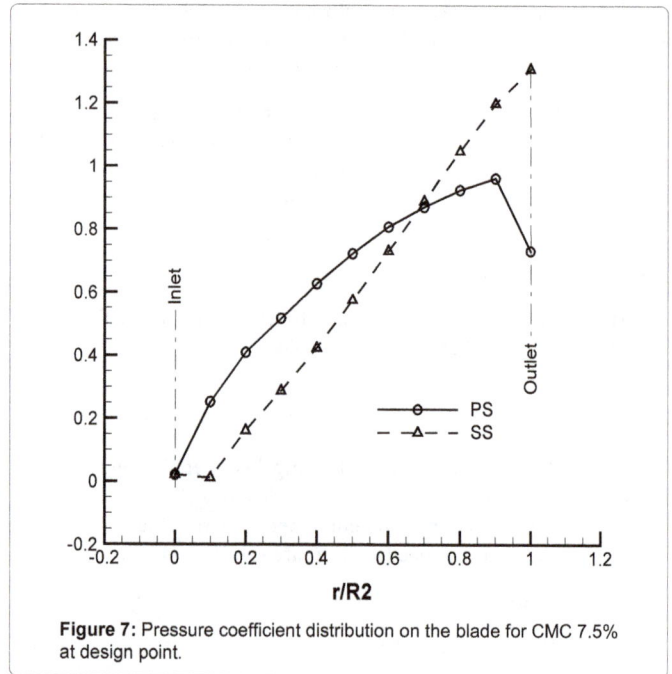

Figure 7: Pressure coefficient distribution on the blade for CMC 7.5% at design point.

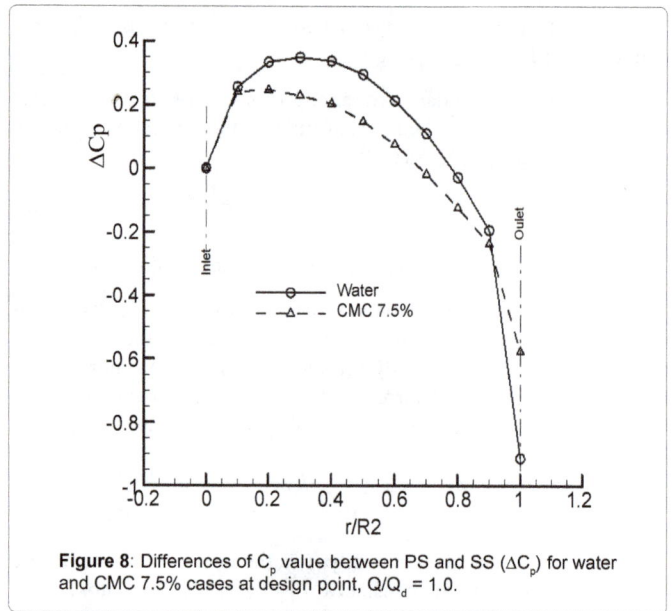

Figure 8: Differences of C_p value between PS and SS (ΔC_p) for water and CMC 7.5% cases at design point, Q/Q_d = 1.0.

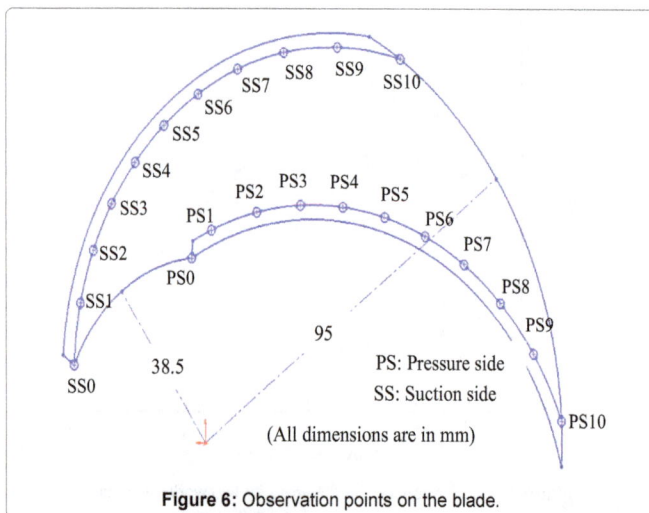

Figure 6: Observation points on the blade.

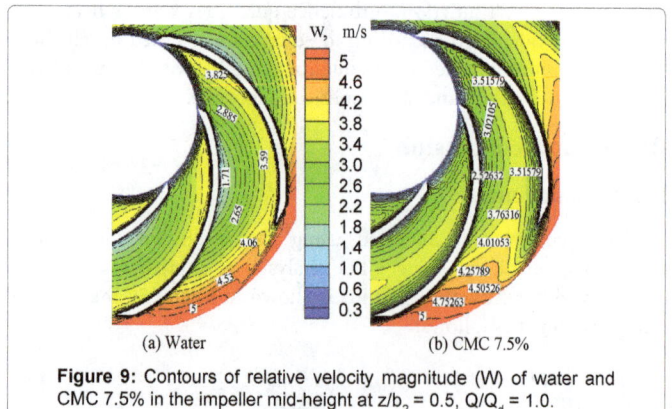

Figure 9: Contours of relative velocity magnitude (W) of water and CMC 7.5% in the impeller mid-height at z/b_2 = 0.5, Q/Q_d = 1.0.

at $r = 0.65\ R_2$. However, the relative velocity near SS is obviously lower than that near PS at $r = 0.9\ R_2$.

The effect of working fluids property on the radial and the tangential velocities at impeller outlet ($r/R_2 = 0.9$) on central section ($z/b2 = 0.5$) is shown in Figure 11. When working fluid is water, the results have a good agreement of velocity between the numerical and experimental data, but for the CMC 7.5%, the results differ quite large, especially in areas near SS.

Analysis of losses

The total flow head in a pump increases as tangential kinetic energy is transferred into the fluid by the rotating blades. However, due to fluid friction (viscosity), the work input to the machine does not equal the isentropic work out. That is, the efficiency is always less than 100% as stated by Klaus Brun et al. in 2005 [11]. Thus, the viscous head

loss is an important parameter to estimate the overall performance of a centrifugal pump. An estimation of this loss can be determined from the viscous dissipation of the internal flow. In fluid mechanics, the viscous dissipation function for incompressible flow is defined as follows:

$$\Phi = \mu_{eff}\left\{2\left[\left(\frac{\partial u}{\partial x}\right)^2 + \left(\frac{\partial v}{\partial y}\right)^2 + \left(\frac{\partial w}{\partial z}\right)^2\right] + \left(\left(\frac{\partial u}{\partial x} + \frac{\partial v}{\partial y}\right)^2 + \left(\frac{\partial v}{\partial y} + \frac{\partial w}{\partial z}\right)^2 + \left(\frac{\partial w}{\partial z} + \frac{\partial u}{\partial x}\right)^2\right)\right\} \quad (10)$$

where $\mu_{eff} = \mu + \mu_t$, μ_t is the turbulent viscosity.

According to Klaus Brun et al. [11] suggestion, the total head loss per passage, ΔH, can be defined as:

$$\Delta H = \int \Phi dV \quad (11)$$

The viscous dissipation function Φ and the total head loss per passage ΔH are used to analyze the energy losses.

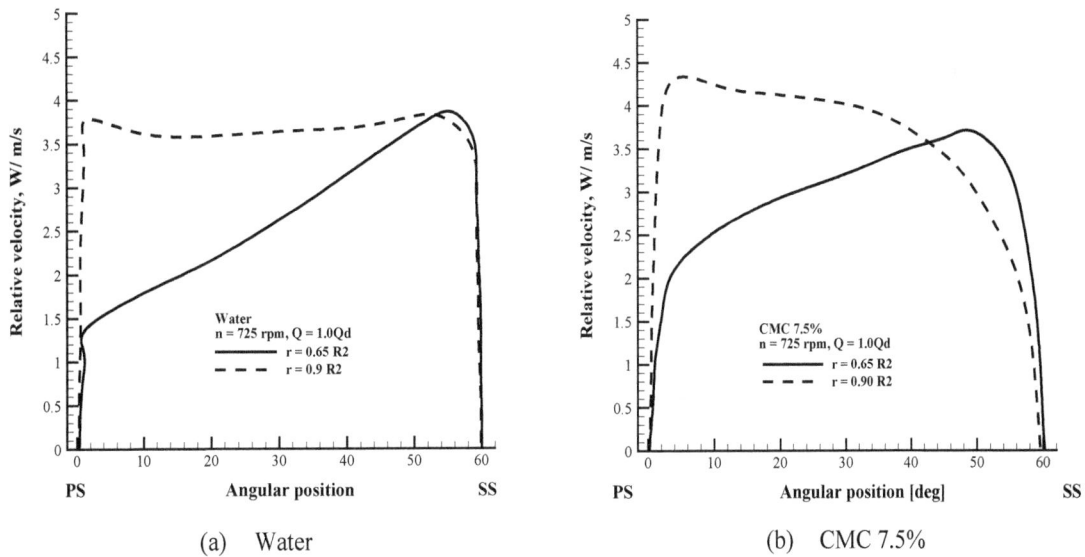

(a) Water
(b) CMC 7.5%

Figure 10: Relative velocity distribution at design condition, $Q/Q_d = 1.0$.

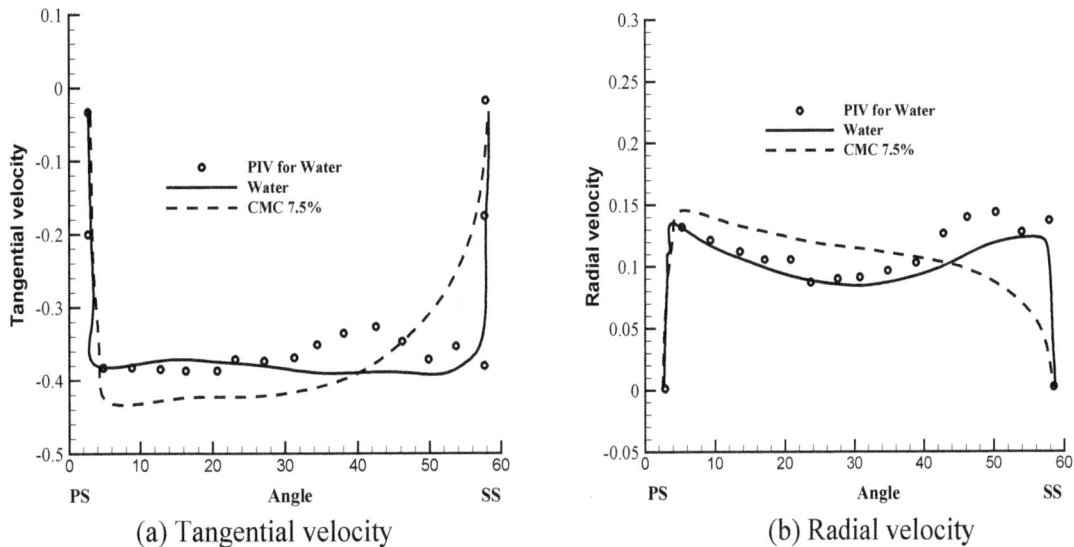

(a) Tangential velocity
(b) Radial velocity

Figure 11: Distributions of tangential and radial velocities at design condition, $Q/Q_d = 1.0$.

Figure 12 shows the distributions of Φ at several radial stations, while Figure 13 is Φ in the mid-height plane of the impeller. In the both cases, the high dissipation region is mainly concentrated near the blade surface and the walls. For the CMC 7.5% case, the values of Φ are on higher levels than those of the water flow case.

Figure 14 shows the total head loss per passage (ΔH) in impeller passage at different flow rate. In the case of non-Newtonian fluid, the total head loss increases rapidly as the flow rate increases. It can also be seen from Figures 12 and 13 that the function dissipation of the CMC 7.5% fluid is much greater than that of the water.

By studying change of dissipation function (Φ) and total head loss (ΔH) in the impeller for two cases (water and CMC 7.5%, respectively), we can clearly see that the loss when handling CMC 7.5% fluid is much greater than those for handling water, and this is also one of the causes for the reduced pump head when pumping CMC 7.5% fluid [12-15].

Turbulent kinetic energy

We now focus on the working fluids effect on the turbulent flow fields. In Figure 15, the distributions of the turbulent kinetic energy k for the two working fluids are shown. For the case of water, we have a strong turbulent kinetic energy at the pressure side in the before of blade. In the case of CMC 7.5% solutions, the magnitude levels are lower. And the contours are relatively evenly distributed along the flow direction. The lower turbulent kinetic energy for CMC 7.5% case may also be a cause of loss when pumping non-Newtonian fluids using this impeller.

Vorticity field analysis

In addition to using viscous dissipation functions to analyze losses,

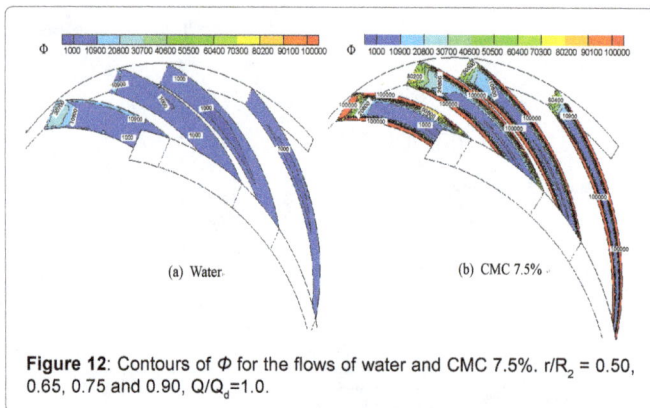

Figure 12: Contours of Φ for the flows of water and CMC 7.5%. r/R_2 = 0.50, 0.65, 0.75 and 0.90, Q/Q_d=1.0.

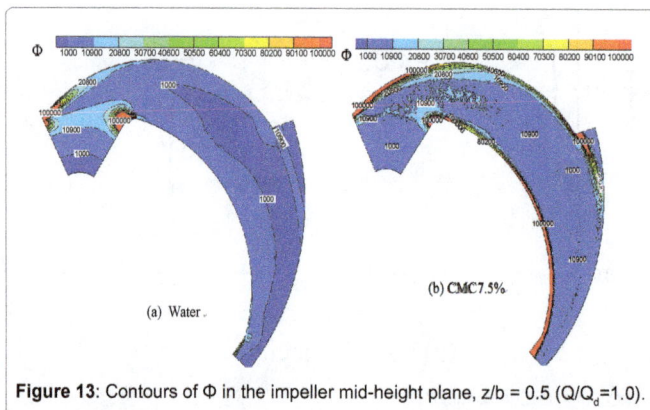

Figure 13: Contours of Φ in the impeller mid-height plane, z/b = 0.5 (Q/Q_d=1.0).

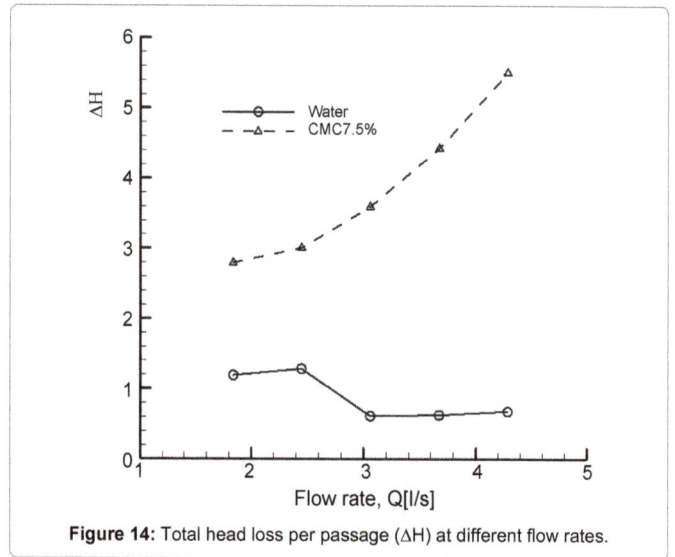

Figure 14: Total head loss per passage (ΔH) at different flow rates.

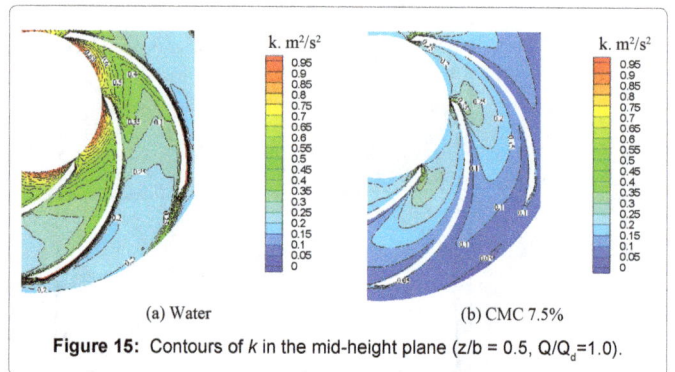

Figure 15: Contours of k in the mid-height plane (z/b = 0.5, Q/Q_d=1.0).

the relationship between losses and the vorticity is also important. Through the analysis of the vorticity, we may find the reason for the difference in performance caused by the working fluids.

The vorticity is defined as follows:

$$\vec{\Omega} = \nabla \times \vec{u} \tag{12}$$

The components of the vector $\vec{\Omega}$ are:

$$\Omega_y = \frac{\partial u}{\partial z} - \frac{\partial w}{\partial x}, \Omega_y = \frac{\partial u}{\partial z} - \frac{\partial w}{\partial x}, \Omega_z = \frac{\partial v}{\partial x} - \frac{\partial u}{\partial y} \tag{13}$$

Where u, v and w are the components of relative velocity in the x, y, and z directions, respectively. The velocities are the dependent variables directly resolved by the numerical simulation. When the calculation reaches convergence, the vorticity can be calculated according to equations (12) and (13). Such calculation must be on the base of convergent solution, and is therefore often referred as a post-processing. The strength of the vorticity is denoted by the norm:

$$\Omega = \sqrt{\Omega_x^2 + \Omega_y^2 + \Omega_z^2} \tag{14}$$

The simulation results of the vorticity magnitude Ω for flows of water and CMC 7.5% are shown in contours and in Figures 16-18.

Figure 16 compares the vorticity distribution in the radial station of r/R_2=0.65 plane, at design condition for water and CMC 7.5% cases. It is found that higher vorticity mainly concentrates near the walls. In near of the PS, the vorticity is much higher than in near the SS. For the

CMC 7.5% case, the zones of higher levels of vorticity near the shroud and hub are larger in area than those in the water flow case.

There are also clear differences in the contours distribution of Ω in the PS (Figure 17) and SS (Figure 18), between the flows of water and CMC 7.5%. In the PS, for the flow of CMC 7.5%, the vorticity intensity tends to rise continuously from entrance to exit. Near the blade leading edge, two low intensity area exist, close to the shroud and hub, respectively. For the water flow case, vorticity seems to increase first and then decrease in the streamwise direction. And an area of low intensity exists in the middle of the flow passage. On the SS shown in Figure 18, the vorticity intensity tends to decrease in the streamwise direction for the flow of CMC 7.5%. While for the water flow case, the vorticity strength tends to rise streamwisely, especially in near of the shroud and the hub walls.

Figure 19 shows vorticity distribution along the blade. Results also indicate that vorticity difference between water flow and CMC 7.5% flow. On the blade surfaces, the reduction or increase of vorticity for the flow of CMC 7.5% has a clear rule, while the water flow is more irregular. This is possibly because the flow of CMC 7.5% is more viscous.

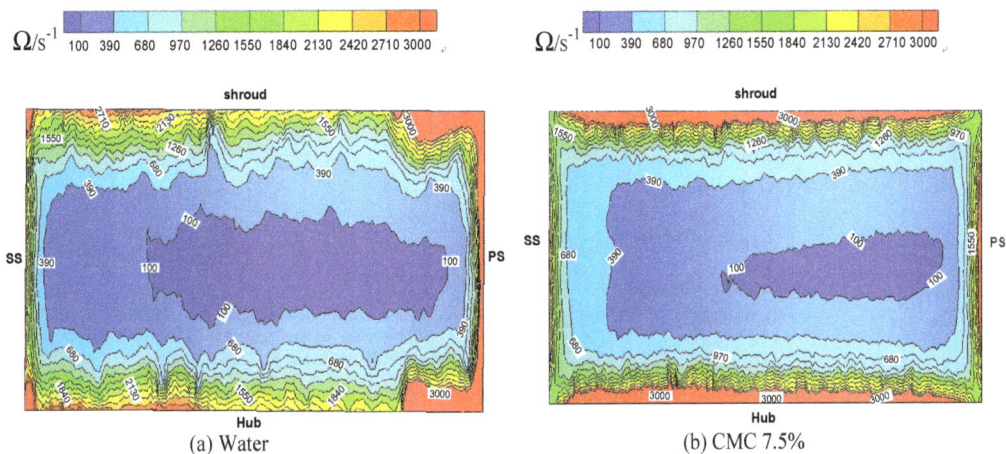

(a) Water   (b) CMC 7.5%

Figure 16: Contours of Ω for flows of water and CMC 7.5%, in the $r/R_2=0.65$ plane, $Q/Q_d = 1.0$.

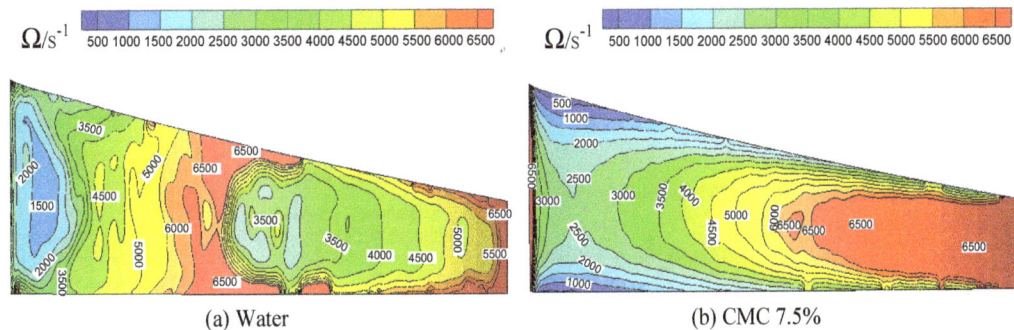

(a) Water   (b) CMC 7.5%

Figure 17: Contours of Ω on the blade pressure surface, $Q/Q_d = 1.0$.

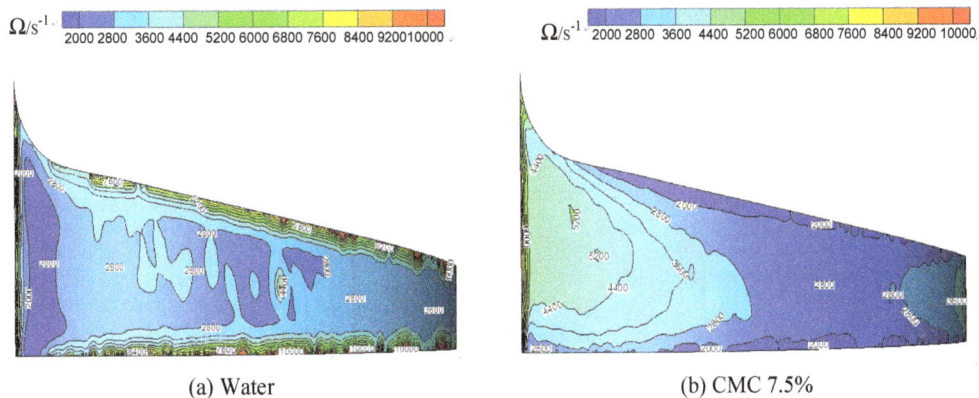

(a) Water   (b) CMC 7.5%

Figure 18: Contours of Ω on blade suction surface, $Q/Q_d = 1.0$.

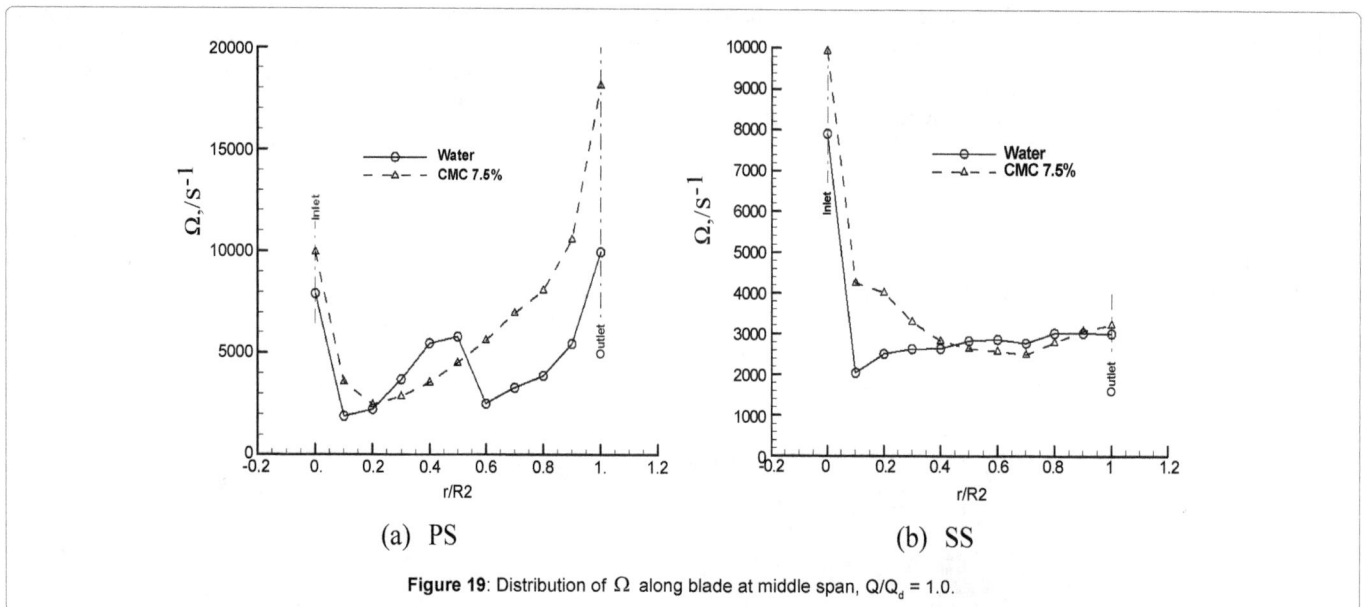

Figure 19: Distribution of Ω along blade at middle span, $Q/Q_d = 1.0$.

Conclusions

In this paper, the Power-law fluid is employed to study the non-Newtonian effects of CMC 7.5% flow in a centrifugal pump impeller. The results are compared with the flow of water, which is a Newtonian fluid. These fluid flows in turbulent regime are compared in details. The following remarks are concluded:

1) The pump head for water flow is higher than that for the flow of CMC 7.5% at the same flow rate. The distributions of total pressure coefficient (C_p), relative velocity and the turbulent kinetic energy (k) are quite different for cases of water and CMC 7.5% flows. The differences in the C_p value between PS and SS, ΔC_p for the pump handling CMC 7.5% are lower than those for handling water.

2) Through the study of the distribution of dissipation function, it is found that the total head loss per passage (ΔH) increases rapidly as the flow rate increases, for the case of CMC 7.5%. And the ΔH for the CMC 7.5% case is much greater than that of the water flow.

3) It can be concluded when this impeller is employed to pump the non-Newton fluid CMC 7.5, the energy efficiency is derated, because of the higher viscosity and energy dissipation.

Acknowledgment

The study in this paper is funded by the National Natural Science Foundation of China under Grants 51176048 and 51576067.

References

1. Kalombo JJN, Haldenwang R, Chhabra RP, Fester VG (2014) Centrifugal pump derating for non – Newtonian slurries. J Fluids Engineering.

2. Ng EYK, Zhou WD, Chan WK (1998) Non-Newtonian effects on mixed-flow water pump using CFD approach. J Hydraulic Machinery and Cavitation Singapore 1: 734-748.

3. Arakawa C, Qian Y, Samejima M (1999) Turbulent flow simulation of Francis water runner with pseudo compressibility. Proceedings of the GAMM Conference on Numerical Methods in Fluid Mechanics: 255-268.

4. Gandhi BK, Singh SN, Seshadri V (2002) Effect of speed on the performance characteristics of a centrifugal slurry pump. J hydraulic Engineering 128: 225-233.

5. Zhao BY, Yuan SQ, Liu HL, Tan MG (2006) Three-dimensional coupled impeller-volute simulation of flow in centrifugal pump and performance prediction. Chinese J Mechanical Engineering 19: 5-10.

6. Graham LJW, Pullum L, Slatter P, Sery G, Rudman M (2009) Centrifugal pump performance calculation for homogeneous suspensions. The Canadian J chemical engineering 87: 526-533.

7. Pullum L, Graham LJW, Rudman M (2007) Centrifugal pump performance calculation for homogeneous and complex heterogeneous suspensions. J Southern African Institute of Mining and Metallurgy 107: 373-379.

8. Yassine YC, Hammoud AH, Khalil MF (2010) Experimental investigation for centrifugal slurry pump performance.

9. Crawford J, Van Sittert F, Vander Walt M (2012) The performance of centrifugal pumps when pumping ultra-viscous paste slurries. J Southern African Institute of Mining and Metallurgy 112: 959-964.

10. Versteeg HK, Malalasekera W (2011) An introduction to computational fluid dynamics the finite volume method Pearson education limited. Harlow England London.

11. Brun K, Kurz R (2005) Analysis of secondary flows in centrifugal impellers . Int J Rotating Machinery 1: 45-52.

12. Pedersen N, Larsen PS, Jacobsen CB (2003) Flow in a centrifugal pump impeller at design and off-design conditions Part I: Particle Image Velocimetry (PIV) and Laser Doppler Velocimetry (LDV) measurements. J Fluids Engineering 125: 61-72.

13. Maiti B, Seshadri V, Malhotra RC (1989) Analysis of flow through centrifugal pump impeller by finite element method. Applied Scientific Research 46: 105-126.

14. Binxin W (2010) Computational fluid dynamics investigation of turbulence models for non-Newtonian fluid flow in anaerobic digesters. Environmental science & technology 44: 8989-8995.

15. Hua-ping L, Huan-long C, Xiao-guang Y, Chen F (2011) Study of loss mechanism in flow speed compressor cascade using dissipation function . J Aerospace Power.

19

Approach of Calculating a Parameter of Ductility in Tensile Test

Regaiguia B[1]* and Abderrazak D[2]

[1]Département de métallurgie et génie des matériaux, Laboratoire de métallurgie et génie des matériaux, Université Badjimokhtar 23000 Annaba, Algeria
[2]Département de métallurgie et génie des matériaux, Laboratoire de mise en forme, Université Badjimokhtar 23000 Annaba, Algeria

Abstract

Ductility by the sum of longitudinal elongation ΔL and reduction of diameter Δd which is the difference of initial diameter and necking diameter assumes that it is represented by the sum of the total longitudinal movement made by ΔL and transverse displacement of necking diameter.

Keywords: Ductility; Longitudinal elongation; Necking diameter; Diameter reduction; Tensile test

Introduction

Ductility by the sum of the distributed elongation and necking with the difference of initial diameter and necking assumes that it is represented by the sum of the total longitudinal movement made by ΔL and transverse displacement Δd (Figure 1).

We purpose to work on the upper half test piece, then we generalize the formula to the whole of the test piece by multiplying the found result of our approach because of the assumptions that we have made suppose that the test piece is perfectly symmetrical compared with the longitudinal and transverse axes passing through its middle point [1-20].

Materials and Methods

Materials

To develop and analyze the second step of our work, we experiment tensile test on 03 different grades of carbon steel. For each grade we use 03 specimens test grades are XC18 carbon steel, XC38 and XC48. Ductility values of the above-mentioned steels is known because of the carbon content, in other words it is known that XC18 is more ductile than XC38 and XC48 because it contains less carbon, and XC38 is more ductile than XC48. Based on this fact we test the ductility approach and we have to prove this order of ductility values of XC18, XC38 and XC48.

The various test pieces in number (03) for each grade were tested in the tensile test; the different values that we identified (final length and final diameter), are used to calculate average ΔL and necking diameter for calculating our approach parameter of ductility [21-40].

We experiment the approach that we called D_2 obtained as we said by the sum of total longitudinal elongation ΔL and reduction of diameter Δd on XC18, XC38 and XC48 after that we compare results.

For proving the task of ductility approach D_2, we must have linear deformation represented by D_2 of XC18 higher than XC38 and XC48; and also linear deformation of XC38 higher than XC48.

Methods

As the sum of the total elongation and the reduction in diameter is obtained:

$$\frac{\Delta L}{2} + \frac{\Delta D}{2} + \frac{\Delta L + \Delta D}{2}$$

From a geometrical point of view the progress of this approach is performed in a linear geometry before starting the initial test point and extending over a right characterizing the uniform elongation in homogeneous deformation then it becomes a geometric form L as soon as appearance of necking.

From Figure 2, we note that the elongation and reduction of diameter whose intersection gives the L-geometric form are perfectly identical on both sides of the axis through the necking which leads us note that the final geometry of the ductility approach D_2 is inverted to T form on the left side in this case because it can also be reversed on the right side. This is the length (mm) of this geometric form which T represents is the ductility approach D_2 characterized by the sum of the total elongation with the difference of initial and necking diameter [41-58].

So we have:

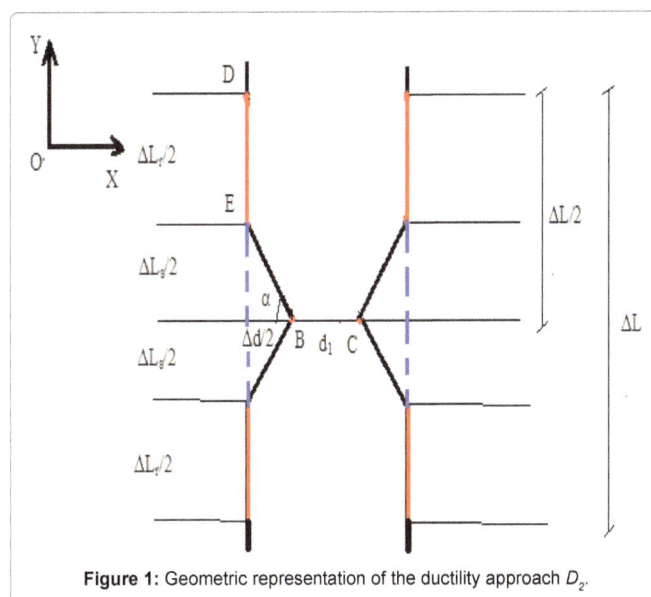

Figure 1: Geometric representation of the ductility approach D_2.

---

*Corresponding author: Regaiguia B, Département de métallurgie et génie des matériaux, Laboratoire de métallurgie et génie des matériaux, Université Badjimokhtar 23000 Annaba, Algeria, E-mail: bregaiguia@yahoo.fr

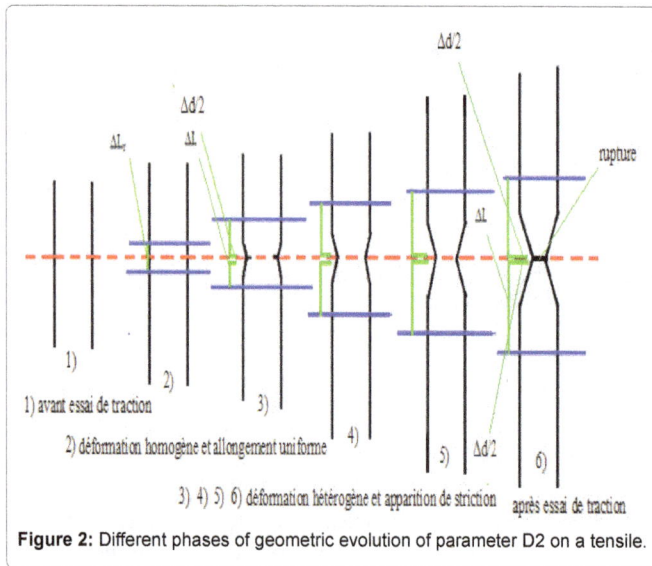

Figure 2: Different phases of geometric evolution of parameter D2 on a tensile.

$D_2(XC18) = 16.1$ mm $> D_2(XC38) = 14.2$ mm and $D_2(XC38) = 14.2$ mm $> D_2(XC48) = 10.9$ mm

So the linear deformation of XC18 symbolized physically by the parameter of ductility approach D_2 is higher than other steels wich is true (Tables 1-4).

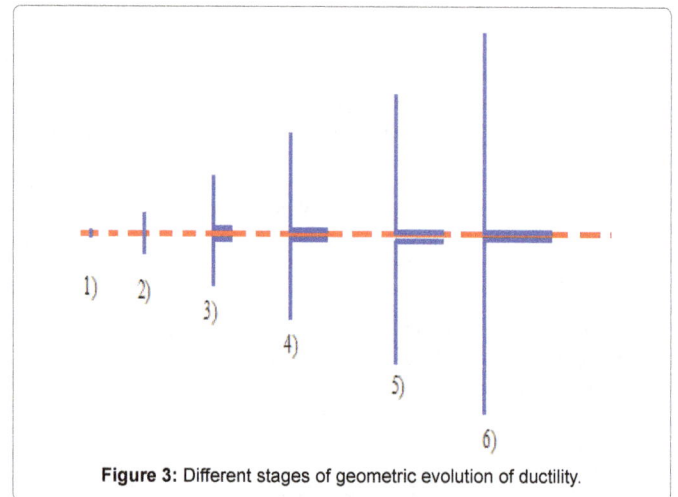

Figure 3: Different stages of geometric evolution of ductility.

Checking D_2 ductility approach

$D_2 = \Delta L + \Delta d$

For a brittle material ductility is zero.

So logically,

$D_2 = 0$

it means that $\Delta L = 0$ and $\Delta d = 0$ which is true for a brittle material.

For a superplastic material ductility is very significant.

So logically $D_2 \gg 0$

Where: $\Delta L \gg 0$ and $\Delta d > 0$, which is true for a superplastic material.

For a plastic material ductility is significant and positive,

So $D_2 > 0$

Where: $\Delta L > 0$ and $\Delta d > 0$, which is true for a plastic material.

So the modeling approach of ductility D_2 offer us à good appreciation of ductility on the other hand its easy to use for calculus.

D_2 also activates simultaneously as a summation key 02 variants influencing ductility: the elongation ΔL and the variation of the diameter Δd. In Figure 3, there is no ductility, it is a brittle material.

Figures I.3.2, I.3.3, I.3.4, I.3.5, I.3.6 show the onset and progression of ductility according to D2 approach.

So,

$DUCT = D_2 = \Delta L + \Delta d$

Finally we note that the ductility approac $D_2 = \Delta L + \Delta d$ h is a measurable quantity of linear dimension (mm). And we can say that this parameter represent a linear deformation with dimension 1 (Figures 4-7).

Results and Discussion

Experimental study of the ductility approach D_2 of XC18

We notice that:

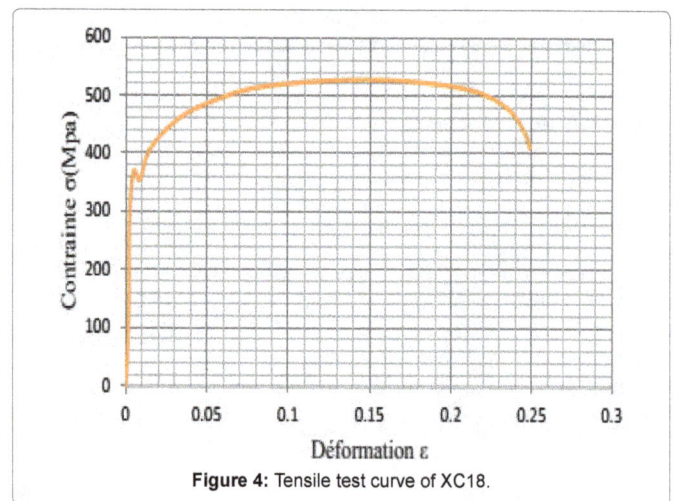

Figure 4: Tensile test curve of XC18.

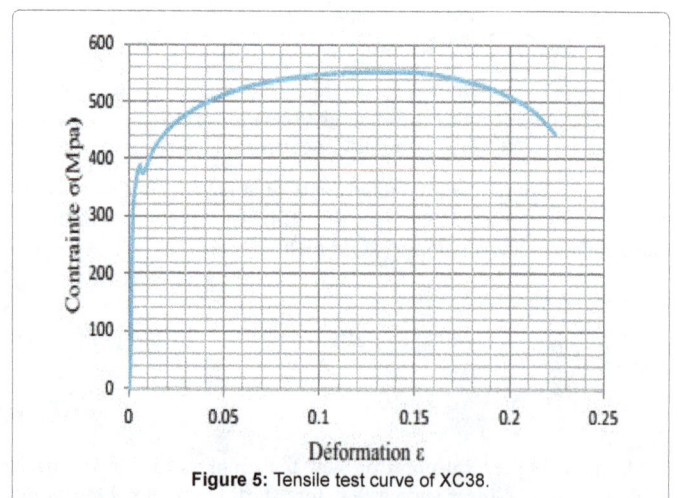

Figure 5: Tensile test curve of XC38.

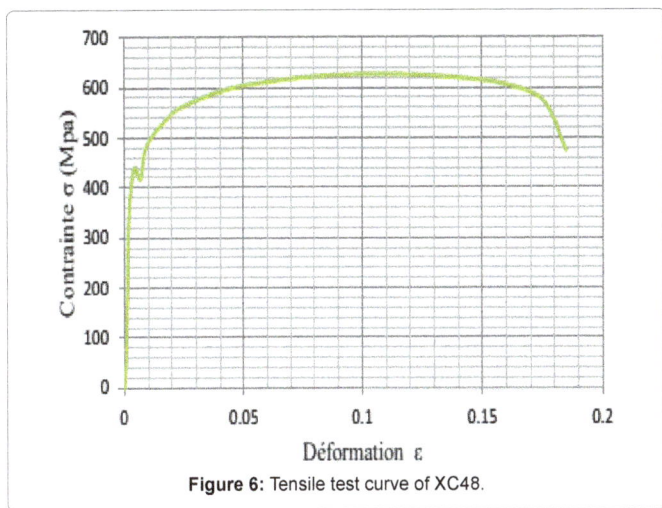

Figure 6: Tensile test curve of XC48.

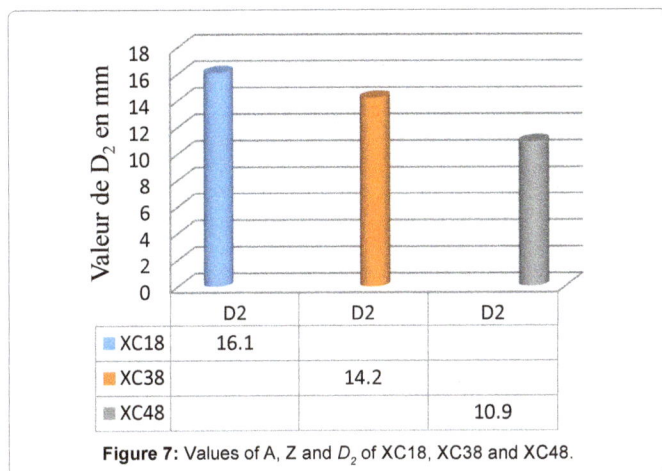

Figure 7: Values of A, Z and D_2 of XC18, XC38 and XC48.

Légend	$A = \dfrac{L_0 - L1}{L_0}$	$Z = \dfrac{\Delta S}{S_0}$	$D_2 = \Delta L + \Delta d$
Unit	%	%	mm
Specimen 1	25.4	61.5	16.5
Specimen 2	25	60.3	16.2
Specimen 3	24	59	15.6
Calcul de la moyenne $\sum \frac{x_i}{3}$	24.8	60.3	16.1

Table 1: Average values of A, Z and D_2 of XC18.

Légende	$A = \dfrac{L_0 - L1}{L_0}$	$Z\% = \dfrac{\Delta S}{S_0}$	$D_2 = \Delta L + \Delta d$
Unité	%	%	mm
Spécimen 1	22.7	57.7	14.8
Spécimen 2	22.2	50.9	14.1
Spécimen 3	21.9	48.2	13.7
Calcul de la moyenne $\sum \frac{x_i}{3}$	22.3	52.3	14.2

Table 2: Average values of A, Z and D_2 of XC38.

Légende	$A = \dfrac{L_0 - L1}{L_0}$	$Z = \dfrac{\Delta S}{S_0}$	$D_2 = \Delta L + \Delta d$
Unité	%	%	mm
Spécimen 1	19.56	45.2	12.4
Spécimen 2	17.24	39.1	9.7
Spécimen 3	18.42	42.1	10.6
Calcul de la moyenne $\sum \frac{x_i}{3}$	18.4	42.1	10.9

Table 3: Average values of A, Z and $D2$ of XC48.

Paramètres de ductilité	A%	Z%	$D_2 = \Delta L + \Delta d$
Unité	%	%	mm
XC18	24.8	60.3	16.1
XC38	22.3	52.3	14.2
XC48	18.4	42.1	10.9

Table 4: Values of A, Z and D_2 of XC18, XC38 and XC48.

Conclusion

Finally the linear and monodimensional approach $D_2 = \Delta L + \Delta d$ (mm) is intersting because it gives the measurement of deformation as a linear deformation easy to calculate.

References

1. Franz G (2008) Prédiction de la limite de formabilité des aciers multiphasés par une approche micromécanique. Engineering Sciences Arts et Métiers Paris Tech.

2. Chastel Y (2006) Matériaux pour l'ingénieur.

3. Degallaix S (2006) Caractérisation des matériaux métalliques.

4. Feng ZQ (2011) Mécanique non linéaire MN91.

5. Tillet P, Mathieu S (1750) Dissertation sur la ductilité des métaux et les moyens de l'augmenter. ETH-Bibliothek Zürich.

6. Mompiou F (2014) Des vermicelles dans les métaux. Centre d'élaboration des matériaux et d'études structurales CEMES-CNRS. Toulouse.

7. Poulain HD (2008) MA 41 Essai de traction. UTBM. Université de technologie de Belfort-Montbéliard.

8. Zhu Y(2013) Mechanical properties of materials. NC State University, USA.

9. Berthaud Y (2004) Matériaux et propriétés.

10. AD Rollett AD, De Graef M (2007) Microstructure-properties: I materials properties: strength ductility.

11. ADMET (2013) Sheet metal testing.

12. Jacquot B (2009) Propriétés mécaniques des biomatériaux utilisés en odontologie. Société francophone de biomatériaux dentaires.

13. ASM (2004) American Society for Materials.

14. Montheillet F, Briottet L (2009) Endommagement et ductilité en mise en forme.

15. Beaucquis S (2012) Propriétés mécaniques des matériaux. Laboratoire SYMME Polytech' IUT Annecy Département Mesures Physiques France

16. Bailon JP (2007) Des matériaux presses internationales polytechniques ca.

17. Chung BJ (2008) Journal of Mechanical Science and Technology.

18. Jardin Nicolas H (2012) Essais des matériaux. Résistance des matériaux.

19. Dorlot JM (2000) Des matériaux. Propriétés mécaniques des métaux.

20. Ghomari F (2014) Sciences des matériaux de construction. Faculté des sciences de l'ingénieur. université aboubekr belkaid.

21. Dupeux M (2005) Aide-mémoire science des matériaux.

22. Zaiser M (2006) Scale invariance in plastic flow of cristalline solid advphysics.

23. Thomas G (2006) Traité Matériaux métalliques.

24. Suquet P (2003) Rupture et plasticité.

25. Dorina N (2009) Matériaux et traitement OFPPT Maroc.

26. Mandel J (1978) Propriétés mécanique des matériaux Editions Eyrolles.

27. Norme NF (2009) EN 10002 Essai de traction.

28. François D, Bailon JP (2005) Essais mécaniques des métaux Détermination des lois de comportement.

29. Charmet JC (1990) Mécanique du solide et des matériaux. Elasticité-Plasticité-Rupture École supérieure de physique et de chimie industrielles de Paris ParisTech - Laboratoire d'Hydrodynamique et Mécanique Physique France.

30. Anduze M (2013) Mécanique des détecteurs. Du détecteur à la mesure. Institut National de Physique Nucléaire et de Physique des Particules LLR - Laboratoire Leprince-Ringuet France.

31. Newey C, Weaver G (1990) Materials Principles and Practice.

32. Symonds J (1976) Mechanical properties of materials.

33. Charvet R (2009) Comportement à la traction de barres d'armature laboratoire de métallurgie mécanique école polytechnique fédérale de Lausanne.

34. Ben Tahar M (2005) Contribution a l'étude et la simulation du procède de l'hydroformage. Engineering Sciences. Ecole Nationale Supérieure des Mines de Paris.

35. Strnadel B, Brumek J (2013) Effect of tensile test specimen size on ductility of R7T steel. Faculty of Metallurgy and Materials Engineering, Ostrava Tchécquie.

36. Brunet M (2011) Mécanique des matériaux et des structures.

37. Altmeyer G, Abed-Meraim F, Balan T (2013) Vue d'ensemble des relations théoriques entre la striction et les critères souches de localisation. Université de Lyon CNRS Laboratoire de Mécanique des Contacts et des Structures UMR5259 Insa de Lyon.

38. Glowacki D, Kozakiewicz K (2013) Numerical simulation of the neck in heterogeneous materials. Third International Conference On Material Modelling. Warsaw University of Technology Varsovie Pologne.

39. Bueno R, Sánchez J, Rodríguez T (2009) New parameter for determining plastic fracture deformation of metallic materials. University of Seville Spain, Avenida Reina Mercedes 2. 41012 Seville Spain.

40. Hosford WF (2010) Mechanical behaviour of materials. University of Michigan, USA

41. Yasuyuki A (2010) Procédé de fabrication de fil d'acier WO 2010101154 A1. General Manager of Corporate Planning & Coordination Group and Senior Managing Executive Officer at Sumitomo Corporation Japan.

42. Maya M (2008) Plasticité mise en forme. Arts et metiers – Paristech Centre d'enseignement et de recherche de Cluny France.

43. Broniewski W (1938) Revue de métallurgie. Allongement de striction et travail de rupture la traction.

44. Blétry M (2007) Méthode de caractérisation mécanique des matériaux.

45. International Standardisation Organisation (ISO) (2009) EN-ISO-6892-1, October 2009 Tensile test - mechanical properties.

46. Degoul PB (2010) Essai de traction. Etude des caractéristiques classiques I 1 Description de l'essai euro-norme 10002-I.

47. Chateigner D (2012) IUT mesures physiques université de caen basse-normandie ecole nationale supérieure d'ingénieurs de caen (ENSI CAEN) Laboratoire de cristallographie et sciences des matériaux (CRISMAT) Caen France.

48. Khalfallah A (2009) Fascicule de atelier de mécanique. Institut supérieur des sciences appliquées et de technologie de sousse. Tunisie.

49. Elias F (2013) Elasticité M2 Fluides Complexes et Milieux Divisés. Université Paris Diderot Paris France.

50. Col A (2011) Emboutissage des tôles aspect mécanique Techniques de l'ingénieur BM7511 : 3-4.

51. Savoie J, Jonas JJ, Mac Ewen SR, Perrin R (1995) Evolution of r-value during the tensile déformation of aluminium textures and microstructures 23: 1-23.

52. Abedrabbo N, Pourboghrat F, Crasley J (2006) Forming of aluminium alloys at elevated temperatures Part-1 Material characterization. Int J Plasticity 22: 314-341.

53. Prensier JL (2004) Les critères de plasticite: compléments.

54. Considère A (1885) Mémoire sur l'emploi du fer et de l'acier dans les constructions. Annales des Ponts et Chaussées 9 : 574.

55. Swift HW (1952) Plastic instability under plane stress. J Mechanics and Physics of Solids 1: 1-18.

56. Hill R (1952) On discontinuous plastic states with special reference to localized necking in thin sheets. J Mechanics and Physics of Solids 1: 19-30.

57. SAE (1989) Society for Automobile Engineers, USA.

58. Chomel P (2001) Selection des matériaux métalliques familles de matériaux. BM 5071.

20

Modeling of Three-Dimensional Soft Tissue Deformation with Dynamic Nonlinear Finite Element

Bahwini T[1]*, Zhong Y[1], Gu C[1] and Smith J[2]

[1]*School of Engineering, RMIT University, Melbourne, Australia*
[2]*Department of Surgery, School of Clinical Sciences at Monash Health, Monash University, Clayton, Australia*

Abstract

Deformation of soft tissue is a significant simulation part in minimally invasive surgery. This paper presents a dynamic nonlinear finite element method for modeling of soft tissue deformation. This method models large-range deformation via the second-order Piola-Kirchhoff stress. It condenses the stiffness matrix to reduce the degrees of freedom of the entire soft body at each node for every time step to improve the computational performance. Simulations and comparison analysis show that the proposed method can predict the nonlinear behaviors of soft tissues and requires a small amount of time.

Keywords: Finite element; Large-range deformation; Surgical simulation; Soft tissue deformation; Second-order Piola-Kirchhoff stress

Introduction

Soft tissue deformation plays a vital role in surgical simulation [1-3]. Surgical simulation requires the mechanical interaction between soft tissues and surgical tools be realistic and in real-time [4]. However, due to the complexity of soft tissues, it is difficult to achieve both conflict requirements, and realistic modeling of soft tissue deformation in real time is still a challenging research problem [4].

The mass-spring model and finite element method (FEM) are the most common modeling methods for soft tissue deformation [5]. The mass-spring model uses springs connected masses to carry out soft tissue deformation. It is simple in computation and easy to implement, but lacks the physical accuracy. The FEM is the exact opposite to the mass-spring model. It carries out soft tissue deformation based on rigorous laws in continuum mechanics, leading to high accuracy for modelling. However, it is expensive in computation. Due to the complexity in computation, the existing FEM models are mainly based on linear elasticity.

In this study, modelling of nonlinear FEM for soft tissue deformation is presented. The Second-order Piola-Kirchhoff stress has been used for modeling of nonlinear soft tissue behaviors. The technique of matrix condensation is also developed to improve the computational performance by reducing the degree of freedom. The rest of the paper is organized as follows: After the literature survey in section 2, section 3 details the proposed method for modelling of soft tissue deformation. Section 4 evaluates the performance of the proposed method. Finally, Section 5 concludes the paper and discusses about the future work.

Literature Review

There have been significant research efforts in soft tissue modeling for surgical simulation and robotic-assisted surgery, and development of virtual surgical schemes for education [6]. Various FEMs have been developed for linear elastic 2D and 3D simulations [7-10]. DiMaio and Salcudean established force distribution over the needle shaft, and developed a 2D finite element simulation using a linear electrostatic material model to measure the tissue deformation path in a phantom tissue [7]. They also applied fast low-rank matrix updates to achieve the real-time contact simulation [11]. Alterovitz et al. developed a dynamic system for needle insertion using the mass-spring model. This system

can simulate the needle insertion process with improved accuracy and computational performance [9].

Okamura et al. developed an empirical force model for soft tissue deformation and penetration, where the needle forces are considered to be a combination of the stiffness force, friction force and cutting force [12]. Webster et al. studied the motion of needle insertion with the use of a bevel-tip needle [13]. Misra et al. reported a mechanics-based method for steering of needle motion by exploring the connections at the tip and the overall bending of the needle with consideration of material properties and the needle tip geometry [14]. Mahvash and Dupont extended Misra's work and developed a dynamic model for characterizing rupture events, showing that the rupture force would decrease when the needle insertion velocity increases [15]. Wang and Hirai developed a dynamic model of needle-tissue interaction by considering needle insertion as a mixture of contact, rupture and friction forces [6]. They also applied a local constraint method to avoid re-meshing, which is generally needed due to the collision between discontinuous finite element structures and continuous needle movement.

In general, the existing FEM methods are mainly dominated by linear elasticity to reduce the computational cost, thus unsuited to handle nonlinear elastic behaviors of soft tissues.

Methodology

Proposed 3D dynamic nonlinear soft tissue model

Modeling of soft tissue behaviors depends on material properties. The mechanical behavior of an object, which results from the object internal structure, can be characterized using constitutive relations. Consider an isotropic and homogenous

*Corresponding author: Bahwini T, School of Engineering, RMIT University, Melbourne, Australia, PhD scholarship is sponsored by the Ministry of Education, Saudi Arabia, E-mail: tamb20@gmail.com

object with nonlinear elastic behavior. For large deformation, the strain is described as

$$
\begin{aligned}
\varepsilon_{x,y,z} &= [\varepsilon_x \varepsilon_y \varepsilon_z \varepsilon_{xy} \varepsilon_{yz} \varepsilon_{zx}]^T \\
\varepsilon_x &= \frac{du}{dx} + \frac{1}{2}[(\frac{\partial u}{\partial x})^2 + (\frac{\partial v}{\partial x})^2 + (\frac{\partial w}{\partial x})^2)] \\
\gamma_{xy} &= \gamma_{yx} = \frac{du}{dy} + \frac{dv}{dx}[\frac{\partial u}{\partial x}\frac{\partial u}{\partial y} + \frac{\partial v}{\partial x}\frac{\partial v}{\partial y} + \frac{\partial w}{\partial x}\frac{\partial w}{\partial y}]
\end{aligned}
\tag{1}
$$

where ε_x is the normal strain, and $\gamma_{xy} = \gamma_{yx}$ is the shear strain. The other terms of the strain can be defined similarly.

In order to describe the large deformation in a global form in terms of nonlinear quadratic strain, let us consider a fixed global coordinate system for the entire simulation. Therefore,

$$
\varepsilon = B_0^e u^e + \frac{1}{2}A\Theta
\tag{2}
$$

Where B_0^e and u^e represent the linear displacement differentiation matrix in size of 6×3 and the nodal displacement in size of 3×1, respectively. $[B_0^e] = [B_1\ B_2\ B_3\ B_4]$ is for each element in the tetrahedron. The second item on the left side represents the nonlinear displacement differentiation matrix and the nodal displacement, where A is a 6×9 nonlinear matrix, and Θ is a 9×1 matrix with three-dimensional identity (I_3).

Under the assumption of nonlinear elastic material, the stress σ and the strain ε are in a non-linear relationship, which can be defined by Hooke's law as

$$
\sigma = D(\varepsilon - \varepsilon_0) + \sigma_0
\tag{3}
$$

where σ_0 and σ are the stresses at times t_0 and t_{n+1} respectively, D is the elasticity matrix, $\lambda = \frac{Ev}{(1+v)(1-2v)}$ and $\mu = \frac{E}{2(1+2v)}$ are the material parameters, and ε_0 and ε are the strains at times t_0 and t_{n+1}. E and v denote the Young's modulus and Poisson's ratio, respectively.

Constitutive relation for nonlinear FEM

The strain energy density function (SEDF) is a common method to describe material behavior. The deformation of a physical object can be characterized by the deformation gradient $F = \frac{\partial x}{\partial X}$, where X and x are the original and the deformed configurations, respectively [16]. In the 3-D nonlinear case, the second-order gradient tensor (internal force) at a given point inside a material has nine strain components, which define the state of stress and strain at the point for the deformation configuration. The material will be hyper elastic when such SEDF exists, from which the stress components can also be derived. Let W be the strain energy per unit volume of the tissue. In a hyper-elastic material, when SEDF (which is derived from the Cauchy stress tensor in the material because of deformation) is known, it can be obtained from the second Piola-Kirchhoff stress

$$
\begin{aligned}
E &= \frac{1}{2}(F^T F - I) = \frac{1}{2}(\nabla U + \nabla U^T - \nabla U^T \nabla U) \\
S &= \frac{\partial W}{\partial E} \\
\tau &= J^{-1}F.\frac{\partial W}{\partial E}.F^T
\end{aligned}
\tag{4}
$$

where E is the Green (nonlinear) strain tensor, S is the second Piola-Kirchhoff stress, $U = x\text{-}X$ is the displacement vector, I is the identity tensor, J is the determinant of F, and τ is the Cauchy stress tensor. For a hyper-elastic material, the SEDF can be derived from Hooke's law as

$$
W = \frac{\lambda}{2}(E)^2 + \mu(E)^2
\tag{5}
$$

Element stiffness matrix

The global deformation is large and involves the entire body. This type of deformation is essential to surgical simulation, and can be mathematically applied to any large motion and deformation. The nonlinear stiffness is

$$
K^e(u^e) = K^e + \frac{1}{2}\int_\Omega B_0^{eT}AG^e d\Omega^e
\tag{6}
$$

where $K^e(u^e)$ is the local nonlinear stiffness matrix, which is dependent on the displacement u^e. A and G^e are the nonlinear deferential matrices. Equation (6) can be further simplified as

$$
K^t(u^e) = B_0^{eT}DB_0^e V + \frac{1}{2}B_0^{eT}AG^e V
\tag{7}
$$

where $K^t(u^e)$ is the nonlinear strain incremental stiffness matrices at time t, and V is the volume of the integral element.

Global equation to nonlinear finite element modelling

The element behavior is characterized by the partial differential equation governing the motion of the material points of a continuum, resulting in the following discrete system differential equation (equation of motion):

$$
M^{t+\Delta t}\ddot{U} + C^{t+\Delta t}\dot{U} + K^{t+\Delta t}(u^e)U = ^{t+\Delta t}R - _t^t F
\tag{8}
$$

Where U is the displacement vector at a node, $\ddot{U}$ and $\dot{U}$ are the acceleration and velocity at time $t +\Delta t$, F is the external force vector at time t, M and C are the time-dependent mass and damping matrices, and R is the external load applied at the nodal at time $t + \Delta t$.

Static condensation method

The static condensation method is employed to reduce the number of degrees of freedom for the element, that is, condense out the internal nodes. This method is to determine a part of the solution to solve the total finite element system equilibrium equations prior to assembling the structure matrices K and U based on the system boundary, leading to reduced computations. The stiffness matrix and corresponding displacement and force vector of the element under consideration can be decomposed into the form

$$
\begin{bmatrix} K_{mm} & K_{ms} \\ K_{sm} & K_{ss} \end{bmatrix}\begin{bmatrix} U_m \\ U_s \end{bmatrix} = \begin{bmatrix} F_m \\ F_s \end{bmatrix}
\tag{9}
$$

where m and s are the degrees of freedom of master (to be returned) and slave nodes (to be condensed out), and U_m and U_s are the desired displacements of the master and slave nodes, and F_m is the load vector.

From (9), we have the condition

$$
K_{sm}U_m + K_{ss}U_s = F_s
\tag{10}
$$

which can be used to eliminate U_s. From (10) we can obtain

$$
U_s = K_{ss}^{-1}F_s - K_{ss}^{-1}K_{sm}U_m
\tag{11}
$$

Substituting (11) into (9) yields

$$
\widehat{K}_m U_m = \widehat{F}_m
\tag{12}
$$

where

$$
\begin{aligned}
\widehat{K}_m &= K_{mm} - K_{ms}K_{ss}^{-1}K_{sm} \\
\widehat{F}_m &= F_m - K_{ms}K_{ss}^{-1}F_s
\end{aligned}
\tag{13}
$$

The new stiffness matrix K_m in the reduced form (12) is obviously denser than the stiffness matrix in the original form (10). This means that the condensed method is mathematically equivalent to the volumetric FEM, keeping the volume characteristic in the solution, but only at the computational expense of the surface FEM.

Model dynamics

The dynamics of the nonlinear FEM is achieved using an implicit Lagrangian formulation. The implicit Lagrangian formulation can be solved with the Newmark's method, leading to unconditionally stable solutions.

By linearizing the motion (8), the nonlinear FEM at each time step after the initial calculation can be described as follows:

1) Calculate effective loads at time $t + \Delta t$

$$
\begin{aligned}
^{t+\Delta t}\widehat{R} = {}_t^t F + M(a_0\,{}^tU + a_2\,{}^t\dot{U} + a_3\,{}^t\ddot{U}) \\
+ C(a_1\,{}^tU + a_4\,{}^t\dot{U} + a_5\,{}^t\ddot{U})
\end{aligned}
\tag{14}
$$

2) Solve for displacements at time $t + \Delta t$.

$$
LDL^T\;{}^{t+\Delta t}U = {}^{t+\Delta t}\widehat{R} \tag{15}
$$

3) Calculate the accelerations and velocities at time $t + \Delta t$:

$$
\begin{aligned}
^{t+\Delta t}\ddot{U} = a_0({}^{t+\Delta t}U - {}^tU) - a_2\,{}^t\dot{U} + a_3\,{}^t\dot{U} \\
^{t+\Delta t}\dot{U} = {}^tU + a_6\,{}^tU - a_7\,{}^{t+\Delta t}U
\end{aligned}
\tag{16}
$$

Where $\delta = 0.5$ and $\alpha = 0.25$ are the parameters to obtain the accuracy and stability, a_0, a_1...a_7 are the coefficient parameters, and $\widehat{K} = LDL^T = K + a_0 M + a_1 C$ is the effective stiffness matrix. Using the de-factorisation (Skyline) method, $\widehat{K}$ can be reduced to an upper triangular form, from which the unknown displacement U can be calculated by back-substitution.

Deformation region selection

With the model dynamics, the topology structure of the deformation for soft tissue modeling using FEM will be reformed and subsequently the stiffness matrix of the deform structure will be updated. Therefore, in order to reduce the computational time and cost, the minimum faces of the model mesh at the deformation region must be determined.

Strictly speaking, a set of triangles from the mesh, which are close to the surgical tool need to be selected. Generally, the selection of the minimum deformation region is according to the interaction between the surgical tool and soft tissues, which is detected based on the information of faces, edges and vertices. Figure 1 shows the three cases for the determination of the minimum deformation region. The interaction process between the surgical tool and soft tissues is highlighted in red and the minimum deformation area is displayed in green. The minimum deformation region is determined based on the following interaction configurations: (a) Vertex: six triangles are adjusted; (b) Edge: two triangles are adjusted; and (c) Face: one triangle is adjusted.

Implementation and result

The prototype system for simulation of soft tissue deformation with the proposed FEM was implemented using Java3D on a PC with Intel Core i7 MacBook Pro (13-inch, Early 2011) at 2.7 GHz with 16 GB 1333 MHz DDR3 RAM memory and Intel HD Graphics 3000 512 MB. Simulations were conducted to investigate the performance of the proposed nonlinear FEM. The values of materials parameters were set according to those reported in [7].

Trials on the interaction between a surgical needle and soft tissues were conducted by the proposed nonlinear FEM. Different shapes of tetrahedron volume models such as the cube, human liver and kidney were tested. Table 1 shows the element numbers after condensation and the average iteration computational times. For example, for the cubic volume model which is made up of 953 elements, it is condensed to 266 tetrahedron elements, and the average time for one iteration of deformation is around 0.174 seconds. As the visual refresh rate should be more than 25 frames per second [11,18]. Yin and Goulette proposed FEM is able to provide real-time visual feedback [11,18]. Figure 2 shows the deformations of the cubic model under a tensile and compression force, respectively. Figure 3 illustrates the deformations of the virtual human liver and kidney models. Figures 4 and 5 show more deformation results on the cubic model as compression and tensile forces.

To further evaluate the performance of the proposed FEM, we further compared the simulation results with experimental data reported in the literature [7]. The simulation results in terms of the relationships between stress and strain as well as force and displacement are shown in Figures 6 and 7, while the corresponding experimental data are shown in Figure 8. It is obvious that the proposed FEM demonstrates the nonlinear deformation behavior remarkably. Therefore, the proposed FEM can achieve large deformations via the nonlinear relationship.

Conclusion

Simulating of soft tissue deformation is discussed in this paper based on dynamic nonlinear FEM in surgical simulation. Soft tissue deformation is carried out by using the second-order Piola-Kirchhoff stress. The model dynamics is conducted based on the

Object Test	Number of elements	Condensation	Average of one iteration time costs (Second)
		Condensed elements	Dynamic
Cubic	953	266	0.174
Liver	4094	873	0.881
Kidney	11494	2188	7.142

Table 1: Matrix condensation and time performance.

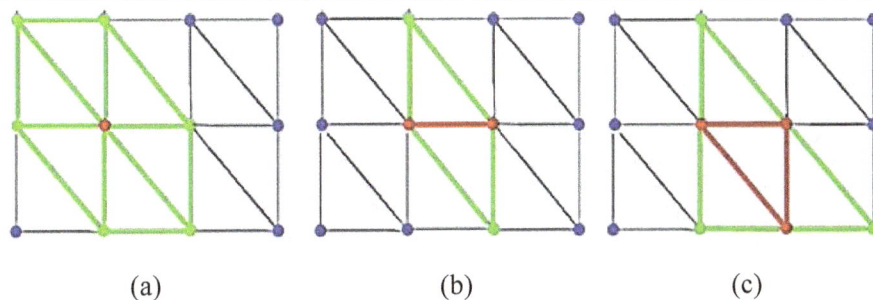

(a)   (b)   (c)

Figure 1: Different interaction cases between the surgical tool and soft tissues (a) Vertex (b) Edge and (c) Face [17].

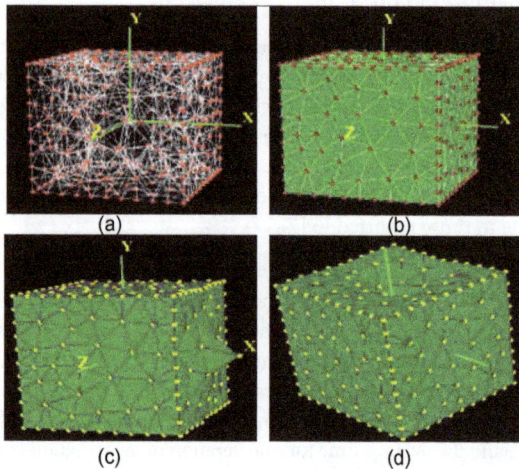

Figure 2: Cubic modeling (a) wireframe (b) shade (c) and (d) deformed force (under a tensile and compression force, respectively).

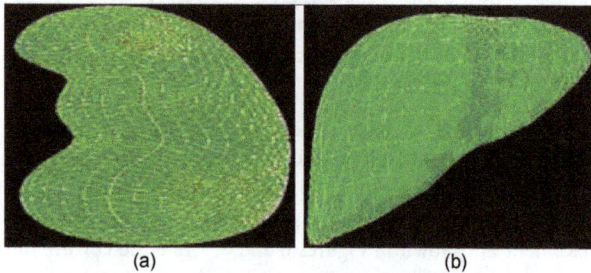

Figure 3: Deformations of the virtual human (a) kidney and (b) liver.

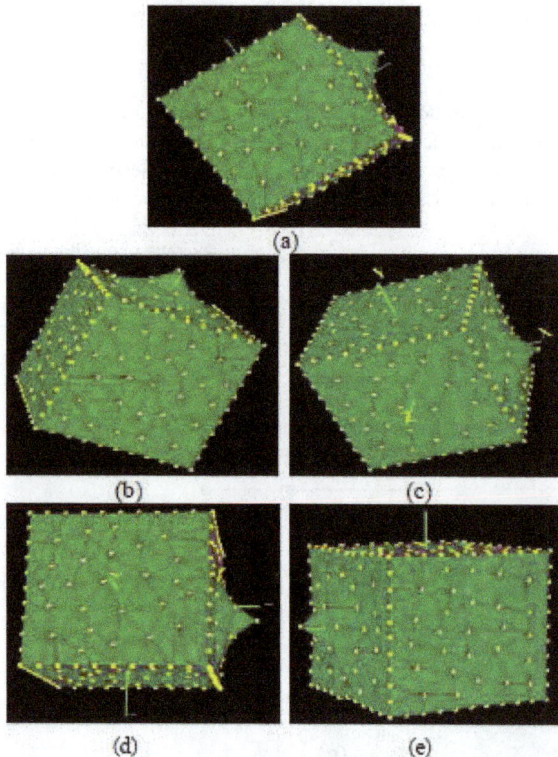

Figure 4: Different views of the deformation of the cubic model under a tensile force.

implicit Lagrangian formulation to define the nonlinear constitutive relationships for large deformations. The Newmark's method under the assumption of isotropic and homogenous materials is established. Moreover, a technique of matrix condensation based on continuum mechanics of solid is also developed to reduce the number of the degrees of freedom for each element.

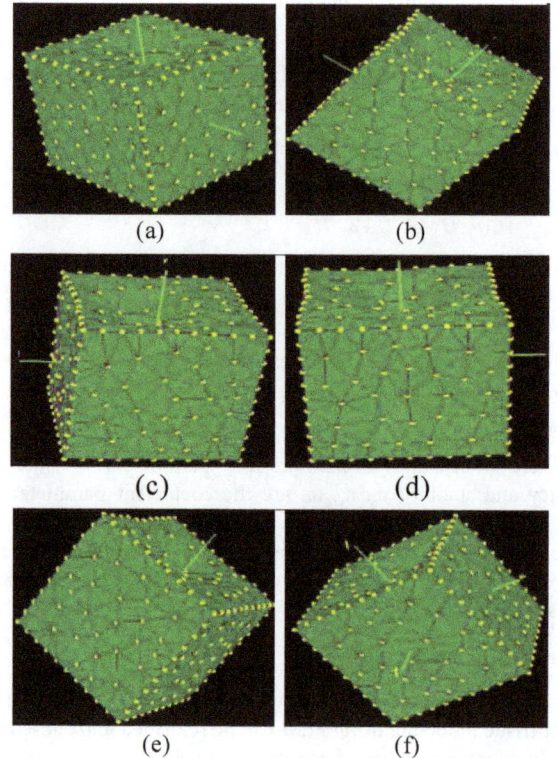

Figure 5: Different views of the deformation of the cubic model under a compressed force.

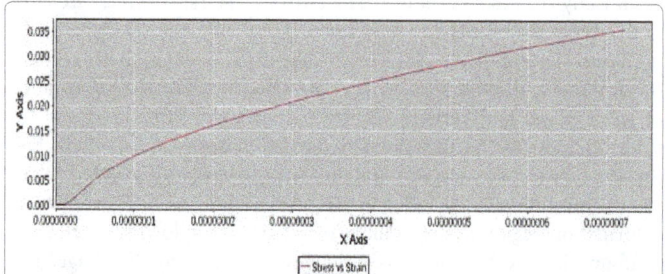

Figure 6: Stress (Y Axis) vs. strain (X Axis) curve.

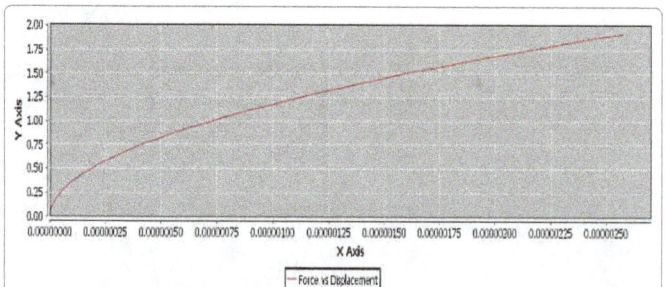

Figure 7: Force (Y Axis) *vs.* displacement (X Axis) curve.

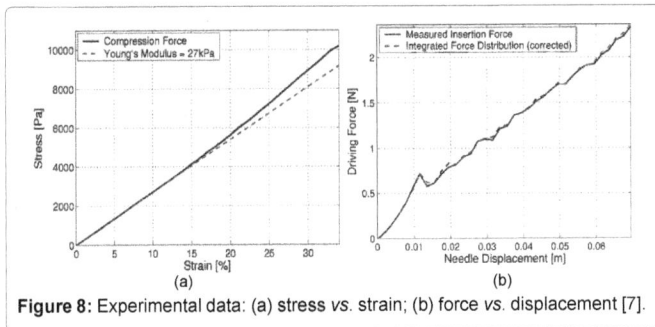

Figure 8: Experimental data: (a) stress *vs.* strain; (b) force *vs.* displacement [7].

Future work will focus on the improvement and application of the proposed FEM for surgical simulation. The proposed FEM will be integrated with a haptic device for achieving force feedback for surgical simulation. It is expected to establish the methods of haptic modeling and rendering for soft tissue deformation with real-time haptic feedback in real time. The proposed FEM will also be applied to model the typical surgical procedure of needle insertion and further develop a surgical simulation system for training and procedure planning of needle insertion.

Acknowledgment

This work was supported by Ministry of Education in Saudi Arabia for financial support for the first author on his research.

References

1. James DL, Pai DK (1999) ArtDefo: accurate real time deformable objects. Proceedings of the 26th annual conference on Computer graphics and interactive techniques: 65-72.

2. Cotin S, Delingette H, Ayache N (1999) Real-time elastic deformations of soft tissues for surgery simulation. IEEE Trans Vis Comput Graph 5: 62-73.

3. Hagemann A, Rohr K, Stiehl HS, Spetzger U, Gilsbach JM (1999o) Nonrigid matching of tomographic images based on a biomechanical model of the human head in Medical Imaging 99: 583-592.

4. Barbé L, Bayle B, de Mathelin M, Gangi A (2007) Needle insertions modeling: Identifiability and limitations. Biomed Signal Process Control 2:191-198.

5. Zhong Y, Shirinzadeh B, Smith J, Gu C (2012) Soft tissue deformation with reaction-diffusion process for surgery simulation. J Visual Languages and Computing 23: 1-12.

6. Wang L, Hirai S (2011)A local constraint method for needle insertion modeling and simulation. IEEE International Symposium on Haptic Audio-Visual Environments and Games. HAVE : 39-44.

7. DiMaio SP, Salcudean SE (2003) Needle insertion modeling and simulation. IEEE Transactions on Robotics and Automation 19: 864-875.

8. Goksel O, Salcudean SE, DiMaio SP, Rohling R, Morris J (2005) 3D needle-tissue interaction simulation for prostate brachytherapy. Med Image Comput Comput Assist Interv 8: 827-34.

9. Alterovitz R, Pouliot J, Taschereau R, Hsu IC, Goldberg K (2003) Simulating needle insertion and radioactive seed implantation for prostate brachytherapy. Stud Health Technol Inform 94:19-25.

10. Dehghan E, Salcudean SE (2007) Needle insertion oint and rientation ptimization in on-linear issue with application to Brachytherapy 25(2): 10-14.

11. DiMaio SP, Salcudean SE (2005) Interactive simulation of needle insertion models. IEEE Trans Biomed Eng 52: 1167-79.

12. Okamura AM, Simone C, O'Leary MD (2004) Force modeling for needle insertion into soft tissue. IEEE Trans Biomed Eng 51: 1707-1716.

13. Webster RJ, Kim JS, Cowan NJ, Chirikjian GS, Okamura AM (2006) Nonholonomic modeling of needle steering. Int J Robot Res 25: 509-525.

14. Misra S, Reed KB, Schafer BW, Ramesh KT, Okamura AM (2010) Mechanics of flexible needles robotically steered through soft tissue. Int J Rob Res 29: 1640-1660.

15. Mahvash M, Dupont PE (2010) Mechanics of dynamic needle insertion into a biological material. Int J Robot Res 57: 934-943.

16. Famaey N, Vander Sloten J (2008) Soft tissue modelling for applications in virtual surgery and surgical robotics. Comput Methods Biomech Biomed Engin 11: 351-66.

17. Yin G, Li Y, Zhang J, Ni J (2009) Soft tissue modeling using tetrahedron finite element method in surgery simulation. In information science and engineering (ICISE) 2009 1st International Conference 3705-3708.

18. Goulette F, Chen ZW (2015) Fast computation of soft tissue deformations in real-time simulation with hyper-elastic mass links. Comput Methods in Appl Mech Engg 295:18-38.

21

A Comparative Study on Wear Behavior of Al6061-6% SiC and Al6061-6% Graphite Composites

Nagaral M[1]*, Auradi V[2], Parashivamurthy KI[3] and Kori SA[4]

[1]Aircraft Research and Design centre, HAL, Bangalore-560037, Karnataka, India
[2]R&D Centre, Department of Mechanical Engineering, SIT, Tumkur-572103, Karnataka, India
[3]Department of Mechanical Engineering, Government Engineering College, Chamarajnagar-571313, Karnataka, India
[4]Department of Mechanical Engineering, Basaveshwara Engineering College, Bagalkot-587102, Karnataka, India

Abstract

This work investigated the influence of SiC and graphite on the microstructure and wear behavior of Al6061- SiC and Al6061-Graphite composites. The investigation reveals the effectiveness of incorporation of SiC and Graphite in the Al6061 alloy for studying wear properties. The composites were fabricated using liquid metallurgy route. The Al6061- SiC and Al6061-Graphite composites were fabricated separately by introducing 6 wt. % of SiC and graphite particulates by two stage melt stirring process. In this reinforcement particulates were added in two steps to increase the wettability. The characterization was performed through Scanning Electron Microscope and Energy Dispersive Spectrum. The particle distribution was uniform in these composites. Pin on disc apparatus was used to conduct the dry sliding wear tests. The experiments were conducted by varying loads and sliding speeds for sliding distance of 2000 m. The results revealed that Al6061-6% SiC and Al6061-6% Graphite composites were shown more resistance to wear. Al6061-6% Graphite composites were shown more resistance to wear as compared to Al6061-6% SiC composites. Further, the volumetric wear loss was found to increase with the load and sliding speed for all materials. Worn surface analysis made by using scanning electron micrographs to know various mechanisms involved in the wear process.

Keywords: Al6061 Alloy; SiC Particulates; Graphite; Wear; Stir casting

Introduction

The metal matrix composites (MMCs) are very attractive materials for several applications. For many researches the term MMCs is often equated with the term light metal matrix composites. Substantial progress in the development of light metal matrix composites has been achieved in recent decades, so that they could be introduced into the most important applications [1]. Aluminum matrix composites (AMCs) are the competent material in the industrial world. Efforts have been made to develop aluminium metal matrix composites in recent years due to their low density, high strength, superior creep resistance and have great potential in automotive and aerospace applications [2,3].

Aluminium (Al) is the second most widely used metal in the world today after iron. It has a low density (2.7 g/cc), superior malleability, excellent corrosion resistance, good thermal conductivity (237 W/mK), very low electrical resistivity (2.65*10-8Ωm) and good formability. It's Young's modulus is 70 GPa and its Vickers hardness is 60 to 70 VHN. Al has a melting point of 660.32°C and at high temperatures, the strength of Al decreases. However, the demand for Al and its alloys having a much higher strength is increasing. Al-matrix composites (AMCs) have been widely used in automobile and aerospace industries due to their excellent physical and mechanical properties. To overcome these shortcomings and to meet the ever increasing demand of modern day technology, composites are one of the most promising materials [4,5].

Many techniques are currently available to fabricate the metal matrix composites (MMCs), such as mechanical alloying, high-energy ball milling, spray deposition, powder metallurgy, sintering and various casting techniques. The powder metallurgy processing method cannot be used for bulk production of large and complex structural MMCs components. The fabrication of MMCs by powder metallurgy route is time-consuming, expensive and energy intensive. The liquid metallurgy route or stir casting process is most commonly used method to fabricate aluminium composites. This technique is the simplest and can able to produce more complex castings by this method. It is economical for bulk production.

Aluminium-based metal matrix composites (AMCs) have been widely used in automotive and aircraft applications due to their low density and concurrent high wear resistance, strength, corrosion resistance, stiffness and thermal conductivity [6]. Ceramic particulates like SiC, B_4C, TiC, WC, ZrO_2 and Al_2O_3 are the most commonly used reinforcements to fabricate composites.

In industrial applications maximum components are subjected to sliding motion, where wear resistance plays important role. Several researchers investigated wear behaviour ceramic particulates reinforced aluminium alloy matrix composites [7,8]. Suresh et al. [9] studied wear prediction of stir cast Al-TiB₂ composites. The wear mechanism of the specimen was studied through scanning electron microscope images. Umanath [10] and co-authors investigated the wear properties of Al6061-SiC-Al_2O_3 reinforced metal matrix composites. The 15% hybrid composite shown better wear resistance compared to 5% composites. The fracture surface of composites shows the ductile tear ridges and cracked SiC and Al_2O_3 particles indicating both ductile and brittle fracture mechanism.

*Corresponding author: Nagaral M, Design Engineer, Aircraft Research and Design centre, HAL, Bangalore-560037, Karnataka, India
E-mail: madev.nagaral@gmail.com

In the present investigation, an attempt is made to develop Al6061-6 wt. % SiC and Al6061-6 wt. % Graphite composites by stir casting technique. Further, a comparative study on wear behaviour of Al6061 alloy, Al6061-SiC and Al6061-Graphite composites has been made at varying loads and sliding speeds.

Experimental Work

Aluminium 6061 alloy with theoretical density of 2.7 g/cm³ was used as the matrix material. Table 1 tabulates the chemical composition used. In the table, the amount of Mg and Si is maximum, which is required for Al6061 alloy matrix.

Micron sized SiC and particulates of Graphite with an average particle size of 90-125 µm were selected as the reinforcements for metal matrix composites. The theoretical density of SiC is 3.10 g/cm³ and graphite is 2.20 g/cm³.

In this present work, stir casting technique was used to fabricate

Element	Wt. Percentage
Magnesium	0.82
Silicon	0.64
Iron	0.23
Copper	0.17
Zinc	0.031
Manganese	0.072
Titanium	0.015
Chromium	0.014
Others	0.15
Aluminium	Bal

Table 1: Chemical composition of Al6061 alloy.

Figure 1: Pin on disc wear testing machine.

Figure 2: Wear test specimens.

Al6061-6 wt. % SiC and Al6061-6 wt. % Graphite particulate composites. Initially the required amount of Al6061 alloy was placed into graphite crucible and heated to a temperature of 750⁰C in an furnace. The temperature of furnace was controlled to an accuracy of ±10°C using a thermocouple. Preheated reinforcement particles were added in novel two step addition method. Ceramic SiC and Graphite particulates were preheated to a temperature of 300°C in an oven to remove the gases from the surface of particles and to avoid temperature drop in molten metal after addition of reinforcements. After degassing of molten Al alloy by using solid hexachloroethane (C_2Cl_6), preheated SiC and Graphite particles were poured into the vortex of the molten Al6061alloy. Vortex was generated with the help of a zirconia coated steel impeller. The extent of incorporation of SiC and Graphite particles in the matrix alloy was achieved in steps of 2. i.e., Total amount of reinforcement required was calculated and is being introduced into the melt 2 times rather than introducing all at once. At every stage before and after introduction of reinforcement, mechanical stirring was carried out for a period of 5 min. The stirrer was preheated before immersing into the melt, and was located approximately to a depth of 2/3 height of the molten metal from the bottom and run at a speed of 300 rpm. Composite mixture was poured into cast iron moulds having diameter 15 mm and length of 125 mm and 710°C was the temperature at the time of pouring.

The prepared composites were machined for microstructure and wear studies as per required dimensions. Samples for SEM/EDS were prepared by diamond grinding and final polishing was done by using 1µm diamond paste. SEM was carried out on HR-SEM (Hitachi S-4800, Japan).

Pin on disc machine (DUCOM, TR-20LE) was used to carry wear tests as per ASTM G99 standard [11-13]. Dry sliding wear tests were performed on specimens of diameter 8 mm and 25 mm height. The counter disc material was of EN31 steel. Prior to testing, the pin and disc surface were cleaned with acetone. The experiments were conducted at a constant sliding speed of 400 rpm and sliding distance of 2000 m over a varying load of 10 N, 30 N and 50 N. Similarly experiments were conducted at a constant load of 50 N and sliding distance of 2000 m over a varying sliding speed of 100 rpm, 200 rpm, 400 rpm and 600 rpm. During testing the pin specimen was kept stationary and perpendicular to the disc while the circular disc was rotated. Electronic weighing machine with the precision of 0.0001 g was used to weigh the initial weight of the specimens. After each test, the counter face disc was cleaned with acetone. The pin was weighed before and after testing to determine the amount of wear loss. The data in the form of weight loss was converted into volumetric wear loss with respect to their corresponding density and from the wear volume. Figure 1 showing the DUCOM made pin on disc wear testing machine used for the experiments. Figure 2 shows the wear test specimens used in the study.

Results and Discussion

Microstructural analysis

Representative SEM micrographs of the synthesized as cast Al6061 alloy and micro SiC and Graphite particulate reinforced aluminium matrix composites are presented in Figures 3a-3c. The dispersion of the reinforcement particles (SiC and Graphite) within the Al6061 alloy matrix is visible and can be considered as homogeneous in the composite.

Figures 3b-3c clearly show and even distribution of SiC and Graphite particles in the Al6061 alloy matrix. In other words, no

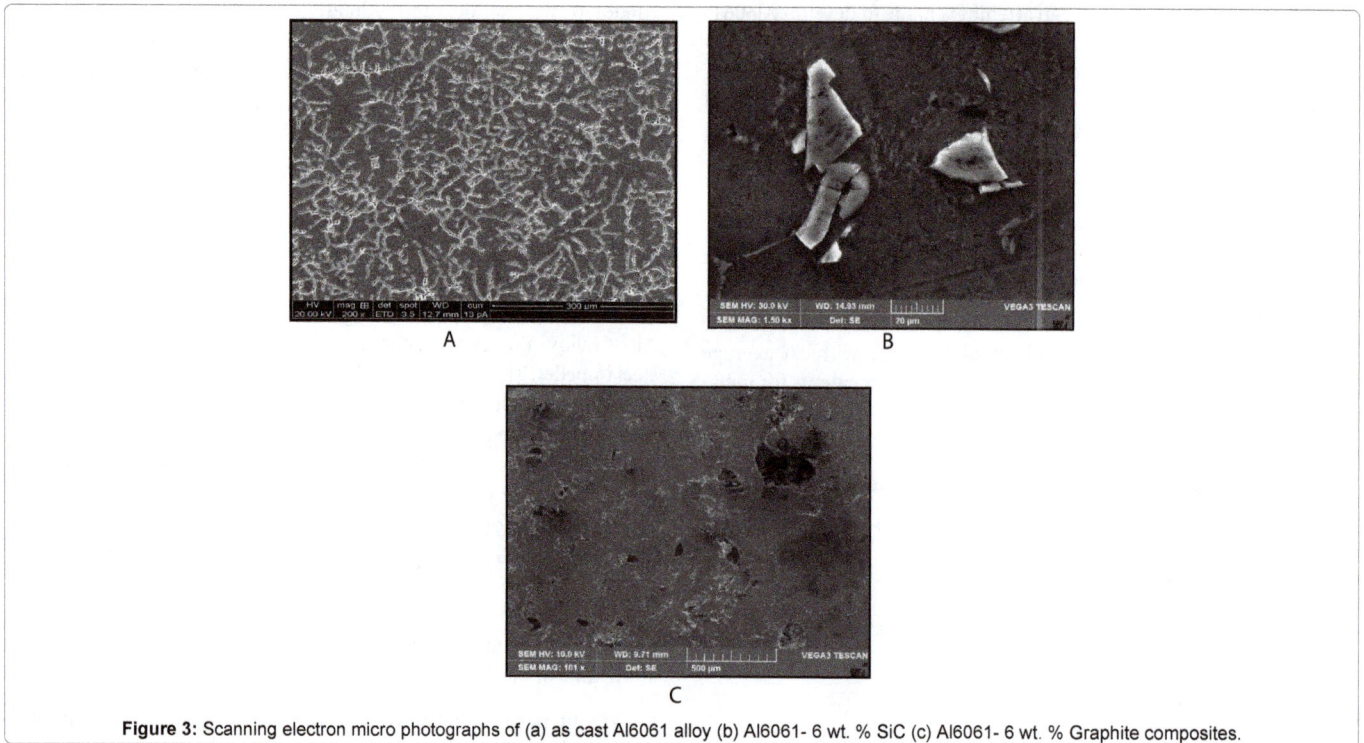

Figure 3: Scanning electron micro photographs of (a) as cast Al6061 alloy (b) Al6061- 6 wt. % SiC (c) Al6061- 6 wt. % Graphite composites.

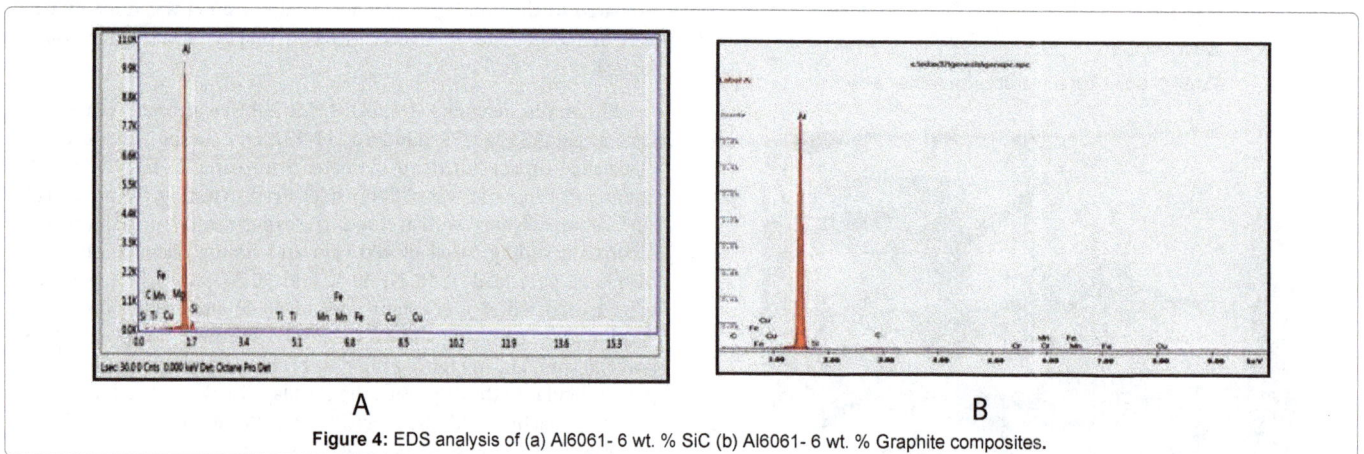

Figure 4: EDS analysis of (a) Al6061- 6 wt. % SiC (b) Al6061- 6 wt. % Graphite composites.

clustering of SiC and Graphite particles are evident. There is no evidence of casting defects such as porosity, shrinkages, slag inclusion and cracks which is indicative of sound castings. This is difficult to achieve in the aluminium matrix by conventional casting process. Two step mixing technique is an ideal process for synthesizing ceramic particulate reinforced composites [14]. In this, wetting effect between particles and molten Al6061 alloy matrix also retards the movement of the SiC and Graphite particles, thus, the particles can remain suspended for a long time in the melt leading to uniform distribution.

Figures 4a and 4b show energy dispersive X-Ray spectrographs of Al6061-6wt. of SiC and Al6061-6 wt. % of Graphite composites respectively. The EDS analysis confirmed the presence of SiC and Graphite in the Al matrix alloy. The presence of SiC shows in the form of Si (Silicon) and Carbon (C), which is evident from the EDS graph (Figure 4a). The presence of graphite shows in the form of Carbon (C), which is evident from the EDS graph (Figure 4b).

Wear studies

The volumetric wear loss of Al6061 alloy and its composites is as shown in Figure 5. The effect of applied load on the wear behavior of Al6061 alloy and its composites is shown in the Figure 5. The volumetric wear loss is increased as the normal load increases from 10 N to 50 N and is y lower in case of SiC and Graphite reinforced composites.

Higher volumetric wear loss is observed for matrix alloy and the composites at higher loads. At maximum loads the temperature of sliding surface and pin exceeds the critical value. So as load increases on the pin ultimately there is an increase in the volumetric wear loss of both the matrix alloy and SiC-Graphite composites.

The variation of volumetric wear loss of the matrix alloy 6061 and its composites with 6 wt. % SiC and 6 wt. % Graphite reinforcement contain are shown in Figure 5. It is observed that the volumetric wear loss of the composites decreases with 6 wt. % SiC reinforcements in the matrix alloy. The improvement in the wear resistance of the composites with 6 wt. % of SiC reinforcements can be attributed to the high hardness of SiC particulates which acts as the barrier for the material loss. There was significant decrease in volumetric wear loss in 6 wt. % Graphite reinforced composites. The volumetric wear loss of Al6061-6 wt. % Graphite composite is lesser than Al6061-6 wt. % SiC and as cast

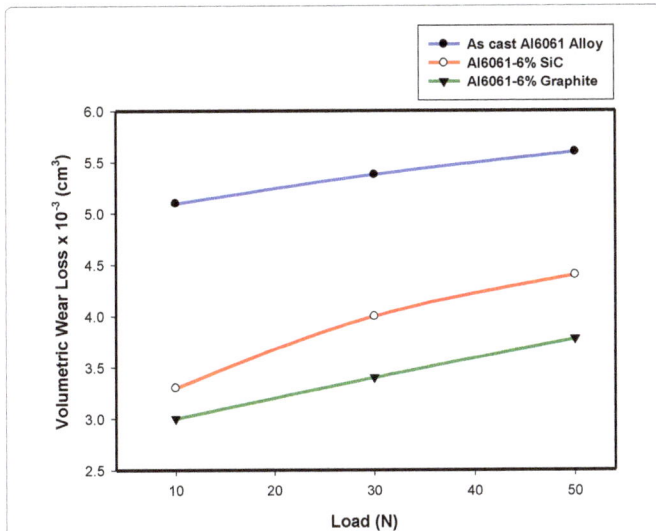

Figure 5: Volumetric wear loss of Al6061 alloy and SiC and Graphite composites at varying loads and 400 rpm constant sliding speed.

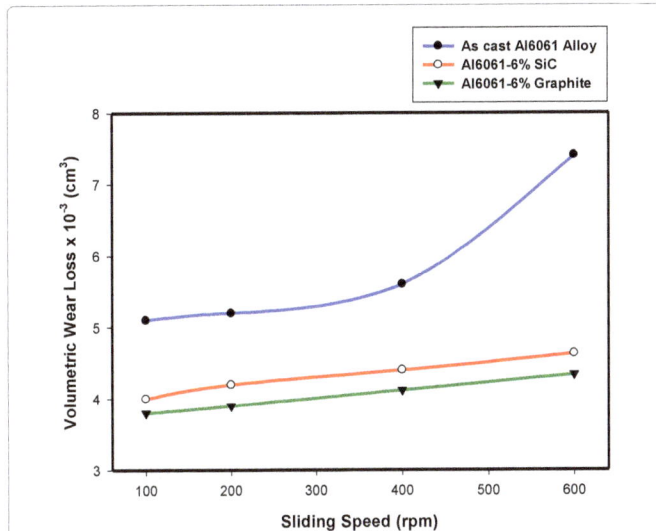

Figure 6: Volumetric wear loss of Al6061 alloy and SiC and Graphite composites at varying sliding speeds and 50 N constant load.

Al6061 alloy matrix. A6061-Graphite composites shown superior wear resistance compared to SiC reinforced composites.

Good lubricating property of graphite makes Al6061-6 wt. % Graphite composites more resistance to wear. When the Graphite content is low, mating parts is largely not covered by a film of graphite and the tri-biological properties are almost similar to or only slightly better than those of the matrix materials. When Graphite content is high, acts as a lubricant in the solid form and smears on the surface of the disc [15,16].

Figure 6 shows the dependence of all the volumetric wear loss of Al-6061 matrix alloy and Al6061-6 wt. % SiC and 6 wt. % graphite composites on sliding speed. With an increasing speed from 100 rpm to 600 rpm, the volumetric wear loss is increased for both Al6061 matrix alloy and fabricated composites. However for all sliding speeds, the wear rate of the composites were much lower, when compared with the Al6061 matrix alloy and is much lesser in the case of Al6061-6 wt. % Graphite composites compared to Al6061 alloy matrix and 6 wt. % SiC composites. Further, as sliding speed increases there is increase in wear due to softening of the composite at high temperature. The increased temperature causes the severe plastic deformation in the specimen at higher sliding speeds can leading to form high strain rate sub-surface deformation. Therefore this leads to enhanced delamination contributing to enhance wear rate.

Worn surface study

Wear surface analysis of Al6061 alloy and SiC-Graphite reinforced composites are studied by using scanning electron micro-photographs. Figure 7 represents the wear worn surfaces of matrix material Al6061 alloy and its composites 6 wt. % of SiC and 6 wt. % of Graphite particles reinforced composite at 50 N load and 400 rpm sliding speed.

Figure 7a shows the wear worn surface of the matrix alloy as cast Al6061 reveals more patches of smashed portions and yawning grooves due to heavy deformation in the plastic form. It can be seen that the wider and deeper grooves in the case of Al6061 alloy compared to its composites.

Figures 7b and 7c show the SEM photographs of the worn surface of Al6061-6 wt. % SiC and Al6061-6wt. % Graphite composites tested at applied load of 50 N and 400 rpm speed. Due to presence of hard SiC particulates in Al alloy wear loss is less and small patches can be seen.

Worn surface shows very minor flaws and cracks mainly due to the presence of SiC particulates (Figure 7b). The smeared particles of graphite from the worn surface of composites form a thin layer between sliding surfaces, which prevents direct metal contact. This acts as tribo-layer between pin and disc, avoids the direct contact between the two surfaces, and causes less volumetric wear loss in graphite composites (Figure 7c).

Conclusion

The investigation made on fabrication and evaluation of wear behavior Al6061-6 wt. % SiC and Al6061-6 wt. % Graphite metal matrix composite by vortex method has led to following conclusions. Al6061 alloy and its composites have been effectively fabricated by stir casting method using two step additions of particulates in the melt. The scanning electron microphotographs of Al6061-SiC and Al6061-Graphite composites were shown uniform distribution of reinforcement particulates in the base matrix material. The wear resistance of Al6061 alloy increased after addition of SiC and graphite particles. The volumetric wear loss is affected by load and sliding speed. There is significant increase in volumetric wear loss as applied load and sliding speed increases. The Al6061-6 wt. % graphite composites have shown lower rate of volumetric wear loss as compared to that observed in as cast Al6061 alloy matrix and Al6061-6 wt. % SiC composites. SEM micrographs of wear worn surfaces indicated the presence of

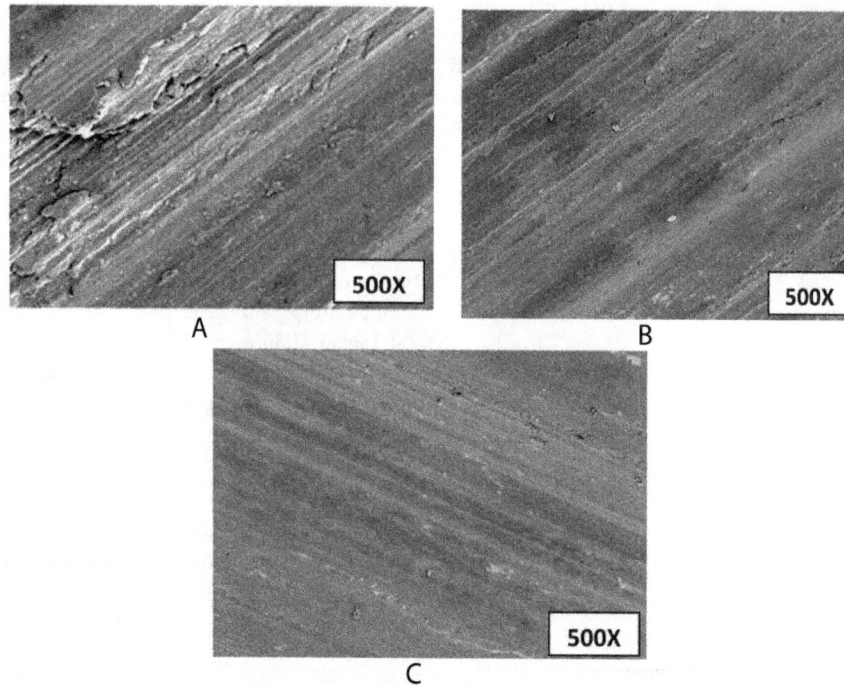

Figure 7: Shows the SEM microphotographs of worn surfaces of (a) as cast Al6061 alloy (b) Al6061-6% SiC (c) Al6061- 6% Graphite composites at 5 kg load and 400 rpm.

soft grooves in the Al6061-SiC and Al6061-graphite composites as observed in the base alloy Al6061.

References

1. Dora Siva P, Chintada S (2014) Hybrid composites a better choice for high wear resistant materials. J Materials Research and Technology 3: 172-178.

2. Rama Rao G, Padmanabhan (2012) Fabrication and mechanical properties of aluminium-boron carbide composites. Int J Materials and Biomaterials Applications 2: 15-18.

3. Nagaral M, Auradi V, Kori SA (2014) Dry sliding wear behaviour of graphite particulate reinforced Al6061 alloy composite materials 592-594: 170-174.

4. Zadra M, Girardini L (2014) High performance low cost titanium metal matrix composites Materials Science & Engineering 608: 155-163.

5. Vijay R, Elanchezhian C, Jaivignesh M, Rajesh S, Parswajinan C (2014) Evaluation of mechanical properties of aluminium alloy alumina boron carbide metal matrix composites. Materials and Design 58: 332-338.

6. Shorowordi KM, Laoui T (2001) Microstructure and interface characteristics of B$_4$C SiC and Al$_2$O$_3$ reinforced Al matrix composites: A comparative study. 4th International Conference on Mechanical Engineering Dhaka Bangladesh: 175-181.

7. Hasim J, Looney L, Hashmi MSJ (1999) Metal matrix composites production by the stir casting method. J Materials Processing Technology 92: 931-937.

8. Nagaral M, Attar S, Reddappa HN, Auradi V, Suresh Kumar et al. (2015) Mechanical behaviour of Al7025-B4C particulate reinforced composites. J Applied Mechanical Engineering 4: 6.

9. Suresh S, Shenbaga VM, Vettivel SC, Selvakumar N (2014) Mechanical behaviour and wear prediction of stir cast Al-TiB2 composites using response surface methodology. Materials and Design 59: 383-396.

10. Umanath K, Palanikumar K, Selvamani ST (2013) Analysis of dry sliding wear behavior of Al6061-SiC-Al$_2$O$_3$ hybrid metal matrix composites. Composites Part-B 53: 159-168.

11. Ramesh CS, Keshavamurthy R, Channabasappa BH, Promod S (2010) Friction and wear behaviour of Ni-P coated Si$_3$N$_4$ reinforced 6061-Al composites Tribiology Int 43: 623-624.

12. Zang J, Alpas AT (1993) Wear regimes and transition in Al$_2$O$_3$ particulate reinforced Al-alloy. Material Science and Engineering 161: 273-284.

13. Naplocha K, Granat K (2008) Dry sliding wear of Al-Saffil-C hybrid metal matrix composites Wear 265: 1734-1740.

14. Nagaral M, Bharath V, Auradi V (2013) Effect of Al$_2$O$_3$ particles on mechanical and wear properties of 6061AL alloy metal matrix composites. Materials Science and Engineering 2: 1.

15. Devaraju A, Kumar A, Kotiveerachari B (2013) Influence of addition of Gr/Al$_2$O$_3$ with SiC on wear properties of aluminium alloy 6061-T6 hybrid composites via friction stir processing. Transactions of Nonferrous Metal Society of China 23: 1275-1280.

16. Nagaral M, Auradi V, Parashivamurthy KI, Kori SA (2015) Wear behaviour of Al2O3 and Graphite particulates reinforced Al6061 alloy hybrid composites. American J Materials Science 5 (3C): 25-29.

22

Deriving the Governing Equations for Shock Damper to Model and Control the Unsteady Flow Caused by Sudden Closer of Valve

Bostan M* and Akhtari AA

Department of Civil Engineering, College of Engineering, Razi University, Kermanshah, Iran

Abstract

This paper focuses on deriving the governing equations for an instrument used in water flow systems known as "shock damper". Shock damper is a tremulous tool with a single degree of freedom system which includes tank, connecting pipe, mass, spring, and a damper. This tool is then applied as a boundary condition for characteristic lines equations. Additional equations will be included by considering conservation of mass, momentum, and energy. This system of equations is then explicitly solved at each time step. In order to illustrate the performance of the shock damper, a gravity-feed system is considered with and without a damper. The control valve of this system will suddenly be shut to impose a water hammer condition. Then, unsteady flow parameters such as minimum/maximum of flow velocity and pressure are evaluated along with a sensitivity analysis. Results demonstrate that shock damper despite being simple and economically efficient, is highly capable of moderating unsteady flow characteristics.

Keywords: Water hammer; Shock damper; Unsteady flow effects; Surge tanks

Introduction

One of the key factors in designing a water distribution system is to consider the effects of possible unsteady flow caused by water hammer. Water hammer is a pressure surge or wave propagating with the speed of sound and is created when a fluid (usually a liquid) in motion is forced to stop or change direction suddenly (momentum change). This phenomenon can occur due to various reasons including rapid opening and/or shutting of a valve, failure of a pump, joint, or pipe, use of check valves. Water hammer causes intensive damages in some cases, making it a significant factor while designing water distribution systems.

Two different approaches have been reported to tackle this problem: a) taking necessary precautions to avoid the occurrence of sudden change of flow velocity (or direction) hence water hammer; and b) equipping the water system with efficient tools to dampen water hammer wrecking impacts. Since water hammer seems inevitable, the second approach is widely employed. Use of such tools (e.g. shock dampers) helps in depreciating consequences of unsteady flow due to water hammer by compensating any change in flow pressure. Equipment used to dampen water hammer can be categorised into surge tanks and safety valves [1]. In order to simulate the dynamic behaviour of surge tanks, they are introduced as a boundary condition for characteristic lines equations. Taking conservation of mass, momentum, and energy into account augments the number of equations forming a system of n equations where n is the number of unknown variables of the surge tank. Detailed discussion on this topic can be found in Bruce et al. [2] and Russell Bent et al. [3]. Safety valves are also applied as a boundary condition, and additional equations are derived to form a system of equations. Grabarczyk et al. [4] and Grabarczyk et al. [5] reviewed the application of safety valves and similar equipment in an unsteady flow condition. Niełacny [6] modelled spring safety valves in water distribution systems with pumps.

High operational cost is the major concerns in utilising surge tanks. Despite being expensive, one-way surge tanks are unable to dampen the positive pressure wave caused by water hammer. They are only capable of preventing the negative waves and column separation by injecting flow to the pipeline [2]. Air surge tanks can handle both positive and negative waves, but they are costly [2]. On the other hand, safety valves have much less installation and maintenance costs, yet they have poor performance for restraining positive waves and incapable of hindering negative waves. Thus, the occurrence of column separation and consequent damages to the system is inevitable [4]. We aim to find an alternative that can handle both negative and positive waves while being simple and economically efficient.

This paper focuses on shock damper in attempt to derive its governing equations and illustrate the effectiveness. Results show that shock dampers are highly capable of moderating unsteady flow characteristics. The structure of the rest of this paper is as follows. In the next section, we define the problem we are dealing with which includes deriving the equation and solving it for the given boundary conditions. Then, we demonstrate the performance of the proposed approach by solving a case study. Results are depicted in the next section followed by a concluding section.

Problem Definition

The current study focuses on the proposed instrument, called shock damper, which is capable of encountering positive and negative waves due to water hammer. In this section, shock damper is demonstrated, the governing equation is derived, boundary conditions are considered, and the resulting system of equations is solved.

Shock Damper

A shock damper consists of a damper, spring, mass, tank, and a joint (Figure 1). The combination of spring, damper, and mass performs as a vibrant system with a single degree of freedom (vertically) when positive or negative waves occur. The spring of this system is pre-pressed to capture enough force. This force is used to balance with the force caused by pressure head. In case of positive waves, the excessive pressure imposed to the spring compress it further, storing potential energy. This potential energy preserves the rest of the water distribution

*Corresponding author: Bostan M, Department of Civil Engineering, College of Engineering, Razi University, Kermanshah, Iran
E-mail: mohamad_bostan@gmx.com

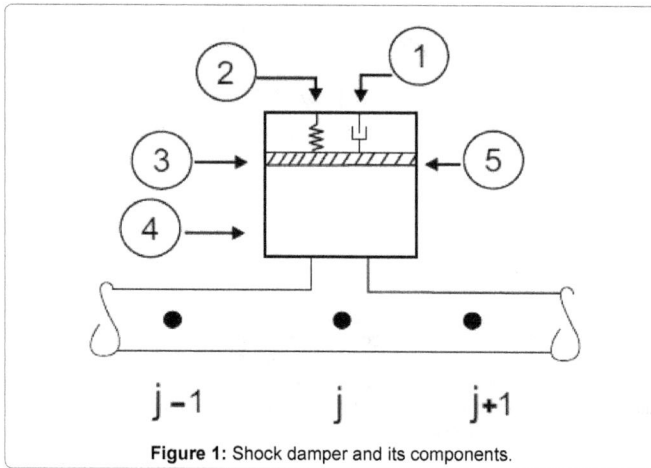

Figure 1: Shock damper and its components.

system from water hammer effects (waves). When a negative wave hits the shock damper, the pre-pressed spring will be elongated and water stored in the shock damper's tank will be injected into the flow to compensate the negative pressure. This will help to prevent column separation and consecutive damage to the system. In theory, due to water hammer, an infinite number of positive and negative waves must propagate in the system. However, this is never the case as a result of friction and depreciation of energy. The role of shock damper is to accelerate the process of depreciation of water hammer waves. Figure 1 shows a shock damper and its components including 1) damper, 2) spring, 3) mass, 4) joint, and 5) tank.

Governing Equations

To simulate the condition of unsteady flow in the system, a pair of partial differential equations as shown in equations 1 and 2 is used. These equations are obtained by considering conservation of mass and momentum for a moving control volume as an element of the pipeline [7].

$$\frac{\partial V}{\partial t} + g\frac{\partial H}{\partial x} + \frac{fV|V|}{2D} = 0 \tag{1}$$

$$\frac{\partial H}{\partial t} + \frac{a^2}{g}\frac{\partial V}{\partial x} = 0 \tag{2}$$

Where H is the piezometric head and V is flow velocity, both are a function of time (t) and space (x); a is the speed of propagation of sound wave in the flow; g is the gravitational acceleration; D is the pipe diameter; and f is the Darcy–Weisbach coefficient. All dimensions follow the standard SI units. One of the most widely-used approaches to solving equations similar to those of equation (1) and (2) is the Characteristic line method [8]. In this method, the solution space is discretized to a finite number of nodes including intermediate and boundary nodes.

Intermediate Nodes

Discretised form of the equations for intermediate nodes under unsteady flow condition is represented below [2].

$$Q_i^{n+1} = \frac{1}{2}\left[\begin{array}{l}(Q_{i-1}^n + Q_{i+1}^n) + \frac{gA}{a}(H_{i-1}^n - H_{i+1}^n) + \frac{g}{a}\Delta t\,(Q_{i-1}^n - Q_{i+1}^n)\sin\alpha \\ -\frac{f\Delta t}{2DA^2}(Q_{i-1}^n|Q_{i-1}^n| + Q_{i+1}^n|Q_{i+1}^n|)\end{array}\right] \tag{3}$$

$$H_i^{n+1} = \frac{1}{2}\left[\begin{array}{l}(H_{i-1}^n + H_{i+1}^n) + \frac{a}{gA}(Q_{i-1}^n - Q_{i+1}^n) + \frac{\Delta t}{A}\,(Q_{i-1}^n + Q_{i+1}^n)\sin\alpha \\ -\frac{a}{g}\frac{f\Delta t}{2DA^2}(Q_{i-1}^n|Q_{i-1}^n| + Q_{i+1}^n|Q_{i+1}^n|)\end{array}\right] \tag{4}$$

In the above equations, H_i^{n+1} and Q_i^{n+1} are the piezometric head and flow rate in node i at time–step $n+1$ respectively; and A is the pipe's cross-section.

Boundary Conditions

In order to derive the governing equations of a shock damper, a gravity-feed system is considered. Components of such system include pipe, tank, valve, and a shock damper, each with a specific equation (boundary condition).

Tank at $i = 1$:

Following equations are considered for a tank with constant head (H^0) located at the beginning of a pipeline [2].

$$H_1^{n+1} = H^0 \tag{5}$$

$$Q_1^{n+1} = Q_2^n + \frac{gA}{a}(H^0 - H_2^n) - \frac{f\Delta t}{2DA^2}Q_2^n|Q_2^n| \tag{6}$$

Valve at $i = m+1$:

For a valve that requires $t=T_c$ to decrease the flow rate to zero and is located at the end of the pipeline, the boundary condition equations are:

$$Q_{m+1}^{n+1} = Q^0(1 - \frac{t}{Tc}) \qquad 0 \le t \le Tc$$
$$Q_{m+1}^{n+1} = 0 \qquad t \ge Tc \tag{7}$$

$$H_{m+1}^{n+1} = H_m^n - \frac{a}{gA}(Q_{m+1}^{n+1} - Q_m^n) - \frac{a}{g}\frac{f\Delta t}{2DA^2}Q_m^n|Q_m^n| \tag{8}$$

Shock damper at $i = B_c$:

As shown in Figure 2, we assume that the shock damper is located at the node $i = B_c$. Expanding the conservation of mass for this node:

$$Q_i^{n+1} = Q_{i+1}^{n+1} + Q_c^{n+1} \tag{9}$$

Where Q_c^{n+1} is the amount of flow rate from the main pipe to the joint of the shock damper at time-step $n+$.

Whenever flow enters the shock damper it will be assumed a positive flow rate, otherwise it is negative.

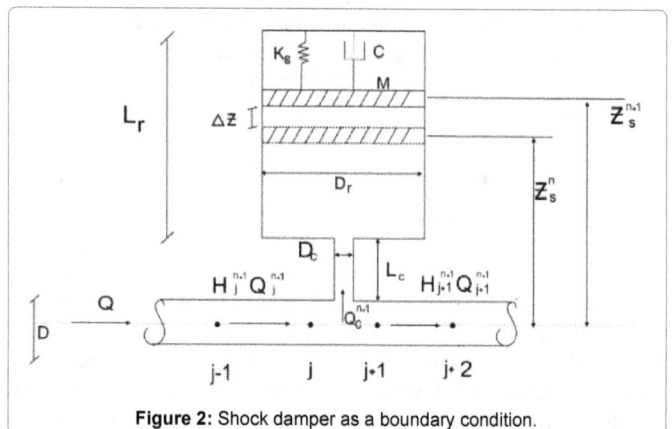

Figure 2: Shock damper as a boundary condition.

Next, conservation of energy is expanded for node i. Since the distance between node i and i+1 is trivial, energy depreciation due to friction is negligible.

By taking into account the conservation of momentum for the joint, we have:

$$\frac{L_c}{gA_c\Delta t}(Q_c^{n+1} - Q_c^n) = \frac{H_i^{n+1} + H_i^n}{2} - \frac{f \, L_c}{4gD_cA_c^2}(Q_c^n\left|Q_c^n\right|$$
$$+Q_c^{n+1}\left|Q_c^{n+1}\right|) - \frac{Z_s^{n+1} + Z_s^n}{2} - \frac{P_0^{n+1} + P_0^n}{2\gamma} \tag{10}$$

Where L_c is the length of the joint; Z_s^n is the mass level proportional to the pipe axis in each time-step; and P_0^n is pressure imposed to the moving mass at each time-step. Using the Continuity Principal, Z_s^n can be calculated as follows:

$$Z_s^{n+1} = Z_s^n + \frac{\Delta t}{2A_r}(Q_c^{n+1} + Q_c^n) \tag{11}$$

Figure 3 shows the free body diagram for any element of mass in the tank. Applying Newton's second law of motion for vertical axis leads to:

$$P_0^{n+1}A_r = M \, \ddot{Z}^{n+1} + C \, \dot{Z}^{n+1} +$$
$$K_s(Z_s^{n+1} - Z^0) + K_s\Delta_0 + Mg \tag{12}$$

Where M is the moving mass; C is damping coefficient; A_r is the cross-section of shock damper's tank; K_s is the spring stiffness factor; Z^0 is the mass level according to the pipe axis in $t=0$; Z^0 is the distance that spring has been compressed at $t=0$. The forces are defined as follow: gravity force is $F_g = M \, g$, spring force is $F_K = K_s(Z_s^{n+1} - Z^0 + \Delta_0)$, damping force is $F_C = C \, \dot{Z}^{n+1}$, and inertia force is $F_a = M \, \ddot{Z}^{n+1}$. As it can be seen, above equations represent the vibration of a system with a single degree of freedom (vertically). $\ddot{Z}$ and $\dot{Z}$ in above equation are acceleration and velocity of flow mass respectively, they can be calculated as:

$$\dot{Z}^{n+1} = \frac{Z_s^{n+1} - Z_s^n}{\Delta t} \tag{13}$$

$$\ddot{Z}^{n+1} = \frac{\dot{Z}^{n+1} - \dot{Z}^n}{\Delta t} \tag{14}$$

The velocity and acceleration of the mass in the tank can be calculated based on flow rate in the joint using the Continuity Principal.

$$\dot{Z}^{n+1} = \frac{Q_c^{n+1} + Q_c^{n+1}}{2A_r} \tag{15}$$

$$\ddot{Z}^{n+1} = \frac{Q_c^{n+1} + Q_c^{n+1}}{A_r\Delta t} \tag{16}$$

Flow rate and head in nodes i and i+1 can be estimated by

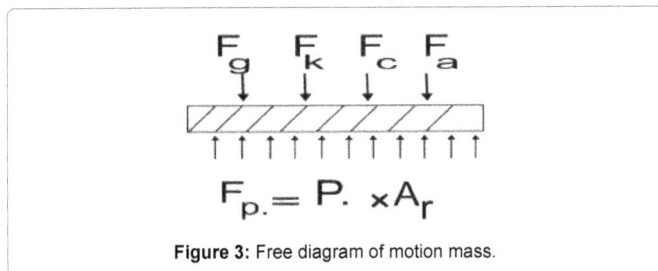

Figure 3: Free diagram of motion mass.

considering the characteristic line equations C⁻ and C⁺, equation 9 and equation 10.

$$Q_i^{n+1} = (C_1^n - \frac{g}{a} H_i^{n+1})A \tag{17}$$

$$Q_{i+1}^{n+1} = (C_2^n + \frac{g}{a} H_{i+1}^{n+1})A \tag{18}$$

$$H_i^{n+1} = \frac{C_1^n - C_2^n - \frac{Q_c^{n+1}}{A}}{\frac{2g}{a}} \tag{19}$$

$$H_i^{n+1} = H_{i+1}^{n+1} \tag{20}$$

In the above equations, C_1^n and C_2^n are derived with the following equations:

$$C_1^n = \frac{Q_{i-1}^n}{A} + \frac{g}{a} H_{i-1}^n - \frac{f\Delta t}{2DA^2}Q_{i-1}^n\left|Q_{i-1}^n\right|$$
$$+ \frac{g}{aA}\Delta t Sin\theta \, Q_{i-1}^n \tag{21}$$

$$C_2^n = \frac{Q_{i+2}^n}{A} + \frac{g}{a} H_{i+2}^n - \frac{f\Delta t}{2DA^2}Q_{i+2}^n\left|Q_{i+2}^n\right|$$
$$- \frac{g}{aA}\Delta t Sin\theta \, Q_{i+2}^n \tag{22}$$

In each time-step the nine variables including H_i^{n+1}, H_{i+1}^{n+1}, $\dot{Z}^{n+1}$, $\ddot{Z}^{n+1}$, Z_s^{n+1}, Q_i^{n+1}, Q_{i+1}^{n+1}, Q_c^{n+1}, and P_0^{n+1} are computed using equations 11 to 13 and equations 16 to 21.

Solving the System of Equations for a Shock Damper

The system of equations defined in the previous section is used to simulate the behaviour of a shock damper when water hammer occurs. The challenge here is to solve this nonlinear and implicit system of equations which cannot be readily solved with current numerical methods. To tackle this issue, we replace the term $\frac{f \, L_c}{4gD_cA_c^2}(Q_c^n\left|Q_c^n\right| + Q_c^{n+1}\left|Q_c^{n+1}\right|)$ in equation 11 with $\frac{f \, L_c}{2gD_cA_c^2}Q_c^n\left|Q_c^n\right|$. In fact, we neglect the numerical error produced here to make equation 11 linear. This error can be diminished by opting a smaller time-steps. By replacing H_i^{n+1}, H_{i+1}^{n+1}, $\dot{Z}^{n+1}$, $\ddot{Z}^{n+1}$, Z_s^{n+1}, Q_i^{n+1}, Q_{i+1}^{n+1}, Q_c^{n+1}, and P_0^{n+1} in equation 11 with their corresponding values from equations 12 and 13 and equations 16 to 21, equation 11 changes into an explicit linear form with Q_c^{n+1} being the only variable:

$$\alpha_1 Q_c^{n+1} = \alpha_2^n Q_c^n + \alpha_3^n \tag{23}$$

In the above equation, new parameters are introduced and defined as follows:

$$\alpha_1 = \begin{bmatrix} \dfrac{L_c}{gA_c\Delta t} + \dfrac{a}{2Ag} + \dfrac{\Delta t}{4A_r} + \\[2mm] \dfrac{M}{2\gamma A_r^2 \, \Delta t} + \dfrac{C}{4\gamma A_r^2} + \dfrac{K_s\Delta t}{4\gamma A_r^2} \end{bmatrix} \tag{24}$$

$$\alpha_2^n = \begin{bmatrix} \dfrac{L_c}{gA_c\Delta t} - \dfrac{\Delta t}{4A_r} + \dfrac{M}{2\gamma A_r^2 \, \Delta t} - \\[2mm] \dfrac{C}{4\gamma A_r^2} - \dfrac{K_s\Delta t}{4\gamma A_r^2} - \dfrac{fL_c}{2gA_c^2 D_c}Q_c^n \end{bmatrix} \tag{25}$$

$$\alpha_3^n = \left[\begin{array}{c} \dfrac{H_i^n}{2} + \dfrac{(C_1^n - C_2^n)a}{2g} - Z_s^n - \\ \dfrac{P_0^n}{2\gamma} - \dfrac{K_s \Delta_0 + Mg}{2\gamma A_r} \end{array} \right] \qquad (26)$$

We can now calculate Q_c^{n+1} from equation 24 for each time-step. Then, rest of the variables can be computed accordingly.

Case Study

A gravity-feed system is considered here which consist of a tank with constant head at the beginning of the pipeline system and a valve at the end of it. This system is solved in two scenarios: with and without using a shock damper. The aim is to demonstrate the effect of installing a shock damper on flow characteristics of such system. Figure 4 shows the system under consideration. In order to create a water hammer condition, the downstream valve is shut in 3 seconds. The initial condition of the system includes a constant flow rate in the pipe.

Scenario 1

In this case, no shock damper is equipped with the system. Piezometric head in the tank is equal 40 m; diameter of the main pipe is 0.3 m; length of the pipe is 1000 m; thickness of the pipe is 0.005 m, and thepipeis made of steel with a modulus of elasticity equal 210000 Mpa. The fluid in the system is water with a density of 1000 Kg/m^3, bulk modulus of 21000000 Mpa, and Darcy–Weisbach coefficient of 0.02 (for both steady and unsteady flow). Δx and Δt are assumed equal 20 m and 0.175 sec respectively.

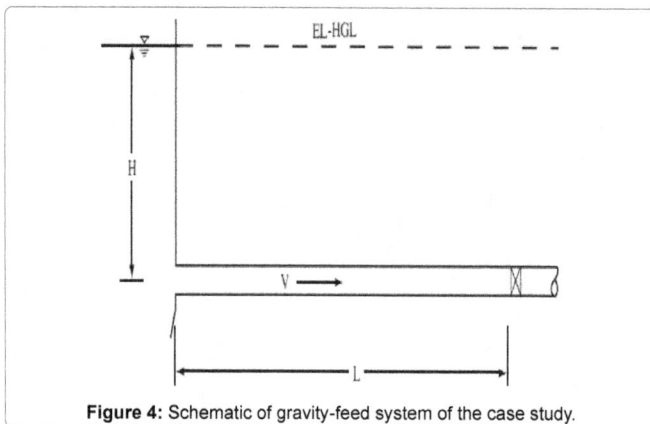

Figure 4: Schematic of gravity-feed system of the case study.

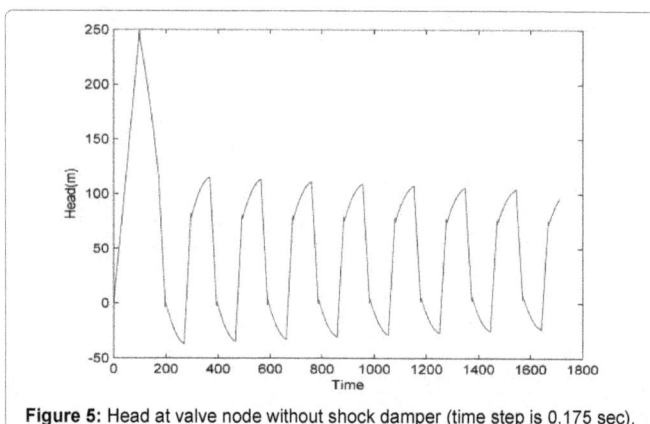

Figure 5: Head at valve node without shock damper (time step is 0.175 sec).

Scenario 2

We use a shock damper in this scenario. The parameters related to the shock damper include M=30 Kg, C=10^7 $N.sec/m$, K_s=30000 N/m, D_r=1 m, D_c=0.1 m, L_r=2 m, and L_c=0.5 m. All other parameters are assumed to be similar to those of scenario 1.

Results and Discussion

The case described in the previous section was solved to illustrate the effect of using a shock damper in a simple water distribution system. The piezometric head at the location of the valve was calculated for both scenarios. Figures 5 and 6 show the fluctuation of piezometric head obtained for scenario 1 and 2 respectively.

As it can be seen, water hammer caused by a sudden closure of the valve can generate piezometric head (and corresponding pressure) as high as 250 m and -50 m where no shock damper was used. This is relatively a severe pressure resulting in extensive damages to the system. The maximum pressure (positive value) can cause pipes to explode while the minimum pressure (negative value) can lead to column separation. Also, the next head peak reached a value of 130 m and -50 m. This is due to friction in the pipeline and depreciation of energy, yet the resulting pressure is high after approximately 5.25 sec. In this scenario, the impact of water hammer is still visible after around 30 sec. When the shock damper described in scenario 2was utilised, maximum and minimum of the piezometric head were calculated as 165 m and -10 m respectively. This is a great reduction (35% for the maximum head and 80% for the minimum head) compared to those of scenario 1. Moreover, the peak heads were significantly depreciated after almost 7 sec, reaching a value of 60 m and 20 m. Herein, we perform a sensitivity analysis of the system. Figures 7 - 10 summarise the outcome of the sensitivity analysis. Results indicate that the performance of a shock damper is more sensitive to two of its parameters, namely the diameter of the main tank and damping coefficient of the shock damper.

Figures 7 and 8 show obtained piezometric head at the location of the valve for damping coefficients of +80% and -60% of its initial value, respectively. It is seen that increase in the damping coefficient decreases the efficiency of the shock damper hence greater piezometric head is generated. Decreasing this parameter improves the performance of the shock damper. However, there is a limit to it as further reduction deteriorates its efficiency. This suggests that an optimal level can be sought. Responds of the system to changes made to the diameter of the main tank is demonstrated in Figures 9 and 10. As it is expected, as the size of the tank increases, the shock damper performs better. Yet

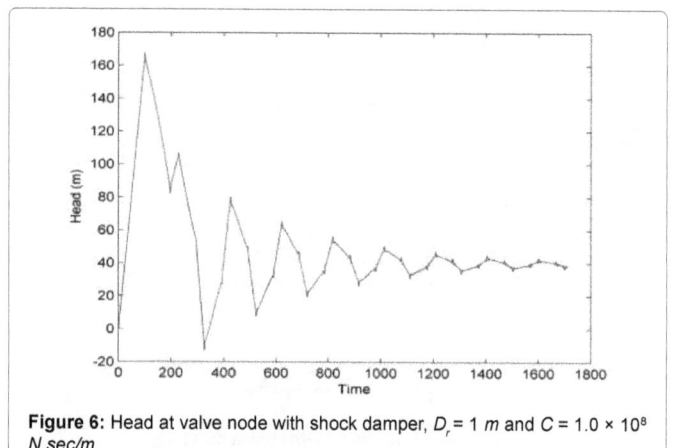

Figure 6: Head at valve node with shock damper, D_r = 1 m and C = 1.0 × 10^8 $N.sec/m$.

there exists a feasible upper limit to expansion due to physical and economic constraints. Also, after a certain level, any further increase in the size shows to have no considerable effect. Again, this implies that there is an optimal level for this capacity expansion. Figure 10 depict a situation where the diameter approaches zero. As we can expect, a system with a shock damper and the main tank diameter of zero will perform similar to the one without any shock damper. Figure 10 closely imitates the behaviour of the system shown in Figure 5 which validates the results.

Conclusion

Water hammer is inevitable in water distribution systems. Because of the extensive damages caused by this phenomenon, its impacts to the system must be efficiently moderated. Shock damper is simple and economically efficient equipment that can be used to handle devastating consequences such as pipe explosion and column separation. This paper focused on deriving the governing equation of shock dampers and using it to model the system. Results demonstrated that installing a shock damper in a gravity-feed system can effectively dampen the pressure waves caused by water hammer. A sensitivity analyses revealed that the performance of shock dampers mostly relies on the diameter of the main tank and damping coefficient of the shock damper. This study also suggests that an optimal value for these parameters can be obtained to maximise the performance.

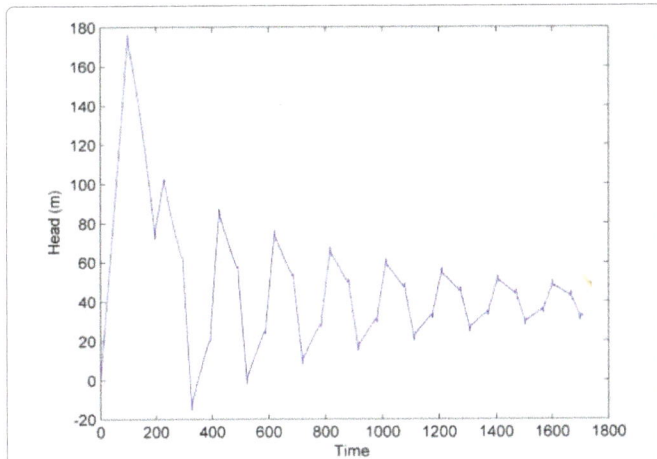

Figure 7: Head at valve node without shock damper with D_r =1 m and C =1.8 × 10^8 $N.sec/m$.

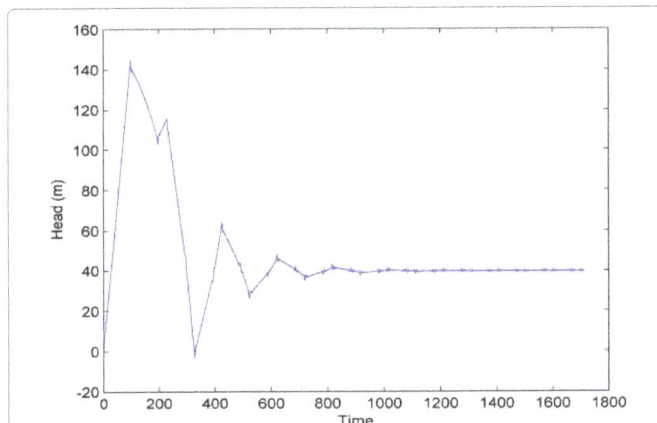

Figure 8: Head at valve node with shock damper, D_r = 1 m and C = 0.4 × 10^8 $N.sec/m$.

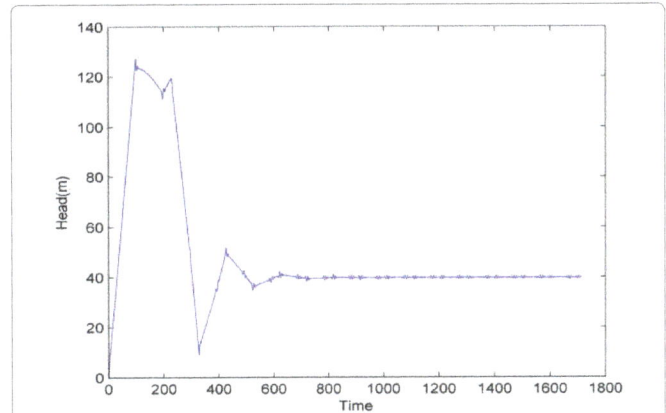

Figure 9: Head at valve node with shock damper, D_r = 4 m and C=1.0 × 10^8 $N.sec/m$.

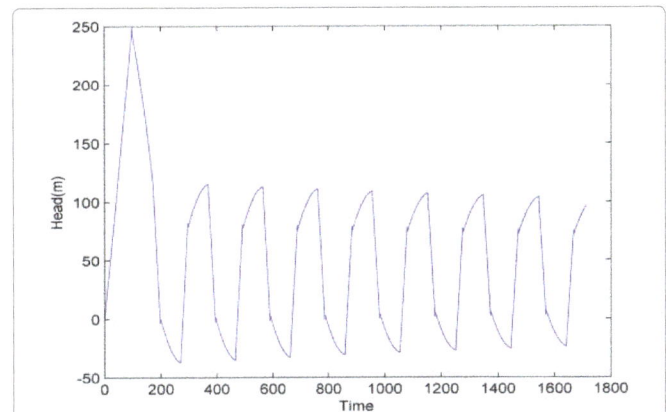

Figure 10: Head at valve node with shock damper, D_r = 0.001 m and C = 1.0 × 10^8 $N.sec/m$.

References

1. Jung BS, Karney BW (2006) ASCE M: Hydraulic optimization of transient protection devices using GA and PSO approaches. J of water resources planning and management 132: 44.

2. Larock BE, Jeppson RW, Watters GZ (1999) Hydraulics of pipeline systems. CRC press boca raton London New York Washington DC.

3. Bent R, Perelman LS (2015) Piece-wise mixed integer programming for optimal sizing of surge control devices in water distribution systems, Water Resources Research.

4. Grabarczyk CZ, Niełacny M, Skiba J (1977–1980) Review of the methods of water-air reservoirs design and construction of valves damping water hammer Part II Departmental problem R-15 Poznan–Krakow (in Polish).

5. Grabarczyk CZ, Niełacny M, Skiba J (1977–1980a) Calculation methodology of water hammer phenomena and design principles of damping equipment departmental problem R-15 Poznan–Krakow (in Polish).

6. Niełacny M (2004) Model of the water-hammer effect considering a spring safety valve. Archives of Hydro-Engineering and Environmental Mechanics 51: 25-40.

7. Wylie EB, Streeter VL (1984) Fluid transients FEB Press, Ann Arbor.

8. Hanif Chaudhry M (1987) Applied hydraulic transients. Van Nostrand Reinhold New York.

23

Analysis of the Wheat Straw/Flax Fiber Reinforced Polymer Hybrid Composites

El Messiry M* and El Deeb R

Textile Engineering Department, Faculty of Engineering, Alexandria University, Alexandria, Egypt

Abstract

In the last decades, the agro wastes in manufacturing different types of composites found the increasing utilization. The world statistics of the Wheat and rice straw indicates that about 710 million metric tons of Wheat straw and 670 million tons of rice straw are produced each year as the agricultural waste, beside the other cereal and forest waste. Therefore, it causes a huge environmental impact and this problem is growing each year. The production of plastic composites from agro waste materials is receiving the substantial consideration currently. In this work, we suggest to use the Wheat straw in its whole shape instead of shredded one for forming the green composite. The hybrid composite was formed of Wheat straw, flax fiber, and jute fabric. This study targets at the mechanical properties of Wheat straw. The results indicate the possibility of manufacturing of the thin plates and shafts from such hybrid combinations after the increasing the crushing force of Wheat straw.

Keywords: Natural fibers; Green composite; Crushing force; Agro waste; Wheat straw

Introduction

Composite materials have been known to mankind for thousands of years, and occur naturally in many living things. The earliest composite materials were straw reinforced bricks [1]. Currently, natural fibers, mainly flax and hemp, are available as commercial products as short fibers for injection molding compounds or as non-woven mats based on short fibers organized in a random array. Enhancing the adhesion between the natural fibers and matrices results in improving composite strength and toughness [2-6].

Green composite combines plant fibers with natural resins to create natural composite materials. Biomaterial composites are made from hemp, kenaf, sisal, soybean, etc. Natural fibers are emerging as low cost, lightweight and apparently environmentally superior alternative to synthetic fibers [7]. The agriculture waste is one of the problems of the modern intensive agriculture which needs innovative methods for its useful industrial applications. The development of composites from renewable raw materials has increased considerably during the last years as they are environmentally friendly materials. Natural fibers are renewable, easily recycled, carbon dioxide neutral, and are available in large quantities. In addition, they have high specific properties and cause less health problems during handling when compared to glass fibers. Hence, apart from the low cost of agro-waste, there is interest nowadays to environment and biodegradability properties of materials to actualize the scale up production of agro waste plastic composites by the formation and synthesis of these filler fibers. Various methods are required for mixing filler husks at different filler blend ratio [1,8-10]. Several researchers [7,8] exploit the facts of value-added and ultra-light composite products from Wheat straw for the implementation in the automotive and transportation industries. With natural fiber composites, car weight reduction up to 35% is possible. This can be translated into lower fuel consumption and the lower environmental impact. Natural fiber based composites also offer good mechanical performance, good formability, high sound absorption, and cost savings due to low material costs. Moreover, their "Green look" as well as ecological and logistical benefits of the natural fiber based technologies looks more attractive. In 2000, more than 23,000 tons of natural fibers have been used in the automotive sector alone furthermore, natural fibers in the automotive industry should experience a sustainable growth as EU regulations regarding recycling and "End of life vehicle" directives set car recycling targets to 95% by 2015 [7].

The world statistics of the Wheat and rice straw indicates that there are about 710 million metric tons of Wheat straw and 670 million ton of rice straw each year. The introduction an innovative environmentally friendly process for turning agricultural residues (agro-waste) into quality products suitable for use for the industrials is always urged [1,11]. The use of straw has been gaining much research attention as a potential alternative lignocellulosic raw material replacing wood for making composites, particularly for particleboards [12-16]. The agro waste may be used individually or in combination with each other to achieve desired structure of agro waste plastic composites. An understanding of the available composites fabrications processes and how they apply to different composites is a necessary requirement for proper selection of composite design for engineering. The characteristic of composite depends on the nature of the reinforcement, the ratio of resin, such as Polyethylene Terephthalate (PETE), High Density Polyethylene (HDPE), PolyvinylChloride (PVC), Low Density Polyethylene (LDPE) and Polypropylene (PP), Polystyrene (PS), to reinforcement and the mode of fabrication [17].

Mechanical properties of the composites with Wheat straw as filler showed better mechanical properties compared to cornstalk and corncob, which is resulted from the good compatibility between the more non-polar surface of Wheat straw and HDPE [18].

The techniques that are used to manufacture the bio-composites are based on existing plastic processing methods. These methods can be used with thermoplastics or thermosetting polymers [1,19].

*Corresponding author: Magdi El Messiry, Textile Engineering Department, Faculty of Engineering, Alexandria University, Alexandria, Egypt
E-mail: mmessiry@yahoo.com

For thermoplastic composite, the manufacturing methods include extrusion, injection moulding and compression moulding. One of the main limitations of compounding and injection moulding is that only relatively short natural fibers can be used [20]. If longer fibers to be involved, alternative methods, such as co-mingling in the form of a non-woven fleece, should be implemented.

The Wheat straw, usually as short lengths, febrile, or after grinding as a micro size particles is used. Literatures not readily accessible to use of whole stem of the Wheat straw to form the composites. This study targets at the mechanical properties of whole stem of Wheat straw as well as the manufacturing methods for Wheat straw/flax fiber to produce low density hybrid composite.

Materials and Methods

Materials

Wheat straw: Wheat straw obtained from Egyptian local farmers was cut into 19–20 cm lengths without nods and used for preparing Wheat straw composite.

Flax fibers: Egyptian Long staple combed flax fibers were used of fiber length110 mm and fiber diameter 20 μm. The fiber breaking strength is 547.1 Mpa, breaking strain is 0.01458, and Modulus of elasticity is 37.52 GP.

Properties of flax fabric: Flax fabric is used with the following specifications: weight 290 g/m2, thickness 0. 43 mm, tensile strength 0.03625 GPA, breaking elongation 9.3%, fabric design plain weave 1/1, and cover factor 0.37 with equal ends/cm and picks/cm 4.5. Warp count and the weft count are 246 tex and 311 tex, respectively. Fabric tensile strength was done on a fabric tensile tester machine with load capacity 5000 N.

Fiber diameter measurement

Flax fiber diameter is measured manually by using an optical microscope. The results reported are averages of 50 samples. The measuring of fiber diameter was done on LEICA DME microscope as shown in Figure 1.

Tensile strength testing: Tensile testing was done on a TITAN2 Universal strength tester machine (Model 710) with load capacity 3000 N. TITAN2 Universal strength tester software was used for data acquisition. The testing for the flax fiber samples was done at gauge length 3 cm, the cross-head displacement rate at 10 mm/min. The results reported are averages of five samples. The specimen ends were clamped into the hydraulic jaws of the testing machine as shown in Figure 2.

Wheat straw crushing force: For the determination of the crushing force, special attachment was designed for measuring the crushing force required to crack the Wheat straw; the straw is pressed against a flat surface by a force at the middle portion till a crack is started under the applied load. Figure 3 shows the principle of the attachment.

Results and Discussion

There are three important physical structural features of Wheat straw which make it an excellent fiber and filler source for thermo-set composite materials, i.e., intricate percolating pore structure of vascular bundles and central void, presence of micro fibrils in the structure, and existence of polymer lignin near the surface of the straw. However, the lack of cross-linking between polymers and Wheat straw

fibers has been a major stumbling block for straw fiber utilization in various composite materials [21].

Figure 4 shows SEM image of a cross sectioned raw Wheat straw from the internode region of the stem with noticeable hollow pith and the scattered placement of the vascular bundles close to the epidermis [5]. The use of the straw in this case will give low strength composites.

Composites preparation

With the anticipation to improve the mechanical properties of the Wheat straw, we suggested infusing risen, protein colloid glues, inside the Wheat straw Lumina. Collagen consists of long protein molecules composed of naturally occurring amino acids that are linked in a specific sequence by covalent peptide bonds. Due to the spatial conformation of some amino acid groups (notably proline and hydroxyproline) and the many ionisable and polar functional groups in the protein chain, the

Figure 1: LEICA DME optical microscope.

Figure 2: TITAN2 Universal strength tester machine.

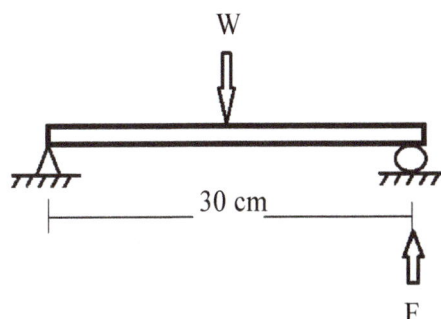

Figure 3: Principle of crushing force attachment.

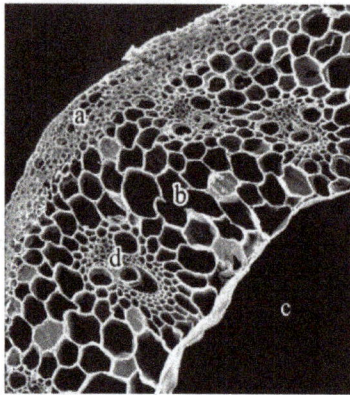

Figure 4: SEM image of a cross sectioned raw Wheat straw from the internode region of the stem [5]. (a) Epidermis, (b) Parenchyma, (c) Lumen, and (d) Vascular bundles.

individual chains form triple-stranded helical coils that are generally believed to be internally stabilized by hydrogen-bonding [22-24].

Animal glues are considered as eco-adhesives. They are non-toxic, environmentally friendly, and biodegradable. Animal glue formulations begin to break down in the natural environment in a matter of weeks when disposed of. It is composed of technical gelatin, usually extracted from collagen found in hides and/or bones of cattle. Other ingredients include Epsom salts, water, natural corn sugars, and glycerin. Chemical composition of hide glue is a protein derived from the simple hydrolysis of collagen which is a principal protein constituent of animal hides. Collagen, hide glue and gelatin are very closely related as to protein and chemical composition [25-27]. Through the infusion of the risen inside the Lumina, the inner wall of the straw will absorb the glue which is water soluble and penetrating in the open areas in the stem cross section. According to the mechanical interlocking theory, adhesive strength is provided by interlocks which are formed when the adhesive penetrates into the pores and cavities of the substrate, causing an increase of wall strength. Formation of the composite using treated Wheat straw will follow this preparation.

Cross-section of Wheat straw treated and untreated with natural glue illustrating glue penetration in parenchyma as black spot areas is shown in Figures 5a and 5b.

Wheat straw preparation

In this work, special technique was used to strengthen Wheat straw through the infusion of the gelatin glue through the internal channel of the straw, using a special attachment as shown in Figure 6.

After the infusion of the gelatin glue inside the Wheat straw, it was dried on the air at room temperature for 24 hours or dried using hot air oven at temperature of 105°C for one hour. The result was a composite with fiber volume fraction of 0.57. The tensile strength of the treated Wheat straw was measured using Titan2 Universal strength tester machine (Model 710) with load capacity 3000 N and with cross-head speed of 10 mm/min. Test specimens were of average diameter 6.0 mm wide and 200 mm long. For each sample, 10 specimens were tested and the average tensile modulus and strengths were estimated. All test samples were preconditioned at 65% relative humidity, 23°C for 48 h.

Tensile properties of treated wheat straw

The tensile properties of Wheat straw/glue are given in Table 1. The

mechanical properties of the straw composite represent the increase of straw strength by about 166%, breaking strain by 125%, and the Young's Modulus by 130%.

The increase in the tensile strength is a direct effect of the absorption of the Wheat straw wall of the glue in the areas b and d in Figure 2. Also the Young's modulus of was increased.

Straw crushing force

The stems of the Wheat straw are solid at the nodes but internodes are hollow. The dimension of the Wheat straw depends on its variety. The crushing force, which is the force required to collapse the cross section of the Wheat straw, depends on its hollowness, the term hollowness as referred to here is the ratio of inner diameter of the wheat straw to the outer diameter expressed in percentage [22] (Figure 7).

The value of the hollowness:

$$H = (D_{in} / D_{out}) * 100 \qquad\qquad\qquad\qquad\dots\dots\dots\dots\dots\dots\dots\dots\dots\dots(1)$$

The hollowness of Wheat straw is laying in the range of 90% to 95% [21].

The deformation of the Wheat straw cross section under the crushing force is illustrated in Figure 8. The study of hollowness effect

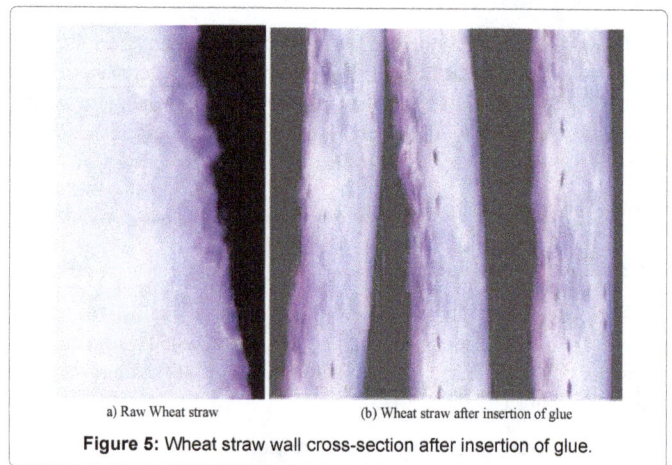

a) Raw Wheat straw   (b) Wheat straw after insertion of glue

Figure 5: Wheat straw wall cross-section after insertion of glue.

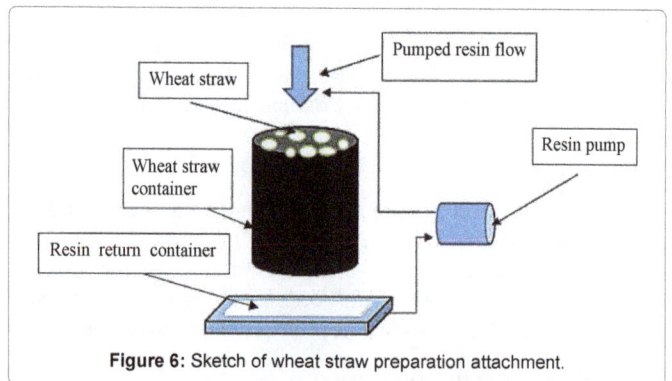

Figure 6: Sketch of wheat straw preparation attachment.

Material	Strain (%)	Strength (MPa)	Young's Modulus (MPa)	Volume fraction Vf (%)
Raw wheat straw	0.0151	14.7	973.899	100
Wheat straw/glue composite	0.02	24.35	1217.685	55.6

Table 1: Mechanical properties of the wheat straw composite.

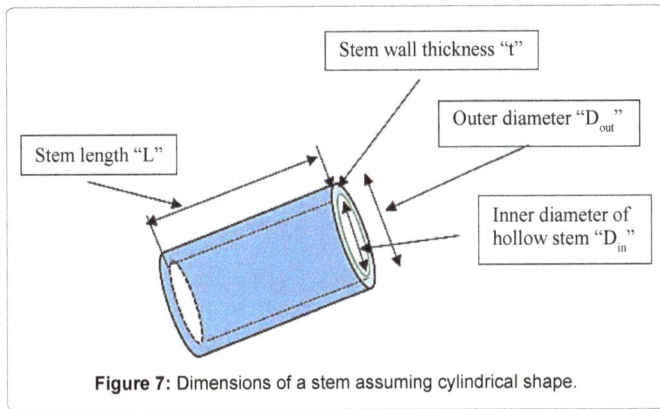

Figure 7: Dimensions of a stem assuming cylindrical shape.

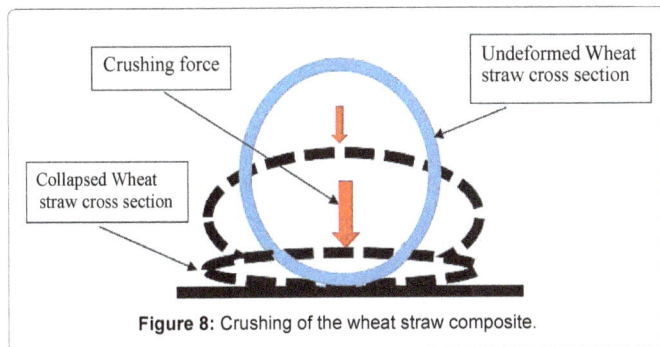

Figure 8: Crushing of the wheat straw composite.

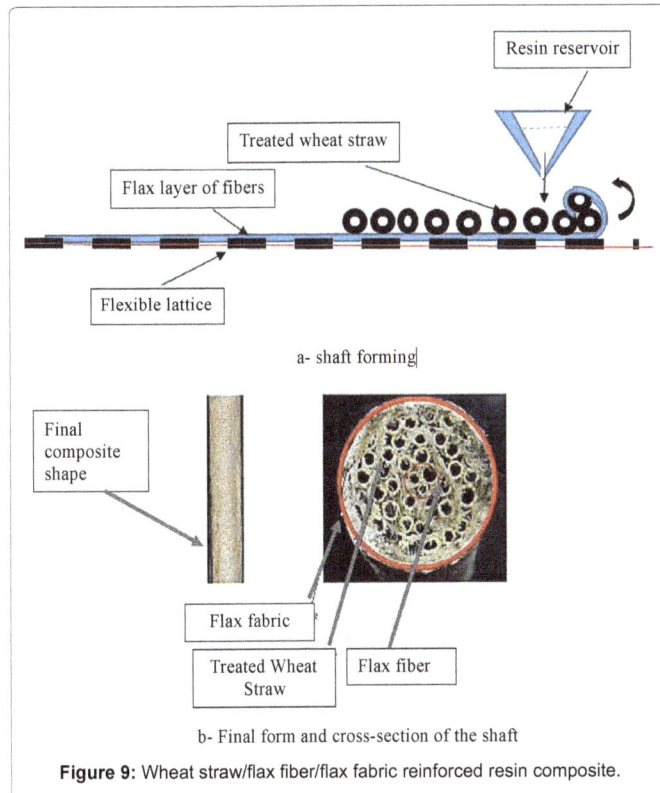

a- shaft forming

b- Final form and cross-section of the shaft

Figure 9: Wheat straw/flax fiber/flax fabric reinforced resin composite.

indicates that bending stress in the wall is continuously increasing as increases of hollowness while the contact pressure decreases. The deformation parabolically increases after hollowness value over 30%, creating more stresses in the wall which leads to crack of the straw

wall and crushing of its cross section. Due to thin wall section very less material is available to resist the force, so von Mises stress increases rapidly after 60% hollowness [23]. Consequently, the maximum crushing force decreases as a function of hollowness.

From the above analysis, one of the drawbacks of the Wheat straw is the low value of crushing force and bending stiffness. Consequently, the infusion of glue will help to increase crushing force. In order to determine the effect of the gelatin glue injection, crushing test was carried out to measure crushing force acting in the middle distance between the straw nodes. Specific crushing force is given in the Equation (2):

$$F_s = F/L \qquad \text{.. (2)}$$

Where: F_s – specific crushing force, $N \cdot m^{-1}$; F – crushing force, N; L – length of straw under pressing jaw, mm.

The wheat straw started to crack under applied load at as low as 16 N. When Wheat straw was covered with glue on the outer surface crushing force was increased up 40 N for untreated straw, while for the treated with infusion by the gelatin glue the cross section was deformed with no cracks in the straw surface. Moreover, the straw wall becomes more flexible; its cross-section is flattened under the applied load, as shown in Figure 8, with a noticeable increase of the crushing force reached 125 N. Hence, the increase of the strength in the thin wall of the Wheat straw reflected directly on the crushing force under which the straw will crack.

The specific crushing force was 4.444 kN/m for composite straw against 0.421 kN/m for raw straw. This increase of specific crushing force is one of the most important advantages of the straw/glue composite, rendering its use possible in various applications.

Application of wheat straw composite

Suitable use of the Wheat straw composite is flat board or a round cross-section shaft. Round shaft hybrid composite was designed using Wheat straw/flax fibers and covered by a layer of flax fabric. The construction of shaft is made of several layers of Wheat straw; first layer (core) consists of three straws winded around by a bundle of flax fibers. Figure 9a illustrates the forming process. The rotation of the laid layer of flax fibers and the stacks around the core structure continues till the final diameter is reached. Flax fabric was wrapped on the formed structure as the outside layer. During the whole process, the resin (PVA) is continuously fed at the rate to insure the required fiber volume fraction of the final composite. In our case, the weight fraction for the straw, flax fibers, and fabric was 51% and composite density was 0.1125 g/cm³. The shaft formation and cross section are shown in Figures 9a and 9b respectively.

Conclusion

The potential use of treated Wheat straw/flax fiber as reinforcement in PVA was investigated. The study focused on whole stems of Wheat straw which were reinforced with injection of the animal glue to improve the strength and crushing force of the Wheat straw. Wheat straw tenacity increased by about 165%, breaking strain by 125%, and the Young's Modulus of elasticity by 125%. The specific crushing force reached 4.444 kN/m for injected Wheat straw, while it was 0.421 kN/m for raw straw. The above supports the possibilities of consuming the whole wheat straw in a composite as an engineering material. Wheat straw stem infused by glue can be used to get low density composites with acceptable mechanical performance. Given method is applicable for fabrication of Wheat straw/flax fiber/flax fabric polymer composite in rod form.

References

1. El Messiry M (2016) Natural fiber textile composite engineering. Apple Academic Press Inc.

2. Mads A, Kristensen J, Felby C, Jorgensen H (2011) Pretreatment and enzymatic hydrolysis of wheat straw (Triticum aestivum L) The impact of lignin relocation and plant tissues on enzymatic accessibility Technology Volume February :1-12.

3. Avella M, La Rota G, Martuscelli E, Raimo M, Sadocco P et al. (2016) Poly (3-hydroxybutyrate-co-3-hydroxyvalerate) and wheat straw fiber composites: thermal mechanical properties and biodegradation behavior. J Composite Materials 35: 829-836.

4. Mantanis G, Nakos P, Berns J, Rigal L. (2008) Turning agricultural straw residues into value added composite products: A new environmentally friendly technology bio-composites from agricultural raw materials.

5. Hui Y, Ruigang L, Dawa S, Zhonghua W, Yong H (2008) Arrangement of cellulose microfibrils in the Wheat straw cell wall. Carbohydrate Polymers 72: 122-127.

6. Pearson C (2003) Animal glues and adhesives engineer material Handbook of 5. Adhesive Technology by Taylor & Francis Group LLC.

7. Shanmugasundaram L (2009) Green composites: Manufacturing techniques & applications the Indian textile journal.

8. Albano C, Karam A, Dominguez N, Sanchez Y, Gonzalez J (2005) Thermal mechanical, morphological thermogravimetric, rheological and toxicological behavior of HDPE/Seaweed residues composites. J Composite Structures 71: 282-288.

9. Yang HS, Kim HG, Son J, Lee BJ, Twang TS (2006) Rice husk flour filled with polypropylene composites, mechanical and morphological studies. J Composite Structure 63: 305-312.

10. Alvarez A, Kenny I, Vazquez A (2005) Influence of twin-screw processing conditions on the mechanical properties of biocomposites. J Composite Materials 39: 2023-2038.

11. Yasina M, Bhuttob A, Karimb A (2010) Efficient utilization of rice-wheat straw to produce value added composite products. Int J Chemical and Environmental Engineering 1: 136-148.

12. Grigoriou AH (2000) Straw–wood composites bonded with various adhesive systems. Wood Sci Technol: 34:355-365.

13. Han G, Zhang C, Zhang D, Umenura D, Kawai S (1998) Upgrading of urea formaldehyde bonded reed and Wheat straw particleboards using silane coupling agents. J Wood Sci 44: 282-286.

14. Wang D, Sun XS (2002) Low density particleboard from Wheat straw and corn pith. Ind Crops Prod 15: 43-50.

15. David AP (1997) Wood process adapted to straw particleboard. Wood Technol 124:20-24.

16. Karr GS, Sun XS (2013) Straw board from vapour phase acetylation of Wheat straw. Ind Crops Prod 11: 31-41.

17. Abba HA, Nur H, Salit S (2013) Review of agro waste plastic composites production. J Minerals and Materials Characterization and Engineering 1: 1-9.

18. Suhara-Panthapulakkal S, Sain M (2007) Agro-residue reinforced high-density polyethylene composites: Fiber characterization and analysis of composite properties. Composites: Part A 38:1445-1454.

19. Amirpouyan Sardashti A (2009) Wheat straw-clay-polypropylene hybrid composites. Waterloo Ontario Canada.

20. Fowler PA, Hughes JM, Elias RM (2006) Biocomposites: Technology environmental credentials and market forces. J the Science of Food and Agriculture 86: 1781-1789.

21. Chevillard A, Angellier-Coussy H, Cuq B, Guillard V, Césa G, et al. (2011) How the biodegradability of Wheat gluten-based agromaterial can be modulated by adding nanoclays. Polymer Degradation and Stability.

22. Lam PS, Sokhansanj S, Bi X, Lim CJ, Naimi LJ et al. (2008) Bulk density of wet and dry wheat straw and switchgrass particles. Applied Engineering in Agriculture 24: 351-358.

23. Darji PH, Vakharia DP (2012) Development of graphical solution to determine optimum hollowness of hollow cylindrical roller bearing using elastic finite element analysis Finite Element Analysis Applications in Mechanical Engineering.

24. Schellmann N (2007) Animal glues: a review of their key properties relevant to conservation. Reviews in conservation 8: 55-66.

25. Von Endt D, Mary T (2007) The chemistry of filled animal glue systems.

26. Landrock A (1987) Adhesives technology. Handbook J Eng Mater Technol 109: 98.

27. Pizzi A, Mittal L, (2009) Handbook of adhesive technology (2ndedn). Hopewell Junction, New York USA.

24

A Review of Computational Fluid Dynamics Simulations on PEFC Performance

Chen Y[1]*, Enearu OL[1], Montalvao D[2] and Sutharssan T[1]

[1]*School of Engineering and Technology, University of Hertfordshire, College Lane, Hatfield, Hertfordshire, AL10 9AB, UK*
[2]*Bournemouth University, Department of Design and Engineering, Faculty of Science and Technology, Poole BH12 5BB, UK.*

Abstract

Among the number of fuel cells in existence, the proton exchange fuel cell (PEFC) has been favoured because of its numerous applications. These applications range from small power generation in cell phones, to stationary power plants or vehicular applications. However, the principle of operation on PEFCs naturally leads to the development of water from the reaction between hydrogen and oxygen. Computational fluid dynamics (CFD) has played an important role in many research and development projects. From automotive to aerospace and even medicine, to the development of fuel cells, by making it possible to investigate different scenarios and fluid flow patterns for optimal performance. CFD allows for in-situ analysis of PEFCs, by studying fluid flow and heat and mass transfer phenomena, thus reducing the need for expensive prototypes and cutting down test-time by a substantial amount. This paper aims at investigating the advances made in the use of CFD as a technique for the performance and optimisation of PEFCs to identify the research and development opportunities in the field, such as the performance of a novel PEFC, with focus on the underlying physics and in-situ analysis of the operations.

Keywords: Proton exchange fuel cell (PEFC); Fuel cell (FC); Computational fluid dynamics (CFD); Validation

Introduction

The world is quickly becoming aware of the need to resolve issues associated with energy and climate changes. Sustainable energy is being developed and is aimed at handling greenhouse effects due to gas emissions and limited availability of fossil fuels by combining social awareness with technologies [1]. PEFCs have been referred to as an effective source of power production with very attractive energy transformation technologies, low operating temperature, high efficiency, zero emissions, quick start-up time and silence during operations [2-5]. Fuel cells have also been identified as a proposed solution to greenhouse emission problems due to operations [6]. Fuel cells come in various forms and are suitable for automotive, stationary, and portable system applications [7].

The versatility of FC applications ranges from small devices capable of supplying just a few watts of electricity to power plants generating power in the megawatts range. However, an individual fuel cell usually delivers low voltages and high currents. Typical voltage and current values range from 0.4 to 0.9V and from 0.5 to 1.0 A/cm^2 respectively after losses [8]. As shown in Figure 1, in order for fuel cells to produce high voltages, a stack is made up by connecting a number of fuel cells in series, separated by bipolar plates [9]. The different fuel cells are categorised according to the type of electrolyte they use. Both advantages and disadvantages can be found with respect to temperature, size, fuel, purity, lifetime and cost. This is because each type of fuel cell requires particular materials and fuels that may be restricted to certain applications [10]. The various categories of fuel cells include:

1. PEMFC or PEFC – Proton Exchange (Membrane) Fuel Cells (also referred to in the literature as Polymer Electrolyte Membrane Fuel Cell).

2. DMFC – Direct Methanol Fuel Cell.

3. PAFC – Phosphoric Acid Fuel Cell.

4. AFC – Alkaline Fuel Cell.

5. SOFC – Solid Oxide Fuel Cell.

6. MCFC – Molten Carbonate Fuel Cell.

The PEFCs are usually made up of polymer electrolyte membranes (commonly Nafion®) as a proton conductor and Platinum as a catalyst [11]. PEFCs have the ability to satisfy vehicular, domestic and larger stationary energy requirements. In the upcoming years, it is envisaged that the fuel cell applications will include automotive powertrains, distributed power generation and portable devices such as batteries [1,12]. Significant research and development efforts have gone into PEFC technologies over the past few years, thus leading to an increase in power density, efficiency and reliability of the existing PEFCs [11,13]. Nevertheless, the increased level of research has given little attention to the underlying physics of the transport phenomena of fluid and gases within the fuel cells. As a result of the highly reactive and compact nature of the fuel cell, in-situ measurements cannot be easily carried out during an operation [14]. Hence, computational fluid dynamics (CFD) modelling and simulations have been preferred to help in providing a better understanding of transportation of fluids and gases within the fuel cell [15].

Fundamentals of the PEFC

The operating principle of a PEFC is relatively simple: it involves the feeding of hydrogen into the cell to be oxidised at the anode, while the oxygen is reduced at the cathode after being carried in by an air feed stream [16]. As shown in Figure 1, a common fuel cell structure

*Corresponding author: Chen Y, School of Engineering and Technology, University of Hertfordshire, College Lane, Hatfield, Hertfordshire, AL10 9AB, UK
E-mail: y.k.chen@herts.ac.uk

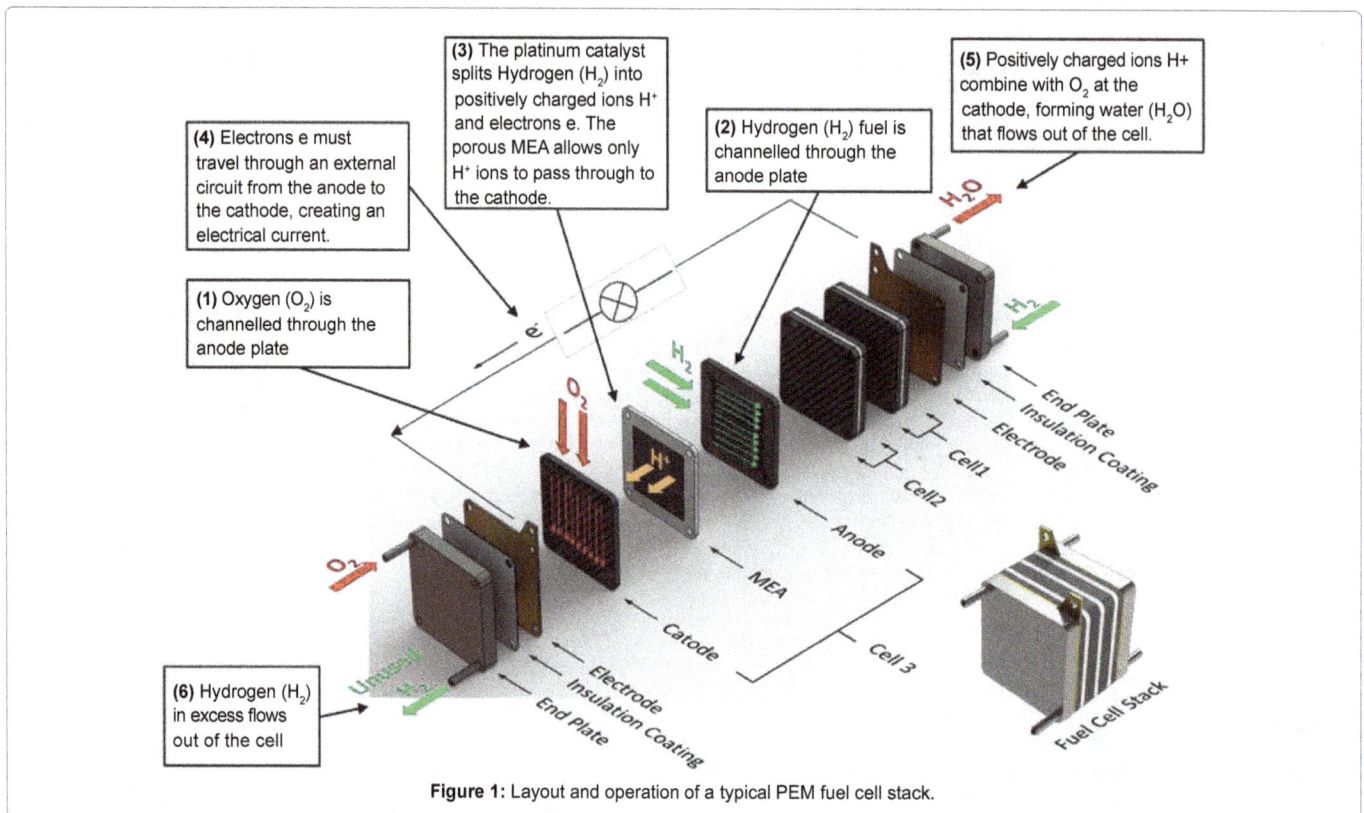

Figure 1: Layout and operation of a typical PEM fuel cell stack.

also known as membrane electrode assembly (MEA) is made up of an ion-conducting electrolyte material layered in between a porous anode and cathode electrodes. The hydrogen is supplied to this porous anode, while the oxygen is supplied to the cathode, commonly by feeding air into the cathode [17]. The input fuel of hydrogen passes over a negatively charged anode electrode, where it is split into H^+ ions and e^-electrons. The electrons are then transported through an external circuit to produce electric current, while oxygen passes over a positively charged cathode electrode. The ions go through the electrolyte to the positively charged electrode and react with one another to produce H_2O (water) and heat as by-products.

Furthermore, the straight transformation of the chemical energy of covalent bonds into electrical energy occurs as a result of the split of the hydrogen and oxygen by the electrolyte also referred to as a separator. The electron transfer in this process takes a long period of time, which allows direct gathering of electrons, thus leading to fuel efficiency more than two times higher than that in ICEs (Internal Combustion Engines) [18]. The direct current electricity that is produced as a result of the flow of electrons from the anode to the cathode within the fuel cell is the main product. It is also worthy to note that the amount of current produced depends on the amount of reactant fuels or gases supplied, chemical activity and power loss within the fuel stack. However, the production of current is constant over time, unlike in batteries where it decreases, because there is no chemical transformation of the fuel cell components involved. Hence, the fuel cell can continue generating power, provided that the supply of reactant gases remains constant for the required period [17]. The ability to continue producing power is one of the many benefits of PEFCs.

Effects of parameters on performance of the PEFC

For the PEFC to perform as intended, it is important that

appropriate operating parameters are observed at all times [5]. These operating parameters can include temperature of the Fuel Cell (FC), operating pressure, heat loss, flow distribution, steam reforming, relative humidity of reactant gases, stoichiometric ratio, water distribution, current density and performance (polarisation) curve [19,20]. Observation of these parameters can be done with the aid of a built-in monitoring system that shows these parameters on a screen and can also shut down the fuel cell in emergency situations. Santarelli and Torchio [21] pointed out the importance of understanding what effects these parameters have on the PEFC so that supplementary costs associated with external systems that are employed to control these variables can be reduced. This is a result of the fuel cell being a complex system, consisting of interconnected parts that depend on each other in order to achieve the aim of energy production. This implies that degradation mechanisms are interconnected and can result from the individual degradation of one part leading to the deterioration of other parts and components of the FC [22].

Water management

The performance of the PEFCs can be influenced by a number of factors and operating conditions such as temperature, pressure and humidification of the reactant gases [23]. However, to a large extent, the performance is dependent on the hydration of the polymer membrane, where the water balance is dependent on the fuel cell current and water content of fuel and oxidant gas stream [24]. When the current density goes beyond a certain threshold, the delivery of water by electro-osmotic drag and oxygen-reduction reaction (ORR) overtakes the water removal from the cathode catalyst [25] hence, the excess water accumulates, flooding the electrode and reducing the rate of oxygen being transported to the catalyst. It is this flooding that limits the mass-transport ability of the electrode which then leads to a quick rise in the cathode overvoltage. A decrease in the FC power output is the end result [26,27].

Without proper water management to keep up the constant energy production rate, especially at the cathode, flooding of the pores in the catalyst and gas diffusion layers is likely to occur. This can lead to a reduction in cell performance as a result of the inability of reactants to reach active catalyst sites. However, a certain quantity of water is required to ensure the optimum operation of the fuel cell because the proton conductivity of the membrane depends on humidification. This level of humidification is required to avoid membrane dehydration and water vapour condensation that could lead to a reduction in proton conductivity and an increase in ohmic losses and overheating of the stack [25,28]. Table 1 illustrates the effects of water content on the parts and operations of a PEFC. It is worthy to note that corrosion can occur as a result of the electrochemical nature of the reactions within the PEFC during operations [29-34].

A number of interacting factors and mechanisms such as flooding, dehydration and corrosion contribute to the degradation and performance of a FC. Therefore, better understanding of the transportation phenomena of liquid water in the fuel cell may lead to substantial performance gains and an enhancement in the lifetime of the fuel cell [35]. Li et al. [36] have reported the following contributing issues that hinder the performance of the cathode; 1) Slow kinetics of the oxygen reduction reaction at the cathode unlike the hydrogen oxidation reaction at the anode, 2) Difficulty in removing water at the cathode leading to a limitation of the mass transport.

The liquid water produced due to the electrochemical reactions in the fuel cell in addition to water from humidified inlet gas, makes it possible to keep the required level of hydration in the membrane

[37] The water generated as a by-product during power-generating operations of the fuel cell is unwelcome. Recent development efforts have led to the production of PEFC models with built-in water management systems to manage water and humidity in the fuel cell. However, the issue with water generated during operation of the fuel cell prevails, eventually leading to unwanted flooding [38] Although the water production within the fuel cell during operation is unavoidable because it is a by-product of the hydrogen fuel reaction during the conversion to electrical energy, it exists and remains in the liquid state as a result of the low operating temperature of the PEFC [39]. Figure 2 shows a schematic illustration of the process of water production within the PEFC.

Liquid water is generated in the cathode and is transported across the electrolyte by the electro-osmotic drag from the anode to the cathode. The oxygen reduction reaction (ORR) is a very important process in fuel cell modelling that occurs at the cathode and the hydrogen oxidation reaction (HOR) at the anode [15,40,41] can be expressed as flows:

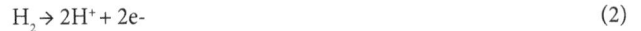

$$1/2 O_2 + 2H^+ + 2e^- \rightarrow H_2O \tag{1}$$

$$H_2 \rightarrow 2H^+ + 2e^- \tag{2}$$

The addition of the two half-cell reactions (1) and (2), which is exothermic, gives the combustion reaction (3) of the fuel cell in:

$$H_2 + 1/2 O_2 \rightarrow H_2O + Electricity + Heat \tag{3}$$

The formation of water in the PEFC actually leads to a 2-phase flow in the channels which consists of liquid water and gas/ vapour, thus increasing the complexity of the problem of water management in PEFCs [25]. Measuring the water content of a conducting PEFC is of high importance because PEFC models are designed specifically

	Dehydration	Humidified	Flooding
Overall performance	Significant drop in cell potential, leading to power loss.	Normal power output based on set operating conditions.	Drop in cell voltage.
			Significant reduction in performance and lifetime.
	Instant and long-term degradation in cell operations.		Reduction in the Electro-chemical active surface area (ECSA).
	Reduction in cell performance		
Catalyst layer	---	----	Reduction in the transport rate of reactants.
			Carbon corrosion, corrosion and degradation of catalyst layers.
Anode	More intense at the cell inlet	----	Fuel starvation.
Cathode	Lower cathode over-potential	----	Increase in mass transport losses.
			Partial drop in pressure of gas.
			Cathode overvoltage.
GDL (Gas Diffusion Layer)	---	Oxygen gas access to cathode catalyst layer and better cell performance.	Reduced pore size, poor diffusivity of gases, thus increasing the concentration and surface over-potential of the fuel cell.
			Blockage of pores, degradation.
			Reactant starvation.
			Dissolution and diffusion of reactant gases into the liquid water flood.
MEA	Dry cell operations over-lengthy period of time can lead to irreversible damage to the membrane.	High proton conductivity	Oxygen concentration decreases.
		Maintenance of mechanical stability of the PEMFC.	Non-uniform current density distribution.
	Become brittle, developing cracks.		
	Shrink in pores leading to low back diffusion rates. Drying of the proton conducting membrane.		Corrosion and degradation of electrodes.
	Decrease in conductivity.		
	Increased ionic resistance and ohmic losses.		
	Increase in voltage loss.		

Table 1: Effects of water on the performance of PEFCs [30-34].

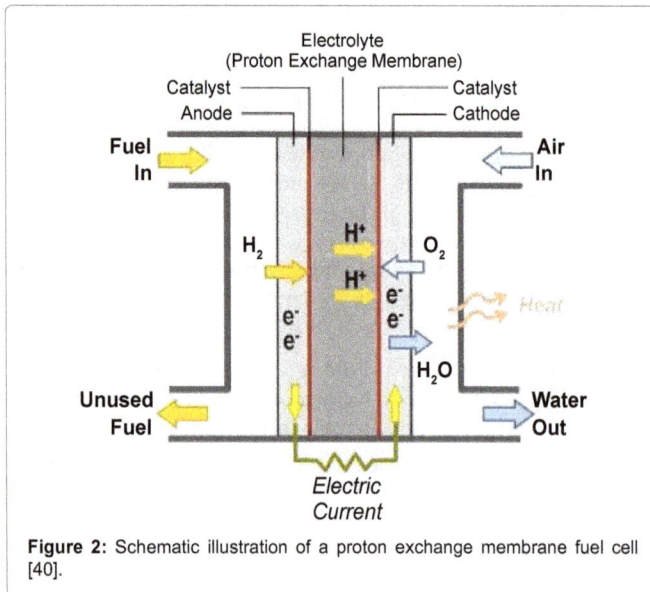

Figure 2: Schematic illustration of a proton exchange membrane fuel cell [40].

for the tests or experiments they are expected to undergo. Therefore, it is necessary to know how much water is in the fuel cell at any point during operation [38]. For the detection of water in the fuel cell, a number of detection techniques have been employed [42,43]. It is worthy to note that many of these techniques have been limited to a qualitative description of flow patterns based on certain operating conditions with respect to the resources available [44]. There are four major visualisation techniques which have been discussed as follows:-

i. Direct visualisation as a result of a transparent cell plate that makes it possible for optical devices to access the channels. These devices include digital and high-speed cameras [45]. The transparent plate also makes it possible to visualise the water directly, without requiring any external device.

ii. Gas chromatography has been used by Mench et al. [46] for the measurement of in situ distribution of water vapour, oxygen and nitrogen, with full humidification within the gas channels of the PEFC during operation. In addition to the current distribution mapping technique, the researchers were able to understand the behaviour of the cathode and anode with regards to water distribution and current density.

iii. Relative humidity (RH) has been measured using a humidity sensor by Nishikawa et al. [43] along the air flow field. Since this experiment only considered water vapour distribution, it implied that a part of the membrane in a working fuel cell was in an over-saturated condition, with the other part in an under-saturated condition which could not be achieved in a typical fuel cell.

iv. Neutron radiography, using a small-angle neutron scattering (SANS) technique [42], where the neutron radiography was used to obtain a 2D image of the liquid water in the channels. The water content of a membrane was observed with this technique, using the resulting spectra made up of the presence of intense scattering at low angles and maximum scattering (ionomer peak), which was directly related to the presence of water pools in the membrane. Neutron radiography was employed for in-situ and non-destructive imagining and also as a measurement method for detecting liquid water in a PEFC during operation [38]. This method is capable of showing the amount of water present in the gas diffusion layer (GDL), rib and flow channel [38,47,48].

However, there are downsides to some of these techniques. For example, the neutron imaging and x-rays are considerably high cost techniques. Also, there are risks associated to the use of radiation, preventing its frequent use [42]. Furthermore, as a result of the rapid weakening of the signal in the carbon layer, the nuclear magnetic resonance imaging is incapable of accessing the water content on the GDL; therefore it can only be used to detect water in the membrane. Presently, a more cost effective method has been used for the visualisation of water. This involves the use of optical devices like thermal imaging cameras or camcorders that can be obtained for prices ranging between 700 and 35,000 GBP. The cost of obtaining a thermal imaging camera is dependent on the model, pixel resolution, thermal sensitivity, field of view, infrared sensitivity, Wi-Fi connectivity and other features. A key advantage of these cameras lies in their ability to determine the presence of water within a fuel cell based on temperature gradients [49]. In addition, these are portable devices that allow fast inspection and repeated use without fear of setbacks from radiation. Finally, the time needed to set up these systems is generally shorter than the time needed to set up the gas chromatography or neutron radiography equipment.

Furthermore, the problems associated with membrane dehydration and flooding phenomena are yet to be overcome. A significant amount of research and development efforts ranging from fluid flow numerical simulations to experimental testing have been put into place in an attempt to find suitable solutions and mitigation approaches to tackle these problems [36]. Numerous studies have been done using theoretical models that help to describe the management and transportation of water in PEFCs. CFD is another tool used to study the saturation, transport and distribution of water within the PEFC [45] CFD is useful in the design process and performance analysis of fuel cells. CFD models that have been developed include zero-dimensional, one-dimensional, single-phase flow, isothermal models (involving all layers in the fuel cell stack), steady-state and single layer models. More recently, two-dimensional, two-phase flow, non-isothermal, transient, multiple layer and three-dimensional models have been proposed [50,51]. These will be further discussed later in section 4.

Heat and gas

The heat losses to the environment during its operation of a PEFC are one of the main factors affecting efficiency. Hence, in order to attain high energy efficiencies, it is important that high levels of thermal integration be reached [52]. Temperature is spatially heterogeneous and under dynamic conditions and its distribution becomes irregular at different time frames. Thus, during the operation of a fuel cell, both the fluctuation on the operating parameters and the physical condition of the FC is responsible for a degree of uncertainty, leading to unstable and difficult to predict its performance. This is as a result of the level of uncertainty associated with temperature, which also is a function of ambient conditions, flow rate of reactant gases and points of operation [53].

A few papers have been published on the topic of heat losses and the effect of performance on PEFCs. Research by Schmittinger and Vahidi [22] showed that when fuel cells were operated under temperatures higher than 100°C there was an increase in degradation on the cell components. This brought a reduction in long-term performance and resilience. Also, the catalyst is chemically unstable at this point, leading to increased movement, sintering and accumulation of particles [54]. However, a number of advantages can be seen with the increase in temperature, such as reduced CO-poisoning, the ability to use reformed hydrogen as an alternative to pure hydrogen, increased catalyst tolerance for contaminants, improvement in electrochemical kinetics and efficiency, improved water management and cooling as a result of difference in temperature between fuel cell and coolant [55].

Computational Fluid Dynamics in PEFC

Computational fluid dynamics (CFD) handles the coupling of multidimensional transport phenomena in addition to electrochemical kinetics and transportation of electrons and ions that help to provide comprehensive understanding of the dynamics of fuel cells [56]. CFD is based on fluid mechanics and Navier-Stokes equations for the conversion of mass, momentum and energy equations. The CFD software carries out iterations of the Navier-Stokes and energy equations for a problem until a convergence of values is attained [57]. Laws of physics are employed in order to derive the equations for the conservation of mass, momentum and energy. These laws include: first law of thermodynamics, mass conservation and the sum of forces that equals the change of momentum.

In fuel cell modelling, CFD models can be seen as rational models that handle the important physio-chemical procedures as well as experimental observations and are a fundamental tool to ensuring that a holistic understanding of the operating of a fuel cell is achieved [16]. However, notable difficulties have been experienced in trying to carry out in-situ analysis of the fuel cell. As a result of the nature of the experimental environment, CFD methods are needed to help to simulate and determine multi-dimensional coupled fluid flow and transport of reactants, heat and charged species [14].

An operation of a PEFC involves a combination of electrochemical reactions taking place simultaneously. These include; multi-component, multi-dimensional and multi-phase fluid flow in addition to heat and mass transfer. Hence, CFD models are necessary in addition to mathematical models that would help to create a better understanding of the physical behaviour and complex phenomena taking place within the fuel cell [58,59]. The CFD modelling methods have been developed over the years from 2D models to the improved computational models that account for fluid flow, thermal and electrochemical transport, 3D geometries and 2-phase transportations [14]. However, the challenges yet to be won to ensure proper functional and predictive capabilities of the CFD models include:

i. Inadequate general models for ionic transportation of water in polymer membranes.

ii. The majority of CFD models for PEFCs consider analytical, semi-empirical 1D and 2D PEFC models, without considering the effects of flooding by liquid-water on performance of the stack [60].

iii. Inadequate macroscopic exhibiting and determination of catalyst layers.

iv. Inadequate information in in-situ data validation for transport and material factors.

v. Wide range of time scales.

Djilali [16] further mentioned that to bring about notable improvements and developments in cost, innovative designs and lead operation times, coupling of computational fluid dynamics (CFD) models with multi-disciplinary optimisation procedures has to take place, in order to provide an opportunity for the development of computational fuel cell engineering (CFCE) tools. In addition, the cost of CFD modelling of a 3D PEFC model is large and can be seen as not practical in certain computing environments because it can be time

consuming and requires powerful computational resources. Modelling can be computationally demanding due to the differences on the geometric dimensions of the various parts and regions of the fuel cell [61]. For example, in a typical fuel cell the thickness of a stack is in the range of 1 – 2 mm and thickness of the anode and cathode catalyst layers is in the range of 10 to 20 μm and the overall height and width dimensions are in the range of 10 to 20 cm.

According to Gurau et al. [62], the governing differential equations for the gas flow channels and gas diffusion electrodes can be joined for both regions because they are similar, thus leading to a computational effect of one domain without any internal boundary conditions. The material properties and source terms acknowledge the two regions and take different values of interest to form one domain; leading to the single-domain approach [63, 64]. Computational fluid dynamics (CFD) has come to play an important part of the design and fabrication of products and systems in recent times. CFD is often said to be a fluid version of finite element analysis, as will be further discussed. This is because it is possible to run simulation tests on a number of design iterations and compare results to see the best course of action. This step greatly reduces the need to build prototypes that can be both costly and time consuming. CFD methodologies are being used in different fields, from the propulsion industry to car, ship and turbo-machinery designs [65,66]. The evolution from the use of 2D flow models, as mentioned earlier, to Euler equations and 3D Navier- Stokes applications can be attributed to the rapid computer power increase and improvements in CFD methodologies and algorithms [57].

Also, CFD models are useful for an analysis of phenomena such as heat and mass transfer, fluid and two-phase flows through a particular cell, based on a finite-element framework [67]. Sivertsen and Djilali [14] suggested that simulations under realistic conditions were important for the continuous development and optimization, commercialisation and the use of cheaper materials for improved design and fabrication of fuel cells in the coming years. This is evident in comparisons between model test experiments and computer-based simulations. CFD software packages can be applied in many other areas, for example, in the flow analysis of wind in a turbine. These software packages are designed in such a way that it is possible to enter parameters such as velocity, pressure and direction of the wind into the program. Hence it is possible to predict the behaviour and performance of the model being studied, and also helps to identify areas of increased temperature to optimise cooling, overall temperature and other required parameters [57].

CFD for the Fuller and Newman [68] modelling of fuel cells was first introduced by Gurau et al. [62] with a 2D calculation of coupling the fluid flow, mass transfer and electro-kinetics, while eliminating the need for artificial boundary conditions at internal interfaces. The revolutionary modelling work was initially done by Bernardi and Verbrugge [69,70] and Springer et al. [71,72], assuming a 1D flow through the layers of the PEFC. Nguyen and White [73] later created a pseudo-2D heat and mass transfer model that also showed the effect of reactant consumption along the flow channels. While Fuller and Newman [68] used a similar 2D model method to understand the relationship between water and thermal management.

Since the beginning of the century, the behaviour of water/vapour in fuel cells has become a central topic in many literature works [74]. Baschuk and Li [75] developed a model showing the effect of water in both the GDL and catalyst layer of the cathode, on overall fuel cell performance. Over the years, it was noted that a number of fuel cell models had been developed, like the single and two-phase models, that considered the effect of the liquid water supplied to the anode and that formed in the cathode. However, as a result of the slow kinetics of oxygen reduction as mentioned earlier, some models were focused on the fuel cell cathode. Jiao and Zhou [76] and others focused on the fuel cell as a whole [77]. One of the most comprehensive 3D multiphase works was done by Berning and Djilali [15]. Their research covered an analysis of the role of the gas diffusion layer (GDL) as the main component of the multiphase treatment and the introduction of the multiphase approach for the cathode and anode in 3D plots, showing cell length and width of the GDL. However, shortcomings of this research lay in the absence of a comparison of simulation results with experimental data on a fuel cell [2].

It is worthy to note that the innovative approach of Bernardi and Verbrugge [69,70] served as a framework that brought about the multi-dimensional and multi-physics models in existence today. However, that research work only involved one-dimensional models, with the assumption that the water in the electrodes was only in vapour form thus providing very narrow applicability to actual PEFCs performance where flooding is concerned [56, 78]. This is also the case for the two-dimensional models developed afterwards by Fuller and Newman [68] and Nguyen and White [73]. Although their pseudo-two-dimensional models revealed the effect of heat and water management on the performance of PEFCs, which led to the concept of modelling proposed by Gurau et al. [62]. There are still limitations with regards to high fuel utilisation and low humidity circumstances in commercial-scale fuel cells.

In addition, CFD simulations are directed at gaining a better understanding of a model PEFC, helping in the prediction of fluid behaviour within the PEFC by carrying out a simulation of its performance [61] to describe in the fluid flow within the PEFC, simulation programs are used more conveniently to develop models, set suitable work conditions, generate mesh and run the calculations. The use of the CFD framework for modelling the PEFC has been developed further from 2D and 3D simulations of coupled electrochemical and transport processes with the aid of CFD codes. Later on, Dutta et al. [79], went on to present a 3D simulation using a commercial CFD package, FLUENT. The subsequent simulations by Nguyen and White [73], represented the MEA (membrane electrode assembly) as an interface without thickness in the computational domain, causing a drop in water transport and Ohmic potential across the membrane [56].

Over recent years, further CFD simulations are being carried out with improved commercial packages, such as FLUENT, FEMLAB, CFX-5, STAR-CD and STAR-CCM+, just to name a few from the various suppliers that give room for users to make use of their own models [51,61]. For example, FLUENT is a finite volume code that can be used to solve the coupled equation of mass conservation, mass diffusion and momentum conservation [80]. These commercial CFD packages are capable of balancing and solving uncommon water equations, as is the nature of PEM fuel cells when modelling. Although STAR-CD is no longer in use and has now been replaced with STAR CCM+ by CD ADAPCO. These software packages need to be selected depending on the specific requirements of a task ranging from heat and mass transfer to electrical power generation and so on.

Complete CFD modelling of PEMFCs is made up of the creation of a 3D model and introducing meshing, mathematical and numerical solving techniques. With the aid of commercial CFD packages, some aspects of the modelling process is simplified, but with the challenge of producing reliable results. Shan and Choe [81] mentioned in

their review that models of PEFCs available now are based on either empirical or 3D CFD, which do not meet the needs that require physical representation of the behaviour of a stack as a result of complexity or even simplicity. The processes that take place within the fuel cell during its operation comprise of many small complex operations that make these models incapable of fully simulating the actual working of the fuel cell at all times.

Validation of CFD Simulations

Validation involves the assessment of simulation results using experimental data as a benchmark for the estimation of errors [82]. In order to confirm accuracy and reliability of CFD results, careful validation is required, where validation serves as a means of ensuring that the correct mathematical model is used as an accurate representation because it meets specific performance criteria [83]. There are many techniques that can be used for the validation of CFD with respect to the needs of the project. The most common type of validation involves a comparison of simulation results with experimental data. A general sense of agreement between the experiment and computation results can provide validation. This method of validation is also useful for creating assurance in the model. Another method involves the use of test rigs that help with the maintenance of control over fuel cells, while ensuring safe and reliable operations. With the aid of a test rig, the conditions such as load, gas flow, gas flow conditioning, humidification of the gases, operating temperature, pressure, gas stoichiometry and utilisation, differential pressure, single cell voltages, gas leakage, overheating and cell reversal can be monitored.

During an operation of a PEFC, a sufficient supply of pure hydrogen is required, as well as oxygen/ air. Contamination by CO_2 in large quantities up to 500 ppm can have a negative impact on the performance of the PEFC. However, trace quantities of CO_2, around 100 ppm can be managed when monitored closely and conditioned by adding a small quantity of oxygen/air. When the fuel cell is operated under increased pressures, the power density also increases, thus elevating problems that may arise from water loss in the membrane. In order to maintain the high conductivity state of the membrane electrolytes, a certain level of humidification of the gases is required. Also, the production of heat as a result of the electrochemical electricity generation from hydrogen needs to be monitored in order to prevent the fuel cell from overheating. Besides comparing simulation results with experimental data, the other methods that can be used to validate CFD simulations include:

i. Comparing simulation results with analytical equations.

ii. Checking mesh convergence. This method allows for assessing the robustness of the model.

iii. Comparing results from two different packages.

iv. Comparing simulation results from existing literature.

The methods for validation listed above (including validation from experimental data) can be performed either independently or concurrently, depending on the resources available and complexity of the problem. For example, a test rig might not be readily available, in another case, there might not be access to another package or data from literature for a specific simulation is non-existent.

Concluding Remarks

One of the major factors that influenced the development of fuel cells is the worldwide concern about the environmental consequences of fossil fuels. These fuels are massively used in the day-to-day production of electricity and propulsion of vehicles for example. Careful consideration needs to be taken for all components and their interactions, from the appropriate cell type to the right fuel and system requirements. Hydrogen and fuel cell systems may be a solution for the sustainable power generation that the world needs today. Benefits can be seen when public incentives and private contributions are put in place to motivate the development of the power generation market in general. However, challenges need to be won before fossil fuels can take a back seat in energy provision.

In addition, CFD serves as an alternative to the modelling of the complex governing equations used to calculate and solve fluid flow problems within PEFCs and other applications spanning across various fields of study, from automotive engineering to aerospace and even medicine. However, there are a number of limitations that can be associated with the use of CFD, for example human, truncation or numerical errors. Inaccuracies can result from the inability of untrained users to spot errors and provide adequate input for the required simulation. CFD contributes to the understanding of underlying physics within fuel cells, involving electrochemical and kinetic variables like current density and water transport in individual cells. CFD models are therefore able to predict failures that may arise from flooding or dehydration of the membrane or other operating conditions that can bring about failure of fuel cells. Also, the ease and speed of running simulations contribute to the positive prospects and advantages of CFD. With the aid of commercial packages the need to constantly make improvements in user interfaces as more complicated designs are made, CFD will continue to undergo advances with benefits for the industries developing fuel cell systems.

In conclusion, the interest in CFD is expanding, as computational modelling makes it possible to evaluate innovative designs and assess the performance of fuel cells. In the long run, with regards to the ability of CFD to predict performance of fuel cells, it will help to improve their marketability, reliability and confidence in designs. However, most of the theoretical models concentrate on steady-state condition analysis, while transient transport behaviours still need to be considered. Also, due to the numerical computation workload required for PEFCs with complex flow fields, more focus has been put into simple flow fields and channels, leaving the problems associated with complex flow fields a challenge to be solved. Proper water management of the PEFC involves maintaining hydration of the membrane without flooding the electrode, while keeping a dynamic balance of water in the membrane during operation. Hence, proper water management and overall performance and marketability can be achieved in the future if and only if these challenges are overcome in order to speed up commercialisation efforts.

References

1. Carrette L, Friedrich AA, Stimming U (2000) Fuel cells: Principles, types, fuels, and applications. Chem Phys Chem 1: 162-193.

2. Cordiner S, Lanzani SP, Mulone V (2011) 3D effects of water-saturation distribution on polymeric electrolyte fuel cell (PEFC) performance. Int J Hydrogen Energy 36: 10366-10375.

3. Dincer I, Rosen MA (2011) Sustainability aspects of hydrogen and fuel cell systems. Energy for Sustainable Development 15: 137-146.

4. Frey HC, Rouphail NM, Zhai H, Farias TL, Gonçalves GA (2007) Comparing real-world fuel consumption for diesel-and hydrogen-fuelled transit buses and implication for emissions. Transportation Research Part D: Transport and Environment 12: 281-291.

5. Yuan W, Tang Y, Pan M, Li Z, Tang B (2010) Model prediction of effects of operating parameters on proton exchange membrane fuel cell performance. Renewable Energy 35: 656-666.

6. Barbir F (2013) PEM fuel cells: Theory and practice. (2ndedn) London: Elsevier Inc. Academic Press.

7. Çelik M, Özışık G, Genç G, Yapıcı H (2014) The effect of microporous layer in phosphoric acid doped polybenzimidazole polymer electrolyte membrane fuel cell. J Appl Mech Eng 3: 2.

8. Gasteiger H, Panels J, Yan S (2004) Dependence of PEM fuel cell performance on catalyst loading. J Power Sources 127: 162-171.

9. Srinivasulu GN, Subrahmanyam T, Rao VD (2011) Parametric sensitivity analysis of PEM fuel cell electrochemical Model. Int j hydrogen energy 36: 14838-14844. PEMFC 2014, Fuel cell today, Technologies.

10. Wang Y, Chen KS, Mishler J, Cho SC, Adroher XC (2011) A review of polymer electrolyte membrane fuel cells: Technology, applications, and needs on fundamental research. Applied Energy 88: 981-1007.

11. Williams RH, Larson ED, Katofsky RE, Chen J (1995) Methanol and hydrogen from biomass for transportation. Energy for Sustainable Development 1: 18-34.

12. Yang WJ, Wang HY, Lee DH, Kim YB (2015) Channel geometry optimization of a polymer electrolyte membrane fuel cell using genetic algorithm. Applied Energy 146: 1-10.

13. Sivertsen B, Djilali N (2005) CFD-based modelling of proton exchange membrane fuel cells. J Power Sources 141: 65-78.

14. Berning T, Djilali N (2003) Three-dimensional computational analysis of transport phenomena in a PEM fuel cell- a parametric study. J Power Sources 124: 440-452.

15. Djilali N (2007) Computational modelling of polymer electrolyte membrane (PEM) fuel cells: challenges and opportunities. Energy 32: 269-280.

16. Song C (2002) Fuel processing for low-temperature and high-temperature fuel cells: Challenges, and opportunities for sustainable development in the 21st century. Catalysis today 77: 17-49.

17. Lepiller C (2013) Fuel cell basics, Pragma Industries.

18. Das PK, Li X, Liu ZS (2010) Analysis of liquid water transport in cathode catalyst layer of PEM fuel cells. Int j hydrogen energy 35: 2403-2416.

19. Kanani H, Shams M, Hasheminasab M, Bozorgnezhad A (2015) Model development and optimization of operating conditions to maximize PEMFC performance by response surface methodology. Energy Conversion and Management 93: 9-22.

20. Santarelli M, Torchio M (2007) Experimental analysis of the effects of the operating variables on the performance of a single PEMFC. Energy Conversion and Management 48: 40-51.

21. Schmittinger W, Vahidi A (2008) A review of the main parameters influencing long-term performance and durability of PEM fuel cells. Journal of Power Sources 180: 1-14.

22. Amirinejad M, Rowshanzamir S, Eikani MH (2006) Effects of operating parameters on performance of a proton exchange membrane fuel cell. J Power Sources 161: 872-875.

23. Liu X, Guo H, Ye F, Ma CF (2007) Water flooding and pressure drop characteristics in flow channels of proton exchange membrane fuel cells. Electrochimica Acta 52: 3607-3614.

24. Anderson R, Zhang L, Ding Y, Blanco M, Bi X, et al. (2010) A critical review of two-phase flow in gas flow channels of proton exchange membrane fuel cells. J Power Sources 195: 4531-4553.

25. Kraytsberg A, Ein-Eli Y (2006) PEM FC with improved water management. J Power Sources 160: 194-201.

26. Houreh NB, Afshari E (2014) Three-dimensional CFD modeling of a planar membrane humidifier for PEM fuel cell systems. Int J Hydrogen Energy 39: 14969-14979.

27. Jiao K, Li X (2011) Water transport in polymer electrolyte membrane fuel cells. Progress in Energy and Combustion Science 37: 221-291.

28. Koulouris K, Konstantopoulos G, Apostolopoulos A, Matikas T, Apostolopoulos C (2016) The Influence of Corrosion Damage on Low Cycle Fatigue Life of Reinforcing Steel Bars S400. J Applied Mechanical Engineering.

29. Deevanhxay P, Sasabe T, Tsushima S, Hirai S (2013) Effect of liquid water distribution in gas diffusion media with and without microporous layer on PEM fuel cell performance. Electrochemistry Communications 34: 239-241.

30. Natarajan D, Van Nguyen T (2003) Three-dimensional effects of liquid water flooding in the cathode of a PEM fuel cell. J Power Sources 115: 66-80.

31. Park S, Lee JW, Popov BN (2012) A review of gas diffusion layer in PEM fuel cells: materials and designs. Int J of hydrogen energy 37: 5850-5865.

32. Wang X, Van Nguyen T (2012) An experimental study of the liquid water saturation level in the cathode gas diffusion layer of a PEM fuel cell. J Power Sources 197: 50-56.

33. Wood DL, Jung SY, Nguyen TV (1998) Effect of direct liquid water injection and interdigitated flow field on the performance of proton exchange membrane fuel cells. Electrochimica Acta 43: 3795-3809.

34. Litster S, Sinton D, Djilali N (2006) Ex situ visualization of liquid water transport in PEM fuel cell gas diffusion layers. J Power Sources 154: 95-105.

35. Li H, Tang Y, Wang Z, Shi Z, Wu S, et al. (2008) A review of water flooding issues in the proton exchange membrane fuel cell. J Power Sources 178: 103-117.

36. Bazylak A (2009) Liquid water visualization in PEM fuel cells: A review. Int J Hydrogen Energy 34: 3845-3857.

37. Chen YS, Peng H, Hussey DS, Jacobson DL, Tran DT, et al. (2007) Water distribution measurement for a PEMFC through neutron radiography. J Power Sources 170: 376-386.

38. Satija R, Jacobson DL, Arif M, Werner S (2004) In situ neutron imaging technique for evaluation of water management systems in operating PEM fuel cells. J Power Sources 129: 238-245.

39. (2015) Fuel Cell Technology: How it works. Catalysis today: Clean power solutions.

40. Zhang X, Gao Y, Ostadi H, Jiang K, Chen R (2014) A proposed agglomerate model for oxygen reduction in the catalyst layer of proton exchange membrane fuel cells. Electrochimica Acta 150: 320-328.

41. Mosdale R, Gebel G, Pineri M (1996) Water profile determination in a running proton exchange membrane fuel cell using small-angle neutron scattering. J Membrane Science 118: 269-277.

42. Nishikawa H, Kurihara R, Sukemori S, Sugawara T, Kobayasi H, et al. (2006) Measurements of humidity and current distribution in a PEFC. J Power Sources 155: 213-218.

43. Hussaini IS, Wang CY (2009) Visualization and qualification of cathode channel flooding in PEM fuel cells. J Power Sources 187: 444-451.

44. Ji M, Wei Z (2009) A review of water management in polymer electrolyte membrane fuel cells. Energies 2: 1057-1106.

45. Mench M, Dong Q, Wang C (2003) In situ water distribution measurements in a polymer electrolyte fuel cell. J Power Sources 124: 90-98.

46. Turhan A, Heller K, Brenizer J, Mench M (2006) Quantification of liquid water accumulation and distribution in a polymer electrolyte fuel cell using neutron imaging. J Power Sources 160: 1195-1203.

47. Zhang J, Kramer D, Shimoi R, Ono Y, Lehmann E, et al. (2006) In situ diagnostic of two-phase flow phenomena in polymer electrolyte fuel cells by neutron imaging: Part B. Material variations. Electrochimica Acta 51: 2715-2727.

48. Scott P (2012) Experimental investigation of a novel design concept of a modular PEMFC stack. UK University of Hertfordshire.

49. Carton J, Lawlor V, Olabi A, Hochenauer C, Zauner G (2012) Water droplet accumulation and motion in PEM (Proton Exchange Membrane) fuel cell mini-channels. Energy 39: 63-73.

50. Young JB (2007) Thermofluid modeling of fuel cells. Annu Rev Fluid Mech 39: 193-215.

51. Uriz I, Arzamendi G, Diéguez P, Echave F, Sanz O, et al. (2014) CFD analysis of the effects of the flow distribution and heat losses on the steam reforming of methanol in catalytic (Pd/ZnO) microreactors. Chemical Engineering Journal 238: 37-44.

52. Noorkami M, Robinson JB, Meyer Q, Obeisun OA, Fraga ES, et al. (2014) Effect of temperature uncertainty on polymer electrolyte fuel cell performance. Int j hydrogen energy 39: 1439-1448.

53. Song Y, Xu H, Wei Y, Kunz HR, Bonville LJ, et al. (2006) Dependence of high-temperature PEM fuel cell performance on Nafion® content. J power sources 154: 138-144.

54. Zhang J, Xie Z, Zhang J, Tang Y, Song C, et al. (2006) High temperature PEM fuel cells. J power Sources 160: 872-891.

55. Wang CY (2004) Fundamental models for fuel cell engineering. Chemical reviews 104: 4727-4766.

56. Hirsch C (2007) Numerical computation of internal and external flows: The fundamentals of computational fluid dynamics: The fundamentals of computational fluid dynamics: butterworth-heinemann 680.

57. Le AD, Zhou B (2008) A general model of proton exchange membrane fuel cell. J power sources 182: 197-222.

58. Ferreira RB, Falcão D, Oliveira V, Pinto A (2015) Numerical simulations of two-phase flow in an anode gas channel of a proton exchange membrane fuel cell. Energy 82: 619-628.

59. Dawes J, Hanspal N, Family O, Turan A (2009) Three-dimensional CFD modelling of PEM fuel cells: an investigation into the effects of water flooding. Chemical Engineering Science 64: 2781-2794.

60. Guvelioglu GH, Stenger HG (2005) Computational fluid dynamics modeling of polymer electrolyte membrane fuel cells. J Power Sources 147: 95-106.

61. Gurau V, Liu H, Kakac S (1998) Two-dimensional model for proton exchange membrane fuel cells. AIChE Journal 44: 2410-2422.

62. Cheddie D, Munroe N (2005) Review and comparison of approaches to proton exchange membrane fuel cell modeling. J Power Sources 147: 72-84.

63. Patankar S (1980) Numerical heat transfer and fluid flow: CRC Press.

64. Blazek J (2001) Computational fluid dynamics: principles and applications. The Netherlands: Elsevier Ltd.

65. Arvay A, Ahmed A, Peng XH, Kannan A (2012) Convergence criteria establishment for 3D simulation of proton exchange membrane fuel cell. Int J hydrogen energy 37: 2482-2489.

66. Weber AZ, Newman J (2004) Modeling transport in polymer-electrolyte fuel cells. Chemical reviews 104: 4679-4726.

67. Fuller TE, Newman J (1993) Water and thermal management model for proton-exchange membrane fuel cells. J Electrochemical Society 140: 1218-1225.

68. Bernardi DM, Verbrugge MW (1991) Mathematical model of a gas diffusion electrode bonded to a polymer electrolyte. AIChE Journal 37: 1151-1163.

69. Bernardi DM, Verbrugge MW (1992) A mathematical model of the solid-polymer-electrolyte fuel cell.

70. J Electrochemical Society 139: 2477-2791.

71. Springer T, Wilson M, Gottesfeld S (1993) Modeling and experimental diagnostics in polymer electrolyte fuel cells. Journal of the Electrochemical Society 140: 3513-3526.

72. Springer TE, Zawodzinski T, Gottesfeld S (1991) Polymer electrolyte fuel cell model. J Electrochemical Society 138: 2334-2342.

73. Nguyen TV, White RE (1993) A water and heat management model for proton-exchange-membrane fuel cells. J Electrochemical Society 140: 2178-2786.

74. Chen YS, Peng H (2008) A segmented model for studying water transport in a PEMFC. J Power Sources 185: 1179-1192.

75. Baschuk J, Li X (2000) Modelling of polymer electrolyte membrane fuel cells with variable degrees of water flooding. J power sources 86: 181-196.

76. Jiao K, Zhou B (2008) Effects of electrode wettabilities on liquid water behaviours in PEM fuel cell cathode. J Power Sources 175: 106-119.

77. Grujicic M, Chittajallu K (2004) Optimization of the cathode geometry in polymer electrolyte membrane (PEM) fuel cells. Chemical Engineering Science 59: 5883-5895.

78. He W, Yi JS, Van Nguyen T (2000) Two-phase flow model of the cathode of PEM fuel cells using interdigitated flow fields. AIChE Journal 46: 2053-2064.

79. Dutta S, Shimpalee S, Van Zee J (2001) Numerical prediction of mass-exchange between cathode and anode channels in a PEM fuel cell. Int J Heat and Mass Transfer 44: 2029-2042.

80. Ramos-Alvarado B, Hernandez-Guerrero A, Juarez-Robles D, Li P (2012) Numerical investigation of the performance of symmetric flow distributors as flow channels for PEM fuel cells. Int J hydrogen energy 37: 436-448.

81. Shan Y, Choe SY (2006) Modeling and simulation of a PEM fuel cell stack considering temperature effects. J Power Sources 158: 274-286.

82. Stern F, Wilson RV, Coleman HW, Paterson EG (1999) Verification and validation of CFD simulations: Iowa Institute of Hydraulic Research, University of Iowa.

83. Oberkampf WL, Trucano TG (2000) Validation methodology in computational fluid dynamics. AIAA paper 2549: 19-22.

25

Numerical Study of Hydrodynamic Characteristics of Gas – Liquid Slug Flow in Vertical Pipes

Massoud EZ[1,2]*, Xiao Q[1], Teamah MA[2] and Saqr KM[2]

[1]*Department of Naval Architecture, Ocean and Marine Engineering, University of Strathclyde, Glasgow G4 0LZ, UK*
[2]*Mechanical Engineering Department, College of Engineering and Technology, Arab Academy for Science, Technology and Maritime Transport, P.O. Box 1029, Abu Qir, Alexandria, Egypt*

Abstract

Multiphase flows occur in wide applications including; nuclear, chemical, and petroleum industries. One of the most important flow regime encountered in multiphase flow is the slug flow which is often encountered in oil and gas production systems. The slugging problems may cause flooding of downstream processing facilities, severe pipe corrosion and the structural instability of pipeline and further induce the reservoir flow oscillations, and a poor reservoir management. In the present study, computational fluid dynamics simulation is used to investigate two phase slug flow in vertical pipe using the volume of fluid (VOF) methodology implemented in the commercial code ANSYS Fluent. The viscous, inertial, and interfacial forces have significant effect on the hydrodynamic characteristics of two-phase slug flow. These forces can have investigated by introducing a set of dimensionless numbers, namely; inverse viscosity number, N_f, Eotvos number, E_o, and Froude number, Fr_{TB}. The simulation accounts for the hydrodynamic features of two phase slug flow including; the shape of Taylor bubble, bubble profile, velocity and thickness of the falling film, wake flow pattern, and wall shear stress distribution. The CFD simulation results are in good agreement with previous experimental data and models available in literature.

Keywords: Gas-liquid slug flow; Taylor bubble velocity; Liquid film; Wake flow pattern; VOF method

Introduction

Multiphase flow is a flow field in which more than one phase exists. It could be two-phase or three-phase or even four-phase flow depending on the number of phases encountered in the flow domain. Different combinations between the phases such as; gas bubbles in liquid, liquid droplets in gas, and solid particles in either gas or liquid are all different types of multiphase flow. Multiphase flow is complex in nature compared to single phase flow. The difficulty of multiphase flow arises from; the need of good knowledge of the chemical, physical and thermodynamics properties of each phase, the variation in the fluid composition with time as it flows in multiple locations, slugging problem, corrosion and erosion, and the variation in pressure and temperature leading to deposit formation.

Multiphase flow occurs in a wide range of applications including; natural processes as rivers and clouds formation, nuclear systems as nuclear waste processing, fluidized beds, boiling and condensation in nuclear reactors, chemical processes as food mineral processing and transportation, transport of cement, grains, and metal powders and petroleum industries as oil processing, oil and gas transport in pipelines, sloshing in offshore separator devices.

Multiphase flow is classified according to the distribution of the phases, which is known as flow regime, or flow pattern. Identifying the encountered flow regime in multiphase flow field is an important aspect prior performing any investigation on the flow. For gas-liquid flow in pipes, one of the common and complex pattern encountered in multiphase flow is known as slug flow. Slug flow is an intermitted flow between stratified and annular flow. It is characterized by an elongated bullet shaped bubble, known as Taylor bubble, and a liquid slug, which is a liquid film flowing downwards between the bubble interface and pipe wall.

Computational fluid dynamics (CFD) has been proven to be a powerful practical tool for the analysis and simulation of the hydrodynamic characteristics of slug flow in pipes. The main complex feature of gas-liquid slug flow is the deformable interface [1]. The volume of fluid (VOF) method originally developed by Hirt CW [2] is often used to simulate complex multiphase flows including slug flow, and is powerful in tracking the interface between fluids [3-7].

The hydrodynamic characteristics of gas-liquid vertical slug flow is governed by three main forces namely; viscous, inertial, and interfacial forces. A dimensionless analysis for two-phase gas-liquid slug flow can be done by applying Pi-Buckingham theorem. This simplifies the problem to three dimensionless groups, which are; Eötvös number, Eo, Froude number, Fr_{TB}, and inverse viscosity number, N_f, defined as follows:

Eötvös number, Eo, is the ratio between gravitational forces, and surface tension forces, and given by; $Eo = \rho_L * g * D^2 / \sigma$.

Froude number, Fr_{TB}, is the ratio between the inertia and gravitational forces, and given by; $Fr_{TB} = U_{TB} / \sqrt{g * D}$

Inverse viscosity number, N_f, is the ratio between Eötvös number, Eo, and Morton number, $M = g \mu_L^4 (\rho_L - \rho) / \rho_L^2 \sigma 3$, and given by; $Nf = \rho_L * (g * D^3)^{0.5} / \mu L$.

Where; ρL, μ_L, g, D, σ, and U_{TB} are the density of the liquid phase (kg/m³), dynamic viscosity of the liquid phase (Pa.s), gravitational acceleration (m/s²), pipe diameter (m), surface tension coefficient

*Corresponding author: Massoud EZ, Department of Naval Architecture, Ocean and Marine Engineering, University of Strathclyde, Glasgow G4 0LZ, UK
E-mail: Enass-zakaria-shafik-massoud@strath.ac.uk

(N/m), and Taylor bubble rise velocity (m/s), respectively [8]. The main hydrodynamic characteristics of gas-liquid vertical slug flow are; the shape of the Taylor bubble, Taylor bubble rise velocity, liquid film velocity, liquid film thickness, and wall shear stress distribution. Despite the fact that hard effort has been done in the modelling process of gas-liquid slug flow, a needs for closure correlations based on experimental data is still required. These relationships included slug characteristics such as; slug frequency, slug length, slug liquid hold up, and slug unit velocity. This is in addition to other relationships as shear force, wall friction, and gas entrainment rate.

In literature, many experimental and numerical studies were done on the rise of single Taylor bubble in both stagnant and flowing fluid in vertical pipes. Taha T [9] investigated the motion of a single Taylor bubble in vertical pipes using the volume of fluid (VOF) method implemented in the CFD software FLUENT. The simulation accounted for the Taylor bubble velocity, bubble shape, wake flow pattern, and wall shear stress distribution for the rise of single Taylor bubble in both stagnant and flowing liquid including both laminar and turbulent flow conditions. Their study involved wide range of Eötvös number, Froude number, and Morton number, and the results showed that the bubble shape is dependent on the liquid viscosity, the surface tension, and independent on bubble length. In addition, the results showed that the inverse viscosity number, N_f, greatly influence the wake structure, and increases the wall shear stress, and finally reduce the liquid film thickness. Similarly, [1] investigated the hydrodynamic characteristics of single Taylor bubble rising in stagnant liquid using the VOF method. The study further included the near wall treatment, and hence accounting for the near wall mass transfer and wall shear stress. The results revealed that the key parameters governing the two-phase flow induced corrosion are; mass transfer coefficient and wall shear stress. Furthermore, (8) simulated the rise of single Taylor bubble through stagnant Newtonian liquid using the VOF method, with particular focus to the liquid wake structure, and analyzing the results in terms of Taylor bubble velocity, flow around the bubble nose, liquid film and wake region. The study accounted for laminar flow regime only. The authors developed a map gathering all information reached while studying the wake structure. This maps could show the existence or absence of wake structure with the knowledge of Morton number, Eotvos numbers, and the type of concavity of the bubble rear. Moreover, Yan K [10] investigated the hydrodynamic characteristics of single Taylor bubble rising in stagnant liquid with further consideration of the small dispersed bubble in the liquid slug zone. The study accounted for the effect of small dispersed gas bubbles in liquid slug zone on the flow hydrodynamics features and CO_2 corrosion rate. It was concluded that the small dispersed gas bubbles result in higher fluctuations in the liquid slug zone, which subsequently increase the mass transfer and wall shear stress. Lastly, Araújo J [11] investigated the rising of two consecutive Taylor bubbles through vertical stagnant Newtonian liquids under laminar regime using the VOF method. The results accounted for bubble-bubble interaction, and showed the dependency of the wake on the separation distance between the bubbles.

The main aim of the present study is to investigate the hydrodynamic characteristics of gas-liquid slug flow in vertical pipe by applying computational fluid dynamics CFD simulation using the volume of fluid (VOF) methodology implemented in the commercial software ANSYS Fluent. The flow consists of single Taylor bubble rising in stagnant fluid. The study includes the investigation of some of the main hydrodynamic features of slug flow, which are; Taylor bubble shape, Taylor bubble rise velocity, velocity fields, liquid film thickness, and liquid film velocity for a range of Eötvös number, Eo,

Froude number, Fr_{TB}, and inverse viscosity number. A validation of the present numerical code is done by performing comparison with both experimental data, and with correlations available in literature.

Model Development

Following the concept of slug unit cell, two-phase vertical slug flow can be characterized by an elongated bullet shaped bubble, known as Taylor bubble, and a liquid slug which is a liquid film flowing downwards between the bubble interface and pipe wall. The computational time is an important issue that need to be considered while performing any multiphase flow simulation. Thus, in order to reduce the computational time, the present simulation is performed in 2D coordinate system, with axial symmetry around the centerline of the pipe. The simulation domain is a vertical pipe of diameter, D, and length, L. The flow domain is constructed and solved using the VOF methodology implemented in the computational dynamic software package, ANSYS Fluent (Release 15.0). In all simulated cases, a uniform grid of quadrilateral control elements is applied. Different grids were tested to check solution convergence. The present simulation has been performed for 2D coordinate system, unsteady, and laminar flow with constant fluid properties. The two phases were assumed as incompressible, viscous, immiscible, and not penetrating each other.

Governing equations

In the present model, the fluids share a well-defined interface and hence the volume of fluid (VOF) method for two phase flow has been selected. The VOF model is a surface-tracking technique applied to a fixed Eulerian mesh. This model is designed for two or more immiscible fluids to track the interface between them. This model solves a single set of momentum equation that is shared by the two fluids, and the volume fraction of each of the fluids in each computational cell is followed throughout the domain. Details of the governing equations and the treatment of the interface can be obtained from the Fluent user guide [12]. The equations being solved in the VOF model models are as given in this section.

One set of continuity and momentum equations are solved for the two-phase system. Firstly, the Reynolds average continuity equation in a VOF model for n number of phases can be expressed as follows;

$$\frac{\partial \rho}{\partial t} + \nabla \cdot (\rho U) = \sum_{q=1}^{n} S_q \tag{1}$$

Where; ρ, t, U, n and S_q and are the volume-fraction-averaged density, time, time average velocity vector of the flowing fluid, number of the phases (for the present two phase flow = 2), and mass source respectively. The mass source, S_q, is set to zero in the present case. In addition, a single Reynolds average momentum equation is solved throughout the computational domain, and all phases share the same resulting velocity field. The general momentum equation can be written as follows;

$$\frac{\partial (\rho U)}{\partial t} + \nabla \cdot (\rho U U) = -\nabla P + \nabla \cdot \left[\mu \left(\nabla U + \nabla U^T \right) \right] + F \tag{2}$$

Where; P, μ, and F and are the pressure in the domain, volume-fraction-averaged viscosity of the flowing fluid, and the additional forces as gravitational term and surface tension, respectively. In other words, the left hand side of equation (2) represents the unsteady term and convection terms. While, the right hand side represents the pressure term, the diffusion term, the body force and other external forces that might act on the system. The continuum surface force (CSF) of Brackbill [13] is used to account for the surface tension effects.

The VOF formulation relies on the fact that two or more fluids (or phases) are not interpenetrating. For each additional phase added to the model, a variable is introduced which is the volume fraction of the phase in the computational cell. In each control volume, the volume fractions of all phases sum to unity. If the volume fraction of the qth fluid in a cell is given by αq, thus the following relationship is valid for each computational cell;

$$\sum_{q=1}^{n} \alpha_q = 1 \tag{3}$$

The fields for all variables and properties are shared by the phases and represent volume-averaged values, as long as the volume fraction of each of the phases is known at each location. Thus the variables and properties in any given cell are either purely representative of one of the phases, or representative of a mixture of the phases, depending upon the volume fraction values [12]. Hence, there are three possible conditions:

1. If $\alpha q = 0$; the cell is empty of the qth fluid.

2. If $\alpha q = 1$; the cell is occupied mainly by the qth fluid.

3. If $0 < \alpha q < 1$; the cell contains the interface between the qth fluid and the other fluid.

The Reynolds average continuity equation (1) and the momentum equation (2) are thus dependent on the volume fractions of all phases through the volume-fraction-averaged properties; ρ and μ. Hence, depending on the local value of as discussed above, the volume-fraction-averaged density and viscosity are calculated as follows;

$$\rho = \sum_{q=1}^{n} \rho_q \alpha_q \tag{4}$$

$$\mu = \sum_{q=1}^{n} \mu_q \alpha_q \tag{5}$$

Where; the subscripts, q, L and G refer to phase, liquid phase (primary phase) and the gas phase (secondary phase). Tracking the interface between the two phases is achieved by the treatment of the volume fraction of the qth fluid, αq through solving a separate continuity equation, given by the AFU Guide [12];

$$\frac{1}{\rho_q} \left[\frac{\partial}{\partial t} \left(\alpha_q \rho_q \right) + \nabla \cdot \left(\alpha_q \rho_q U_q \right) \right] = S_{\alpha q} + \sum_{p=1}^{n} \left(\overset{\circ}{m}_{pq} - \overset{\circ}{m}_{qp} \right) \tag{6}$$

Where; $S\alpha q$, $\overset{\circ}{m}_{pq}$ and $\overset{\circ}{m}_{qp}$ are the mass source term, mass transfer from phase p to phase q and the mass transfer from phase to phase and the mass transfer from phase p to phase q, respectively. According to ANSYS Fluent user guide (12) the source term on the right-hand side of equation (6) is by default set to zero. The volume fraction equation will only be used to solve the volume fraction of the qth fluid, αq, and not the primary phase (liquid phase). The gas phase is computed according to the constraint given in equation (3).

Model geometry and boundary conditions

As mentioned earlier, the solution domain is a vertical pipe with diameter, D, and length, L, with symmetry along the centerline. The length of domain is eleven times larger than pipe diameter. Figure 1 shows the boundary conditions and the initial Taylor bubble shape used in the simulation. The initial bubble shape is a cylinder connected to a hemisphere with the same radius giving an overall bullet shape of Taylor bubble. The length and radius of the Taylor bubble are given by; L_{TB}, and R_{TB} respectively. This initial shape is simulated until a steady bubble shape is reached. Different bubble shapes were tested and final

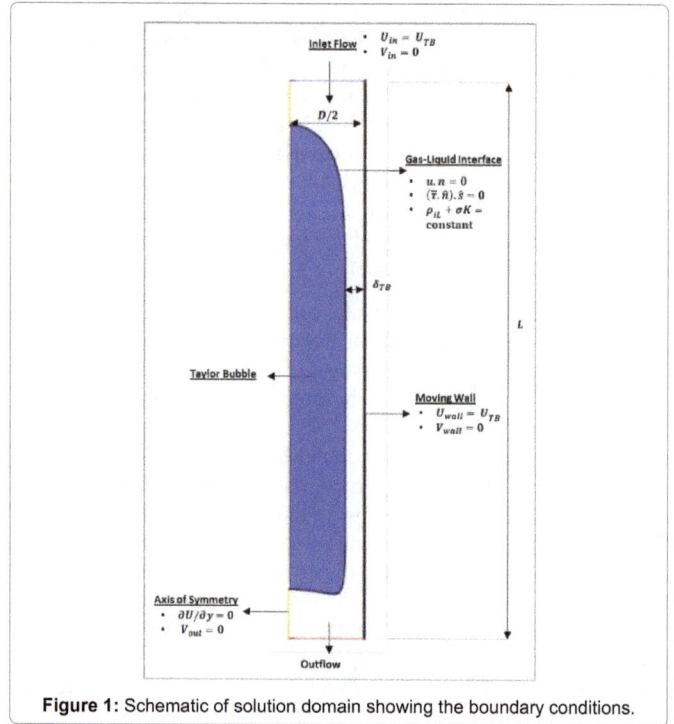

Figure 1: Schematic of solution domain showing the boundary conditions.

steady shape of bubble was similar but this only affects the solution convergence.

The simulation is performed by attaching a reference frame to the rising Taylor bubble. Enabling moving reference frame (MRF) in the simulation, causes the rising Taylor bubble to be stationary and the pipe wall moves downwards with velocity equal to that of the bubble [14]. The initial guess of Taylor bubble velocity, U_{TB}, is estimated according to the general correlation of Wallis GB [15], which is given by,

$$Fr = \frac{U_{TB}}{\sqrt{g * D}} = 0.345 \left(1 - e^{\frac{-0.01N_f}{0.345}} \right) \left(1 - e^{\frac{3.37 - E_0}{m}} \right)$$

$$m = \begin{cases} 25, & N_f < 18 \\ 69 N_f^{-0.35}, & 18 < N_f < 250 \\ 10, & N_f > 250 \end{cases} \tag{7}$$

Once the Taylor bubble ceases moving up or down in the axial direction, and hence pseudo-steady solution is reached, the velocity is then adjusted.

The initial guess of the liquid film thickness, δ_{LF}, is estimated using the following equation [16];

$$\delta_{LF} = \left[\frac{3 * v}{2 * g * \left(R - \delta_{LF} \right)} * U_{TB} * \left(R - \delta_{LF} \right)^2 \right]^{1/3} \tag{8}$$

Referring to Figure 1, the boundary conditions could be summarized as follow;

At the top of domain, the inlet flow boundary condition is applied with liquid entering at average uniform velocity equal to velocity of Taylor bubble, $U_{in} = U_{TB}$, $V_{in} = 0$. This uniform profile assumption is due to the investigation of the rise of Taylor bubble in stagnant liquid.

At the bottom of the domain, the outflow boundary condition is applied as the liquid phase is the only phase available.

The symmetry boundary condition is applied at the pipe centerline; $\partial U / \partial y = 0, V_{out} = 0$.

At the wall, the no-slip condition is applied with wall moving downwards with the following velocities; $U_{wall} = UTB, V_{wall} = 0$.

The pressure variation in the gas phase is assumed to be constant. The boundary conditions at the gas-liquid interface are given by;

1. The kinematic condition assuming full slip at the interface is applied; $u. n = 0$.

2. The dynamic boundary condition, which is also known by stress jump condition can be divided into two separate boundary conditions; the tangential stress balance assuming zero interfacial shear stress along the interface; $(\overline{\tau} \hat{n}). \hat{s} = 0$, and the normal stress balance; $\rho_{iL} + \sigma K$ = constant, where; $\overline{\tau}$, $\hat{n}$, $\hat{s}$, σ, ρ_{iL} and K and are the shear stress, unit normal vector at the interface, unit tangential vector at the interface, surface tension, liquid phase side pressure, and curvature of the interface, respectively.

According to Mao [14], the curvature of the interface is expressed in terms of radii of the curvature of the bubble surface, as follows;

$$K = \frac{1}{r_1} + \frac{1}{r_2} \tag{9}$$

Solution strategy and convergence criterion

As mentioned earlier, a transient simulation is carried out in the present case in order to consider the transient behavior of two phase flow. The simulation was carried out using the explicit VOF model. The PISO pressure-velocity was selected. The spatial discretization scheme used are as follows; Green Gauss Node Based for gradient, PRESTO for pressure, Compressive for volume fraction, Quick scheme for momentum, and first order implicit for transient formulation. The scaled absolute values of the residual of the calculated values of mass, velocity in x, and y directions were monitored including a convergence criterion of 10^{-3} for each time step.

Results and Discussion

As mentioned earlier, the aim of the present investigation is to

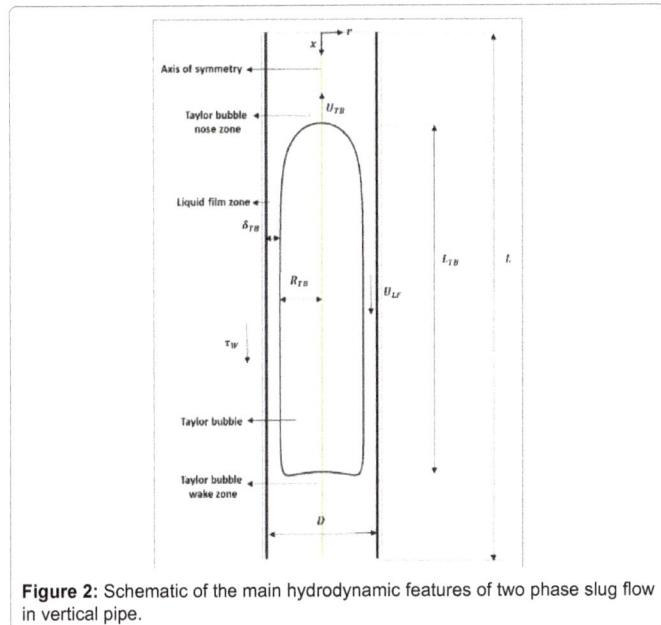

Figure 2: Schematic of the main hydrodynamic features of two phase slug flow in vertical pipe.

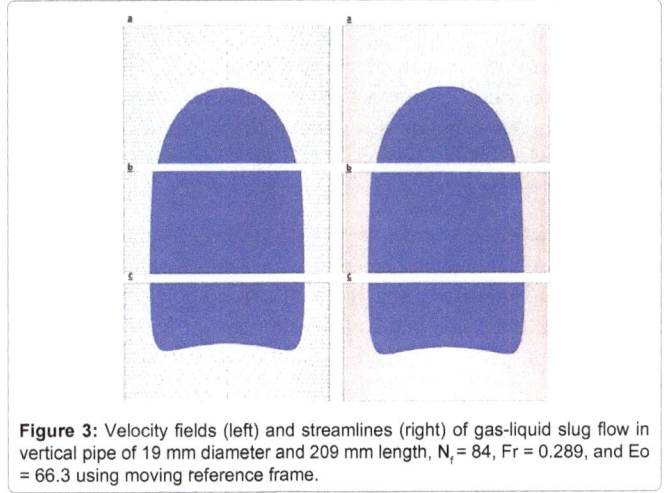

Figure 3: Velocity fields (left) and streamlines (right) of gas-liquid slug flow in vertical pipe of 19 mm diameter and 209 mm length, N_f = 84, Fr = 0.289, and Eo = 66.3 using moving reference frame.

Fluid Properties	Case 1	Case 2	Case 3	Case 4
ρL	1223	1206	1202	1190
vL	9.7653 E-05	4.66071 E-05	4.00139 E-05	2.52395 E-05
σ	0.0656234	0.066733445	0.067247627	0.067536505
Eo	66	64	63.3	62.4
Nf	84	176	205	325

Table 1: Properties of the fluid systems used in the present study.

study the hydrodynamics characteristics of single Taylor bubble rising in a stagnant Newtonian liquid using set of dimensionless number, which are; Eötvös number, E_o, Froude number, Fr_{TB}, inverse viscosity number, N_f, and Morton number, M. Figure 2 shows a schematic of a single Taylor bubble rising in stagnant liquid with some of the main hydrodynamic features of slug flow, including; Taylor bubble shape, Taylor bubble velocity, U_{TB}, liquid film thickness, δ_{LF}, liquid film velocity, U_{LF}, and wall shear stress distribution, τ_W.

Figure 3 provide the velocity fields and streamlines of the liquid phase in two phase slug flow in vertical pipe. As shown in the figure, the flow could be divided into three zones a, b, and c namely; Taylor bubble nose zone, falling liquid film zone, and Taylor bubble wake zone (liquid slug zone). In the Taylor bubble nose zone, the Taylor bubble moves up with velocity, due to buoyancy, pushing the liquid sideways where liquid film zone starts to develop. In the falling liquid film zone, the liquid moves downwards with velocity, U_{LF}, and decreasing liquid film thickness, δ_{LF}. Once a balanced between the gravitational and friction forces is reached, a constant terminal liquid film velocity and thickness is developed. In the Taylor bubble wake zone, the falling liquid film starts to plugs into the liquid slug ending with highly mixing zone in the wake structure of the bubble.

Taylor bubble shape

This section introduces the effect of inverse viscosity number, N_f, on Taylor bubble shape and its profile. Four cases are simulated according to the experimental work by Campos J [17], and their relevant properties are given in Table 1. In all simulation cases, the gas phase used is air with density =1.225 kg/m³, and viscosity =1.7894E-05 Pa. s. The liquid phase is aqueous glycerol solution with varying viscosity range. The simulation domain is a vertical pipe of 19 mm diameter, and 209 mm (11D) length. Figure 4 shows the effect of on Taylor bubble shape in a glycerol solution. The increase in gradually increases the concave shape of the Taylor bubble bottom surface. The simulated results are in good agreement with the experimental observations of Goldsmith H [18]; in highly viscous flow (viscosity dominated flow) the Taylor bubble has spheroid shape where the top end of bubble is prolate, and the bottom end is oblate. While, in low viscosity flow the flattening or concaving shape of Taylor bubble bottom end is observed. There simulation results are as well in good agreement with the numerical work [8,9,19].

To validate the current simulation results, Figure 5 gives a comparison of the bubble shape profile for different values of inverse viscosity number, (cases 1 and 2) with the results of (9). As indicated in the figure, there is a good agreement between both results. The small deviations in the results are due to the different methods used to estimate the initial guess of the liquid film thickness prior simulation.

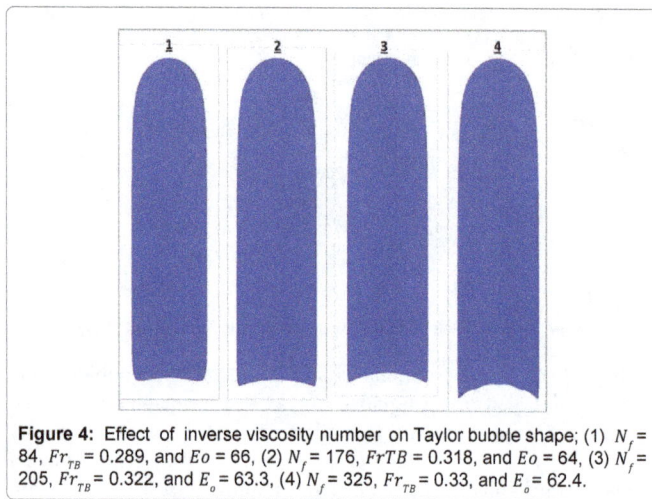

Figure 4: Effect of inverse viscosity number on Taylor bubble shape; (1) N_f = 84, Fr_{TB} = 0.289, and Eo = 66, (2) N_f = 176, $FrTB$ = 0.318, and Eo = 64, (3) N_f = 205, Fr_{TB} = 0.322, and E_o = 63.3, (4) N_f = 325, Fr_{TB} = 0.33, and E_o = 62.4.

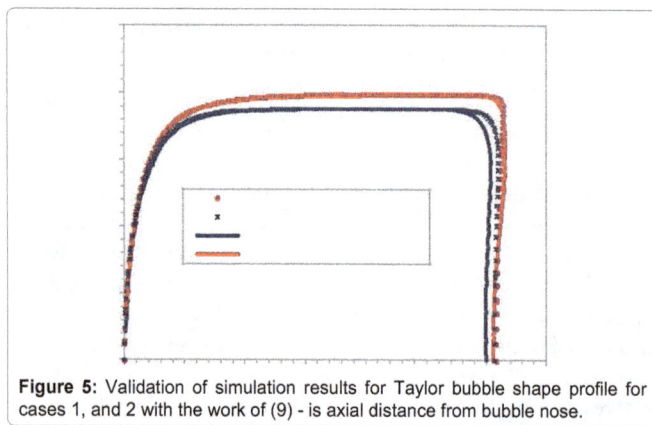

Figure 5: Validation of simulation results for Taylor bubble shape profile for cases 1, and 2 with the work of (9) - is axial distance from bubble nose.

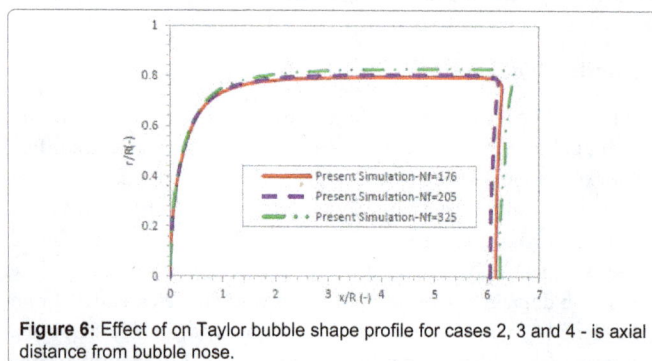

Figure 6: Effect of on Taylor bubble shape profile for cases 2, 3 and 4 - is axial distance from bubble nose.

Cases	Simulation	Ref [22]	Wallis Correlation [15]
Case 1			
UTB	0.1251	0.1381	0.134
Error (%)	...	9.41	6.62
Case 2			
UTB	0.1374	0.1467	0.1473
Error (%)	...	6.35	6.71
Case 3			
UTB	0.139	0.1467	0.148
Error (%)	...	5.24	6.02
Case 4			
UTB	0.1425	0.1424	0.1485
Error (%)	...	-0.07	4.0

Table 2: Numerical and experimental values of Taylor bubble velocity, for all simulation cases given in Table 1 with corresponding deviations.

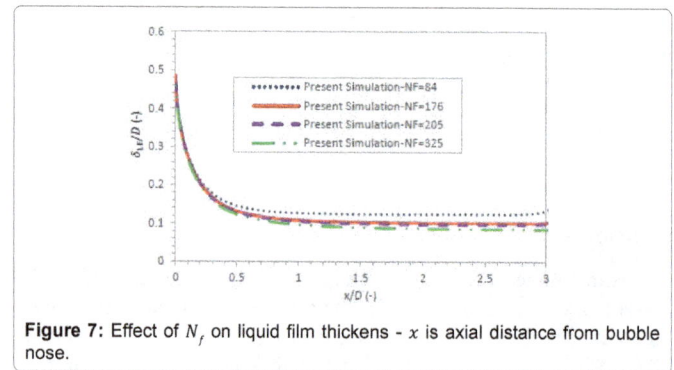

Figure 7: Effect of N_f on liquid film thickens - x is axial distance from bubble nose.

Figure 6 shows the effect of inverse viscosity number, on the bubble shape profile for wide range of N_f. The findings suggest that the increase in increases the bluntness of the bubble nose, and the bubble bottom edge is more flattened. In addition, the increase in decreases the liquid film thickness, δ_{LF}. The results are in good agreement with the work of Zheng et al. [19].

Taylor bubble rise velocity

The Taylor bubble rise velocity, U_{TB}, is one of the main hydrodynamic features frequently used in the description of two phase slug flow. In literature, many experiments were done to estimate the Taylor bubble rise velocity [15,20-22]. Table 2 gives the numerical and experimental results for Taylor bubble velocity for all simulation cases with the deviation between results. All deviations obtained are below 10%, which is an accepted range.

Liquid film

The flow in the liquid film is investigated by studying the liquid film thickness, δ_{LF}, and the liquid film axial velocity, U_{LF}. Figure 7 shows the effect of inverse viscosity number, N_f, on the liquid film thickness, for the four cases even in Table 1. At constant value of the dimensionless thickness of the liquid film, δ_{LF}/D, decreases with the increase in the dimensionless distance from bubble nose, x/D. It then remains constant at around x/D =1.

Increasing the inverse viscosity number, N_f, decreases the liquid film thickness. On the other hand, the increase in increases the liquid film axial velocity, as shown in Figure 8. Due to gravity, the falling liquid film is accelerated along the Taylor bubble with a decrease in the liquid film thickness. Thus, at film axial velocity, U_{LF}, increases with the increase in the dimensionless distance from bubble nose, x/D. In conclusion, once a balanced between the gravitational and friction forces is reached, a constant liquid film thickness, and velocity is established. The results are in good agreement with the work of Zheng et al. [19].

Wake flow structure

The wake structure is one of the vital hydrodynamic characteristics of vertical slug flow especially in describing the interaction and coalescences between successive Taylor bubbles (8). Prior studding the wake structure, it is important to identify the flow behavior in the liquid film, whether it is laminar or turbulent. This is done by estimating the value of critical Reynolds number, $Re_{LFcritical} = \rho_L * U_{LFavg} * \delta/\mu_L$, where; U_{LFavg} is the average liquid velocity in the liquid film.

To ensure laminar flow regime in the liquid film in the present study, the selected simulation cases are based on the experimental work

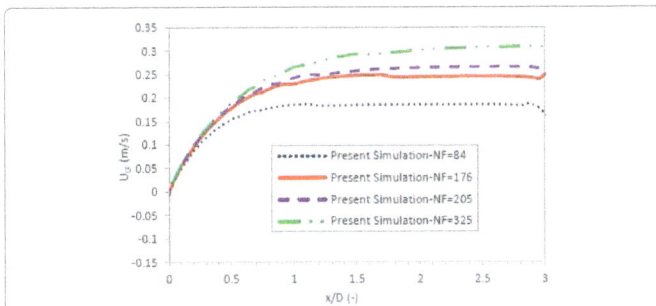

Figure 8: Effect of N_f on liquid film axial velocity - x is axial distance from bubble nose.

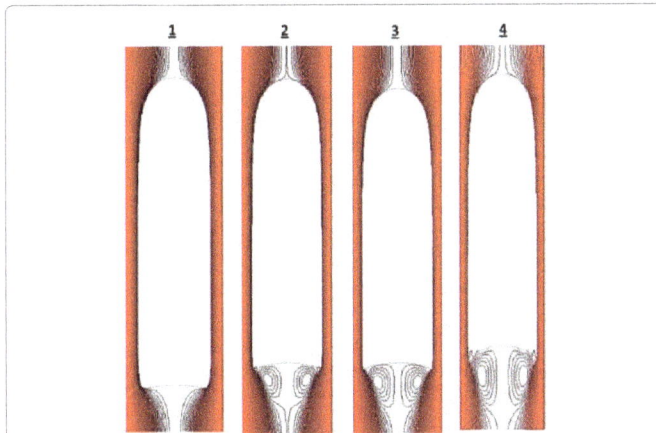

Figure 9: Effect of N_f on the wake flow pattern; (1) N_f = 84, (2) N_f = 176, (3) N_f = 205, (4) N_f = 325.

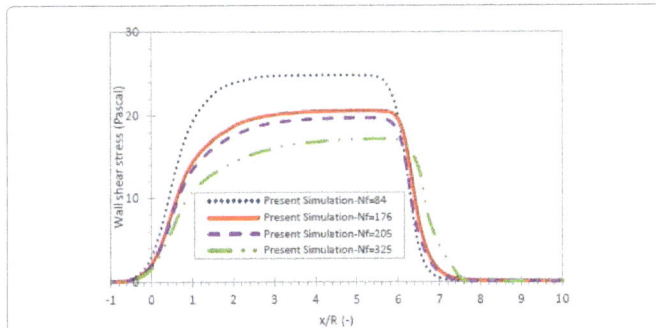

Figure 10: Effect of N_f on the wall shear stress distribution around a slug unit- x is axial distance from bubble nose.

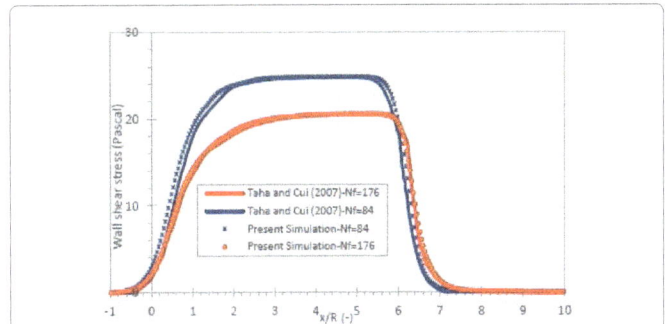

Figure 11: Validation of simulation results for the wall shear stress distribution around a slug unit with the work of (9) - x is axial distance from bubble nose.

of Campos et al. [17], where the authors concluded that for laminar flow regime Nf should be less than 500.

Figure 9 shows the wake flow structure for different value of Nf. In all cases the flow is characterized by closed axisymmetric wake as indicated by Campos et al. [17], and no gas entrainment in the liquid slug zone is noticed. At low values of Nf, the liquid from the liquid film zone expands directly and smoothly in to bubble wake zone, which is noticed by parallel streamlines in case 1. Increasing the values of Nf leads to the development of circulatory vortex in the bubble wake as liquid plugs into the Taylor bubble bottom. The scale and intensity of the vortex increases with higher values of Nf. The wake flow pattern agrees as well with the work of [8,9,19].

Wall shear stress

If the two phase slug flow problem is involved in heat or mass transfer, then the wall shear stress becomes a main significant hydrodynamic parameter. This processes are often referred to as slug flow induced corrosion [1, 8,19,20-24]. The main problems results from slug flow corrosion are; pipeline damage, decrease pipeline lifetime and may possibly lead to shut down of the pipeline.

Figure 10 illustrates the wall shear stress distribution, τW, around a slug unit for different values of N_f. It is obvious that the wall shear stress has the same distribution for all values of which starts with an increase in the wall shear stress near the bubble nose then it reaches a maximum positive value with the formation of a constant liquid film characteristics (thickness and velocity). The wall shear stress then starts to decrease once again till it reaches zero value in the bubble tail or wake zone. The results show that the wall shear stress is dependent on the distance between the Taylor bubble and the pipe wall, known as the liquid film thickness, δ_{LF}, which decreases with the increase in value of N_f. This is shown in Figure 10 where the maximum peak of the wall shear stress profile is decreased from N_f = 84 to higher values of N_f = 325. In addition, Figure 11 shows that there is a good agreement between the present results and the work of Taha [9].

Conclusions

In this work, a detailed numerical simulation of the hydrodynamic characteristics of gas-liquid slug flow in vertical pipe is developed using the volume of fluid (VOF) methodology implemented in the commercial software ANSYS Fluent. The main hydrodynamic characteristic studied are; the Taylor bubble shape and profile, the Taylor bubble rise velocity, the liquid film thickness and velocity, the wake flow structure, and the wall shear stress distribution. The simulation was done using air as the gas phase, and aqueous glycerol solution as the liquid phase. The following conclusions can be pointed out;

1. The Taylor bubble bottom depends on the liquid viscosity where the increase in the inverse viscosity number, N_f, increases the concave shape of the Taylor bubble bottom surface.

2. The calculated Taylor bubble rise velocity, U_{TB}, is in an acceptable range when compared with experimental values and commonly used correlations in the literature.

3. The liquid film zone can be describes using the liquid film thickness, δ_{LF}, and the liquid film axial velocity, that are both directly affected by the inverse viscosity number N_f.

4. The wake flow structure has a closed axisymmetric nature for all the simulation cases with the development of circulatory vortex in the bubble wake with the increase in N_f.

5. The wall shear stress, τ_W is mainly dependent on the liquid film thickness, δ_{LF}, and has a peak positive value in the stabilized liquid film zone.

Recommendations for Future Work

This study set out to develop a basic simulation model for gas-liquid slug flow in vertical pipe under laminar flow regime. It is recommended that further research be undertaken in the following areas; investigating the hydrodynamic characteristics of slug flow for different fluid system including the effect of viscosity and density ratios, investigating the hydrodynamic characteristics of slug flow under turbulent regime with > 500, exploring the hydrodynamic characteristics of slug flow including the flow of two consecutive Taylor bubbles in vertical pipe, and studying the wake flow pattern of single Taylor bubble or two consecutive Taylor bubbles under turbulent flow regime in terms of wake volume and length.

Acknowledgment

Results were obtained using the Engineering and Physical Sciences Research Council (EPSRC) funded ARCHIE-West high performance computer (www.archie-west.ac.uk) with EPSRC grant EP/K000586/1.

References

1. Zheng D, Che D (2007) An investigation on near wall transport characteristics in an adiabatic upward gasliquid two-phase slug flow. Heat and Mass Transfer 43: 1019-1036.

2. Hirt CW, Nichols BD (1981) Volume of fluid (VOF) method for the dynamics of free boundaries. J computational Physics 39: 201-225.

3. Desamala AB, Dasamahapatra AK, Mandal TK (2014) Oil-water two-phase flow characteristics in horizontal pipeline a comprehensive CFD study.

4. Desamala AB, Dasari A, Vijayan V, Goshika BK, Dasmahapatra AK (2013) CFD simulation and validation of flow pattern transition boundaries during moderately viscous oil-water two-phase flow through horizontal pipeline. WASET J 73: 1150-1155.

5. Febres M, Nieckele AO (2010) Slug flow prediction with the volume of fluid model. 13th Brazilian Congress of Thermal Sciences and Engineering Uberlandia MG, Brazil.

6. Rahimi R, Bahramifar E, Sotoodeh MM (2013)The indication of two-phase flow pattern and slug characteristics in a pipeline using CFD method.

7. Razavi S, Namin M (2011) Numerical model of slug development on horizontal two-phase flow.

8. Araújo J, Miranda J, Pinto A, Campos J (2012) Wide-ranging survey on the laminar flow of individual Taylor bubbles rising through stagnant Newtonian liquids. Int J Multiphase Flow 43: 131-148.

9. Taha T, Cui Z (2006) CFD modelling of slug flow in vertical tubes. Chemical Engineering Science 61: 676-687.

10. Yan K, Che D (2011) Hydrodynamic and mass transfer characteristics of slug flow in a vertical pipe with and without dispersed small bubbles. Int J multiphase flow 37: 299-325.

11. Araújo J, Miranda J, Campos J (2013) Flow of two consecutive Taylor bubbles through a vertical column of stagnant liquid A CFD study about the influence of the leading bubble on the hydrodynamics of the trailing one. Chemical Engineering Science 97: 16-33.

12. AFU Guide (2013) Ansys. Inc. Release 15.

13. Brackbill J, Kothe DB, Zemach C (1992) A continuum method for modeling surface tension. J computational Physics 100: 335-354.

14. Mao ZS, Dukler A (1990) The motion of Taylor bubbles in vertical tubes I A numerical simulation for the shape and rise velocity of Taylor bubbles in stagnant and flowing liquid. J computational physics 91: 132-160.

15. Wallis GB (1969) One-dimensional two-phase flow.

16. Brown R (1965) The mechanics of large gas bubbles in tubes: I Bubble velocities in stagnant liquids. The Canadian J Chemical Engineering 43: 217-223.

17. Campos J, De Carvalho JG (1988) An experimental study of the wake of gas slugs rising in liquids. J Fluid Mechanics 196: 27-37.

18. Goldsmith H, Mason S (1962) The movement of single large bubbles in closed vertical tubes. J Fluid Mechanics 14: 42-58.

19. Zheng D, He X, Che D (2007) CFD simulations of hydrodynamic characteristics in a gas liquid vertical upward slug flow. Int J heat and mass transfer 50: 4151-4165.

20. Bendiksen KH (1985) On the motion of long bubbles in vertical tubes. Int J multiphase flow 11: 797-812.

21. Davies R, Taylor G (1950) The mechanics of large bubbles rising through extended liquids and through liquids in tubes. Proceedings of the Royal Society of London A: Mathematical Physical and Engineering Sciences The Royal Society.

22. White E, Beardmore R (1962) The velocity of rise of single cylindrical air bubbles through liquids contained in vertical tubes. Chemical Engineering Science 17: 351-361.

23. Yan K, Zhang Y, Che D (2012) Experimental study on near wall transport characteristics of slug flow in a vertical pipe. Heat and Mass Transfer 48: 1193-1205.

24. Zheng D, Che D (2006) Experimental study on hydrodynamic characteristics of upward gas–liquid slug flow. Int J multiphase flow 32: 1191-1218.

New Approach of Metals Ductility in Tensile Test

Badreddine R*[1], Abderrazak D[1] and Kheireddine S[2]

[1]Departement de metallurgie et génie des materiaux, laboratoire de metallurgie et genie des materiaux, universite badjimokhtar 23000 Annaba
[2]Research Center in Industrial Technologies, CRTI P. O. Box 64, Cheraga 16014 Algiers, Algeria

Abstract

Ductility is the ability of a material to deform plastically before rupture. This is an important feature in shaping because it helps to define the behavior of materials. Ductility is therefore essential to know and thus determine to anticipate the behavior of materials in various situations of stress. Ductility is commonly defined by the two parameters A elongation (in percent) or necking Z (in percent) with:

A(%) = ΔL/L_0(%) = (L_1-L_0)/L_0(%) and Z(%) = ΔS/S_0(%) = (S_0-S_1)/S_0(%)

These two parameters are determined from tensile tests on standard specimens.

We will focus on the study and analysis of ductility using the tensile test.

However these two indicators (A) and (Z) of the ductility may present deficiencies (contradictions) in the interpretation of the ductility in case where for two samples (1) and (2) with same original dimensions (Lo) and (So) and different composition we could have : A1>A2 and Z1<Z2 or A1<A2 and Z1>Z2.

These two cases show the anomaly between A and Z in the assessment of the ductility, in fact in the first case the sample (1) is more ductile than the sample (2) in terms of elongation (A) is less ductile necking in terms of (Z) against the 2nd case we find the opposite behavior; it is this inconsistency that we will approach the ductility by introducing a parameter which will be called ductility (D) which takes into account the elongation and necking in a single formulation. In fact, (D) could remedy this deficiency involving computational approaches by activating the settings of the length (L) and Section (S) across the diameter (d) together in a first approach and to other computational approaches that take into account the elongation A and the neck.

Keywords: Ductility; Elongation; Necking; Tensile; Approach

Introduction

Ductility is the ability of a material to deform plastically before rupture. This is an important feature in shaping because it helps to define the behavior of materials. Ductility is therefore essential to know and thus determine to anticipate the behavior of materials in various situations of stress. Ductility is commonly defined by the two parameters A elongation (in percent) or necking Z (in percent) with:

$$A(\%) = \frac{\Delta L}{L_0}(\%) = \frac{L_1 - L_0}{L_0}(\%) \quad and$$

$$Z(\%) = \frac{\Delta S}{S_0}(\%) = \frac{S_0 - S_1}{S_0}(\%)$$

These two parameters are determined from tensile tests on standard specimens.

We will focus on the study and analysis of ductility using the tensile test.

However these two indicators (A) and (Z) of the ductility may present deficiencies (contradictions) in the interpretation of the ductility in case where for two samples (1) and (2) with same original dimensions (Lo) and (So) and different composition we could have:

$A_1 > A_2$ and $Z_1 < Z_2$

Or $A_1 < A_2$ and $Z_1 > Z_2$

These two cases show the anomaly between A and Z in the assessment of the ductility, in fact in the first case the sample (1) is more ductile than the sample (2) in terms of elongation (A) and less ductile in terms of necking (Z); in the 2nd case we find the opposite behavior; it is this inconsistency that we will approach the ductility by introducing a parameter which will be called ductility (D_1) which

takes into account the elongation and necking in a single formulation. In fact, (D_1) could remedy this deficiency involving computational approaches by activating the settings of the length (L) and Section (S) across the diameter (d) together [1-20].

Materials and Methods

Ductility modeling and approaches of metals

Highlighting the contradiction of the ductility value between the parameters A(%) and Z(%): The anomaly of appreciation of ductility that we expose, concerned the contradiction between the percent elongation parameter A(%), and the percent necking Z parameter (%). This anomaly is confirmed by numerous examples; among metals and alloys defined according to the American standard AISI and ASTM Table 1, some of them confirmed the contradiction between A(%) and Z(%) (Table 1) [21-40].

So this contradiction leads us to propose modeling approaches of ductility to remedy this inconsistency and thus give a more meaningful assessment of the ductility by inter reactive factors such as the length, section and through which the diameter during deformation (Table 2).

*Corresponding author: Badreddine R, Department de métallurgie et genie des materiaux, laboratoire de métallurgie et genie des matériaux, université badjimokhtar 23000 Annaba, E-mail: bregaiguia@yahoo.fr

	E (Gpa)	Rp (Mpa)	Rm (Mpa)	A%	Z%
Ductile Iron A536 (6545-12)	159	334	448	15	19.8
Rolled AISI 1020	203	260	441	36	61
ASTM A514, T1	208	724	807	20	66
Ni Maraging Steel (250)	186	1791	1860	8	56
Aluminium 2024 –T4	3.1	303	476	20	35

Table 1: Mechanical properties of some metals [57].

Metals	A	Z
Ductile Iron A536 (65-45-12)	15	19.8
Ni Maraging Steel (250)	8	56
Rolled AISI 1020	36	61
ASTM A514 , T1	20	66
Ni Maraging Steel (250)	8	56
Aluminium 2024-T4	20	35

Table 2: Examples of metals with a contradiction between the ductility parameters A (%) and Z (%).

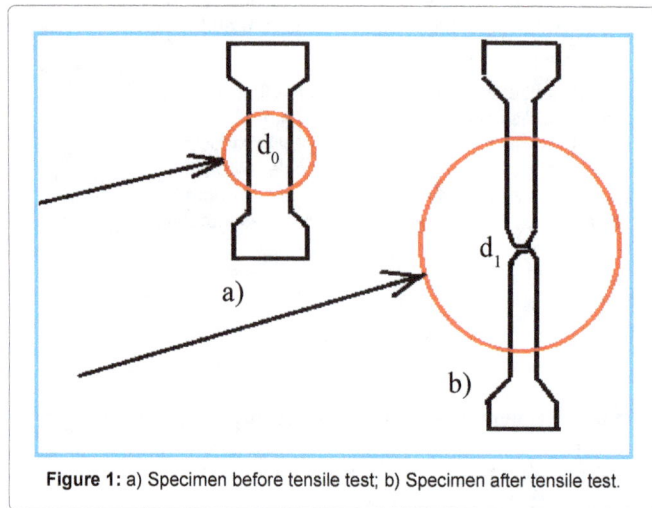

Figure 1: a) Specimen before tensile test; b) Specimen after tensile test.

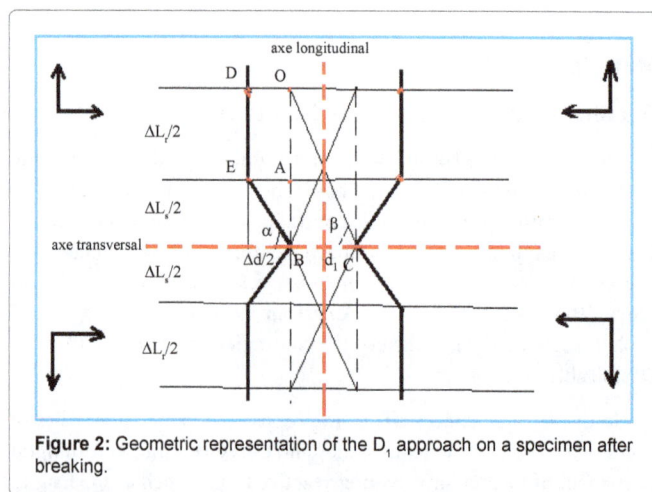

Figure 2: Geometric representation of the D_1 approach on a specimen after breaking.

Approach 1: Geometric modeling approach ductility according to the elongation and diameter necking

This is a standardized tensile specimen with respectively Lo, So length and initial section L_1 and S_1 length and section after tensile test. It is known that the elongation consists of two separate elongations including one distributed almost uniformly over the entire length of the test piece, while the other is located at the point of necking [41,42]. Uniform elongation is calculated by expressing the volume of the specimen between the landmarks of the test piece is not changed by this elongation, then:

$$\left[\left(\frac{d_0}{d_1}\right)^2 1\right] x 100\% \tag{1}$$

Or: d_0 is the diameter of the specimen before the tensile test and d_1 after breaking.

b(%): lengthening of necking, established by the difference:

$$b(\%) = A(\%) - a(\%) \tag{2}$$

A(%): Total elongation measured on standard test specimen.

Assuming that $A = A_R + A_S$.

We focus on the study of uniform elongation and elongation of necking during the tensile deformation.

It discusses this approach by a geometric representation formulation of these parameters enabled.

We consider a standard tensile bar with respectively:

L_o and S_o: the length and the section before tensile test.

L_1 and S_1: the length and the section after tensile test.

By analyzing necking area, it is noted that:

Longer necking which is ΔL_S is a line that develops in the direction of the loading axis.

Necking which is the ratio of ΔS and initial section S_0 is represented in our approach by the difference of initial diameter d_0 and diameter after breaking d_1.

In the work that follows, we will focus our approach on the specimen deformation area (Figure 1).

According to Figure 2, it is assumed that:

- The test piece is perfectly symmetrical on both sides of the longitudinal and transverse axes.

- After the tensile test piece halves of both sides with respect to the break line (necking) are symmetrical.

- The breaking line (necking) is mingled with the transverse axis.

- The variation of the section profile in the necked area is made along a straight right from the beginning of the necking to failure.

- We introduced the concept of necking bearing represented by the oblique line EB represents the profile of the evolution of reduced diameter.

- We introduce the notion of necking angle (tgα) which is the angle resulting from the intersection of the transverse confused with BC and EB bearing.

- We introduced the concept of 1/2 ductility triangle whose base is necking diameter and height is total elongation.

- We introduced the concept of 1/2 ductility angle β formed by the intersection of the base of the triangle (diameter necking) and the hypotenuse OC.

We note in Figure 2 that we have:

04 geometric representations highlighting $\frac{\Delta L_r}{2}$ $\frac{\Delta L_S}{2}$ et $\frac{\Delta d}{2}$.

Symmetrically, 02-02 relative to longitudinal and transverse axes.

However, we will focus our approach of calculating the approach D_1 on geometric representation of the left upper half specimen. In the Figure 2 we see the necking profile across the necking bearing EB that means we see the profile of the evolution of strain in term of elongation and diameter reduction necking d_1 until the breaking [43-47].

Calculation Method of D_1 Ductility Approach

Either the diagram of portion 1(3);

Note from the geometry:

$\frac{\Delta L_r}{2}$ is the uniform elongation of the portion 1.

$\frac{\Delta L_s}{2}$ is the extension of the constriction portion 1.

With: ΔL_s total elongation at necking.

$\frac{\Delta d}{2}$: is the difference between initial and final diameter of the portion 1.

According to Figure 3 we have,

The point O:

$$\frac{\Delta L}{2} = 0 \text{ et } \frac{\Delta d}{2} = 0$$

With $\frac{\Delta L}{2}$: the total elongation of the portion 1 of the specimen and Δd: Δd the variation of the diameter.

This is the initial state at time t = 0 before tensile test.

At point A: There, there's uniform elongation but there is no necking as $\Delta d = 0$.

Consequently the total elongation of the portion 1 is: $\frac{\Delta L}{2} = \frac{\Delta L_r}{2}$

At point B:

We have uniform elongation and necking because there's $\Delta d \neq 0$

Consequently the total elongation of the portion 1 is:

$$\frac{\Delta L}{2} = \frac{\Delta L_r}{2} + \frac{\Delta L_S}{2}$$

With: $\frac{\Delta L_r}{2}$ et $\frac{\Delta L_S}{2}$ respectively: uniform elongation and necking elongation.

We note already in Figure 2 that: the necking elongation ΔL_S can be calculated as follows:

$$tg\alpha = \frac{\frac{\Delta L_S}{2}}{\frac{\Delta d}{2}} = \frac{\Delta L_S}{\Delta d} \tag{3}$$

The angle β is introduced between the transverse which coincides with the necking diameter and the slant segment passing through the point O and the end of the necking diameter in point C introducing the tangent of the angle β:

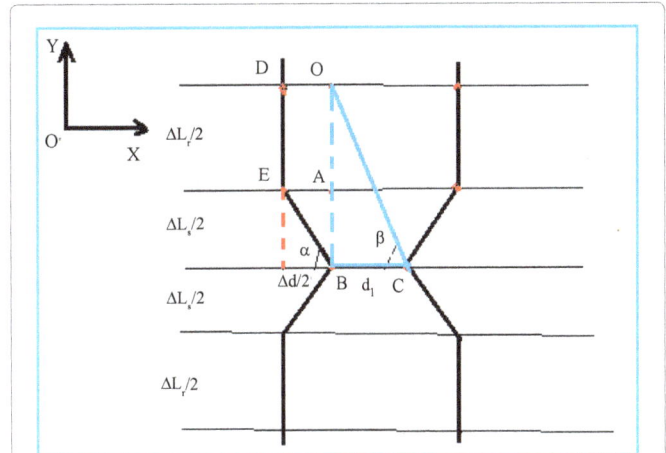

Figure 3: Representation of the D_1 approach of the upper left half specimen (portion 1).

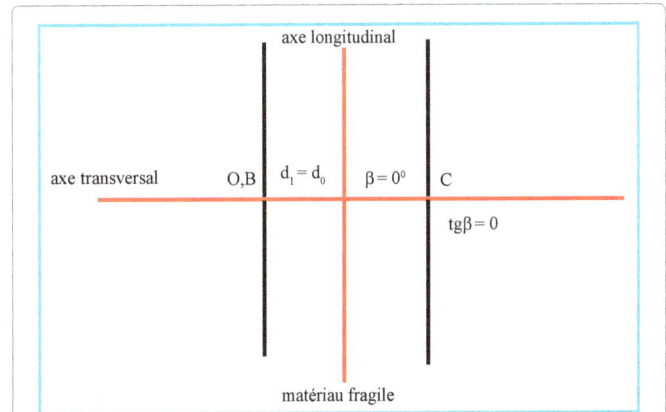

Figure 4: Geometric representation of the D_1 approach on a test specimen of brittle material.

$$tg\beta = \frac{OB}{BC} = \frac{OB}{d_1} = \frac{OA + AB}{d_1} = \frac{\frac{\Delta L_r}{2} + \frac{\Delta L_S}{2}}{d_1} = \frac{\Delta L}{2d_1}$$

So $tg\beta = \frac{\Delta L}{2d_1}$ (4)

Note that the β angle is between 0° and 90°: $0^0 \leq \widehat{OCB} \leq 90^0$

$$tg0^o = \frac{0}{1} = 0 \text{ et } tg90^o = \frac{1}{0} = \infty$$

Indeed 0° corresponds to the initial state before tensile test or the ductility of the material is zero.

At 90° is the condition for which the ductility of the material is infinite.

Interpretation

We note from Figure 4 that:

When $\beta = 0°$ this means that the material is brittle because its ductility expressed $tg\beta$ is zero.

When $\beta = 90°$ (Figure 5), it means that the material is ideally plastic that is to say the material is infinitely superplastic because its ductility expressed by $tg\beta$ approaches infinity.

We see that $tg\beta$ is a credible indicator of ductility because it activates simultaneously the elongation ΔL and the final necking diameter d_1 (Figure 6).

This formulation is interesting because it is based on the values of the elongation ΔL and the final necking diameter d_1 which are essential variables in the determination of the ductility of a material.

Among others we note that conventional formulas of ductility represented by the percent elongation A and the necking percent Z are dependent on these 02 values, in fact elongation percent A is a function of only ΔL and necking Z is a function of the necking section S_1 it means that Z is a function of necking diameter d_1 [48-57].

We shall agree to say, therefore, that our approach of ductility through the tangent of the angle β gives a better description of the state of ductility of material than conventional parameters A and Z.

Checking the formula of ductility approach $tg\beta = \dfrac{\Delta L}{2d_1}$ of portion

When, $tg\beta = 0$ this corresponds to $\beta = 0° \Rightarrow \dfrac{\Delta L}{2d_1} = 0$

We deduce: $\Delta L = 0$ et $d_1 = d_0$.

This is the case of a brittle material whose ductility is zero.

When $tg\beta = \infty$ it corresponds to $\beta = 90° \Rightarrow \dfrac{\Delta L}{2d_1} \to \infty$ we deduce: $\Delta L >> 0$ et $d_1 \to 0$.

This is the case of a ductile material with ideally infinite ductility it

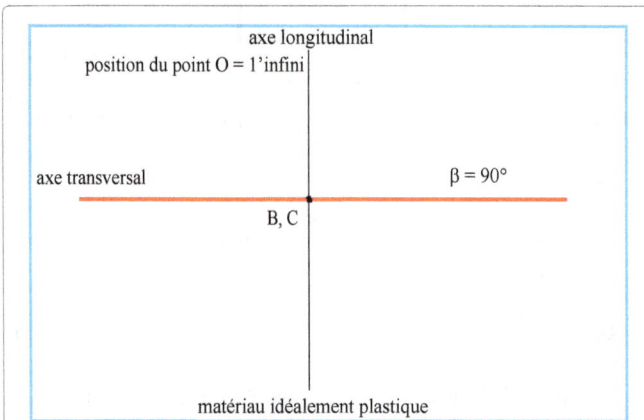

Figure 5: Geometric representation of the D_1 approach on a test specimen of an ideal plastic material.

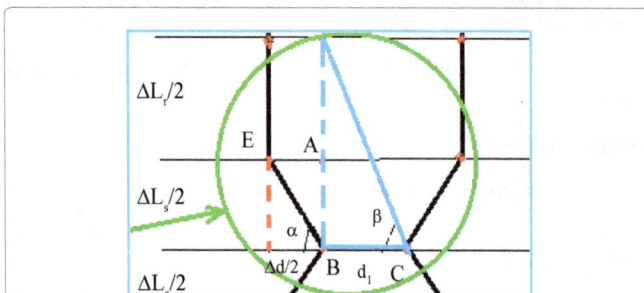

Figure 6: Angle β representation in function of the elongation and diameter necking d_1.

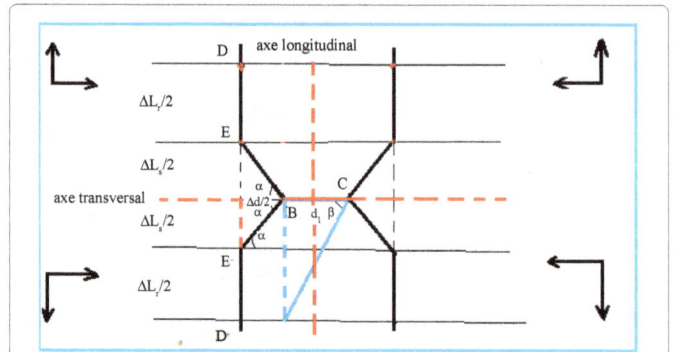

Figure 7: Representation of the D_1 approach of the lower left half specimen.

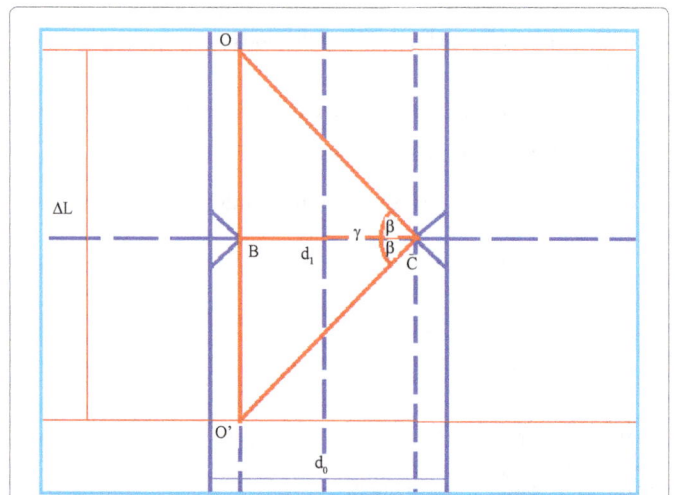

Figure 8: Representation of the D_1 approach for left side specimen.

is concluded from our approach $tg\beta = \dfrac{\Delta L}{2d_1}$ the ductility of superplastic metallic materials and plastic is between 0 and infinity.

We conclude $tg\beta = \dfrac{\Delta L}{2d_1}$ may be representative of the ductility of any material.

To finalize our approach of ductility formula, we notice that $tg\beta$ is representative of the ductility of the upper half of specimen (Figure 3).

Based on the assumptions mentioned above, the test specimen is perfectly symmetrical on both sides of the axes, thus to have the total ductility, it is appropriate to add the ductility of the lower half specimen (Figure 7) or multiply our parameter $tg\beta = \dfrac{\Delta L}{2d_1}$ by 2.

So $D_1 = 2tg\beta = 2\dfrac{\Delta L}{2d_1} = \dfrac{\Delta L}{d_1}$ ⠀⠀⠀⠀⠀⠀(5)

With ΔL: total elongation and d_1: necking diameter.

So we see that the total ductility triangle is isosceles shape, with basic ΔL and total height the extension of necking diameter d_1 (Figure 8).

We also note that the total angle of ductility is the angle $\gamma = 2\beta$.

We note that the ductility parameter approach $D_1 = \dfrac{\Delta L}{d_1}$ is an interesting and promising contribution in the interpretation of ductility. It has been confirmed by audits on 03 types of materials (brittle, plastic and superplastic); in the other hand it is easy to use and involves the elongation ΔL and the necking Z throughout the diameter d_1 under a single formulation.

The ductility approach $D_1 = \dfrac{\Delta L}{d_1}$ effectively solves the problem of the contradiction between A and Z concerning quantification of ductility and that is the problem targeted by the work (Figure 9).

Analysis of approach $D_1 = \dfrac{\Delta L}{d_1}$ for ductile metallic materials

To simplify the geometric representation of the D_1, we preferably used approach for the upper 1/2 specimen (Figure 10) or lower 1/2 specimen; while noting that the geometric shape relative to D_1 approach is an isosceles triangle that is the sum of 02 right triangles (rectangle triangles) perfectly identical and symmetrical and having a common base the diameter of the specimen.

According to Figure 10a shows the evolution of our approach ductility parameter from a simple tensile test on the 1/2 upper specimen.

In Figure 10, D_1 approach parameter before the beginning of the tensile test is shown by the initial state $\Delta L_r = \Delta L_s = 0$; Indeed, the elongation is zero and $\Delta d = 0$ which implies $d_1 = d_0$.

In this case: $D_1 = \dfrac{\Delta L}{d_1} = \dfrac{\Delta L}{d_0} = \dfrac{0}{d_0} = 0$

This is the typical case of brittle materials whose ductility is zero.

Figure 10b represents the beginning of the test; we note that there is a linear deformation along the longitudinal axis therefore representing the homogeneous deformation of the uniform elongation of the specimen.

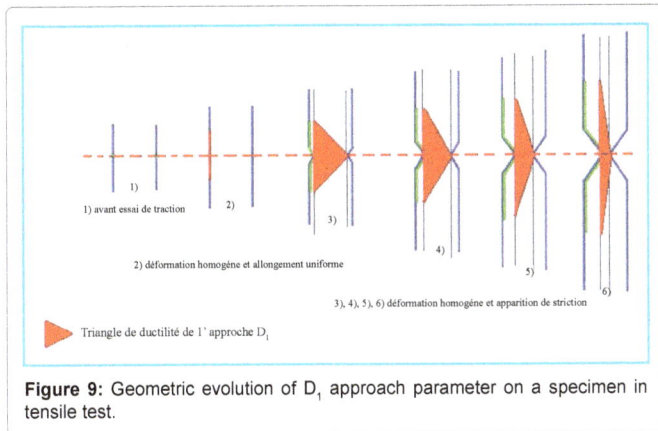

Figure 9: Geometric evolution of D_1 approach parameter on a specimen in tensile test.

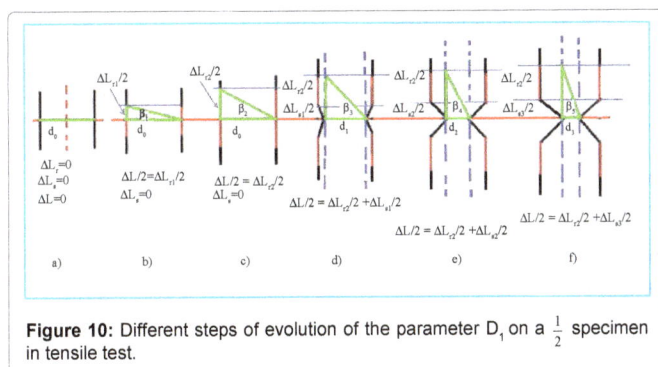

Figure 10: Different steps of evolution of the parameter D_1 on a $\frac{1}{2}$ specimen in tensile test.

The D_1 parameter in this case is based only on the uniform elongation because it is homogeneous deformation and the initial diameter d 0 because there is no necking.

$D_1 = \dfrac{\Delta L_{r1}}{d_0}$

In Figure 10c the homogeneous deformation representing uniform elongation is more pronounced than the previous but still remains in the field of homogeneous deformation, the uniform elongation ΔL_r increased passing ΔL_{r2} with $\Delta L_{r2} > \Delta L_{r1}$ therefore only the uniform elongation occurs in changing the quantization ductility because necking elongation is zero and the diameter $d_1 = d_0$.

What gives us the setting $D_1 = \dfrac{\Delta L_{r2}}{d_0}$

In Figure 10d we notice the onset of necking one enters the area of the heterogeneous deformation indeed it there's diameter reduction and lengthening necking. The total elongation in this case is the sum of 02 elongations: distributed elongation and elongation necking ΔL_{s1} and the diameter d_1 which is less than d_0.

Note that the uniform elongation is constant and equals to ΔL_{r2} because it is no longer homogeneous deformation.

Where: $D_1 = \dfrac{\Delta L_{r2} + \Delta L_{s1}}{d_1}$

In Figure 10e necking is more pronounced, as necking elongation and diameter reduction are enabled, there is a slight increase in the value of the necking elongation and a decrease of the diameter from d_1 to d_2; $d_2 < d_1$.

This increase is induced by the increase in the value of the diameter reduction, that we confirm through the formula that we presented $\Delta L_s = \Delta dx \, tg\alpha$.

Indeed when Δd increases, ΔL_s also increases, which means $\Delta L_{s2} > \Delta L_{s21}$

Therefore the approach of parameter D_1 becomes equal to: $D_1 = \dfrac{\Delta L_{r2} + \Delta L_{s2}}{d_2}$

According to Figure 10f the breaking phase we are witnessing is the end of the necking so necking elongation increases $\Delta L_{s3} > \Delta L_{s2}$ and necking diameter decreases with d_3 less than d_2.

So the approach parameter will be: $D_1 = \dfrac{\Delta L_{r2} + \Delta L_{s3}}{d_3}$

We note that in general the ductility parameter $D_1 = \dfrac{\Delta L}{d_1}$ is growing throughout the tensile test and interprets perfectly plastic behavior through its 02 variables that are elongation ΔL and necking diameter d_1.

Ductility Triangle and Ductility Angle Concepts of Approach D_1

It is a specimen of a material subjected to a simple tensile test. Changes in ductility through the D_1 setting approach has been described above, however we will try to study the evolution of ductility triangle through the various phases of tensile strain. Either the 1/2 upper specimen, the mapping of the ductility is proposed to be made through the triangle by the projection on the longitudinal axis in Figure 11 which gives us the following:

It is seen (FIG II.11) that D_1 approach parameter relating to the 1/2 specimen changes according to a right triangle whose height characterizing the elongation increases at the expense of shrinkage of diameter.

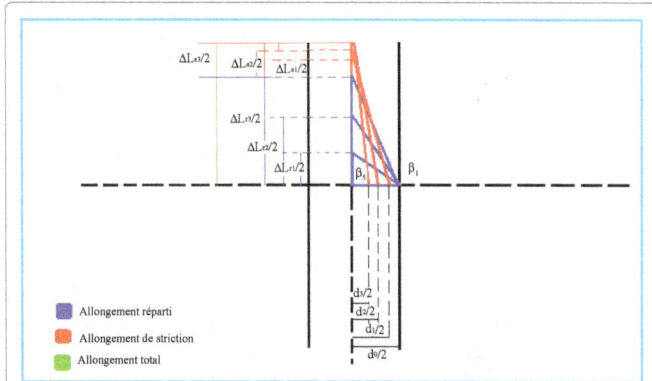

Figure 11: Evolution of shape of ductility triangle and ductility angle of the upper half specimen.

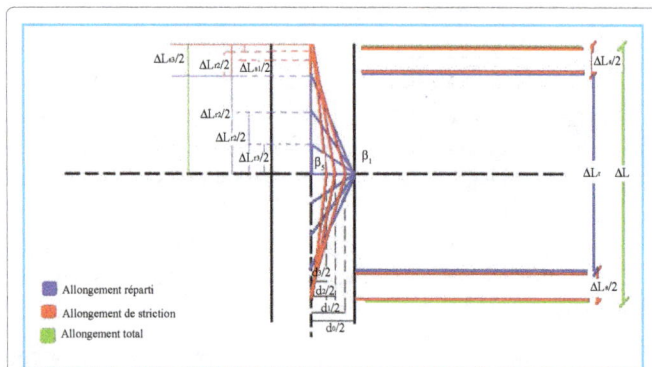

Figure 12: Evolution of shape of the ductility triangle and the ductility angle of the specimen.

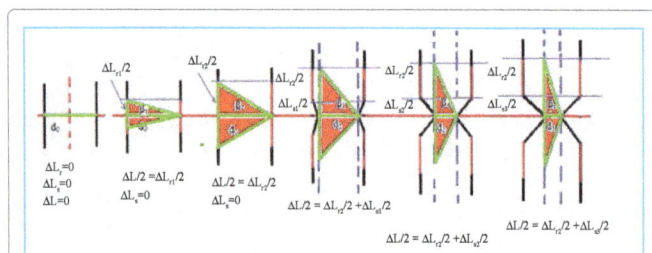

Figure 13: Geometric distribution of the D_1 approach ductility triangle.

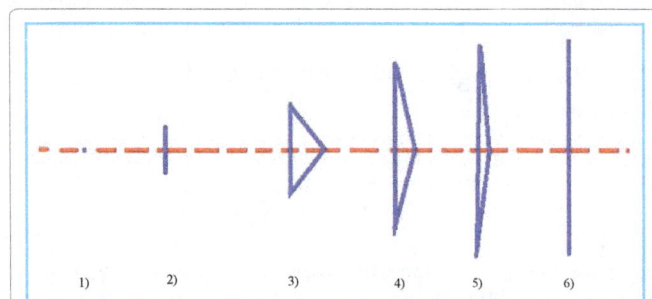

Figure 14: Different stages of evolution of ductility triangle.

This change in shape of the right triangle called ½ ductility triangle is expressed by the variation of angle of inclination β called 1/2 ductility angle.

We also note from the Figure 11, that as β increases, the ductility formulated by D_1 approach also increases and vice versa true.

Indeed it is clear that: $\beta_6 > \beta_5 > \beta_4 > \beta_3 > \beta_2 > \beta_1$.

Therefore it is concluded that the ductility of approach $D_1 = \dfrac{\Delta L}{d_1}$ on the ½ specimen is geometrically represented by a variable right triangular D_1 and this parameter is a function of the ½ β ductility angle whose value is only determining ductility relative to our approach.

Note that the parameter $D_1 = \dfrac{\Delta L}{d_1}$ is finally dimensionless.

For the entire test the geometric representation of the D_1 approach is as follows in Figure 12.

Note on Figure 13, the evolution of the isosceles triangle ductility characterizing D_1 approach; Indeed, starting from t=0 to the beginning of the homogeneous deformation ductility scales linearly along the axis of stress is the uniform elongation and to the onset of necking there is formation of a triangle isosceles which characterizes the heterogeneous deformation according to the approach D_1. This form of this triangle changes gradually as the tensile test is carried out until breaking of the specimen.

Finally we conclude that changes in the geometry of the D_1 approach operate generally as follows in the Figure 14.

Figure 14.1: No deformation, it is the initial state before tensile testing; ductility is confused to a point.

Figure 14.2: Homogeneous deformations, it is the phase of the distributed linear expansion, there is no ductility triangle.

Figures 14.3 - 14.5: Heterogeneous deformation, it is the phase of the emergence and evolution of ductility triangle according D_1 approach.

Figure 14.6: The ductility triangle coincides with the vertical and corresponds to an ideal state of ductility that does not exist in reality. It operates between other than the area of the isosceles triangle is equal to $\dfrac{bh}{2}$ that is the product of base of the triangle and its 1/2 height.

Therefore the area of the triangle ductility characterizing D_1 ductility approach is equal to $\dfrac{\Delta L x d_1}{2}$.

Note that the parameter $D_1 = \dfrac{\Delta L}{d_1}$ is dimensionless.

Results and Discussion

To study and analyze the contradiction between the assessment parameters of ductility A(%) and Z(%) we used in our experiment iron annealed copper annealed. These (02) grades are delivered in rolled state by the precision machining company located in El-Hadjar (Annaba city).

The various test pieces in number (03) for each grade were tested in the tensile test; the different values that we identified (final length and final diameter), are used to calculate average ΔL and necking diameter for proving the contradiction.

For iron annealed, elongation is between 40 and 50, the constriction is 80 to 93, the hardness HRB is from 45 to 55 and its modulus of elasticity (Young) is 206000 MPa.

To develop and analyze the second step of our work, we experiment tensile test on 03 different grades of carbon steel. For each grade, we use 03 specimens. Test grades are XC18 carbon steel, XC38 and XC48. Ductility values of the above-mentioned steels is known because of the carbon content, in other words it is known that XC18 is more ductile than XC38 and XC48 because it contains less carbon, and XC38 is more ductile than XC48. Based on this fact we test the ductility approach and we have to prove this order of ductility values of XC18, XC38 and XC48 (Figure 15).

Experimental study of the ductility annealed iron and annealed copper

The tensile tests were performed on 03 samples of annealed iron

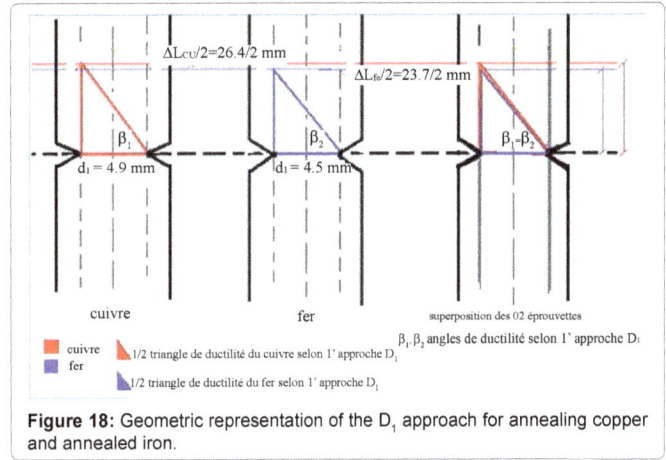

Figure 18: Geometric representation of the D_1 approach for annealing copper and annealed iron.

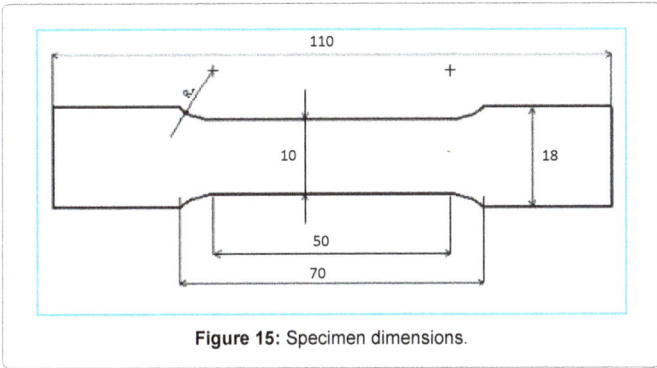

Figure 15: Specimen dimensions.

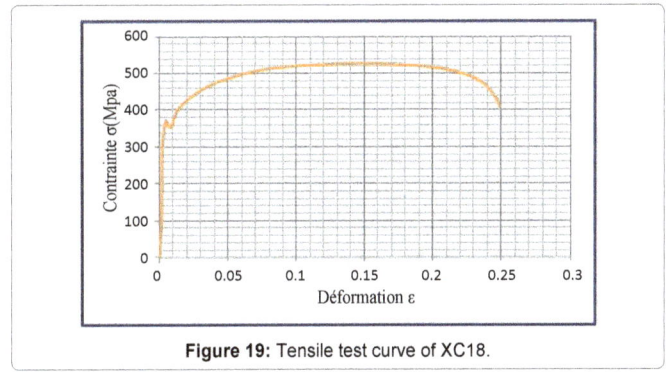

Figure 19: Tensile test curve of XC18.

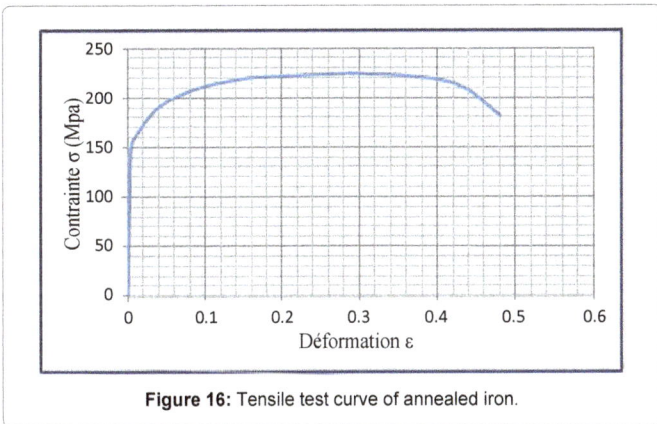

Figure 16: Tensile test curve of annealed iron.

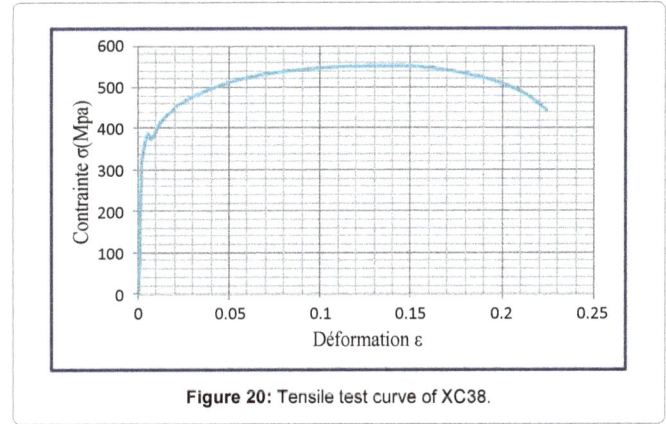

Figure 20: Tensile test curve of XC38.

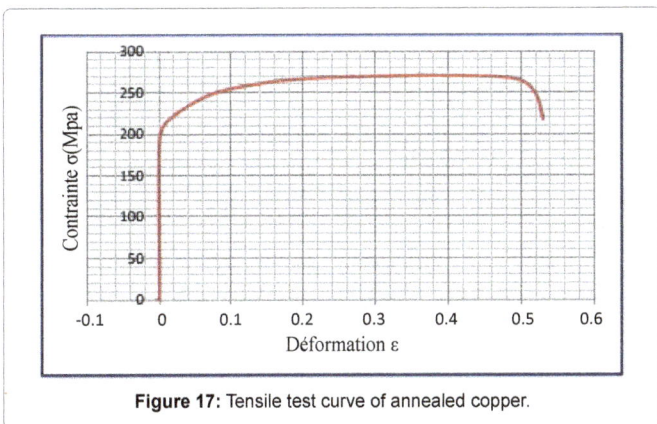

Figure 17: Tensile test curve of annealed copper.

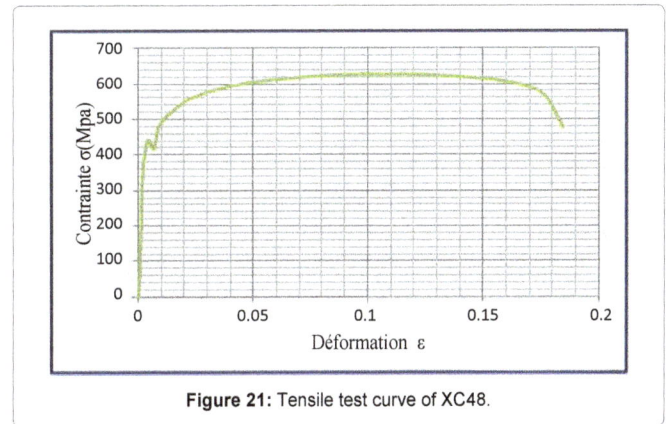

Figure 21: Tensile test curve of XC48.

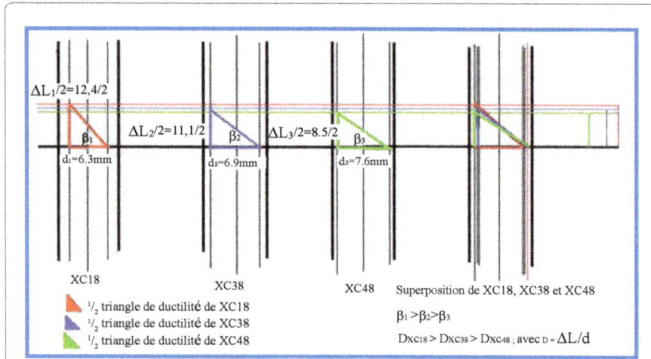

Figure 22: Geometric representation of the D_1 approach for XC18, XC38 and XC48.

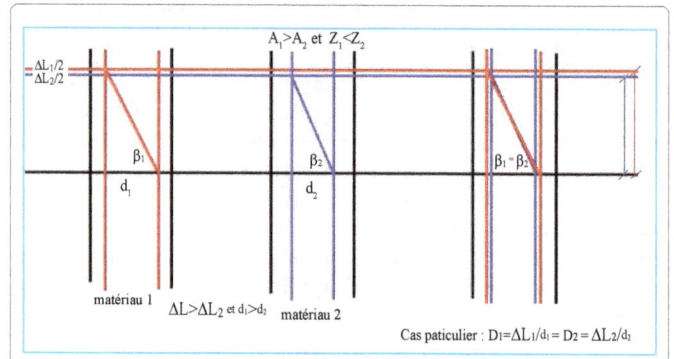

Figure 26: The 3rd possibility is a particular case.

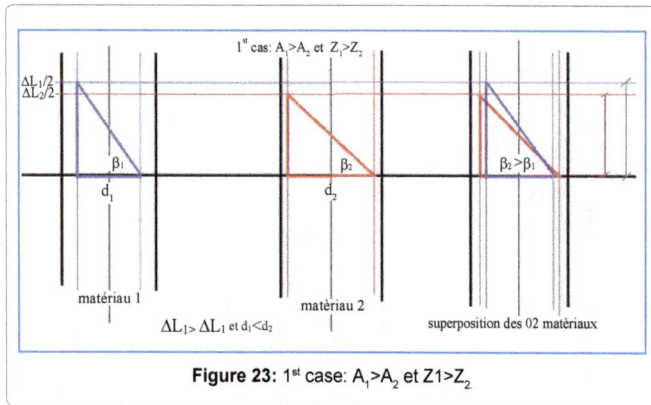

Figure 23: 1st case: $A_1 > A_2$ et $Z1 > Z_2$.

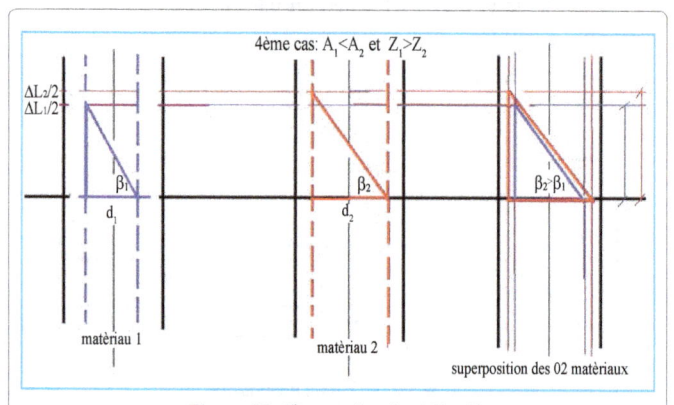

Figure 27: 4th case: $A_1 < A_2$ et $Z_1 > Z_2$.

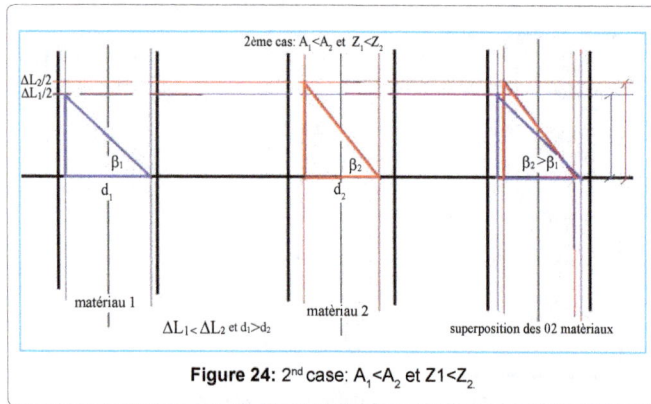

Figure 24: 2nd case: $A_1 < A_2$ et $Z1 < Z_2$.

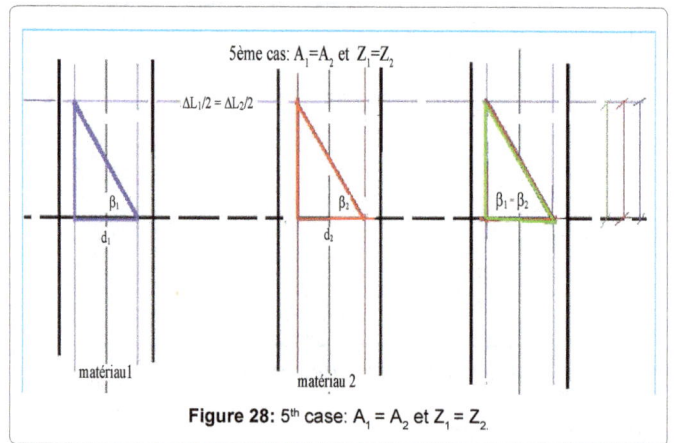

Figure 28: 5th case: $A_1 = A_2$ et $Z_1 = Z_2$.

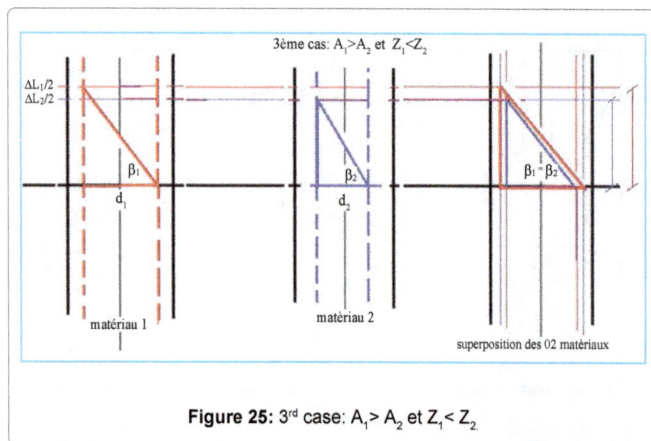

Figure 25: 3rd case: $A_1 > A_2$ et $Z_1 < Z_2$.

Average Ductility $\sum \frac{X_i}{3}$	A%	Z%
Annealed Iron	47.5	79.9
Annealed Copper	52.8	76.2

Légend	Annealed Copper	Annealed Iron
$\Delta L_{moy} = \sum\limits_{i=1}^{i=3} \frac{\Delta L_i}{3}$	26.4 mm	23.7 mm
$d_{1moy} = \sum\limits_{i=1}^{i=3} \frac{d_{1i}}{3}$	4.9 mm	4.5 mm

Table 3: Comparative A, Z annealed iron and annealed copper.

Average Ductility	D1
Annealed Iron	5.3
Annealed Copper	5.4

Table 4: Experimental application of ductility modeling approach of annealed iron and annealed copper.

Légende	$A\% = \dfrac{L_0 - L_1}{L_0}$	$Z\% = \dfrac{\Delta S}{S_0}$	$D_1 = \dfrac{\Delta L}{d_1}$
Unit	%	%	Sans
Specimen 1	25.4	61.5	2
Specimen 2	25	60.3	1.9
Specimen 3	24	59	1.8
Calcul De La Moyenne $\sum \frac{X_i}{3}$	24.8	60.3	1.9

Table 5: Average values of A, Z and D_1 of XC18.

Légende	$A\% = \dfrac{L_0 - L_1}{L_0}$	$Z\% = \dfrac{\Delta S}{S_0}$	$D_1 = \dfrac{\Delta L}{d_1}$
Unité	%	%	Sans
Specimen 1	22.7	57.7	1.7
Specimen 2	22.2	50.9	1.6
Specimen 3	21.9	48.2	1.5
Calcul De La Moyenne $\sum \frac{X_i}{3}$	22.3	52.3	1.6

Table 6: Average values of A, Z and D_1 of XC38.

Légende	$A\% = \dfrac{L_0 - L_1}{L_0}$	$Z\% = \dfrac{\Delta S}{S_0}$	$D_1 = \dfrac{\Delta L}{d_1}$
Unité	%	%	Sans
Specimen 1	19.56	45.2	1.3
Specimen 2	17.24	39.1	0.9
Specimen 3	18.42	42.1	1
Calcul De La Moyenne $\sum \frac{X_i}{3}$	18.4	42.1	1

Table 7: Average values of A, Z and D_1 of XC48.

Paramètres De Ductilité	A%	Z%	D_1
Unité	%	%	Sans
XC18	24.8	60.3	1.9
XC38	22.3	52.3	1.6
XC48	18.4	42.1	1

Table 8: Values of A, Z and D_1 XC18, XC38 and XC48.

and annealed copper; we use the results of 03 curves we got, and then we exploited the Curve Unscan program to have the following average tensile curve for each metal (Figures 16-28).

Experimental study of the ductility of XC18, XC38 and XC48

It is shown in Tables 3-8.

General application of the approach, $D_1 = \dfrac{\Delta L}{d_1}$

In this case, we have 3 possibilities; most intéressant are:

Possibility 1 and 2: $\dfrac{\Delta L_1}{d_1} > \dfrac{\Delta L_2}{d_2}$

The 3rd possibility is a particular case:

With $\Delta L_1 \neq \Delta L_2$ et $\Delta d_1 \neq \Delta d_2$, we obtain:

$$\frac{\Delta L_1}{d_1} = \frac{\Delta L_2}{d_2}$$

Conclusion

Through this approach, we notice that D_1 is a valid and credible general parameter because its formula takes into account the longitudinal deformations using ΔL and the transverse deformations and transverse deformation across the necking diameter d_1. On the other hand the geometric representation of D_1 is interesting because it schematizes the ductility of materials using ductility triangle and ductility angle that we have presented.

References

1. Franz G (2008) Prédiction de la limite de formabilité des aciers multiphasés par une approche micromécanique. Engineering Sciences. Arts et Métiers Paris Tech.

2. Chastel Y (2006) Matériaux pour l'ingénieur.

3. Degallaix S (2006) Caractérisation des matériaux métalliques. 4. Feng ZQ (2011) Mécanique non linéaire MN91.

4. 4.Mathieu T (1750) Dissertation sur la ductilité des métaux, et les moyens de l'augmenter ETH-Bibliothek Zürich.

5. Mompiou F (2014) Des vermicelles dans les métaux. Centre d'élaboration des matériaux et d'études structurales CEMES-CNRS. Toulouse.

6. Poulain H D (2008) MA 41 Essai de traction. UTBM. Université de technologie de Belfort-Montbéliard.

7. ZhuY (2013) Mechanical properties of materials. NC State University. USA.

8. Berthaud Y (2004) Matériaux et proprieties.

9. Rollett AD, De Graef M (2007) Microstructure properties: I materials properties: strength ductility.

10. ADMET (2013) Sheet metal testing.

11. Jacquot B (2009) Propriétés mécaniques des. Biomatériaux utilizes en. Odontologie. Société Francophone de Biomatériaux Dentaires.

12. ASM (2004) American society for materials.

13. Montheillet F, Briottet L (2009) Endommagement et ductilité en mise en forme.

14. Beaucquis S (2012) Propriètés mécaniques des matériaux. Laboratoire SYMME Polytech'. IUT Annecy Département Mesures Physiques, France.

15. Bailon JP (2007) Des Matériaux, presses internationales polytechniques.

16. Chung BJ (2008) Film-wise and drop-wise condensation of steam on short inclined plates. J Mechanical Science and Technology 22: 127-133.

17. Jardin Nicolas H (2012) Essais des matériaux. Résistance des matériaux.

18. Dorlot JM (2000) Des matériaux. Propriétés mécaniques des métaux.

19. Ghomari F (2014) Sciences des matériaux de construction. Faculté des sciences de l'ingénieur. université aboubekr belkaid.

20. Dupeux M (2005) Aide-mémoire science des matériaux.

21. Zaiser M (2006) Scale invariance in plastic flow of cristalline solids. advphysics.

22. Thomas T (2006) Traité Matériaux métalliques.

23. Suquet P (2003) Rupture et plasticité.

24. Dorina N (2009) Matériaux et traitement. OFPPT Maroc.

25. Mandel J (1978) Propriétés mécanique des matériaux editions eyrolles.

26. Norme NF (2009) EN 10002, Essai de traction.

27. François D, Bailon JP (2005) Essais mécaniques des métaux Détermination des lois de comportement.

28. Charmet JC (1990) Mécanique du solide et des matériaux elasticité-plasticité-rupture école supérieure de physique et de chimie industrielles de paris paris tech - laboratoire d'hydrodynamique et mécanique Physique. France.

29. Anduze M (2013) Mécanique des détecteurs Du détecteur à la mesure. Institut National de Physique Nucléaire et de Physique des Particules, LLR - Laboratoire Leprince-Ringuet France.

30. Newey C, Weaver G (1990) Materials principles and practice.

31. Symonds J (1976) Mechanical properties of materials.

32. Charvet R (2009) Comportement à la traction de barres d'armature. Laboratoire de métallurgie mécanique école polytechnique. Fédérale de Lausanne.

33. Ben Tahar M (2005) Contribution a l'étude et la simulation du procède de l'hydroformage Engineering Sciences. Ecole Nationale Supérieure des Mines de Paris.

34. Strnadel B, Brumek J (2013) Effect of tensile test specimen size on ductility of R7T steel faculty of metallurgy and materials engineering Ostrava Tchécquie.

35. Brunet M (2011) Mécanique des matériaux et des structures.

36. Altmeyer G, Abed-Meraim F, Balan T (2013) Vue d'ensemble des relations théoriques entre la striction et les critères souches de localisation. Université de Lyon CNRS Laboratoire de Mécanique des Contacts et des Structures UMR5259 Insa de Lyon

37. Glowacki D, Kozakiewicz K (2013) Numerical simulation of the neck in heterogeneous material. Third international conference on material modelling. Warsaw University of technology, Varsovie Pologne.

38. Bueno R, Sánchez J, Rodríguez T (2009) New parameter for determining plastic fracture deformation of metallic materials. Department of Mechanics and Structures. University of Seville Spain Avenida Reina Mercedes 2 41012 Seville Spain.

39. Hosford WF (2010) Mechanical behaviour of materials. University of Michigan, USA.

40. Yasuyuki A (2010) Procédé de fabrication de fil d'acier japan.

41. Maya M (2008) Plasticité mise en forme. Arts et metiers – Paristech Centre d'enseignement et de recherche de Cluny. France.

42. Broniewski W (1938) Revue de métallurgie. Allongement de striction et travail de rupture la traction.

43. Blétry M (2007) Méthode de caractérisation mécanique des matériaux.

44. International standardization organization (ISO) (2009) EN-ISO-6892-1 October 2009 tensile test: mechanical properties.

45. Degoul PB (2010) Essai de traction. Etude des caractéristiques classiques I 1 Description de l'essai euro-norme 10002-I.

46. Chateigner D (2012) IUT Mesures Physiques Université de Caen Basse-Normandie Ecole nationale supérieure d'ingénieurs de Caen (ENSI CAEN) Laboratoire de cristallographie et sciences des matériaux (CRISMAT) Caen France.

47. Khalfallah A (2009) Fascicule de atelier de mécanique Institut supérieur des sciences appliquées et de technologie de sousse Tunisie.

48. Elias F (2013) Elasticité M2 Fluides Complexes et Milieux Divisés. Université Paris Diderot Paris France.

49. Col A (2011) Emboutissage des tôles, aspect mécanique Techniques de l'ingénieur BM7511.

50. Savoie J, Jonas JJ, Mac Ewen SR, Perrin R (1995) Evolution of r- value during the tensile déformation of aluminium textures and Microstructures 23: 1-23.

51. Abedrabbo N, Pourboghrat F, Crasley J (2006) Forming of aluminium alloys at elevated temperatures Part 1: Material characterization. Int J Plasticity 22: 314-341.

52. Prensier JL (2004) Les critères de plasticité: compléments.

53. Considère A (1885) Annales des Ponts et Chaussées 9: 574.

54. Swift HW (1952) Plastic instability under plane stress. J Mechanics and Physics of Solids 1: 1-18.

55. Hill R (1952) On discontinuous plastic states, with special reference to localized necking in thin sheets. J Mechanics and Physics of Solids 1: 19-30.

56. SAE (1989) Society for Automobile Engineers, USA.

57. Chomel P (2001) Sélection des matériaux métalliques Familles de matériaux.

27

Assessment of the Homogeneous Approach to Predict Unsteady Flow Characteristics of Sheet and Cloud Cavitation

Cote P and Dumas G*

LMFN Lab, Mechanical Engineering Department, Laval University, Quebec, QC G1V 0A6, Canada

Abstract

In this work, the homogeneous approach, frequently used to simulate cavitation in hydraulic machinery, is used to compute unsteady cavitating flows for two simplified geometries. After a quick review of the literature and a rigorous presentation of the proposed methodology, the detailed computed physics of sheet and cloud cavitation are compared with experimental observations and with theory. Results suggest that the assumption of a homogeneous medium is not suitable to predict the fine physics of attached cavitation and thus to predict its precise unsteady characteristics. However, the inhomogeneous approach, in which a momentum equation is solved for both phases under a volume of fluid (VOF) approach, is shown to be more promising. Although it is numerically less stable, such an approach allows the effective body to be modified by the presence of vapor in contrast with the homogeneous approach. The resulting flow topology around the vapor cavity is found to better agree with the experimental observations, and thus the inhomogeneous approach offers the potential to better predict the unsteady characteristics of attached cavitation.

Keywords: CFD; Cavitation modelling; Homogeneous approach; Unsteady flows; Sheet cavitation; Cloud cavitation

Introduction

In an era of emergence of new renewable energy technologies, hydraulic turbines become the corner stone of a complex energy market. As a quick, reliable source of renewable energy, they are operated more frequently in transient and off-design operating conditions to secure the network. As documented in Dörfler et al. [1] and further demonstrated by the works of Yakamoto [2] and Lewys [3], in off-design operating conditions, cavitation may occur and play a leading role in the dynamics of the fluid flow inside the turbine runners.

The state of the art in the simulation of cavitation relies on the assumption of a homogeneous medium which is simply characterized by a mixture volume density. The physics of vaporization and condensation are then governed by different cavitation models. The literature is rich in studies assessing the capacity of numerous cavitation models to predict steady and unsteady characteristics of cavitation with a variable degree of success. For example, the works of Arndt and Song [4], Coutier-Delgosha [5,6], Frikha [7], Ducoin [8] and Zwart [9] and many others have all proposed promising avenues in simulating sheet and cloud cavitation with the homogeneous approach under various assumptions for phase change. However, none of those works has focused on the fine flow physics associated with the homogeneous assumption close to the vapor cavity. As recalled by Brennen [10], this homogeneous assumption is only reasonable if one considers that the dispersed phase is formed of small bubbles, well mixed with the liquid and mainly transported by its convection. However, in the case of attached cavitation where a significant vapor cavity is present along the body, it is not clear if such an assumption is well suited.

The objective of the work presented in this paper is thus to assess the capabilities of the homogeneous approach implemented in a widely used commercial solver to predict the unsteady characteristics of attached cavitation on two experimentally tested setups. The results are compared to those obtained from simulations using the inhomogeneous approach on a case of attached cavitation.

Numerical Methodology

To perform the computations, the commercial solver ANSYS CFX 14.5 [11,12] is used on 1 cell thick pseudo 2-D meshes of the flow fields. In this work, cavitation is modeled through a mass transfer ideology via the Rayleigh-Plesset model included in CFX, whereas the continuity equation states that:

$$\frac{1}{\rho_l}\left(\frac{\partial(\rho_l\alpha_l)}{\partial t} + \nabla.(\alpha_l\rho_l\mathbf{u}_l)\right) + \frac{1}{\rho_v}\left(\frac{\partial(\rho_v\alpha_v)}{\partial t} + \nabla.(\alpha_v\rho_v\mathbf{u}_v)\right) = \Gamma_{lv}\left(\frac{1}{\rho_l}+\frac{1}{\rho_v}\right) \quad (1)$$

where subscripts v and l correspond respectively to the vapor and liquid phases and Γ_{lv} is the mass transfer by unit volume which is being vaporized. It is calculated at each time step through a simplified Rayleigh-Plesset equation $(\Gamma_{lv} = \dot{m}^+ + \dot{m}^-)$ assuming a shared pressure field, adapted both for vaporization $(\dot{m}^+)$ and condensation $(\dot{m}^-)$:

$$\dot{m}^+ = C_{vap}\cdot\frac{3\alpha_{nuc}(1-\alpha_v)\rho_v}{R_{nuc}}\cdot\left(\frac{2}{3}\cdot\frac{p_v\text{-}p}{\rho_l}\right)^{\frac{1}{2}} \quad (p < p_v) \quad (2)$$

$$\dot{m}^- = C_{cond}\cdot\frac{3\alpha_v\rho_v}{R_b}\cdot\left(\frac{2}{3}\cdot\frac{p\text{-}p_v}{\rho_l}\right)^{\frac{1}{2}} \quad (p > p_v) \quad (3)$$

where, C_{vap}, C_{cond}, α_{nuc}, R_{nuc}, R_b are constants that are defined in the solver's documentation [11]. In the inhomogeneous approach, interphase momentum transfer is accounted for via additional terms in the Navier-Stokes equations, shown in Equations 4 and 5:

$$\frac{\partial(\rho_l\alpha_l\mathbf{u}_l)}{\partial t} + \nabla.(\alpha_l(\rho_l\mathbf{u}_l;\mathbf{u}_l)) = -\alpha_l\nabla p + \nabla.\left(\alpha_l\mu_{\text{eff}}(\nabla\mathbf{u}_l + (\nabla\mathbf{u}_l)^T\right) + \mathbf{M}_l - \Gamma_{lv}(\mathbf{u}_l - \mathbf{u}_v) \quad (4)$$

*Corresponding author: Dumas G, LMFN Lab, Mechanical Engineering Department, Laval University, Quebec, QC G1V 0A6, Canada
E-mail: gdumas@gmc.ulaval.ca

$$\frac{\partial(\rho_v\alpha_v\mathbf{u}_v)}{\partial t} + \nabla.(\alpha_v(\rho_v\mathbf{u}_v;\mathbf{u}_v)) = -\alpha_v\nabla p + \nabla.\left(\alpha_v\mu_{eff v}(\nabla\mathbf{u}_v + (\nabla\mathbf{u}_v)^T)\right) \quad (5)$$
$$+ \mathbf{M}_v - \Gamma_{lv}(\mathbf{u}_l - \mathbf{u}_v)$$

where μ_{eff} is the effective viscosity ($\mu_{eff} = \mu + \mu_t$), as expressed by the Boussinesq assumption and modeled in this work via a standard $k - \omega$ SST turbulence model. In Equations (4) and (5), $\mathbf{M}$ is the force acting on a phase due to the presence of the other (i.e. drag; $\mathbf{M}_v = -\mathbf{M}_l$). By using a volume of fluid (VOF) approach to represent transport phenomena at the interface, one can use the interfacial area $A_{lv} = |\nabla\alpha_v|$ to calculate the drag force exerted at the interface:

$$\mathbf{M}_v = C_D\rho_m A_{lv}|\mathbf{u}_l - \mathbf{u}_v|(\mathbf{u}_l - \mathbf{u}_v) \quad (6)$$

where $C_D = 0.44$, which corresponds to the drag coefficient of spherical particles in Newton's regime, independent of the Reynolds number [11].

A homogeneous approach to simulate multiphase flows relies on the assumption of a mixture, simply characterized by a volume density $\rho_m = \alpha_v\rho_v + (1 - \alpha_v)\rho_l$ in which the velocity, turbulence fluctuations and temperature are shared homogeneously. With such assumptions, one can rearrange equations (4) and (5) to obtain the Navier-Stokes equation of the mixture:

$$\frac{\partial(\rho_m\mathbf{u})}{\partial t} + \nabla.(\rho_m\mathbf{u};\mathbf{u}) = \nabla p + \nabla.\left(\mu_{eff}(\nabla\mathbf{u} + (\nabla\mathbf{u})^T)\right) \quad (7)$$

Experimental Cases and their Numerical Representation

Two different 2-D geometries are proposed for comparison with simulations (Figure 1). The first geometry corresponds to a cavitation tunnel studied by Leroux and Astolfi [13-15], in which a c chord NACA hydrofoil is positioned at a 6° angle of attack. The second case consists of an 8° throat venture geometry studied by Barre and Aeschlimann [16,17]. For both cases, space and time variables are hereafter normalized respectively with reference length and convective time-scale as x/L and $t* = t/(L/U_\infty)$.

Boundary conditions and numerical representation

To correctly represent both experimental setups, boundary conditions are set with great care after relevant validations [18]. As shown at the left of Figure 1, 20 chord length extensions are set upstream and downstream of the hydrofoil. The walls of the tunnel are modeled as slip free walls, without considerations for viscous effects. A total pressure condition is set at the inlet along with an averaged absolute pressure outlet, controlling the value computed which is defined as:

$$\sigma = \frac{p_\infty - p_v}{\frac{1}{2}\rho U_\infty^2} \quad (8)$$

where the actual velocity, and thus Reynolds number, are results of the simulation. For the venturi geometry, a velocity inlet is used along with an averaged absolute pressure outlet. The inlet total pressure therefore becomes part of the solution.

In this work, regimes of both sheet and cloud cavitation are simulated on the foil geometry (cases "Sheet-foil" and "Cloud-foil"), while the venturi geometry is simulated only in an unstable manner (case "Cloud-venturi"). For a quick review of the regimes of cavitation, one can refer to the experimental works of Arndt or Leroux [13]. The specific details of the experimental and numerical setups are given below in Table 1. Both numerical setups along with their respective boundary conditions are shown in Figure 1.

For the foil geometry, the proposed numerical mesh is composed of nearly 146 000 elements and allows for a good resolution in the cavity area and in the wake region. For the venturi geometry, the retained

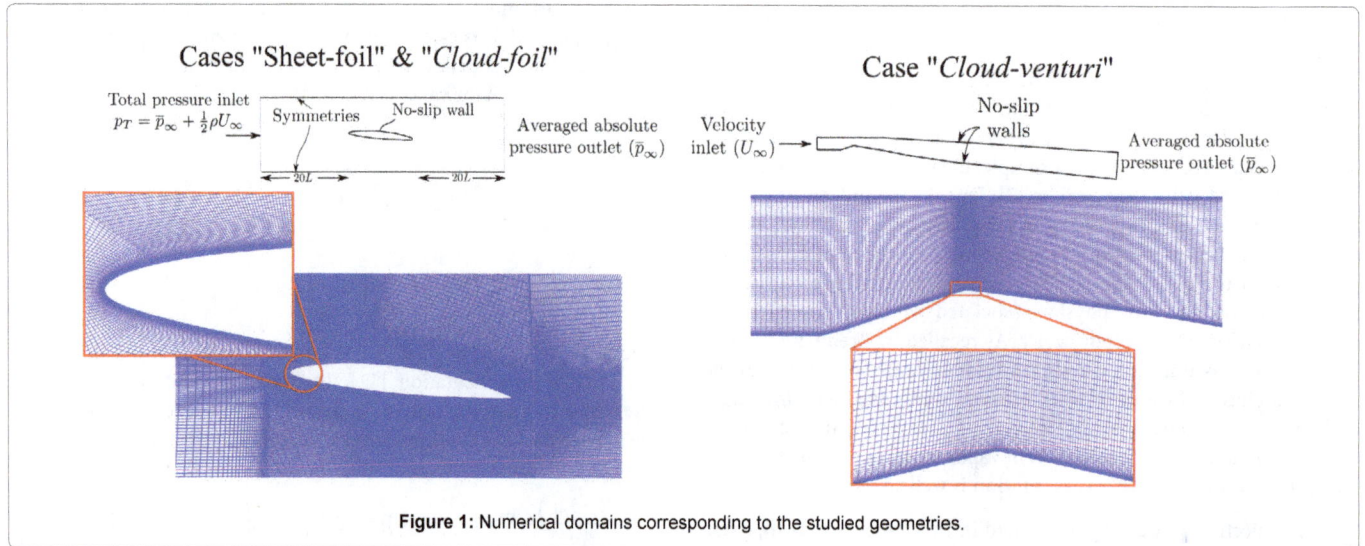

Figure 1: Numerical domains corresponding to the studied geometries.

Parameter	Sheet-foil	Cloud-foil	Cloud-venturi
Inlet velocity U_∞ (m/s)	4.76	4.76	7.04
Exp. cavitation number σ (-)	1.34	1.25	2.15
Exp. cloud shedding frequency (Hz)	n/a	3.5	45
Reference length (chord/venturi throat length) (m)	0.15	0.15	0.224
Reynolds number Re (-)	0.8×10^6	0.8×10^6	1.6×10^6

Table 1: Experimental conditions of the studied cases.

structured mesh contains a total of 50 000 elements and is also properly adapted in the vapor cavity area. To assess the unsteady behavior of the cavitating flow, pressure probes are positioned in both cases as in the experiments.

The unsteady simulations progress in time at a reduced discrete time step of $dt^* = 0.01$. Each unsteady simulation is initialized with the steady state, single phase flow solution in absolute pressure conditions. During the simulation, probes capture an absolute pressure signal that is transformed into a pressure coefficient. The time-averaged and RMS values are then calculated by eliminating the transient time of the cavity formation [18]. For all cases, the proper statistical convergence of unsteady flow characteristics was systematically verified.

The simulations performed under the venturi geometry use a "High resolution" advection scheme while calculations done on the foil geometry use a 2nd order upstream advection scheme. Depending on the ease of convergence, adjustments were done on the residuals convergence criteria. A thorough validation of the methodology proposed here has been carried out and is presented in [18].

Homogeneous Simulations of Sheet and Cloud Cavitation

When simulating sheet cavitation with a homogeneous approach, one rapidly notices that certain aspects of the physics associated to the re-entrant jet underneath the vapor cavity are discarded due to the underlying assumptions of a shared momentum field. In Figure 2 for example, the flow field surrounding the sheet cavitation vapor cavity is shown both with velocity streamlines (left) and with the reduced vorticity (right).

To facilitate interpretation of Figure 2, arrows have been added to point to the areas of interest. Near the leading edge of the foil at the left, the flow separates slightly because of the incidence which leads to the formation of a region of pure vapor inside the separation bubble ($\alpha_v \approx 1$). In the closure region of the cavity, the re-entrant jet is visible as the liquid flowing near the wall moves toward the leading edge of the cavity. At that particular location, an important velocity gradient is created between the low velocity re-entrant jet, near the wall, and the free stream velocity over the cavity.

One can see at the right of Figure 2 that in the homogeneous approach, the boundary layer develops on the hydrofoil wall as it would in a non-cavitating simulation. In the closure region of the cavity,

where an adverse pressure gradient allows the pressure to reach non-cavitating conditions, the shear layer detaches from the foil. Finally, one can note that the region under the separated shear layer contains positive vorticity ($\omega_z > 0$), related to the presence of the re-entrant jet.

One could argue that the physics simulated here is not representative of what would be observed in experiments. The presence of vapor at the leading edge would indeed modify the effective body of the foil. The boundary layer, forming at the leading edge, would then develop on top of the liquid-vapor interface and contribute to the formation of the re-entrant jet by the diffusion of vorticity in the cavity closure region. This would indeed be closer to the experimental observations of Franc and Michel [19,20], Callenaere [21] and Kawanami [22]. However, even if the local flow field surrounding the cavity is not precisely simulated with respect to the existence of a modified effective body, the time-averaged pressure distribution and the RMS distribution of the pressure fluctuations for both sheet and cloud cavitation on the foil are found to match fairly well with experiments, as shown in Figure 3.

It can be observed in Figure 3 that the time-averaged pressure distributions match quite well with the experimental data obtained by Leroux for both cases. As is the case with experiments for sheet cavitation ($\sigma = 1.34$), inside the cavity the pressure is mostly constant and equal to the vapor pressure ($C_p \approx -\sigma$).

When going further toward the trailing edge, an adverse pressure gradient allows the pressure to reach non-cavitating conditions. Again, as was observed experimentally, this recompression is associated with an increase in the pressure fluctuations amplitude, which is greatest in the cavity closure region. Regarding the "Cloud-foil" case, one can notice that even though the shape of the pressure distribution matches fairly well with experiments, it is slightly closer to the non-cavitating pressure distribution than to experiments.

At the bottom of Figure 3, one can see for the "Sheet-foil" case that the amplitude of the fluctuations is over predicted from the cavity closure ($x/L = 0.5$) to the trailing edge of the foil. As mentioned by Leroux, inside the cavity, the pressure fluctuations are mostly constant and equal the fluctuations measured in non-cavitating conditions. These are well predicted by the simulations. However, for the case of cloud cavitation, the pressure fluctuations are greatly over predicted in the vicinity of the leading edge ($x/L < 0.4$). From mid-chord to the trailing edge, the fluctuations distribution is again over predicted but the shape of the latter better matches the experimental data.

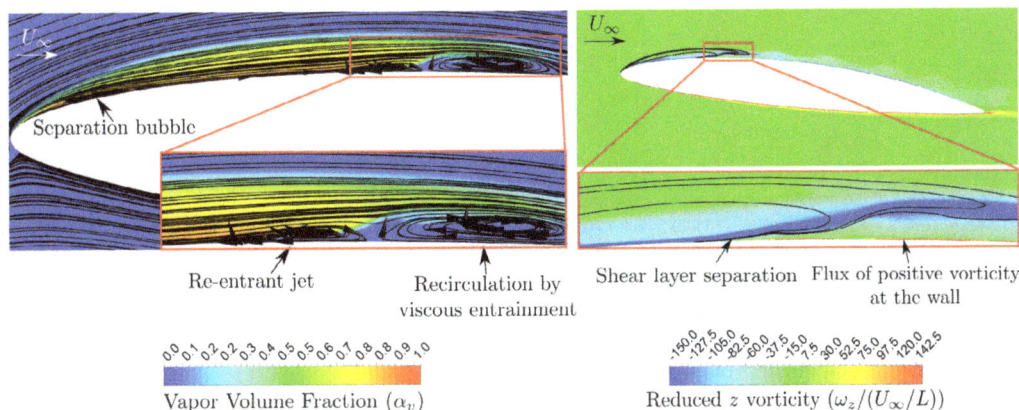

Vapor Volume Fraction (α_v)

Reduced z vorticity ($\omega_z/(U_\infty/L)$)

Figure 2: Flow field surrounding the vapor cavity (shown with the vapour volume fraction) illustrated with velocity streamlines (left) and with the reduced vorticity (right).

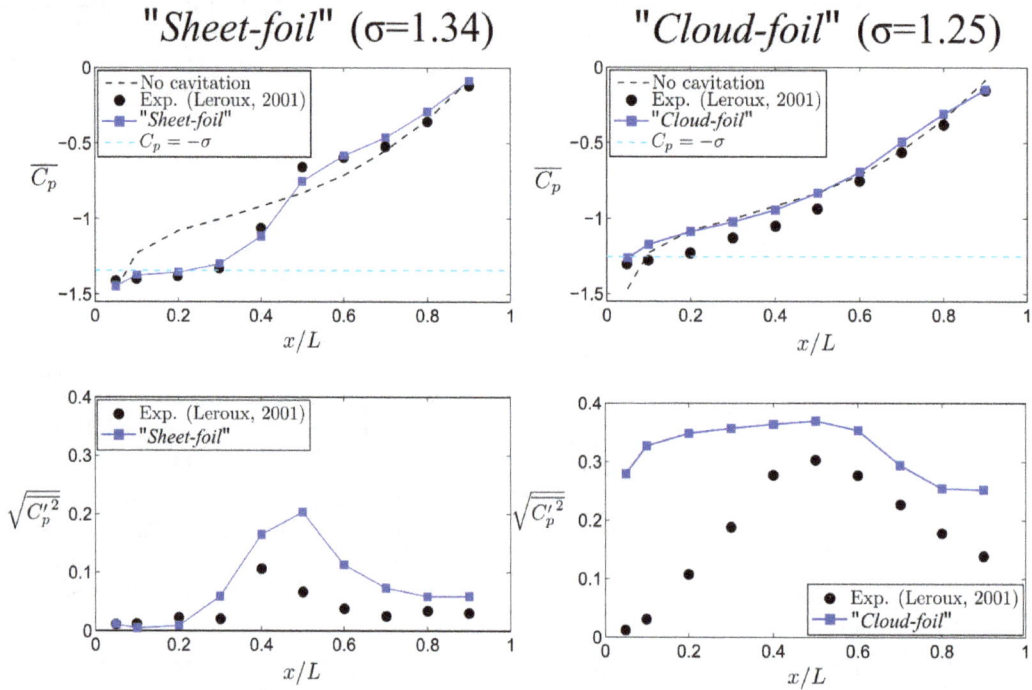

Figure 3: Time-averaged pressure distribution and RMS distribution of the pressure fluctuations on the suction side of the hydrofoil for both sheet (left) and cloud (right) cavitation case.

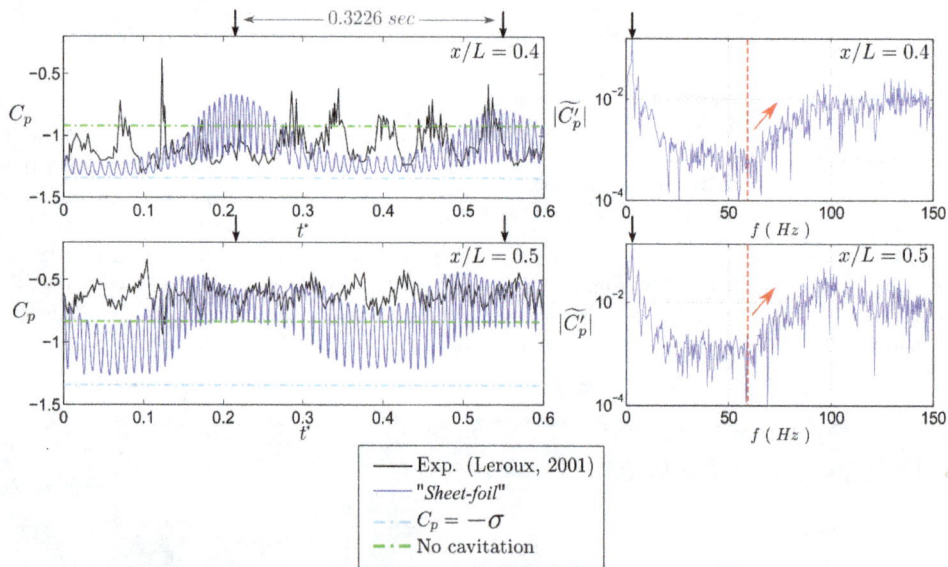

Figure 4: At the left, time evolution of the pressure signals at x/L = 0.4 and x/L = 0.5 for the case of sheet cavitation on the foil. At the right, corresponding frequency domain is analysed. Black arrows show the low frequency fluctuations peaks on the time evolution signal and their corresponding frequency at the right. The red lines and red arrows show the increase in energy caused by the higher frequency fluctuations.

For both cases, experimental pressure signals were measured on the foil, which helps to gain a better understanding of the computed and experimental physics. For the simulated cases, FFT analyses allow the frequency domain content of the time domain signals to be obtained. The resulting signals and frequency contents for the case of sheet cavitation on the foil, at resulting signals and frequency contents

for the case of sheet cavitation on $x/L = 0.4$ and $x/L = 0.5$ in the cavity closure region, are presented in Figure 4.

It appears from Figure 4 that the unsteady flow that is simulated is not in agreement with what is being observed experimentally. The pressure signal obtained in the simulation contains high amplitude, low frequency content that is shown in the left of Figure 4 with black arrows. In the right of Figure 4, the black arrow shows the frequency corresponding to this movement at $f = 3.10$ Hz. We can also observe that the experimental data does not possess this low frequency behavior. In experiments, the signal contains energy at medium frequencies ($f = 18.75$ Hz) plus weaker fluctuations at higher frequencies. As shown in the right of Figure 4, the numerical signal also contains energy at higher frequencies with great amplitudes. This tends to create a camber in the frequency spectrum from around $f = 60$ Hz and above, as shown with the red lines and arrows.

One can also note on the upper left plot of Figure 4 that the pressure coefficient at $x/L = 0.4$ varies between the value without cavitation (shown with the dashed green line) to a value slightly above the vapor pressure ($C_p = -\sigma$, in the dashed blue line). This suggests a movement of the cavity closure caused by the instability of attached cavitation. It also suggests that the simulated cavity possesses two

different pulsating behaviors. First, the cavity shows a large movement of its closure position, generating the fluctuations of lower frequencies. Secondly, it appears that the pressure fluctuations of higher frequencies are not caused by the movement of the cavity closure region itself but rather by the whole flow around the foil. Unsteady visualizations of the numerical simulation help to validate this last point. The same time-frequency signal analysis is proposed below in Figure 5 for the case of cloud cavitation on the foil.

As one can see in the middle of Figure 5 ($x/L = 0.5$), the self-oscillating behavior of the vapor cavity is easily visible as the pressure oscillates from the saturated pressure value ($C_p = -\sigma$, blue dashed line) to the value without cavitation (dashed green line). The phenomenon repeats itself at a frequency of $f = 2.96$ Hz (shown with the black arrow at the right of Figure 5) and is associated with the collapse of the cavity which generates a strong pressure pulsation (pointed with a black arrow, left of Figure 5). In the experiments, this behavior is characterized by

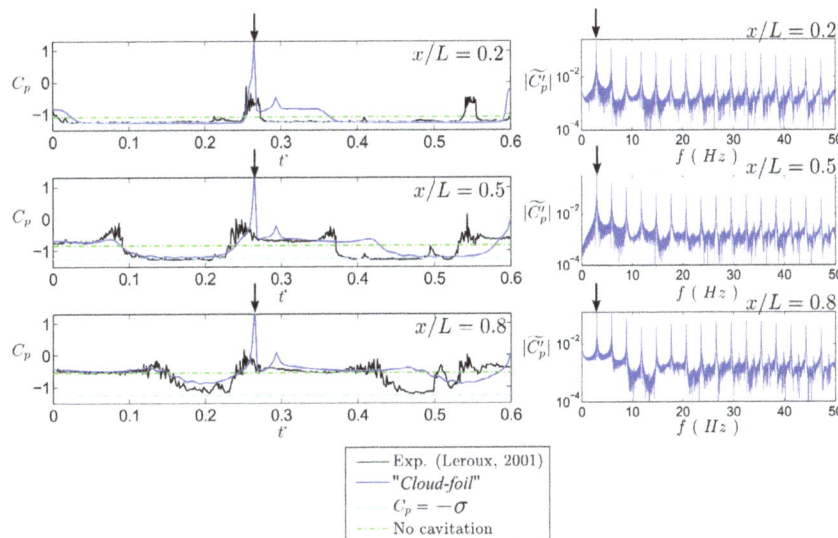

Figure 5: At the left, time evolution of the pressure signals at x/L = 0.2; x/L = 0.5 and x/L = 0.8 for the case of cloud cavitation on the foil. At the right, corresponding frequency domain is analysed. Black arrows show the low frequency fluctuations peaks on the time evolution signals and their corresponding frequency at the right.

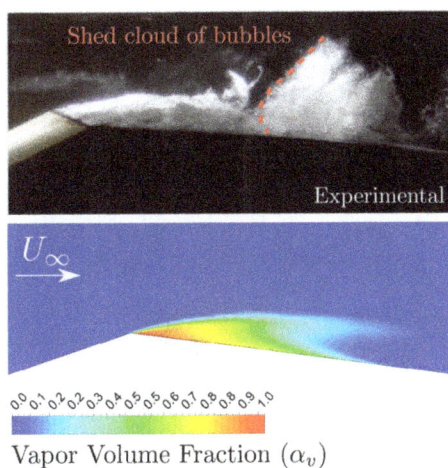

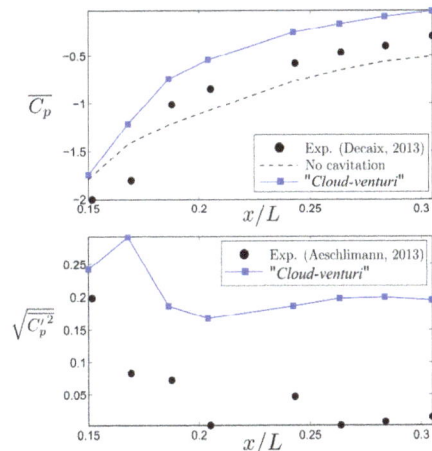

Figure 6: At the left, comparison of the instantaneous vapor cavity between case "Cloud-venturi" and experiments. At the right, time-averaged pressure distribution and RMS distribution of the pressure fluctuations on the lower wall of the diffuser part of the venturi geometry.

a frequency of $f = 3.5$ and leads to a strong fluid-structure interaction phenomenon. The latter can be explained by the important quantity of energy that is contained in the pressure fluctuations and their harmonics, as shown at the right of Figure 5.

Regarding the computations done on the venturi geometry, the flow field surrounding the instantaneous vapor cavity of case "Cloud-venturi" is shown in Figure 6.

One can see at the left of Figure 6 that the location, shape, and amount of vapor appears to be in relative agreement with the experimental observation, if one excludes the cloud of bubbles that was just shed by the main cavity. However, one can also see at the right of Figure 6 that calculations are not as successful in predicting the time-averaged and unsteady characteristics of the cloud cavitation regime in the venturi geometry as it is in the case of the foil geometry.

The unsteady shedding behavior of cloud cavitation was reproduced in the simulations at a frequency of 19.2 Hz, well below the measurements (45 Hz). The phenomenon rapidly appeared very energetic, which caused numerical instability problems. As was the case with "Cloud-foil", but now with a greater importance, the pressure pulsation that periodically appears (as pointed with a black arrow, left of Figure 5), leads to the over prediction of both the time-averaged and the RMS of the pressure shown in Figure 6 (as it was the case for "Cloud-foil", bottom right of Figure 3). A certain amount of effort in assessing the best practices from the literature to produce more accurate results was done, as proposed by Zwart [9] and Coutier-Delgosha [5,6,15], but without any success [18].

For all cases studied in this work, despite being able to predict the shape and location of vapor and the physical mechanisms in cause, the methodology proposed does not allow the accurate prediction of the unsteady flow characteristics caused by attached cavitation. The model's inability might cause this to reproduce the underlying physics of the re-entrant jet, as it was discussed in relation to Figure 2. The next sections now present simulations performed with the inhomogeneous approach on a case of sheet cavitation to identify what physics the consideration of an interface may induce.

Comparisons with the Inhomogeneous Approach

By using the inhomogeneous model included in ANSYS CFX (described in equations 4 and 5), a higher σ case ($\sigma = 1.72$) of sheet cavitation is simulated on the foil geometry. Numerical stability rapidly became problematic with the inhomogeneous approach. It was indeed not possible to simulate lower σ cases because of numerical divergence. For the two new cases of the flow around the foil, the time step is set to $dt* = 0.05$ and a lighter mesh of 50 000 elements is used, both to improve numerical convergence. For both cases, the statistical convergence of unsteady flow characteristics was validated. The comparison of both homogeneous ("HOM") and inhomogeneous ("INH") cases of the foil at $\sigma = 1.72$ are presented below in Figure 7 with the reduced vorticity and vapor volume fraction contours. It appears that with the homogeneous approach, only a small region of the leading edge is filled with pure vapor. This is in contrast with the inhomogeneous approach in which a larger amount of vapor is found at the leading edge of the foil, which also better fits the qualitative experimental observations

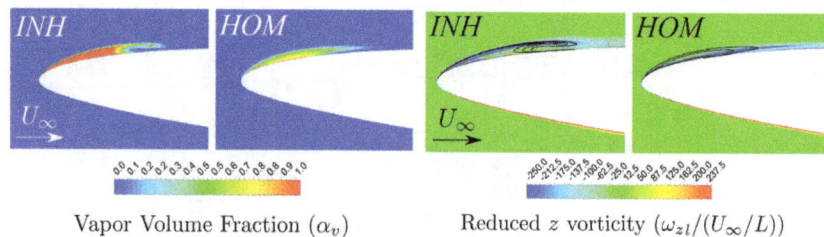

Figure 7: Flow field computed with both the inhomogeneous ("INH") and homogeneous ("HOM") approaches on the foil geometry at = 1.72. At the left is the instantaneous vapour volume fraction and at the right, the reduced vorticity.

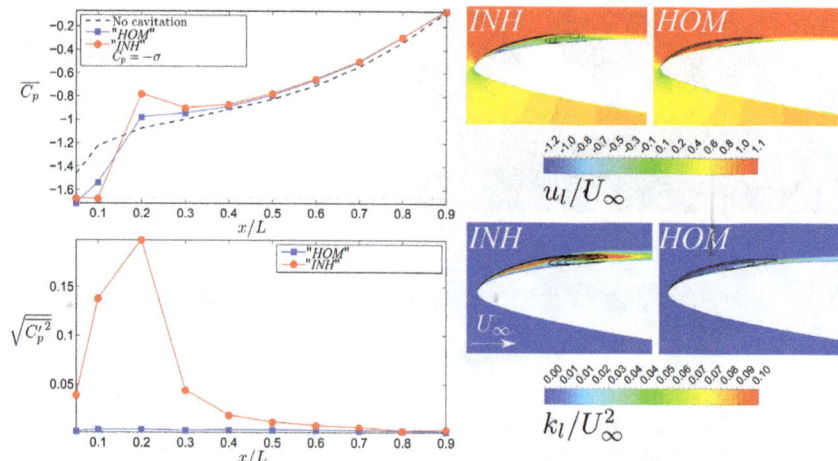

Figure 8: At the left, time-averaged pressure and RMS of the pressure fluctuations distributions for both homogeneous ("HOM") and inhomogeneous ("INH") cases. At the right, flow field surrounding the vapor cavity, shown by the liquid velocity and turbulent kinetic energy.

of Leroux [13]. For reminder, one of the highlighted problematics of the homogeneous approach, as identified in this work, is the lack of modification to the effective body by the presence of vapor. As observed at the right of Figure 7, the vorticity contours clearly show that with the inhomogeneous approach, the presence of vapor modifies the effective body encountered by the liquid flow, as the boundary layer develops at the water-vapor interface. As shown below in Figure 8, even though the resulting time-averaged pressure distributions are quite similar, the resulting different flow topology around the cavity leads to a significantly different RMS distribution of the pressure fluctuations.

One can see in the left of Figure 8 that both cases "*HOM*" and "*INH*" give slightly different time-averaged pressure distributions inside the vapor cavity. Resulting from this, the adverse pressure gradient, which allows the flow to return to non-cavitating conditions, is stronger when using the inhomogeneous approach. One can finally observe that the flow simulated with the homogeneous approach is very calm and generates almost no pressure fluctuations while considerable fluctuations are induced in the closure region with the inhomogeneous approach, possibly caused by the greater adverse pressure gradient.

It can be seen at the top right of Figure 8 that inside and downstream of the vapor cavity, the liquid velocity is greatly reduced as the effective body of the foil is modified. As shown in the bottom right of Figure 8, the resultant effective body leads to the detachment of the boundary layer, which rolls up and diffuses over the vapor cavity, as was experimentally observed by Gopalan and Katz [23]. On the other side, it appears that the small and late separation of the shear layer, downstream of the cavity in case "*HOM*", only leads to a slight increase in the turbulent energy level.

While no experimental data are available for validation at σ= 1.72, the present investigation is quite revealing when considering the simulated physics. It was demonstrated that the modification of the body is only effective when using the inhomogeneous approach as per Figure 7, on the vorticity plot, showing the shear layer developing on the interface rather than at the wall. In the homogeneous approach, the two-phase mixture is indeed considered as a continuous medium. Although no detailed time evolutions of the pressure signal were available, the vorticity injected by the detached shear layer is expected to induce a richer dynamic behavior in the closure region of the cavity, which is suspected to play a considerable role in the spectral response of unsteady cavitation.

This also suggests that locally in the cavity area, the hypothesis of a continuous medium may not be appropriate. In fact, close to the foil, the presence of vapor tends to modify the effective body encountered by the liquid flow. The unsteady physics resulting from this phenomenon may therefore only be characterized with an inhomogeneous approach as they are partly associated with the vortical motion caused by the resulting modified body. Thus, it appears that the consideration of an interface might be critical in assessing the fine characteristics of unsteady cavitating flows.

Conclusion

Through this work, the capacity of the homogeneous approach to solve cavitating flows and predict the resulting unsteady flow characteristics, when used with the Rayleigh-Plesset cavitation model, has been assessed. By performing simulations on two relevant geometries, it was shown that the location, the shape and amount of vapor was always qualitatively close to experimental observations. For the foil geometry, the proposed methodology resulted in accurate time-averaged results for both sheet and cloud cavitation regimes.

However, for the venturi geometry, both time-averaged and unsteady pressure characteristics were amplified by the numerical pressure pulsations associated with the collapse. It was further observed that the homogeneous approach greatly simplifies the flow field surrounding the cavity, whereas the body encountered by the flow is not modified by the appearance of vapor at the leading edge. It is suggested that the vortical motions associated with the modified body would contribute to the proper unsteady response of cavitation.

By using a more complex numerical implementation, solving the flow as an inhomogeneous medium, it was found that with such a model, both phases share an interface. Therefore, the effective body of the foil becomes altered by the presence of vapor, which leads to a more complex but more physically relevant flow field around the vapor cavity. Numerical instability issues however limited the possibility to compare with experimental data and thus to further study the computed physics. More algorithmic developments in the simulation of inhomogeneous cavitation (with the free surface or VOF approach) should aim at rendering the method more numerically robust. This would allow comparisons of simulations of attached cavitation with detailed experimental data, and thus, the capabilities of the inhomogeneous approach could be clearly evaluated. However, from the physics at play, it clearly appears that inhomogeneous simulations of cavitation could lead to better predictions of unsteady characteristics of cavitation. However, for applications into hydraulic machinery, it appears not possible now to accurately predict unsteady flow characteristics caused by cavitation in a robust and general manner.

Acknowledgments

The authors would like to thank the FRQNT and NSERC for their financial support of this research project. Computations were performed on the Colosse supercomputer at Laval University, under the auspices of Calcul Québec and Compute Canada.

References

1. Dörfler P, Sick M, Coutu A (2013) Flow-induced pulsation and vibration in hydroelectric machinery. Springer.

2. Yamamoto K, Müller A, Favrel A, Landry C, Avellan F (2014) Pressure measurements and high speed visualizations of the cavitation phenomena at deep part load condition in a Francis turbine. In 27th IAHR Symposium on Hydraulic Machinery and Systems, Number 22 in IOP Conference Series: Earth and Environmental Science.

3. Lowys PY, Paquet F, Couston M, Farhat M, Natal S, et al. (2002) Onboard measurements of pressure and strain fluctuations in a model of low head francis turbine - part 2: measurements and preliminary analysis results. In Proceedings of the Hydraulic Machinery and Systems 21st IAHR Symposium: 873-880.

4. Arndt REA, Song CCS, Kjeldsen M, Keller A (2000) Instability of partial cavitation: A numerical/experimental approach. In The Proceedings of the Twenty-Third Symposium on Naval Hydrodynamics.

5. Coutier-Delgosha O, Stutz B, Vabre A, Legoupil S (2007) Analysis of cavitating flow structure by experimental and numerical investigations. J Fluid Mechanics 578: 171-222.

6. Coutier-Delgosha O, Fortes-Patella R, Reboud JL (2003) Evaluation of the turbulence model influence on the numerical simulations of unsteady cavitation. J. Fluids Eng 125: 38-45.

7. Frikha S, Coutier-Delgosha O, Astolfi JA (2008) Influence of the cavitation model on the simulation of cloud cavitation on 2d foil section. Int J Rotating Machinery 2008: 12.

8. Huang DB, Young YL (2012) Numerical modeling of unsteady cavitating flows around a stationary hydrofoil. Int J of rotating machinery 2012: 17.

9. Zwart P, Gerber AG, Belamri T (2004) A two-phase flow model for predicting cavitation dynamics. In ICMF 2004 International Conference on Multiphase Flow.

10. Brennen C (2005) Fundamentals of multiphase flows. Cambridge University Press.

11. ANSYS (2012) ANSYS CFX-Solver Modelling Guide. ANSYS.

12. ANSYS (2012) ANSYS CFX-Solver theory guide. ANSYS Inc release.

13. Leroux JB, Astolfi JA, Billard JY (2001) An experimental investigation of partial cavitation on a two-dimensional hydrofoil. In the Proceedings of the Fourth International Symposium on Cavitation.

14. Leroux JB, Coutier-Delgosha O, Astolfi JA (2005) A joint experimental and numerical study of mechanisms associated to instability of partial cavitation on two-dimensional hydrofoil. Physics of fluids 17: 052101.

15. Coutier-Delgosha O, Fortes-Patella R, Reboud JL, Stutz B (2002) Test case number 30: Unsteady cavitation in a venturi type section (pn).

16. Barre S, Rolland J, Boitel G, Goncalves E, Fortes Patella R (2009) Experiments and modeling of cavitating flows in venturi: Attached sheet cavitation. European J Mechanics B/Fluids 28: 444-464.

17. Aeschlimann V, Barre S, Djeridi H (2013) Unsteady cavitation analysis using phase averaging and conditional approaches in a 2D venturi flow. Open J Fluid Dynamics 3: 171-183.

18. Côté P (2015) Investigation of the self-excited vibrations in a Francis runner in transient conditions of load rejection. Master's thesis, Département de génie mécanique, Université Laval.

19. Franc JP, Michel JM (1985) Attached cavitation and the boundary layer: Experimental investigation and numerical treatment. J Fluid Mechanics 154: 63-90.

20. Franc JP (2001) Partial cavity instabilities and re-entrant jet. In The Proceedings of the Fourth International Symposium on Cavitation

21. Callenaere M, Franc JP, Michel JM (1998) Influence of cavity thickness and pressure gradients on the unsteady behavior of partial cavities. In The Proceedings of the Third International Symposium on Cavitation.

22. Kawanami Y, Kato H, Yamaguchi H, Tanimura M, Tagaya Y (1997) Mechanism and control of cloud cavitation. J Fluids Eng 119: 788-794.

23. Gopalan S, Katz J (2000) Flow structure and modeling issues in the closure region of attached cavitation 12: 895.

28

Miniaturized Gas Sensing Assembly for Automobile Exhaust

Kakade IN*, Sanghavi L, Sharma P and Mishra V

Centre for Nano-Science and Engineering, Indian Institute of Science, Bangalore, Karnataka, India

Abstract

Vehicle emissions are composed of a plethora of toxic, non-toxic and greenhouse gases. These consist of Nitrogen Oxides (NOx), Sulphur Oxides (SOx), Carbon Monoxide (CO) and Carbon Dioxide (CO_2) and other Hydrocarbons (HCs). The detection and analysis of these gases is vital in the fight against climate change. With the advent of MEMS technology, solid state sensors have become more and more common in sensor modules used to detect various gases. These sensors used to detect the aforementioned gases are highly delicate and must be placed in protected, safe and suitable environments to perform accurately. This mechanical investigation has lead into the designing and analysis of a miniaturised chamber that is mounted in proximity to the exhaust gases emanating from the tail pipes of automobiles. The chamber protects the sensors from high temperatures, humidity, prevents back pressure or back flow and allows the sensors a sampling time, all while being economical, compact and easy to use. Data from the sensors may be used to actively monitor emissions from the internal combustion engines and hence allow authorities to take off the roads, vehicles that do not meet the emission standards if they exceed them or send the vehicle for servicing. This would mean an overall improvement in the environmental quality of automobiles on the roads.

Keywords: Automobile exhaust; Exhaust gases; Sampling chamber; Solid state sensor

Introduction

The detection and analysing the level of various gases like Nitrogen Oxides (NOx), Sulphur Oxides (SOx), Carbon Monoxide (CO), Carbon Dioxide (CO_2) and other Hydrocarbons (HCs) in vehicular emissions [1] from automobile exhaust has been vital to compete with climatic change. Solid state sensors have become more common in detecting these toxic, non-toxic and greenhouse gases [2,3]. However, these sensors are highly delicate and must be placed in a protected, safe and suitable environment while performing accurately and being compact. For this purpose, a miniaturized chamber was designed. The various aspects considered in the design of the chamber were sampling of the exhaust gas, efficient removal of this sampled gas and intake of fresh air while removing the measure and gas to bring the sensors back to the normal state if solid state sensors with high response time are used. Since the sensors operate at lower temperatures, the hot exhaust gases are cooled in the miniaturized sensing chamber to avoid erroneous readings. To check humidity a suitable filter was provided. The chamber was designed such that there was no backflow or backpressure [4]. The dimensions were such that it occupied less space and could be easily mounted in proximity to the exhaust gases emanating from the tail pipes of automobiles while inducing negligible drag. The sensing chamber provides these advantages all while being economical and safe. The data from this assembly can be actively used to monitor emissions from the internal combustion engine and allow authorities to keep a check on the quality of automobiles (Table 1).

Selection of Material

The miniature sensor chamber was made up of Alumina as it can be easily fabricated and has a comparatively low density. Alumina is chemically inert to most of the chemicals as shown in Table 2. Material used for no return valve was Brass. Since Brass and Alumina have high operating temperatures, it does not get damaged in normal use. Properties of brass are shown in Table 3.

Mechanical Design

The various aspects that affected the design of the chamber are minimum backflow, cooling capacity, heat transfer rate, gas sampling

mechanism, drawing in of fresh air during exhaust post sampling. The design of the interior of the sensing chamber was made so as to maintain zero or negligible backflow or back pressure of inlet gas to maintain laminar flow to the furthest extent possible. The sensing chamber is 40 mm × 90 mm and walls made up of Aluminium are of thickness 3-4 mm that conducts the heat to the Peltier plate which acts as a heat sink by maintaining one of its sides cooler than the other by applying a DC voltage across the plate (Peltier effect) [5-7]. The cooler side is placed in perfect contact with the sensing chamber and the other side is exposed to ambient conditions through fins and a fan. This

Property	Value	Unit
Density	3.98	g/cm³
Bulk Modulus	324	GPa
Compressive Strength	5500	MPa
Elastic Limit	665	MPa
Endurance Limit	488	MPa
Fracture Toughness	5	MPa.m$^{1/2}$
Hardness	22050	MPa
Modulus of Rupture	800	MPa
Poisson's Ratio	0.33	-
Shear Modulus	165	GPa
Tensile Strength	665	MPa
Young's Modulus	413	GPa
Maximum Service Temperature	2114	K
Thermal Expansion	10.9×10^{-6}	K

Table 1: Mechanical and thermal properties of alumina ceramic oxide [5].

*Corresponding author: Kakade IN, Centre for Nano-Science and Engineering, Indian Institute of Science, Bangalore, Karnataka, India
E-mail: kakadeishaan@gmail.com

Environmental Properties	
Resistance Factors: 1=Poor 5=Excellent	
Flammability	5
Fresh Water	5
Organic Solvents	5
Oxidation	5
Sea Water	5
Strong Acids	5
Strong Alkalis	5
Ultraviolet Rays (UV)	5
Wear	5
Weak Acid	5
Weak Alkali	5

Table 2: Chemical properties of alumina [5].

Property	Value	Unit
Thermal Conductivity	121	W/m.K
Specific Heat	0.402	KJ/kg.K
Melting Point	940	°C
Thermal Expansion	20.2×10^{-6}	°C

Table 3: Thermal properties of brass [6].

arrangement cools the incoming gas and avoids erroneous readings or damage of sensor or PCB. The Peltier cooling region of the chamber was designed such that it decreases in area thus increasing the velocity which increases the heat transfer coefficient [8] of the inner wall of the chamber. A 2-way 2-position (2/2) valve is placed at the inlet and the exhaust. The inlet valve should have a higher heat resistant rating to withstand high exhaust temperature. A brass non-return valve (NRV) is placed in the fresh air inlet pipe and the gas inlet pipe. The NRV stops the charge to flow out through it and only allows fresh charge to enter when needed. The sensing chamber is connected to a frame to take all the vibrations and provide dynamic stability. Before entering the chamber, the gas passes through a thermoelectric generator. The gas is filtered out before entering the sensing unit. The filter membrane is made up of materials like Polytetrafluoroethylene (PTFE) and the filter substrate is made up of materials like Polyethylene Terephthalate (PET). This filter can be placed either in the chamber or on the sensor packaging material according to convenience. It is preferred to place the filter on the sensing cap for ease of design. The exhaust suction can be provided by 3 processes: a vacuum servo assembly, a vacuum pump or a diaphragm pump (Figures 1 and 2).

Operation of the Sensing Chamber

The schematic diagram of the sensing unit assembly and the Peltier cooler cooling the gas and the analysis is shown in Figures 3 and 4. A part of the exhaust gas is taken into the sensing chamber through the inlet pipe. This exhaust gas is considered to be uniform due to high turbulence [9]. This gas passes through a thermoelectric generator and converts heat of the exhaust gas to electricity which charges a battery thus reducing the power required to supply across the Peltier plates. The measure and gas passes through the 2-way valve and the filter and enters the sensing chamber. The Peltier cooler cools the gas and the analysis is shown in Figure 3. The backpressure at the fresh air non-return valve increases and fresh air is prevented from entering in the sensing chamber and the gas is sampled when the inlet valve is closed. The sampling time depends on the solid state sensor's response. When the output of the sensor is obtained, the exhaust valve opens and the gas is removed in a way depending on the exhaust mechanism used. If a vacuum pump (already present in the automobile) or a diaphragm

pump [10] is used, the sampled gas is removed by suction and during this the backpressure at the inlet pipe non return valve reduces and the valve opens and fresh air is drawn in due to suction. If a vacuum servo mechanism is used as shown in Fig. the fresh air is pumped in which drives the sampled gas out. In the vacuum servo mechanism, similar to the vehicle braking system [11], when the valve is pushed in the upward position, one side of the piston is exposed to atmosphere and the other side is exposed to vacuum from the inlet manifold [12] through a vacuum reservoir and non-return valve mechanism and the piston moves due to pressure difference thus providing the driving force to pump fresh air into the sensing unit and bring the sensors back to the initial normal condition. If the solid state sensors have a quick response time then a continuous flow of exhaust gas can be maintained and the 2-way 2-position valves, fresh air inlet pipe assembly and the exhaust assembly can be avoided in the design and the chamber consists only of the thermoelectric generator, filter, sensing chamber and frame connection. Thus, the response time of the sensor plays a crucial role in the chamber design, cost involved and the size of the chamber (Figures 3 and 4).

Mechanical Analysis

Analysis for the miniature chamber was carried out by using the Fluid Flow (CFX) and Transient Thermal Analysis systems. Boundary conditions are set to extreme conditions to ensure that a level of design safety is maintained. The model that was used in the analysis was built in such a way that it represents the inner wall of the miniature sensing chamber throughout. Thermal and flow analysis were done to check the temperature distribution over the sensing chamber and the flow of exhaust gas into the sensing chamber. To perform the thermal fluid

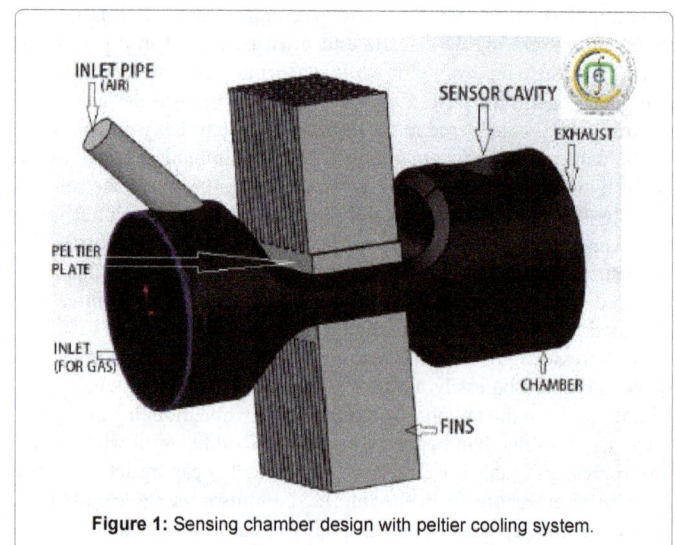

Figure 1: Sensing chamber design with peltier cooling system.

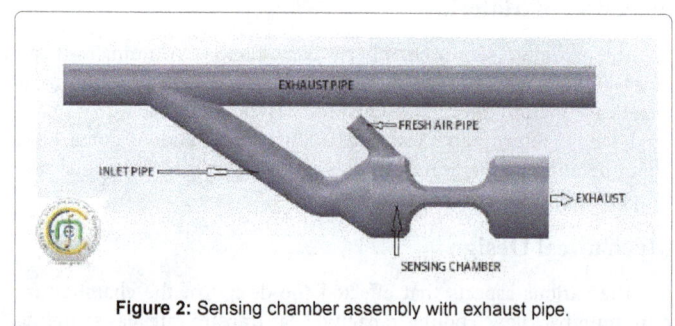

Figure 2: Sensing chamber assembly with exhaust pipe.

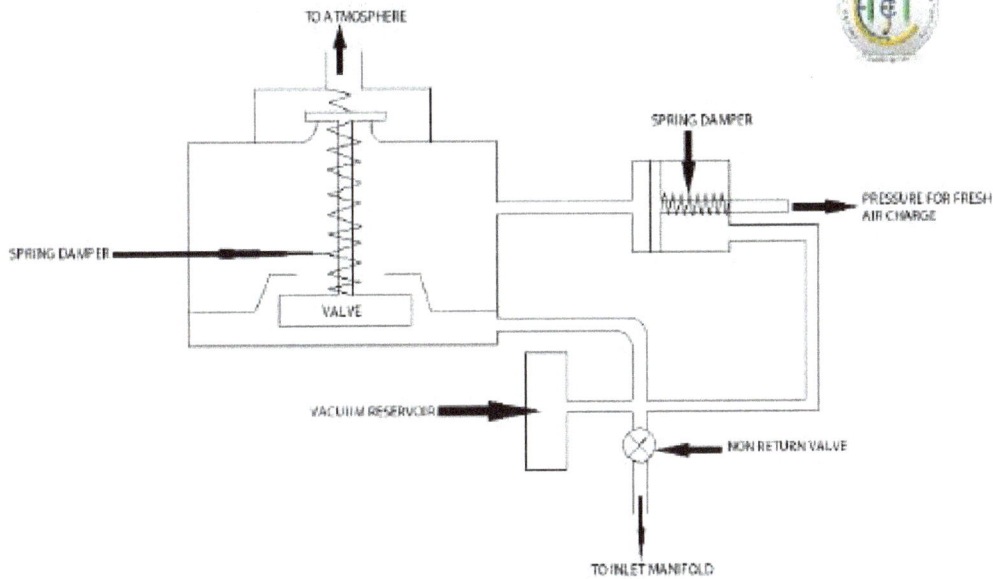

Figure 3: Schematic diagram of vacuum servo mechanism to pump in fresh air.

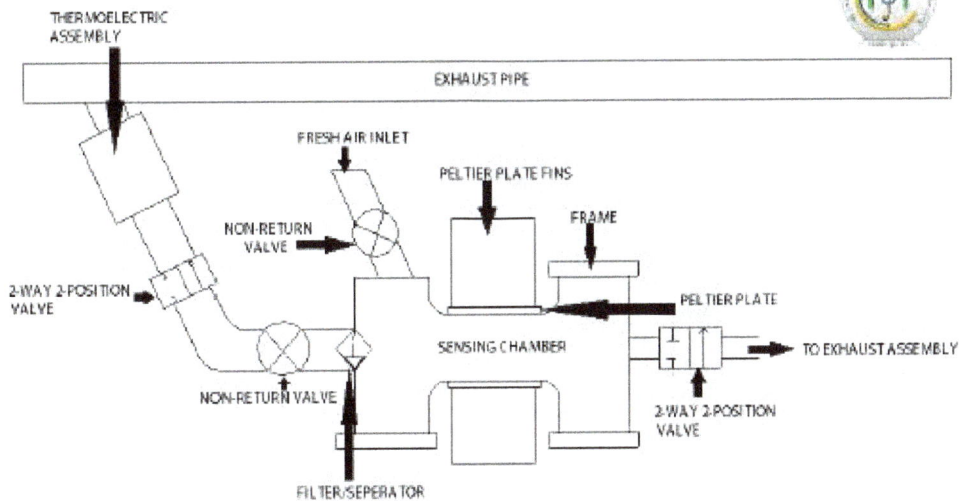

Figure 4: Schematic diagram of sensing chamber assembly.

flow analysis, the inlet gas temperature was maintained at 120°C and the inlet, outlet and walls are defined. The inlet gas velocity was kept at 60 m/s whose value was acquired by measuring the exhaust gas velocity of a single cylinder petrol engine. The medium was chosen as atmospheric air. The pressure difference across the chamber was maintained zero. For the transient thermal analysis, the Peltier plates were maintained at -10°C, internal convection was assigned to the domain with the inlet temperature 120°C. The modes of heat transfer that are being simulated are conduction (stainless steel), convection (stagnant air, thermal conductive coefficient = 5W/m²K), as well as radiation (emissivity = 0.5) with an ambient air temperature of 22°C. The number of time steps is 5 with each step being 1 second.

Results and Conclusion

For the fluid flow analysis, a major determining factor for the miniature chamber was the absence of any back flow. This was a major factor as the use of this chamber must not affect the performance and operation of the source of the measured gas. As seen in Figure 5, the streamlines seen all flow generally in the forward direction, thus resulting in negligible back-flow. For the thermal fluid flow analysis, we look at reducing the temperature of the hot inlet gas as far as possible for the aforementioned reasons. As seen in Figure 6, we see that the temperature of the gas falls substantially in a rather small span. The graph shown in Figure 7 gives us the outlet temperature as the exit, which after 5 seconds of flow at 60 m/s of air at 120°C is around 47°C,

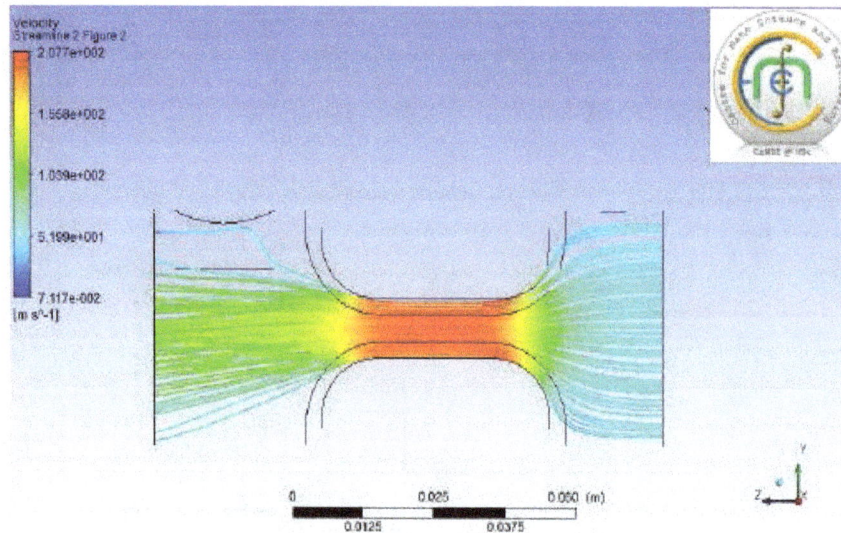

Figure 5: Flow analysis for miniature sensing chamber.

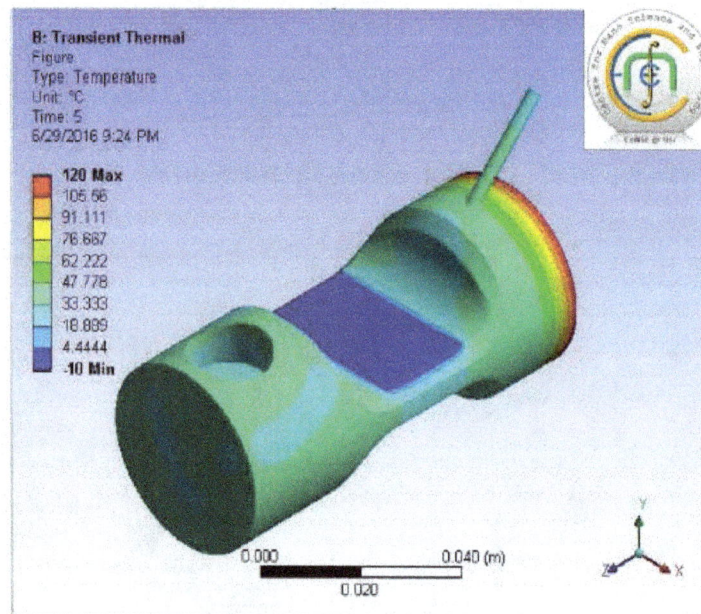

Figure 6: Thermal flow analysis for miniature chamber.

which is well below the operating temperature of most of the solid state sensors. This proves that the sensor packaging material and the sensor itself is safe from the harmful effects of excessive temperatures. Figure 8 shows the laminar flow of the exhaust gases from the exhaust pipe to the sensing chamber. Other applications where such a sensing chamber can be used is:

Automobile

The sensor and sensing chamber assembly can be used in automobiles not only for exhaust gases but to check oxygen level in other parts of the engine too.

Health and medicine

A simple sensing chamber can be used in a breath analyser to detect diseases like CO poisoning and asthma. It could also act like a diet monitor while also checking alcohol intoxication level by using few solid state sensors. Such a Breathalyzer uses an NRV (non-return valve) at the inlet and a diaphragm pump cum valve at the exhaust to regulate the flow of the breath and fresh air.

Agricultural robot

The sensor and sensing chamber assembly can be used in an agricultural robot which could in turn be used to inspect whether the plant has a particular disease. Plants release a particular gas at a certain level which can be drawn in the sensing chamber of the robot and measured thus informing the user about the plant's condition. This chamber could be used in grading of various agricultural products like coffee and spices too.

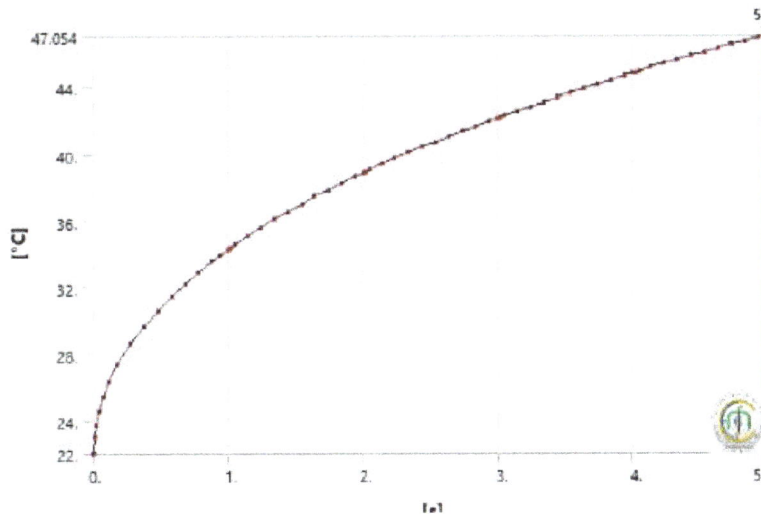

Figure 7: Outlet temperature versus time plot.

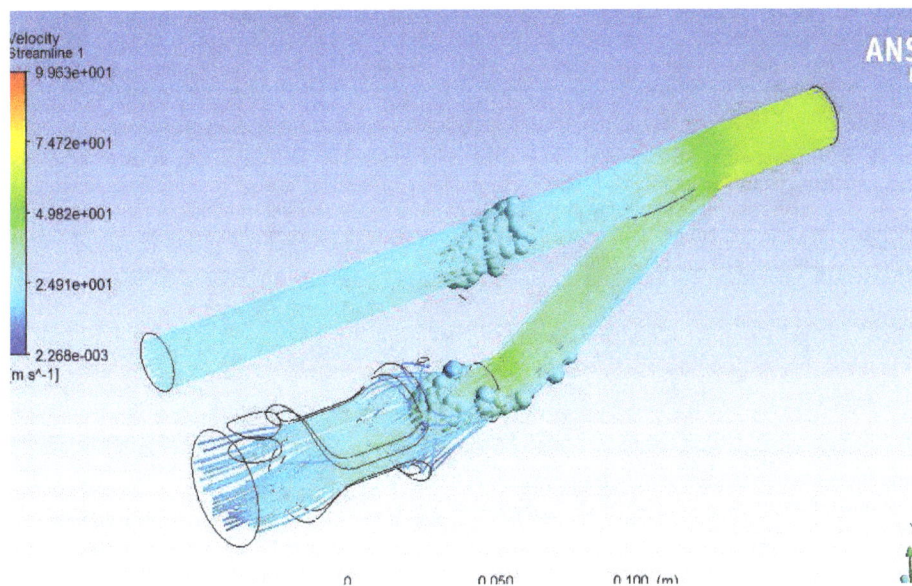

Figure 8: Flow analysis with exhaust pipe of an automobile.

Air pollution sensing drone

An environment monitor or air pollution sensing drone could adopt this assembly of the miniaturized chamber and sensor to detect the level of particular pollutants in air.

Safety: Carbon Monoxide and other harmful gas detection devices could adopt the same assembly. These can be either used in homes or in mines.

References

1. United States Environmental Protection Agency Air and Radiation (2000) EPA420-F-00-013 Office of transportation and air quality average annual emissions and fuel consumption for passenger cars and light trucks.

2. Capone S, Forleo A, Francioso L, Rella R, Siciliano P, et al. (2003) Solid state gas sensors: state of the art and future activities. J Optoelectronics and Advanced Materials 5: 1335-1348.

3. Fernando Garzon H, Mukundan R, Eric Brosha L (2000) Solid-state mixed potential gas sensors theory experiments and challenges. Electronic and electrochemical materials and devices group Los Alamos National Laboratory Los Alamos NM 87545 USA 136: 633-638.

4. Hield P (2011) The effect of back pressure on the operation of a diesel engine. Maritime platforms division defence science and technology organisation.

5. AZO Materials (2000) Mechanical and thermal properties of alumina.

6. Engineers Edge (2010) Engineering and Manufacturing Solutions. Thermal properties of brass.

7. [No authors listed] (1972) Peltier-effect heat pump. Illelmut Hubert Erda Germany United States Patent No 3,635,037.

8. Salwe M, Pande SS, Khan JS (2014) Effect of velocity and particle size on the coefficient of heat transfer in fluidized bed heat exchanger Nagpur University, Nagpur.

9. Jha SJ, Sharma A (2013) Optimal automobile muffler vibration and noise analysis. ASET Amity University Noida, India.

10. KNF Group International (2016) Diaphragm pump technology, KNF Neuberger, Inc., NJ, USA.

29

Mechanical Performance of Woodwork Joinery Produced by Industrial Methods

Altunok M*, Kureli I, Doganay S and Onduran A

Faculty of Technology, Gazi University, Turkey

Abstract

In this study, the improvement of the mechanical properties of wood material were investigated to determine the effects of heat treatment (thermo-process) and impregnation processes held in open-air conditions on the mechanical performance of woodwork joinery. To this end, pine (*Pinus sylvestris* L.) and chestnut (*Castanea sativa*) species grown in our country were used which have significant natural and commercial value. Actual sizes of the examples of woodworks were set out diagonally.

In the double-tenon construction and glue type polyurethane (PU) and D4 adhesives used. Diagonal woodwork and the test samples prepared by class at 185°C temperature with a thermo-vapor process protections apply. Thus, thermo-treated wood joinery was obtained. After application of the heat treatment by immersion, the test samples, and a 5% solution of tannin and oak with a natural pine cone resin made impregnation process. Then, the impregnation of natural wood (control) only heat-treated and heat treatment+impregnated samples kept for one year in outdoor conditions. At the end of the waiting period, on samples prepared by the principles stated in TS 2472 and TS 7251 EN 107 standard "Windows-Test Methods-Mechanical Experiments," according to the principles specified in standard performance tests (diagonal compression test, windows hang test) and microscopic observations and investigations to determine the cellular changes were made. These observations suggest that the natural form treated and untreated samples and comparison of window wings holding the external environment gained strength during the degradation and the reasons for this deterioration was detected against the impregnated material. Experimental density measurements on samples without waiting in outdoor conditions before, after waiting and waiting, as periodic process was determined by performing the difference.

In this study, experiments were carried out on diagonal examples and especially examples in actual size woodwork. In addition, natural oak tannin solution impregnation process was used. From previous research literature diagonal examples have been used not only in instances of life-size woodwork. In addition, thermo-transaction and impregnation process used in conjunction studies, natural pine resin is used, and there is no such study created a solution. This also shows that the original value of our work.

Keywords: Thermo process; Wood material; Wood preservation; Wood modification; Mechanical properties; Industrial woodwork; Microscopic examination

Introduction

Wood material occupies a significant place in human life thanks to the developments in woodworking industry. Wood material is preferred for its many favorable properties and advantages such as easy processing, isolation, being lightweight, strength, naturality and aestheticity Özalp [1]. However, inflammability, insect infestation induced degradation, fungi induced decomposition, change in its size depending on balance humidity altered by ambient temperature and relative humidity, and fading in color and decomposition due to sunlight exposure, resulting from being an organic material, are considered as its disadvantages Kurtoğlu [2].

It is important to protect, increase longevity and physical durability of the wood material which is exposed to factors such as fungi, insects, ultraviolet rays from the sun in the outer environment. The methods applied to protect the wood material from these factors are called as "modification", the most prevalent ones being impregnation Ayar [3] and thermo-process.

Wood and wood products, being of the conventional material group, are used after being treated with various modification processes, as massive material or wood-derived products. Wood material is widely used chiefly in the flooring and walls of the wooden houses and in wood constructions, and also in the roof, moulding and scaffolding works. Particularly the producers who refrain from PVC material prefer again wood material especially for the production of window and door frames in constructions. Different techniques and corner joinings are applied in the production of door and window frames. These corner joinings come under more diagonal traction – pressure tension influence when faced with the door and window frames' self-weights and other external loads Tokgöz, Kap, and Özgan [4].

In the course of the use of the wooden frames, problems such as deformation in the casement geommetry arise, and these bring about difficulties in opening and closing Arslan, Subaşı, and Altuntaş [5]. The studies related to this problem are not found to be sufficient in the literature.

Thermo-process is an important wood modification method in which the size stabilization is established without the use of any chemical substance. By this way, volumetric swelling can be reduced approximately by 50% and thermo-processed wood can be safely used in the outer environment, without any requirement for the

---

*Corresponding author: Altunok M, Gazi University, Faculty of Technology, Turkey
E-mail: altinok@gazi.edu.tr

chemicals that are hazardous to the environment and human health. However, thermo-process completely destroys the natural preservative substances (i.e. tannin, resin) intrinsic to the wood material Altınok, Perçin, and Doruk [6].

Altinok et al. in 2009, tested the Scotch pine and Uludag fir woods gluing with PVAc-D3 and PVAc-D4 in single and double mortise wooden frame corner joinings, and reported Scotch pine and PVAc-D4 glue as having the highest diagonal tension performance, while Uludağ fir and PVAc-D3 were reported to have the lowest Altınok, at al. [7]

In the study of Alen et al. in 2002, thermo-process was applied to spruce wood (*Picea abies* L.) at 180-225°C temperature for 2-8 hours, and it was found in the chemical examinations of thermo-processed samples that carbohydrates had been more decomposed than lignin during the thermo-process procedure Alen et al. [8]

Aydemir and Gündüz in 2009 stated that in the thermo-processes and above 150°C, the color of the wood had darkened, and biological resistance and constancy in terms of size had improved. Additionally in the same study, it was reported that a reduction in the mechanical properties and alteration in the chemical structure of the wood had occured, and for this reason the utilization areas of the wood material were curtailed Aydemir and Gündüz [9].

Korkut et al. in 2008, subjected beech and maple (Acer trautvetteri Medw.) woods to thermo-process at different temperatures (120°C, 150°C and 180°C) and for different times (2 h, 6 h, 10 h). They stated that as the temperature and the duration of the thermo-process increase, technological properties of the wood material are decreased Korkut at al. [10].

Bekhta and Niemz in 2003, determined that in beech wood (Fagus orientalis L.) mechanical properties and bending strength had decreased by 5%-40% and modulus of elasticity (MOE) by 4%-9%, stabilization in terms of size had increased and wood color had darkened Bekhta and Niemz [11].

Edlundl and Jermer in 2004 stated that there had not been any decomposition for 2 years in the samples of Scotch pine and spruce woods thermo-processed at 220°C and for 4 hours (Edlundl and Jermer) [12].

In the study of Follrich et al. in 2006, it was determined that the adhesion on the surfaces brought together by polyethylene in thermo-processed spruce woods had broken, and for this reason the contact angle had increased. It was determined in mechanical experiments that the bonding strength between the polyethylene and wood surfaces was higher than non-thermo-processed wood material Follrich et al. [13].

In the study of Hakkoul 2005, it was established that there had been a sharp change in the water retension capacity of fir, poplar wood, beech and pine species of woods -kept in steam environment for 8 hours- after the thermo-process in between the temperature of 100°C-160°C, and the bonding angle had reached an average of 90 degrees. Changes in wettability had been seen in between 100°C-160°C without any detected weight loss, and it was detected that the wettability of the wood had been unaffected at high temperatures Hakkoul [14].

In the study of Ishikawa et al. in 2004, changes in the humidity content of the sugi wood (*Cryptomeria japonica* D. Don) at high temperature was investigated. It was detected that the loss in the wood structure resulting from the thermo-process at 160°C was significant,

and this loss had affected the humidity values. It was seen that the losses which might occur at 120°C were not significant. Thermo-process and temperature degree were found to be effective on the balance humidity of the wood. In the closed thermo-process systems with vapor is done, the saturated vapor was detected to increase the deterioration on the surface and interior parts of wood Ishikawa et al. [15].

Johansson and Mor'en in 2006 investigated the changing in color and resistance properties of birch wood at 175°C and 200°C temperatures and at 1 h, 3 h, and 10 h. Results showed that color and balance humidity were not related with resistance properties Johansson and Mor'en [16].

The aim of this study was to determine the effects of wood modification (thermo-process and impregnation with natural wood preservatives) and weathering on the mechanical performance of industrial woodwork joinery samples. At the same time, since the carbohydrates and resins (natural preservatives), which are of the wood components, decompose before lignin and cellulose and move away from the wood in the form of gas during thermo-process, the wood becomes slack and deprived of natural preservative substances (tannin and resins). With regard to this situation, a new dimension is brought to the research by the impregnation of tannin and pine resin solution into both woods of research after thermo-process.

Material and Methods

Wood materials and glues

In the experiments, Scotch pine (*Pinus sylvestris* L.) and chestnut (*Castanea Sativa*) wood types were used. Wood materials were obtained by the method of random selection in the site of Woodworking Ankara. Wood materials' being natural in color, clean, smooth parallel to grain and part of the sapwood, free of decay, growth defects, insect and fungi damages were ensured. For gluing of the experimental samples, two-component polyvinyl acetate (PVAc-D4) and polyurethane based Desmodur VTKA were used. Two-component polyvinyl acetate (PVAc-D4) was hardened by the addition of 5% hardener (Turbo hardener 303.5) [17]. According to BS EN 204, PVAc-D4 can be aligned with the pasting quality of D4 by increasing the durability to moisture by adding 5% hardener to the Klebit 303 glue solution [18].

Impregnation solution and process:

As impregnation material, pine resin (95% liquid) and tannin (5% liquid) solution was used. Liquid pine resin was obtained by dissolving the solid resin in the fresh pine cones in hot water at 80-90°C. The tannin was joined up to 5% of the weight of pine resin solution. After dipping in pine solution, the draft samples were impregnated by keeping in the solution for 45 minutes. At the end of each period, the samples were removed from the solution and their surfaces were air-dried. The impregnated samples were kept in air-conditioning fridge under 20 ± 2°C temperature and 65% ± 5% relative humidity conditions until reaching moisture balance. Impregnated samples, retention rate of which to be determined, were dried in the oven at 55°C until reaching constant weight, cooled in desiccator and weighed. Thus, the amount of net impregnation of each sample, whose exact dry weight was determined after impregnation, was calculated from the equation below.

$$ANI = ((M_s - M_o)/M_o) \times 100$$

Where ANI is Amount of Net Impregnation (%) Ms [g] is the oven dry weight after impregnation and M_0 [g] is the oven dry weight before impregnation of the same sample.

Preparation of frame samples

The diagonal test samples were prepared as 60 frame In first place.

120 L-Type samples were obtained by cutting four samples from each of the 60 frame second stage.

To be used as the L-Type samples for diagonal compression performance, 120 test samples were prepared as a $2 \times 2 \times 3 \times 10$ combination of 2 wood species (pine, chestnut), 2 glue types (PVAc D4, VTKA PUR), and 3 processes (control, heat treated, heat treated+impregnated).

The preparation of test samples was initiated with heat treatment. The heat treatment was applied to draft-size wood pieces via thermowood method (175°C - 36 h). These pieces of heat-treated samples were cut with 0.1 mm precision and opened by automatic control of the machine to obtain corner bridle joint components which were then assembled and was impregnated (Figure 1).

The corners were mounted by applying each type of glue on a 150 g/m² basis with a brush to the surfaces of test samples. The pressure was applied by pressing to corner joints samples and they were allowed to harden for 24 hours.

Weathering of test samples

The control samples, heat treated samples and heat treated+impregnated samples were waited between 22.04.2012 - 22.04.2013 (for 12 months), according to the principles of ASTM G7 standard [19]. They were placed, in Ankara-Turkey, facing to the south and in 45° oblique position to the ground (Figure 2).

The lowest level of the test sample was 50 cm in height. The attention was paid to the hay etc. around the stand with organic residues in soil, which increase water content. These months' climatic conditions, such as the following, have been identified for each month in Graphics by the Center for Meteorology (Figure 3) [20].

Methods

The diagonal compression test experiments were held on the basis of ASTM 1037 in 800 Kp stage of Universal Testing Machine with a capacity of 5 tons in the material laboratory of the Faculty of Technology of Gazi University [21]. Experimental device was set to take forward speed of 2 mm/min. The maximum force, read from the machine, was recorded in N (Figure 4).

Evaluation of data

Multiple analyses of variance were made in order to determine the effects of wood type, adhesion type, heat treatment and impregnation in wood frame corner joints. In case of mutual interactions of sources of variance being significant according to α=0.05, for which factors are the differences important was identified by Duncan test. Duncan test was used as the post-hoc analysis.

Results and Discussions

Density and retention amount

The densities of the L-type samples were determined according to TS 2472 (1976) [22]. Density and retention amount of L-type samples are given below (Table 1).

As it can be inferred from Table 1, the heat treatment reduced the density of both wood species. In case of increase in the moisture content, although density rises again in control samples, this increase was not observed in the heat treated samples. As compared to chestnut wood, more retention occured in pine wood due to its low density.

Diagonal compression performance on L-Type samples

The diagonal compression performances of the L-type samples were determined according to TS 7251 EN 107 (2003) [23]. Statistical means of the diagonal compression performances at the levels of wood type, glue type and process type of the L-type samples before and after the natural aging are summarized in Table 2.

According to Table 2, when all samples are examined at the wood and glue levels before and after the natural aging (weathering) process, the interaction of chestnut wood with PVA-D4 glue is found to have the highest performance.

Figure 1: Illustrated corner bridle joint components.

Figure 2: Samples placed in oblique position.

Multiple analysis of variance results for the diagonal compression performances of L-Type samples obtained with variations in wood type, glue type and process type are given in Table 3.

According to Table 3; wood type, adhesive type, process type and the differences between these dual and triple interactions' effects on the diagonal compression were statistically significant ($\alpha=0.05$). Duncan test was performed to determine which groups are significantly different. Diagonal compression performance values (N) by triple interactions of wood type, glue type and process type of L-type samples are given in Figure 5.

According to Figure 5; the triple interaction of control-chestnut with D4 glue was found to have the highest performance value (5726.71 N), while the triple interaction of heat treated-pine with D4 glue samples had the lowest value (1822.82 N). Excluding the control samples, the highest performance was obtained in the chesnut-heattreated-D4 combination. Others performed close to each other.

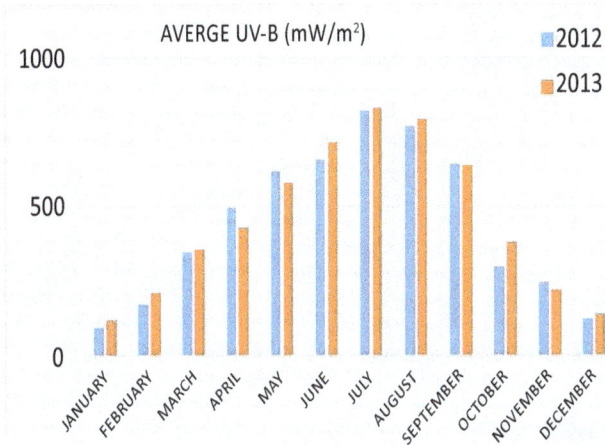

Mounths	1.	2.	3.	4.	5.	6.	7.	8.	9.	10.	11.	12.
Avr. Temperature (C°)	0,4	1,9	6,1	11,3	16,2	20,2	23,6	23,3	18,7	13,1	7,0	2,6
Avr. Highest Temperature(C°)	4,4	6,5	11,7	17,2	22,3	26,7	30,2	30,2	25,9	19,9	12,9	6,6
Avr. Lowest Temperature(C°)	-3,0	-2,2	1,0	5,6	9,7	13,1	16,0	16,0	11,7	7,3	2,5	-0,6
Avr. Sunny Time (h)	2,5	3,5	5,2	6,3	8,4	10,2	11,4	11,0	9,2	6,5	4,4	2,3
Avr. Rainly Day	12,2	11,0	10,9	11,9	12,5	8,6	3,7	2,8	3,9	6,8	8,5	11,8

Figure 3: Graphic and table of UV-B radiation between 2012-2013 years and metrological data [20].

Figure 4: Universal test machine and test sample.

Wood types	0% MC (gr/cm³)		12% MC (gr/cm³)		Retention amount (kg/m³)
	control	Heat treated	control	Heat treated	Heat treated + impregnated
Pine	0.48	0.44	0.50	0.44	11.26
Chestnut	0.54	0.53	0.59	0.55	10.96

Table 1: Density and retention amount of L-type samples.

Conclusions

This study was performed in order to determine the effect of wood modification (heat treatment and Short-term -45 min- dip impregnation method with natural pine resin and tannin solution) on the specific gravity, diagonal compression performance and deformation by aging. Following conclusions can be drawn from this study.

Density and retention amount

The heat treatment, one of the wood modification techniques, reduced the density of both wood types. After heat treatment and impregnation were hold the situation of the both wood types in the natural aging. This decrease of 9% in the pine samples, 7% in the chestnut sample realized. During the aging in natural environment

(weathering) and heat treatment especially natural woods heavy loss; degradation to such as lignin, cellulose and extractives of wood components is reportedly due [8,9].

With the wood types impregnated in a solution of pine cone resin and tannin, the maximum retention was obtained in the pine wood. The reason for this may be the higher degree of degradation in pine and more amount of space.

Diagonal compression performance on the L-type samples

The element that makes up the advanced performance of L-type samples is its ability to resist diagonal test forces. This resistance constituent factor is the adequacy of the sticking of samples' corner joining components. Specifically, the adhesive's resistance to natural aging is the

Process	Wood Types	L-Type Samples	
		PVAc-D4	D-VTKA
Control (Before Weathering)	Pine	2554.31	3116.41
	Chestnut	6813.94	5217.69
Control (After Weathering)	Pine	2111.23	2597.84
	Chestnut	5726.71	4384.26
Heat Treated (After Weathering)	Pine	1822.82	2761.22
	Chestnut	3316.38	2886.69
Heat Treated + İmpregnated (After Weathering)	Pine	2844.28	2862.61
	Chestnut	2884.05	2797.75

Table 2: Statistical means of the performance of diagonal compression performance (as N).

Source of variance	SD.	Sum of Squares	Average of Squares	Value of F	P ≤ 0.05
Wood type (A)	1	38845630.208	3884530.208	160.5797	0.0000
Process type (B)	2	19626690.017	9813345.008	40.5663	0.0000
Interaction (AB)	2	26744116.217	13372058.108	55.2773	0.0000
Glue type (C)	1	1481851.875	1481851.875	6.1257	0.0149
Interaction (AC)	2	12407971.408	12407971.408	51.2920	0.0000
Interaction (BC)	2	7516243.550	3758121.775	15.5353	0.0000
Interaction (ABC)	2	5567584.017	2783792.008	11.5076	0.0000
Error	108	26126141.300	241908.716		
Sum	119	138316228.592			

Table 3: Multiple analyses of variance for the diagonal compression performances of L-Type samples.

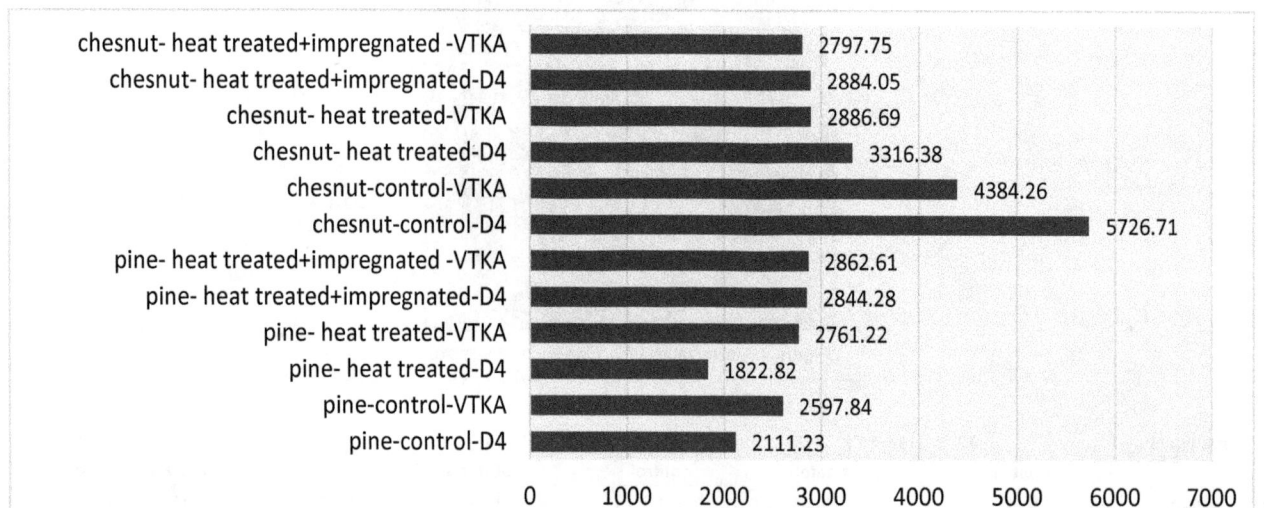

Figure 5: Diagonal compression performance values (N) by triple interactions of L-type samples.

Wood and glue type	Before aging			After aging			
	Control	Untreated	Change (%)	Heat-treated	Change (%)	Heat-treated+impregnated	Change (I%)
Pine-D4	2554.31	2111.23	-18	1822.82	-29	2844.28	+11
Pine-VTKA	3116.41	2597.84	-18	2761.22	-22	2862.61	-8
Chesnut-D4	6813.94	5726.71	-16	3316.38	-52	2884.05	-58
Chestnut-VTKA	5217.69	4384.26	-16	2886.69	-45	2797.75	-46

Table 4: Changes in performance by aging – heat treatment – impregnation.

most important determinant of performance in the L-type samples.

In this regard when the performance of the diagonal compression in the L-type samples is examined; the diagonal compression performance generally attained the highest value in the control experiments of both types of wood and glue before aging (Figure 3). Heat treatment decreased the performance of both types of wood. During the heat treatment process, wood materials became brittle and fragile. Therefore, some heat-treated samples were cracked or broken during the test.

As compared to the untreated (control) and non-aged samples of both wood and glue types; the changes in performance in the aged control samples, heat-treated samples and heat-treated+impregnated samples are given in Table 4.

In this case, the performance of untreated samples decreased between 16% and 18% by the aging acording control samples, in untreated samples of both wood and glue types decreased between 22% and 52% by the heat treatment, but increased only in heat-treated+impregnated pine samples.

In this case, in terms of performance; firstly Chesnut as the wood and VTKA glue as adhesive and secondly pine wood and D4 glue are recommended for the manufacture of joinery.

This study, a scientific research project under the code 18/2012-03, has been supported by "Gazi University - Scientific Research Project Unit".

References

1. Ozalp M (2003) Investigation of physical mechanical and chemical characteristic change of water repellent (Protim WR230) and preservative (Wolmanit-cb) treated pine specimens (Pinus sylvesteris, Pinus Nigra and Pinus brutia) and their use waiter cooling toweras. Zonguldak Karaelmas University Zonguldak Turkey 10-12

2. Kurtoglu A (2000) Wood material finishings general information. Forest Faculty İstanbul University Publication İstanbul Turkey 463: 31-32.

3. Ayar S (2008) Determination of the effect to the wood material penetration of impregnate materials of pressure and Waiting Time. Karabük University Karabük Turkey 1-2.

4. Tokgöz H, Kap T, Özgan E (2005) Diagonal load analysis in wood frame joinery construction created with combining of different wood types and corner bridle joining. J Technology Ankara Turkey 8: 36-376.

5. Arslan M, Subaşı S, Altuntaş C (2006) Mechanical properties of joints at the wooden window casement. J of the Faculty of Engineering and Architecture of Gazi University Ankara Turkey 21: 265-273.

6. Altınok M, Perçin O, Doruk Ş (2010) Effects of heat treatment on technological properties of wood material. J Institute of Science Dumlupınar University Kutahya Turkey 23: 71-83.

7. Altınok M, Söğütlü C, Döngel N, Doruk Ş (2009) Determination of the diagonal tensile strength performances of the wooden window corner joints with single or double mortise and tenon. J of the Polytechnic of Gazi University Ankara Turkey 12: 107-112.

8. Ale'n R, Kotilainen R, Zaman A (2002) Thermochemical behavior of Norway spruce (Picea abies L.) at 180-225°C Wood Science and Technology 36:163-171.

9. Aydemir D, Gündüz G (2009) The Effect of heat treatment on physical chemical, mechanical and biological properties of wood. J the Bartın Faculty of Forestry Bartın Turkey 11: 71-81.

10. Korkut S, Kök MS, Korkut DS, Gürleyen T (2008) The Effects of heat treatment on technological properties in red-bud maple (Acer Trautvetteri Medw.) Wood Bioresource Technology 99: 1538-1543.

11. Bekhta P, Niemz P (2003) Effect of high temperature on the change in color dimensional stability and mechanical Properties of spruce wood. Holzforschung 57: 539-546.

12. Edlund ML, Jermer J (2004) Durability of heat-treated wood final workshop COST action E22-environmental optimisation of wood protection. Lisbon, Portugal.

13. Follrich J, Uller UM, Gindl W (2006) Effects of thermal modification on the adhesion between spruce wood (Picea Abies Karst.) and a thermoplastic polymer. Holz als Roh- und Werkstoff 64: 373-376.

14. Hakkou M, Pe'trissans M, Zoulalian A, Ge'rardin P (2005) Investigation of wood wettability changes during heat treatment on the basis of chemical analysis. Polymer Degradation and Stability 89:1-5.

15. Ishikawa A, Kuroda N, Kato A (2004) In Situ measurement of wood moisture content in high-temperature steam. J Wood Science 50: 7-14.

16. Johansson D, Mor'en T (2006) The potential of colour measurement for strength prediction of thermally treated wood Holz als Roh- und Werkstoff 64: 104-110.

17. Söğütlü C, Döngel N (2007) Tensile shear strengths of same local woods bonded with polyvinyl acetate and polyurethane adhesives. J of the Polytechnic Faculty of Technical Education Gazi University 10: 287-293.

18. BS EN 204 (2001) Classification of thermoplastic wood adhesives for non-structural applications. British Standards Institution UK.

19. ASTM G7-05 (2005) Standard practice for atmospheric environmental exposure testing of nonmetallic materials. ASTM International West Conshohocken PA.

20. Center for Meteorology, Ankara Turkey.

21. ASTM D1037-12 (2012) Standard test methods for evaluating properties of wood-base fiber and particle panel materials. ASTM International West Conshohocken PA.

22. TS 2472 (1976) Wood determination density of physical and mechinacal tests. Turkish Standards Institution Ankara, Turkey.

23. TS 7251 EN 107 (2003) Methods of testing windows mechanical test. Turkish Standards Institution Ankara, Turkey.

30

Effect of Variable Valve Timing to Reduce Specific Fuel Consumption in HD Diesel Engine

Afshari D and Afrabandpey A*

Department of Mechanical Engineering, University of Zanjan, Iran

Abstract

Applying variable valve Actuation systems is one of the most effective ways to improve specific fuel consumption in an engine, which largely affect the pumping work. In this article determination of optimum valve timing angels, using approximation of discrete data and nonlinear regression analysis is investigated for a HD diesel engine to minimize SFC. In the first part of this study a model of compression ignition engine (OM457) in GT-SUITE software are applied for optimization. Then the indicated best angels for EVO, IVO, EVC and IVC were added to the model as lookup tables to shape the VVT system. Eventually, results indicated that using VVT angles the SFC parameter decreases more than 2% in average. Furthermore, to compare differences in emissions rate, the European stationary cycle (ESC) was applied and generated NOx pollutant was reduced 7.4%.

Keywords: OM457 diesel engine; Variable valve timing; Optimization; Specific fuel consumption; Specific emissions

Abbreviations: VVT: Variable Valve Timing; VVA: Variable Valve Actuation; EVO: Exhaust Valve Opening; EVC: Exhaust Valve Closing; IVO: Intake Valve Opening; IVC: Intake Valve Closing; SFC: Specific Fuel Consumption; ESC: European Stationary Cycle; DOE: Design of Experiments; MEP: Mean Effective Pressure; CO: Carbon Monoxide; NOx: Nitrogen Oxides; HC: Hydro Carbon; ELR: European Load Response

Introduction

Diesel engines are favored in heavy-duty commercial and military applications as they have high performance in terms of fuel economy, torque at low speed, and power density [1]. For that purpose, the intake and exhaust valve timing of an engine greatly influence the fuel economy, emissions, and performance of an engine. Conventional valve train systems can only optimize the intake and exhaust valve timing for one given operational condition. Thus, the optimized valve timing can either improve fuel economy and reduce emissions at low engine speeds or maximize engine power and torque outputs at high speeds. With the development of continuously variable valve timing (VVT) systems, the intake and exhaust valve timing can be modified as a function of engine speed and load to obtain both improved fuel economy and reduced emissions at low engine speeds and increased power and torque at high engine speeds [2]. As one of the most promising control strategies for diesel engine, variable valve actuation (VVA) had attracted increasing attention due to its fast-response characteristics. Recently, several different VVA mechanisms have been implemented into diesel engines for realization of low emissions and high thermal efficiency [3].

VVT is computer-controlled and typically uses oil pressure to change the position of a phaser mechanism on the end of the camshaft to advance or retard cam timing [4]. The first VVT systems came into existence in the nineteenth century on early steam locomotives. In early 1920s VVT was developed on some airplane radial engines with high compression ratios to enhance their performance [5] and in automotive applications, the VVT was first developed by Fiat in late 1960 [6]. Considering the ability of the system it was soon used by other companies like Honda [7]. General motors, Ford and other automobile manufacturers [8]. The mechanisms currently available on the market allow, as a function of the engine operating conditions, variations in the timing (VVT) or, in addition, in the lift (VVA). The aim of both the solutions is the adjustment of the load through reduced valve throttling; this adjustment provides a significant positive influence on the pumping work [9].

The aim of this paper is to find optimum angles for EVO, EVC, IVO and IVC in each revolutionary speed of 'OM457' engine to minimizing SFC parameter, regardless of any specific mechanism to operate VVT system. To evaluation of optimum angels, a one-dimensional model of 'OM457' engine was used. To summarize the calculation, EVC and IVO angels are linked together as overlap of valves. Consequently, only EVO, overlap and IVC are three independent parameters to calculate. At the second phase of this paper, a control unit will be adding to the 1-D model of engine to change optimum angels in different revolutionary speeds and finally, at third phase the ESC test will be operate to compare the engine emissions with and without VVT system. It must be noted that regarding the very stringent emissions limit for nitrogen oxides (NOx) and particulate matter (PM) (which will take effect in the EU and the US from 2008 and 2010 respectively), [10] these are unlikely to be met by using current VVT system.

Methodology

The specific fuel consumption parameter measures how efficiently an engine is using the fuel supplied to produce work and it can be obtained by equation (1). The theoretical approach of the simulation is supplied by [11-13].

$$sfc = \frac{\dot{m}_f}{P} \tag{1}$$

In this equation $\dot{m}_f$ is fuel flow rate and P is engine power. The power of a four-stroke engine can be expressed as

---

*Corresponding author: Afrabandpey A, Department of Mechanical Engineering, University of Zanjan, Iran, E-mail: arian.afrabandpey@znu.ac.ir

$$P = \frac{mep \, A_p \overline{S}_p}{4} \tag{2}$$

where mep is mean effective pressure, A_p is area of the piston head, and $\overline{S}_p$ is average speed of piston. The mean effective pressure or mep is given by

$$mep = \eta_f \eta_v Q_{HV} \rho_a \, (F/A) \tag{3}$$

Where η_v is volumetric efficiency, η_f is fuel conversion efficiency, Q_{HV} is the heating value of the fuel, ρ_a is the inlet air density and F/A is fuel to air ratio. The important parameter of volumetric efficiency in equation (3) is defined as the volume flow rate of air into the intake system divided by the rate at which volume is displaced by the piston.

$$\eta_v = \frac{2\dot{m}_a}{\rho_a V_h N} \tag{4}$$

In equation (4) $\dot{m}_a$ is mass flow rate of air, N is engine speed and V_h is engine displacement or swept volume. The mass flow rates are adequately represented by standard expressions for steady, adiabatic and reversible flow. For gas flows through restriction (throttle, valves), a discharge coefficient C_D is introduced to give the effective flow area. The general form of the mass flow rates for an un-choked flow is given by equation (5),

$$\dot{m}_a = \frac{C_D A_R p_0}{\sqrt{RT_0}} \left(\frac{p_T}{p_0}\right)^{\frac{1}{\gamma}} \left\{ \frac{2\gamma}{\gamma-1} \left[1 - \left(\frac{p_T}{p_0}\right)^{\left(\frac{\gamma-1}{\gamma}\right)} \right] \right\}^{\frac{1}{2}} \tag{5}$$

And for a choked flow

$$\dot{m}_a = \frac{C_D A_R p_0}{\sqrt{RT_0}} \gamma^{\frac{1}{2}} \left(\frac{2}{\gamma+1}\right)^{\frac{\gamma+1}{2(\gamma-1)}} \tag{6}$$

where A_R is the reference area, p_0 and T_0 are the upstream stagnation pressure and temperature, p_T the downstream stagnation pressure, R the gas constant for the gas and γ the specific heat ratio [14]. The flow velocity in this model can be expressed as,

$$v_{ps} = \frac{1}{A_m} \cdot \frac{dv}{d\theta} = \frac{\pi B^2}{4A_m} \cdot \frac{ds}{d\theta} \tag{7}$$

Where V is cylinder volume, B is cylinder bore, s is distance between crank axis and wrist pin and A_m is the area of valves.

To find the best EVO, IVC and overlap degrees, GT-suite software was applied to study and calculate the optimum parameters related to SFC. A one-dimensional model of 'OM457' engine was used for optimization of proposed parameters and Figure 1 is representing this model.

'OM457' is a Diesel engine with maximum Torque of 1598 (Nm) in 1200 (rpm) and maximum power of 350 (HP) in 2000 (rpm). Some characteristics of this engine are listed in Table 1.

To achieve a reliable model, with reducing the error between the models output and experimental results, some parameters were achieved in DESA Company and they were added to software model. The software also performs the calculation several times to achieve converged results. For this problem, the work will be followed in three stages:

Stage 1: Optimization of EVO, IVC and overlap degrees to achieve minimum value for SFC in each speed.

Stage 2: Adding a control unit to the valves to change timing and shape the VVT system.

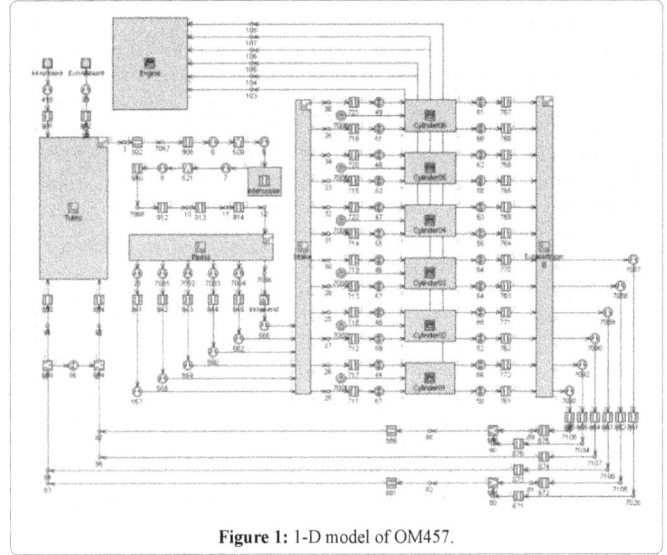

Figure 1: 1-D model of OM457.

No. of cylinders/Location	6/In-line
Engine Displacement [Liter]	12
Bore [mm]	128
Stroke [mm]	155
Connecting rod length [mm]	247
Compression ratio	17.5:1
Aspiration	Single Turbocharger

Table 1: Characteristics of 'OM457'.

Stage 3: Operating the ESC test to illustrate the difference of engine emissions with and without VVT system.

Stage one

For these types of optimization problems, different methods like sensitivity analysis [15] or variation methods, genetic algorithm [16] neural networks [17] and the cascade model [18] have been applied. In this paper, the EVO, IVC and overlap degrees considered as input variables and the SFC is the output.

In some cases, due to the decoupled response of parameters respect to the EVO, IVC and overlap degrees, it is possible to find best value for EVO, IVC and valves overlap as follow:

1. Finding the best EVO, considering a fixed value, equal to primary engine, for IVC and overlap.

2. Finding the best amount for IVC, considering the best amount of EVO but maintaining the overlap value.

3. Find the best value for engine overlap, considering the optimum amounts of EVO and IVC.

4. After choosing best values for EVO, IVC and overlap values, it is necessary to shift them to check the authenticity of the results.

For 'OM457', the DOE method used to shape the standardized effects and check the dependency of elements in responses. Utilizing Minitab software, it was specified that the EVO, IVC and valves overlap degrees haven't any interaction in the proposed range.

Stage two

The obtained values for best EVO, IVC and overlap degrees in each revolutionary speed have been collected in four lookup tables as showed in Figure 2. In this scheme, VVT system uses the feedback of revolutionary speed of engine and changes the EVO, IVC and overlap degrees to achieve the minimum SFC.

In this stage, after deploying VVT system which causes additional MEP and reduction of wasted power in pumping cycles, it is possible to modify the injected fuel in each cycle. Respect to other features of engine such as maximum MEP, the injected fuel into the cylinder in each cycle could be modified. After fuel modification, an alternation in emission rate is expected.

Stage three

The following table contains a summary of the emission standards and their implementation dates for HD diesel engines [19]. The proposed model of engine in this paper comply the Euro IV standard, so to realize engine emissions, the ESC and ELR standards could be executed. Since only software analysis is considered in this paper, the ESC test will be implemented to assess emissions. European stationary cycle (ESC): The test cycle consists of a number of speed and power modes which cover the typical operating range of diesel engines. It is determined by 13 steady and modes (Table 2).

The engine is tested on an engine dynamometer over a sequence of steady-state modes as illustrated in Table 3 and Figure 3. Emissions are measured during each mode and averaged over the cycle using a set of weighting factors. Particulate matter emissions are sampled on one filter over the 13 modes. The final emission results are expressed in g/kWh [20].

In accordance with the ESC testing procedure, the summed average emission will be calculated in the following way:

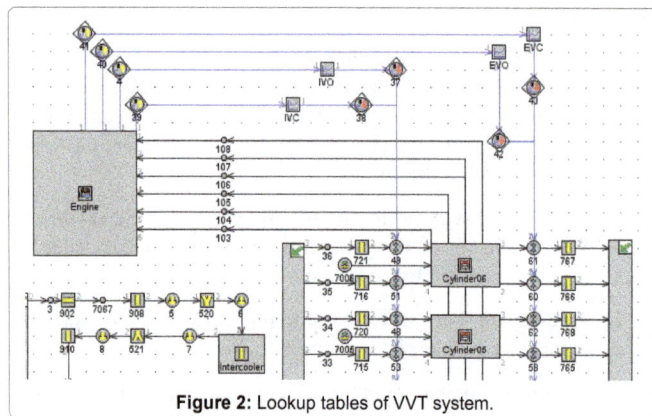

Figure 2: Lookup tables of VVT system.

$$\overline{e}_{g/kWh} = \frac{\sum_{i=1}^{13} e_i * WF_i}{\sum_{i=1}^{13} P_i * WF_i} \tag{8}$$

Which e_i is emission in mode i (g/h), P_i is engine power in mode i (kW) and WF_i is weighting factor in mode i.

The catalysts used in the after-treatment system consist of catalytically active transition metal compounds, which are fixed onto ceramic carriers. The after-treatment system in modern Euro VI engines cause up to 95% reduction of NOx. Poor activity of the SCR[1] after-treatment system due to inactive catalysts may cause an increase in NOx emission and cause secondary damage in the engine itself due to an exhaust gas pressure increase [21].

Modelling Results

The OM457 has many applications, including trucks, marine, military, municipal, and agricultural vehicles, as well as stationary settings. The engine has differing trim and power levels [22].

Here are some of the features of this engine:

- In this engine, the dedicated times for exhaust and intake valves are illustrated in Figure 4. Respect to crank angle degree, both exhaust valves open at 118 and close at 387. Moreover, both intake valves open at 336 and close at 576 degrees.

[1] Selective Catalytic Reduction

Mode	Engine Speed	Load [%]	Weight [%]	Duration [min]
1	Low idle	0	15	4
2	A	100	8	2
3	B	50	10	2
4	B	75	10	2
5	A	50	5	2
6	A	75	5	2
7	A	25	5	2
8	B	100	9	2
9	B	25	10	2
10	C	100	8	2
11	C	25	5	2
12	C	75	5	2
13	C	50	5	2

Table 3: ESC test modes [20].

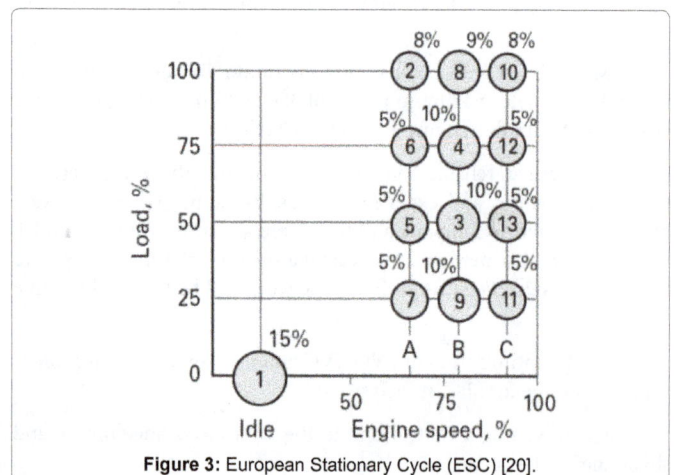

Figure 3: European Stationary Cycle (ESC) [20].

Stage	Date	Test	CO [g/kWh]	HC [g/kWh]	NOx [g/kWh]
Euro I	1992, ≤ 85 kW	ECE R-49	4.5	1.1	8.0
	1992, > 85 kW		4.5	1.1	8.0
Euro II	1996.10		4.0	1.1	7.0
	1998.10		4.0	1.1	7.0
Euro III	1999.10	ESC & ELR	1.5	0.25	2.0
	2000.10		2.1	0.66	5.0
Euro IV	2005.10		1.5	0.46	3.5
Euro V	2008.10		1.5	0.46	2.0
Euro VI	2013.01	WHSC	1.5	0.13	0.40

Table 2: EU emission standards for heavy-duty diesel engines [19].

- In this model, the injected fuel in each cycle is illustrated in Figure 5. This picture indicates the climax of injected fuel at 1100 and 1200 (rpm).

- The generated power and torque in 'OM457' engine, respect to the previous valve timing and injected fuel are illustrated in Figure 6.

Optimized angles

As mentioned before, DOE determined that in 'OM457' the EVO, IVC and valves overlap are Non-dependent parameters. Then, SFC of several angles were calculated to achieve optimum EVO for each speed. Afterward, Chebyshev approximation formula employed to connect the discrete data by a curve which leads to yield the optimum EVOs. This method was applied to calculate the IVC and valves overlap. The evaluated results of EVO, IVC and valves overlap for minimum SFC are gathered in Table 4.

Results of deploying VVT system

After indicating the best angels in each revolutionary speed, these numbers inserted to 1-D model of engine as lookup tables. The

evaluated torque in both 'OM457' and 'VVT-OM457' is illustrated in Figure 7.

Engine Speed [rpm]	Optimized Parameter [Crank Angle Deg]		
	EVO	IVC	Valves Overlap
800	149	528	58
1000	148	541	55.5
1200	146	567	48
1400	144	590	46
1600	142	610	44
1800	139.5	621.5	40
2000	137	631	39

Table 4: Optimum valve timing for minimum SFC.

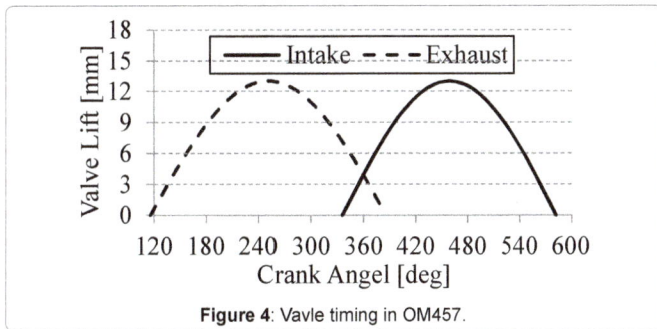

Figure 4: Vavle timing in OM457.

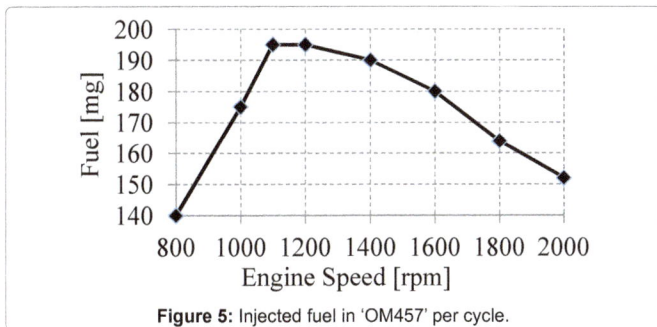

Figure 5: Injected fuel in 'OM457' per cycle.

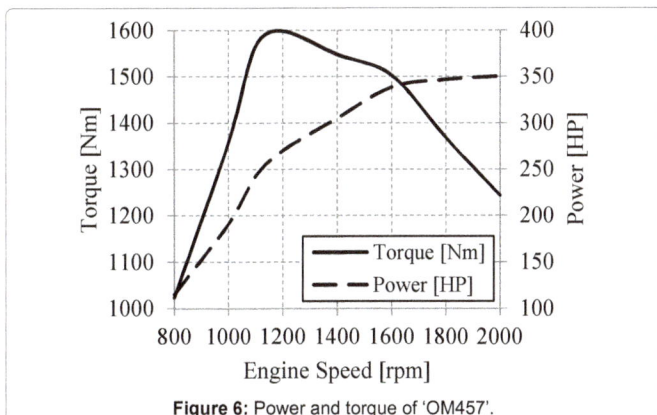

Figure 6: Power and torque of 'OM457'.

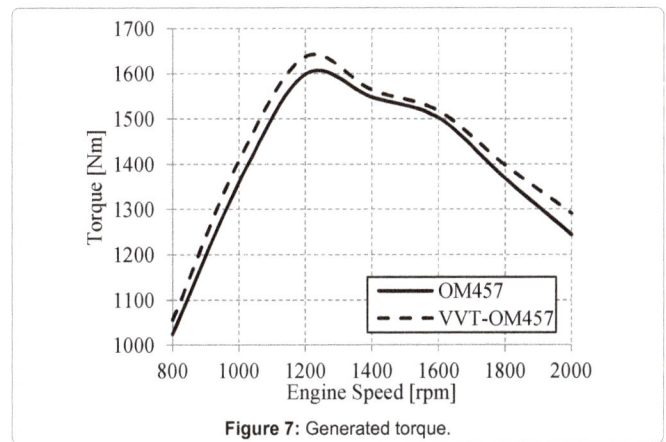

Figure 7: Generated torque.

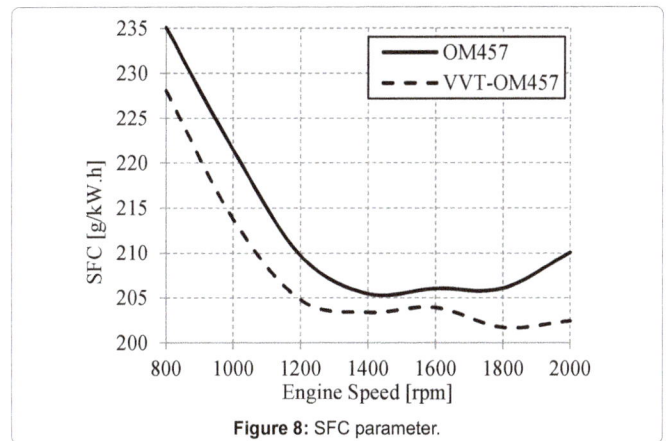

Figure 8: SFC parameter.

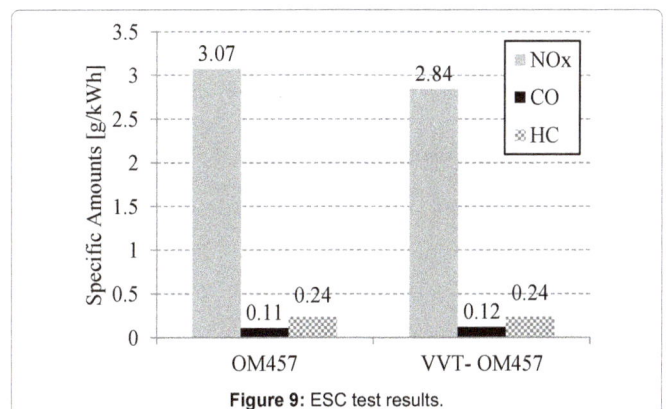

Figure 9: ESC test results.

As expected, the generated torque in 'VVT-OM457' is always greater than primary engine and this increase was more significant in higher and lower range of speed, equal to 3%. But in middle range of speed it was just about 1%.

Due to the constant amount of injected fuel in both modes, the same event is expected for SFC as illustrated in Figure 8. All charts are drawn in full load condition with the injected fuel of Figure 5.

Similar to Figure 7, the greatest improvement for SFC of 'VVT-OM457' took place in higher and lower range of speed and in middle range of speed it is close to SFC of 'OM457'.

According to these results, it can be state that the manufacturer set the valves timing in order reach optimum parameters in these speeds which is the average speed of most standard cycles like ESC, WHSC[2] and even NRTC[3].

Results of European stationary cycle

Figure 9 illustrates the difference of emissions in primary engine and VVT engine. According to ESC test, the NOx pollutant rate decreases from 3.07 in 'OM457' to 2.84 in 'VVT-OM457' (considering the same after treatment system for both). In this test the CO and HC pollutants also had slight variations, away from standard bounds.

By applying VVT system in this engine the major emission of NOx decreases 7.4% but CO and HC levels during ESC test didn't show any significant changes.

Conclusion

This paper indicates that average generated torque in VVT mode increased 2% respect to primary 'OM457' engine while average of SFC parameter decreased 2.3% and average of NOx pollutant decreased 1.6% from 800 to 2000 rpm.

In this paper the evaluated results for EVO, IVC and valves overlap for minimum SFC are based on different speeds of engine. In another approach, it is possible to determine the optimum valve timing as a function of eider speed and load of engine Appendix 1.

Acknowledgment

The authors would like to thank DESA Company for providing an appropriate model.

Declaration of conflicting interests

The authors declared no potential conflicts of interest with respect to the research, authorship, and/or publication of this article.

References

1. Asmus T (1999) A manufacturer's perspective on IC engine technology at century's end. Paper presented at the Daimler Chrysler ASME ICE Division Fall Technical Conference Ann Arbor MI.

2. White AP, Zhu G, Choi J (2013) Linear parameter-varying control for engineering applications Springer.

3. Jia M, Xie M, Wang T (2013) Numerical investigation of the influence of intake valve lift profile on a diesel premixed charge compression ignition engine with a variable valve actuation system at moderate loads and speeds. Int J Engine Research 14: 151-179.

4. Carley L (2014) The inner workings of variable valve timing. Engine Builder.

5. Snell JB (1971) Mechanical engineering: railways: Longman.

6. Altmann W (1975) Valve adjustment mechanism for internal combustion engine. Google Patents.

7. Honda Motor Co. (2006) The VTEC Beakthrough solving a century-old dilemma.

8. Torazza G, Giacesa D (1972) Valve-actuating mechanism for an internal combustion engine: Google Patents.

9. Gimelli A, Muccillo M, Pennacchia O (2014) Study of a new mechanical variable valve actuation system: Part I-valve train design and friction modeling Int J Engine Research.

10. Milovanovic N, Turner J, Kenchington S, Pitcher G, Blundell D (2005) Active valvetrain for homogeneous charge compression ignition. Int J Engine Research 6: 377-397.

11. Pulkrabek WW (2004) Engineering fundamentals of the internal combustion engine. Pearson Prentice Hall New Jersey.

12. Heywood JB (1988) Internal combustion engine fundamentals Mcgraw-hill New York.

13. Ehsani M, Gao Y, Emadi A (2009) Modern electric, hybrid electric, and fuel cell vehicles: Fundamentals theory and design: CRC press.

14. Wu H, Collings N, Etheridge J, Mosbach S, Kraft M (2012) Spark ignition to homogeneous charge compression ignition mode transition study: a new modelling approach. Int J Engine Research 13: 540-564.

15. Kleuter B, Menzel A, Steinmann P (2007) Generalized parameter identification for finite viscoelasticity. Computer methods in applied mechanics and engineering 196: 3315-3334.

16. Alonge F, D'ippolito F, Raimondi F (2001) Least squares and genetic algorithms for parameter identification of induction motors. Control Engineering Practice 9: 647-657.

17. Nyarko EK, Scitovski R (2004) Solving the parameter identification problem of mathematical models using genetic algorithms. Applied mathematics and Computation 153: 651-658.

18. Ponthot JP, Kleinermann JP (2006) A cascade optimization methodology for automatic parameter identification and shape/process optimization in metal forming simulation. Computer methods in applied mechanics and engineering 195: 5472-5508.

19. Diesel Net. Emission Standards.

20. Diesel Net. European Stationary Cycle (ESC).

21. Khan U, Sen S, Dey S, Banerjee A (2016) Development of multi cylinder CRD-I engine to meet euro VI emission norms. Technology 7: 26-36.

22. Wikipedia. Mercedes-Benz OM457 engine.

---

[2]World Harmonized Stationary Cycle
[3]Non Road Transient Cycle

Multi Scale Modeling and Failure Analysis of Laminated Composites

Uniyal P*, Gunwant D and Misra A

Department of Mechanical Engineering, Gbpuat Pantnagar, India

Abstract

In present study a multi scale modeling and failure analysis of laminated composites is performed. for micro level study Rule of Mixtures and Halphin-Tsai equations are used to determine lamina properties. Off-axis failure strength of lamina for different volume fractions are calculated using Finite Element software ANSYS. Finite element analysis results are compared with analytical results and published experimental results. In macro level study of laminates first ply failure load of laminates is calculated using ANSYS and compared with analytical results. Various failure theories i.e. maximum stress theory, maximum strain theory, Tsai-Wu, Tsai-Hill and Puck failure criteria are implemented. First ply failure load for different lamination schemes are calculated for uni-axial and Bi-axial loading conditions.

Keywords: Composites; Macromechanical analysis

Introduction

Laminated composites are made by stacking various layers of unidirectional lamina at different angles to provide required stiffness and strength in particular direction. Each lamina is made up of unidirectional fibers arranged in a matrix. Hence study of laminated composites can be performed at different scales. In micromechanical analysis of lamina properties of individual constituents, interaction between fiber and matrix, distribution of fibers in matrix all these factors are considered. In macromechanical analysis lamina is considered as a homogenous and orthotropic body. In macromechanical analysis of laminates entire stack of lamina is considered as a single body and integration of individual properties of lamina are used for analysis. Micromechanics of lamina is helpful while developing new fiber matrix system. Macro level analysis of lamina can be used when determine stress and strains in lamina. Further off axis strengths can be calculated when loading is not along the direction of fibers. Macromechanical analysis of laminates is capable to predict behavior of laminates under different loading conditions.

Experimental study of laminated composites is expensive and time consuming work. Hence so many theoretical models have been proposed to determine properties of laminated composites. For micromechanical analysis of lamina Halphin and Tsai proposed model to predict elastic properties of lamina. Huang developed a formulae based on micromechanics to predict strength properties of lamina [1]. Off-axis strength of lamina is important when direction of loading is not along the direction of fibers. Pipes performed a benchmark experimental study on boron-epoxy, aramid-epoxy and graphite epoxy lamina to determine off-axis failure strength [2]. First ply failure load is important parameter while designing laminated composite structure. Theoretical and finite element analysis procedures have been developed to determine first ply failure of laminated composites. Reddy et al. performed a benchmark study to determine first ply failure load of laminated composites using finite element method [3]. Kam et al. determined first ply failure load of laminated composites using analytical and experimental methods [4]. Rahimi et al. determined first ply failure and last ply failure loads for $[\Theta_4/0_4/-\Theta_4]_s$ lamination scheme using finite element software ANSYS [5,6].

In present work modeling and failure analysis of laminated composites is performed using finite element software ANSYS. Finite element results are compared with theoretical results. Theoretical procedure is coded in MATLAB program.

Methodology

In present study investigation is carried out at three different scales. In micromechanics of lamina Rule of Mixtures and Halphin-Tsai equations are used for to determine elastic properties of lamina. Haug's micromechanics approach is used to determine strength properties of lamina. For macromechanics of lamina generalized Hook's law is used to determine stress and strains in local axis and various failure theories are used to determine failure strength. For macromechanical analysis of laminates ANSYS software is used to determine first ply failure load. Various failure theories are coded in MATLAB to determine first ply failure load of laminates.

Micromechanics of lamina

In present investigation semi-empirical method Halphin-Tsai model is used to determine elastic properties of lamina. Halphin-Tsai proposed following relations to determine elastic properties of lamina-

Longitudinal Young's Modulus-

$$E_1 = E_f V_f + E_m V_m$$

Transverse Young's Modulus-

$$E_2 = \frac{1 + \xi \eta V_f}{1 - \eta V_f} E_m$$

$$\eta = \frac{(E_f / E_m) - 1}{(E_f / E_m) + \xi}$$

---

*Corresponding author: Uniyal P, Department of Mechanical Engineering, Gbpuat Pantnagar; E-mail: piyu.uniyal@gmail.com

In plane Poisson's Ratio-

$$\nu_{12} = \nu_f V_f + \nu_m V_m$$

In plane Shear Modulus-

$$G_{12} = \frac{1 + \xi \eta V_f}{1 - \eta V_f} G_m$$

$$\eta = \frac{(G_f / G_m) - 1}{(G_f / G_m) + \xi}$$

$$\xi = 1 + 40 V_f^{10}$$

Macromechanics of lamina

Hook's law for 2-D Lamina: Laminates are generally made up of stacking various layers at different angles because lamina is weak in the direction perpendicular to fibers. Hence to provide transverse stiffness and strength lamina are placed at various angles. So it is necessary to develop a stress-strain relationship for an angle lamina.

For an angle lamina it required to define a different co-ordinate system which is known as local co-ordinate system. Axis along the direction of fibers is known as longitudinal local axis whereas axis perpendicular to the fiber is known as transverse local axis.

The hook's law for a 2D angle lamina can be written as:

$$\begin{bmatrix} \sigma_x \\ \sigma_y \\ \tau_{xy} \end{bmatrix} = [T]^{-1}[Q][R][T][R]^{-1} \begin{bmatrix} \varepsilon_x \\ \varepsilon_y \\ \varepsilon_{xy} \end{bmatrix}$$

Where,

$$[T] = \begin{bmatrix} c^2 & s^2 & 2sc \\ s^2 & c^2 & -2sc \\ -sc & sc & c^2 - s^2 \end{bmatrix}, c = \cos(\theta) \, and \, s = \sin(\theta)$$

$$[Q] = Stiffness \, matrix$$

$$[R] = \begin{bmatrix} 1 & 0 & 0 \\ 0 & 1 & 0 \\ 0 & 0 & 2 \end{bmatrix}, Reuter's \, matrix$$

Above equation can be written as-

$$\begin{bmatrix} \sigma_x \\ \sigma_y \\ \tau_{xy} \end{bmatrix} = \begin{bmatrix} \bar{Q}_{11} & \bar{Q}_{12} & \bar{Q}_{16} \\ \bar{Q}_{12} & \bar{Q}_{22} & \bar{Q}_{26} \\ \bar{Q}_{16} & \bar{Q}_{26} & \bar{Q}_{66} \end{bmatrix} \begin{bmatrix} \varepsilon_x \\ \varepsilon_y \\ \varepsilon_{xy} \end{bmatrix}$$

Or

$$\begin{bmatrix} \varepsilon_x \\ \varepsilon_y \\ \varepsilon_{xy} \end{bmatrix} = \begin{bmatrix} \bar{S}_{11} & \bar{S}_{12} & \bar{S}_{16} \\ \bar{S}_{12} & \bar{S}_{22} & \bar{S}_{26} \\ \bar{S}_{16} & \bar{S}_{26} & \bar{S}_{66} \end{bmatrix} \begin{bmatrix} \sigma_x \\ \sigma_y \\ \tau_{xy} \end{bmatrix}$$

Where,

$[\bar{Q}]$=Reduced Stiffness matrix.

$[\bar{S}]$=Reduced Compliance matrix (inverse of reduced striffness matrix).

Failure Theories

Maximum stress failure theory

This theory is based on maximum stress theory of Rankine and maximum shear stress theory of Tresca. According to this theory failure occurs when any one of stress in material axis exceeded the failure value of stress.

Lamina is considered to be failed if any one of the following conditions violates-

$$-(\sigma_1^C)_{ult} < (\sigma_1) < (\sigma_1^T)_{ult}$$
$$-(\sigma_2^C)_{ult} < (\sigma_2) < (\sigma_2^T)_{ult}$$
$$-(\tau_{12})_{ult} < \tau_{12} < (\tau_{12})_{ult}$$

Maximum strain failure theory

This theory is based on max. Strain theory of St. Venant. According to this theory lamina fails when values of strains in material axis exceed limiting values of strains.

Lamina is considered to be failed if any one of the following conditions violates-

$$-(\varepsilon_1^C)_{ult} < (\varepsilon_1) < (\varepsilon_1^T)_{ult}$$
$$-(\varepsilon_2^C)_{ult} < (\varepsilon_2) < (\varepsilon_2^T)_{ult}$$
$$-(\gamma_{12})_{ult} < \gamma_{12} < (\gamma_{12})_{ult}$$

Tsai-Wu failure theory

This interactive failure theory is based on strain energy theory for isotropic materials.

Tsai-Wu failure theory when applied to a lamina states that, a lamina is considered to be safe if:

$$H_1\sigma_1 + H_2\sigma_2 + H_6\tau_{12} + H_{11}\sigma_2^2 + H_{66}\tau_{12}^2 + 2H_{12}\sigma_1\sigma_2 < 1$$

$$H_2 = \frac{1}{(\sigma_1^T)_{ult}} - \frac{1}{(\sigma_1^C)_{ult}},$$

$$H_{11} = \frac{1}{(\sigma_1^T)_{ult}(\sigma_1^C)_{ult}}$$

$$H_2 = \frac{1}{(\sigma_2^T)_{ult}} - \frac{1}{(\sigma_2^C)_{ult}},$$

$$H_{22} = \frac{1}{(\sigma_2^T)_{ult}(\sigma_2^C)_{ult}}$$

$$H_6 = 0, \; H_{66} = \frac{1}{(\tau_{12})_{ult}^2}$$

$$_{12} = -\frac{}{2(\;)_{ult}}, \text{ as per Tsai} - \text{Hill failure theory}$$

$$H_{12} = -\frac{1}{2\sigma_{1\,ult}^{T}\sigma_{1\,ult}^{C}}, \text{ as per Hoffman criterion}$$

$$H_{12} = -\frac{1}{2}\sqrt{\frac{1}{(\sigma_1^T)_{ult}(\sigma_1^C)_{ult}(\sigma_2^T)_{ult}(\sigma_2^C)_{ult}}}, \text{ as per Mises} - \text{Hencky criterion}$$

Tsai-Hill failure theory

Based on distortion energy theory for isotropic materials. Lamina is considered fail when following condition violates-

$$(G_2 + G_3)\sigma_1^2 + (G_1 + G_3)\sigma_2^2 + (G_1 + G_2)\sigma_3^2 - 2G_3\sigma_1\sigma_2 - 2G_2\sigma_1\sigma_3 - 2G_1\sigma_3\sigma_2 + 2G_4\tau_{23}^2 + 2G_5\tau_{13}^2 + 2G_6\tau_{12}^2 < 1$$

$$G_1 = \frac{1}{2}\left(\frac{2}{[(\sigma_2^T)_{ult}]^2} - \frac{1}{[(\sigma_1^T)_{ult}]^2}\right)$$

$$G_2 = G_3 = \frac{1}{2}\left(\frac{1}{[(\sigma_1^T)_{ult}]^2}\right)$$

$$G_6 = \frac{1}{2}\left(\frac{1}{[(\tau_{12})_{ult}]^2}\right)$$

Modeling and Finite Element Analysis of Laminates in ANSYS

Steps involve in finite element analysis (ANSYS)

Finite element analysis involves three stages of activity: pre-processing, processing and post processing. A complete finite element analysis is a logical interaction of the three stages.

Steps of finite element analysis in ANSYS-

- Preprocessing
- Specify element types to be used
- Specify options for element behavior
- Specify real constants
- Specify material model

- Specify material properties
- Create geometry
- Specify meshing options
- Mesh model
- Apply boundary conditions
- Solve problem
- Post processing (reviewing results)

Material properties

Shown in Tables 1-4.

Results and Discussion

Micromechanical analysis of lamina

In this section three different fibers Graphite, Boron and Aramid fibers with epoxy as a matrix system has been analyzed. Elastic properties such as Young's modulus, Poission'ratio, shear modulus and strength properties has been calculated and compared at different fiber volume fractions. Volume fraction of fibers is varied from 30% to 65%.

Figure 1 shows a linear variation of longitudinal Young's modulus with fiber volume fraction. Boron/epoxy composition shows maximum value of longitudinal and transverse Young's modulus, and longitudinal tensile strength among aramid/epoxy and graphite/epoxy compositions whereas aramid/epoxy exhibits minimum values among all compositions. Transverse Young's modulus shows a nonlinear trend with fiber volume fraction. It can be observed that up to 0.4 volume fraction there is no significant difference in transverse Young's modulus for three different fiber and values are diverging as volume fraction increases. Stiffness and strength of lamina increases with increase in number of fibers.

Macromechanical analysis of lamina

In this section strength of lamina is calculated when orientation of fibers is not along the direction of loading. This strength is also known as off-axis strength of lamina. Off axis strength of lamina is calculated using FEA software ANSYS and stress-strain relationship and failure theory are

	E_1(GPa)	E_2(GPa)	v_{12}	v_{23}	G_{12}(GPa)	X_t(MPa)	Y_t(MPa)	S(MPa)
Graphite	213.7	13.8	0.2	0.25	13.8	2250	-	-
Epoxy	3.45	3.45	0.35	0.35	1.3	-	62.9	108

Table 1: Properties of graphite and epoxy [15].

	E_1(GPa)	E_2(GPa)	v_{12}	v_{23}	G_{12}(GPa)	X_t(MPa)	Y_t(MPa)	S(MPa)
Boron	400		0.2		166.7	2566		
Epoxy	3.45	3.45	0.35	0.35	1.3	-	62.9	108

Table 2: Properties of boron and epoxy [15].

	E_1(GPa)	E_2(GPa)	v_{12}	v_{23}	G_{12}(GPa)	X_t(MPa)	Y_t(MPa)	S(MPa)
Aramid	124.1	4.1	0.35	0.35	2.9	2031		
Epoxy	3.45	3.45	0.35	0.35	1.3	-	62.9	108

Table 3: Properties of aramid and epoxy [15].

E_x(GPa)	E_y(GPa)	v_{xy}	G_{xy}(GPa)	X_t(MPa)	X_c(MPa)	Y_t(MPa)	Y_c(MPa)	S(MPa)
204	185	0.23	5.59	1260	2500	61	202	67

Table 4: Elastic and strength properties for boron-epoxy lamina [15].

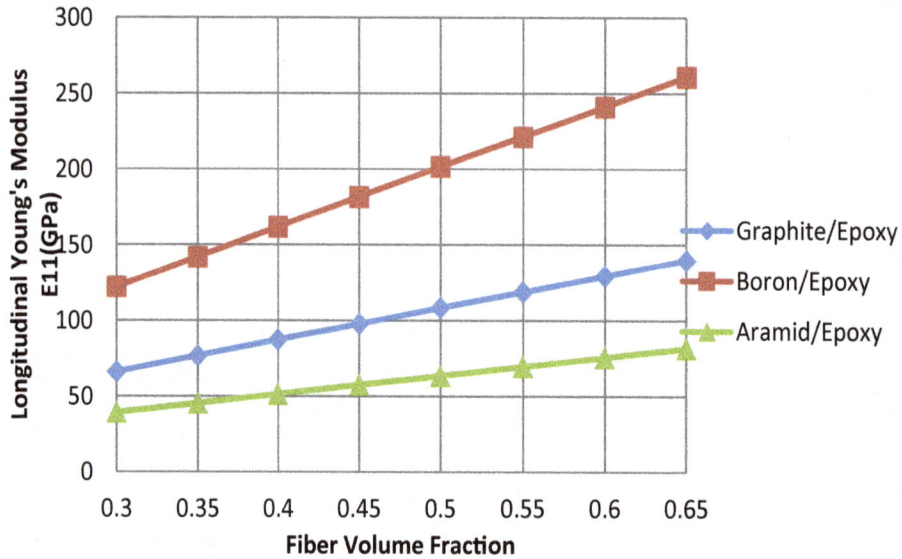

Figure 1: Variation of longitudinal Young's modulus with fiber volume fraction.

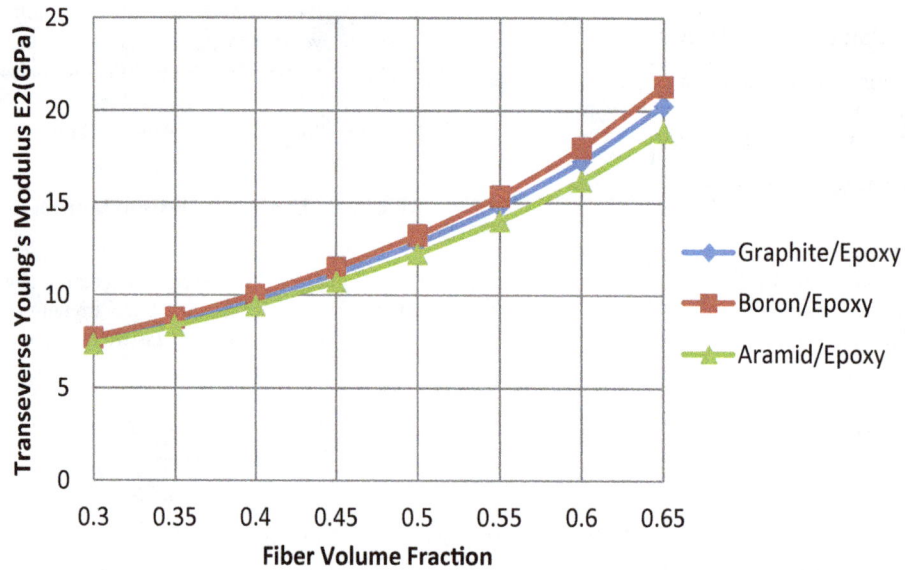

Figure 2: Variation of transverse young's modulus with fiber volume fraction.

Fiber Orientation Angle	Off-axis Failure Strength (MPa)		%Error
	FEM(ANSYS)	Experimental [14]	
15°	217.68	231.28	5.88
30°	113.07	123.16	8.13
45°	79.64	84.87	6.16
60°	64.72	63.94	1.21

Table 5: Comparison of FEM results with experimental results [14].

coded in MATLAB. Results obtained using ANSYS and MATLAB are compared with available experimental results. Finally variation of off-axis strength with fiber orientation angle is plotted for different volume fraction of boron fiber in epoxy matrix (Figures 2 and 3).

Validation of FEA model (ANSYS)

Comparison of FEM results with Experimental Results [7] shown in Table 5.

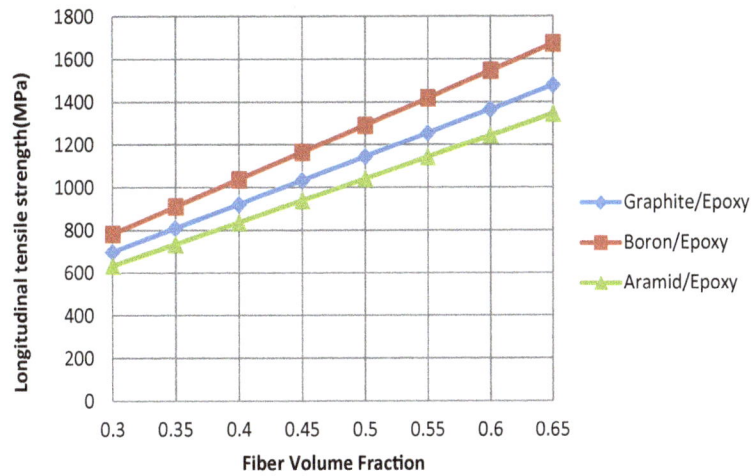

Figure 3: Variation of longitudinal tensile strength with fiber volume fraction.

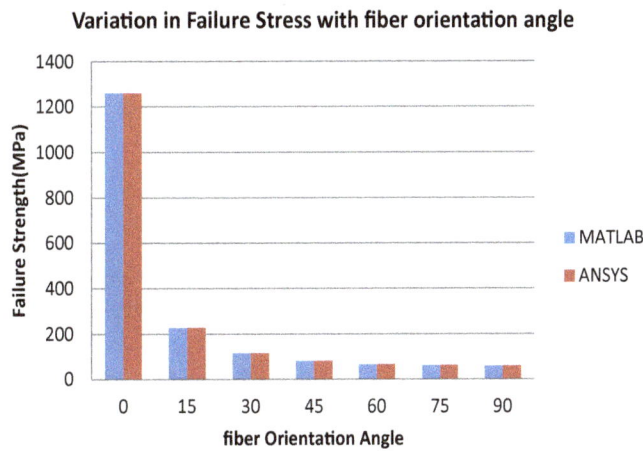

Figure 4: Comparison of ANSYS results with theoretical results.

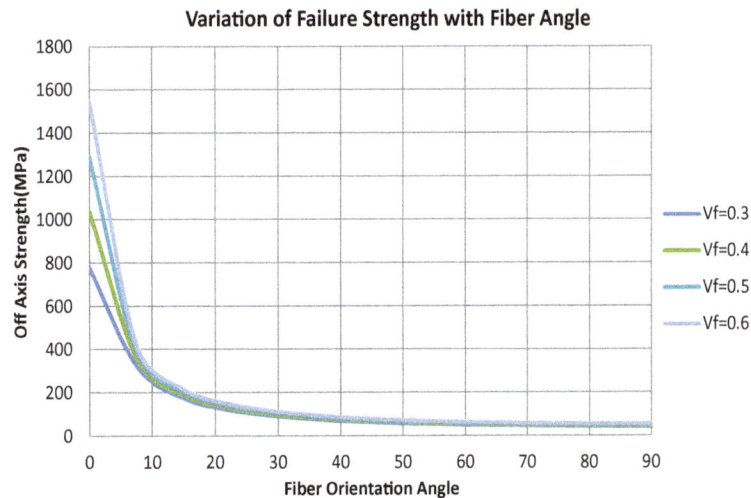

Figure 5: Variation of off axis failure strength with fiber orientation for different volume fractions.

In Figure 4 comparison between FEA (ANSYS) results and theoretical (MATLAB) results is presented. A very good agreement is found between ANSYS and MATLAB results. Failure strength at different fiber orientation angles and local stress results are calculated using ANSYS and MATLAB. From above observations it can be concluded that analysis of lamina can be performed accurately using ANSYS.

Variation of off-axis strength with fiber orientation angle for different volume fractions

It has been observed from Figure 5 that failure strength decreases with increase in fiber orientation angle. Maximum strength is obtained when loading is along the direction of fibers and minimum when loading is perpendicular to the direction of fibers. Higher strength values are associated with higher fiber volume fractions. It is also noticeable that effect of fiber volume fraction is only significant for only low fiber orientation angles. Beyond 10° fiber orientation angles failure strength of lamina is almost same for all volume fractions. Since cost is directly associated with fiber volume fractions. Higher the fiber volume fraction, higher will be cost of lamina. Hence it is suggested to use lamina with lower fiber volume fraction when off axis angles are higher than 10°.

Macro-mechanical Analysis of Laminates

Validation of FEA (ANSYS) model

Comparison of FEA (ANSYS) results with Experimental data [8] for $(0_2/90_2)_s$ laminate shown in Table 6.

Comparison of FEA (ANSYS) results with Experimental data [8] for $(0/90/0/90)_s$ laminate shown in Table 7.

Uniaxial loading

Case 1: $(0/\Theta\,\Theta/0\,\Theta/\Theta\,\Theta)_s$: Shown in Table 8. From Figure 6 it can be concluded that failure strength is maximum when all layers are orientated at 0° and minimum for $(0/90/0/90)_s$ laminates. First ply failure strength decreases as Θ increases for $(0/\Theta/0/\Theta)$ laminates.

Failure strength drastically decreases from 0° to 15° which is from 1260MPa to 760.75MPa according to maximum stress theory. Results obtained using Tsai-Wu and Puck failure criteria are very close whereas maximum stress theory predicts higher failure load in range of 15° to 60°.

Case 2: $(\Theta°/90°-\Theta/\Theta°/90°-\Theta)_s$: For $(\Theta°/90°-\Theta/\Theta°/90°-\Theta)_s$ lamination scheme from Figure 7 it can be concluded that results for failure stress are symmetric about $\Theta= 45°$. Failure strengths predicted using different failure criteria are very close except maximum stress theory at 45°. Maximum stress theory predicts higher failure load at 45° compare to other failure theories. $(45°/45°/45°/45°)_s$ lamination scheme shows minimum failure strength as compare to other combinations [9-15].

Case 3: $(\Theta°/-\Theta/\Theta°/-\Theta)_s$: Figure 8 shows variation of failure strength with layer orientation angle Θ for angle ply lamination scheme. It can be observed from figure that failure strength decreases drastically as ply angle from 15° to 30°. Beyond 45° curve become asymptotic to X axis and there is no significant variation in failure strength. All failure criteria provide almost same results. Maximum stress theory predicts higher failure loads compare to other failure theories (Figure 9).

Case 4: $(\Theta°/\Theta°/\Theta°/\Theta°)_s$: Failure strength for unidirectional laminate is maximum when all layers are oriented along the direction of loading. For unidirectional laminates failure strength decreases drastically from 0° to 15°. Beyond 15° ply angles there is no significant difference in failure strength. Results obtained using all failure criteria is very close (Figures 10-13).

Biaxial loading

Case 1: $(0°/\Theta°/0°/\Theta°)_s$: Under bi-axial loading condition for $(0°/\Theta°/0°/\Theta°)$ failure strength increases as Θ increases. Maximum failure strength is obtained when ply angle Θ is 90°. Failure strength increases rapidly beyond when ply angle Θ is greater than 60°. For lamination scheme $(0°/\Theta°/0°/\Theta°)$ when Θ is 90° laminate becomes cross ply laminate. Under bi axial loading cross ply laminates can take load in both directions. For unidirectional laminates under bi axial loading laminates is strong in longitudinal direction whereas weak in transverse direction.

	Theoretical [3]	ANSYS	Experimental [3]	Error %
Max. stress	229.11	225.63		11.02
Max. Strain	267.55		253.60	
Tsai-Wu	238.60	274.30		7.54
Tsai-Hill	224.19			

Table 6: Comparison of FEA (ANSYS) results with experimental data [3] for $(0_2/90_2)_s$ laminate.

	Theoretical [3]	ANSYS	Experimental [3]	Error %
Max. stress	290.12	281.77		11.32
Max. Strain	355.47		317.74	
Tsai-Wu	304.13	301.70		5.04
Tsai-Hill	280.81			

Table 7: Comparison of FEA (ANSYS) results with Experimental data [3] for $(0/90/0/90)_s$ laminate.

Failure Criteria	First Ply Failure Stress (MPa)						
	0°	15°	30°	45°	60°	75°	90°
Max. Stress	1260	760.75	778.07	714.20	534.15	413.57	369.53
Tsai-Wu	1260	597.77	544.77	521.76	464.84	399.22	369.80
Tsai-Hill	1260	675.67	610.37	565.66	485.20	403.89	369.43
Puck	1260	666.67	588.23	526.31	476.19	400	370.37
Max Strain	1260.5	758.58	778.29	737.65	538.77	413.19	368

Table 8: Uniaxial loading.

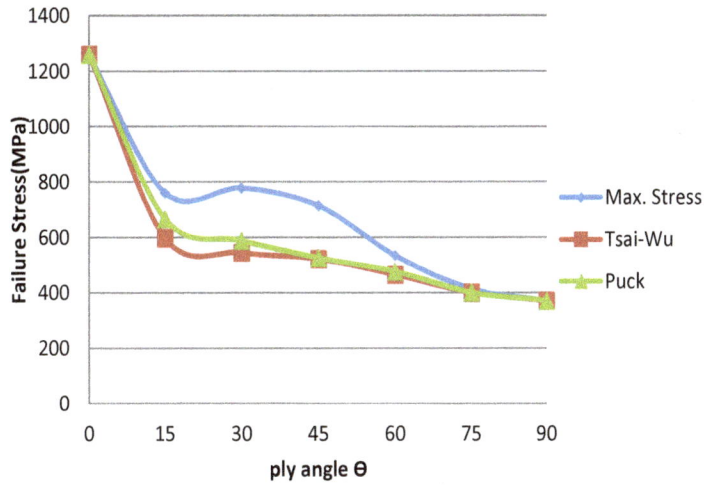

Figure 6: Variation of failure stress with ply angle Θ for (0˚/Θ˚/0˚/Θ˚) laminate.

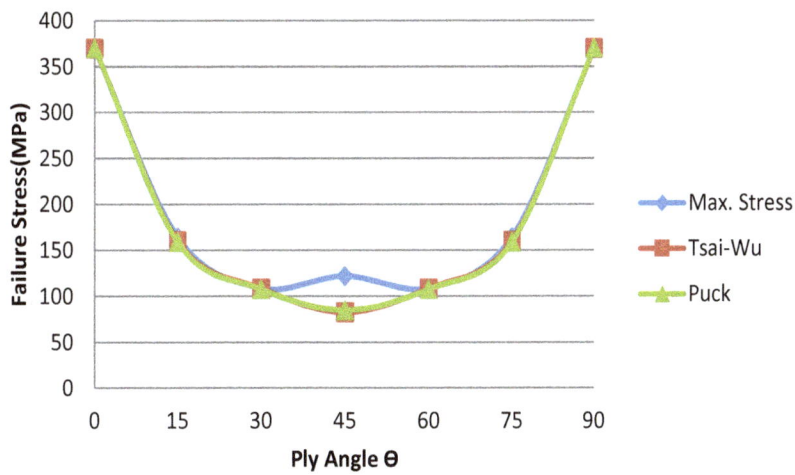

Figure 7: Variation of failure stress with ply angle Θ for (Θ˚/90˚-Θ˚/ Θ˚/90˚-Θ˚) laminate.

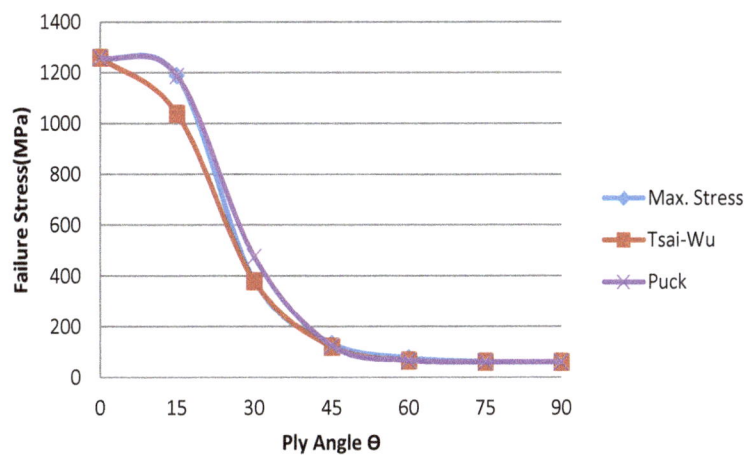

Figure 8: Variation of failure stress with ply angle Θ for (Θ˚/-Θ˚/ Θ˚/-Θ˚) laminate.

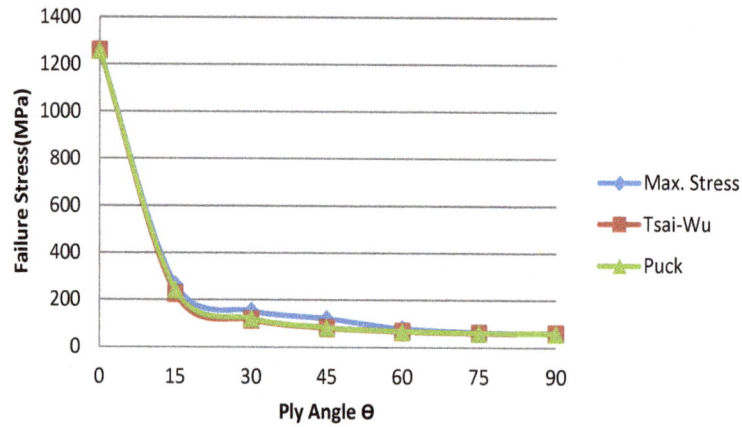

Figure 9: Variation of failure stress with ply angle θ for (θ°/θ°/ θ°/θ°) laminate.

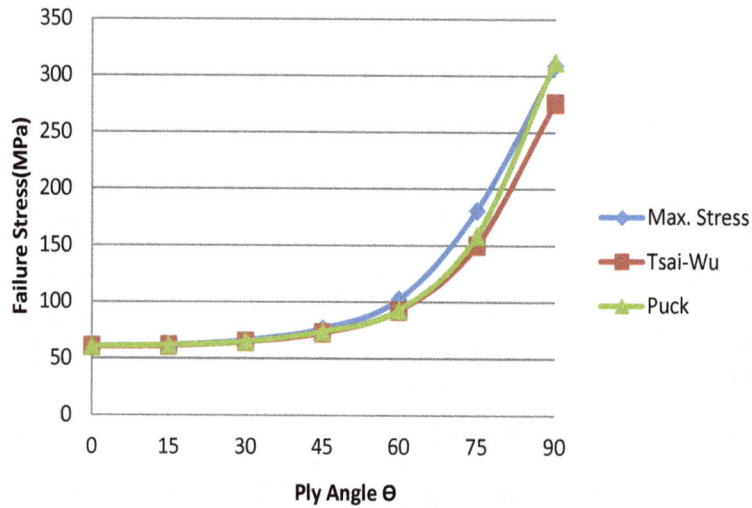

Figure 10: Variation of failure stress with ply angle θ for (0°/θ°/0°/θ°) laminate.

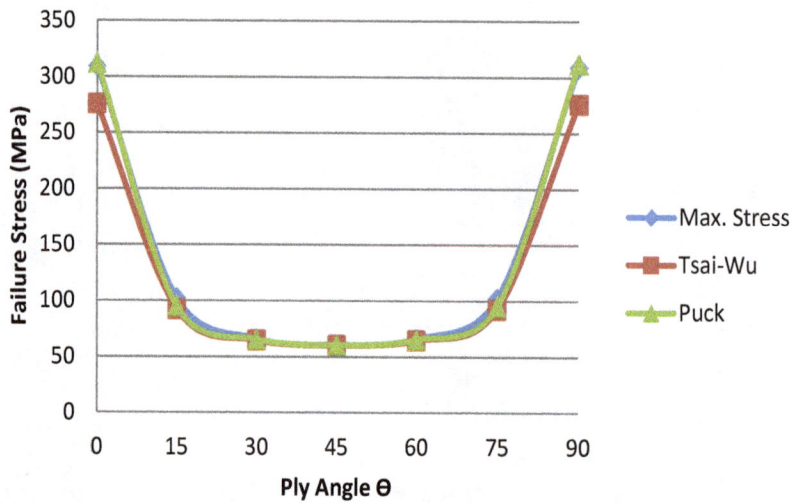

Figure 11: Variation of failure stress with ply angle θ for (θ°/90°-θ°/ θ°/90°-θ°) laminate.

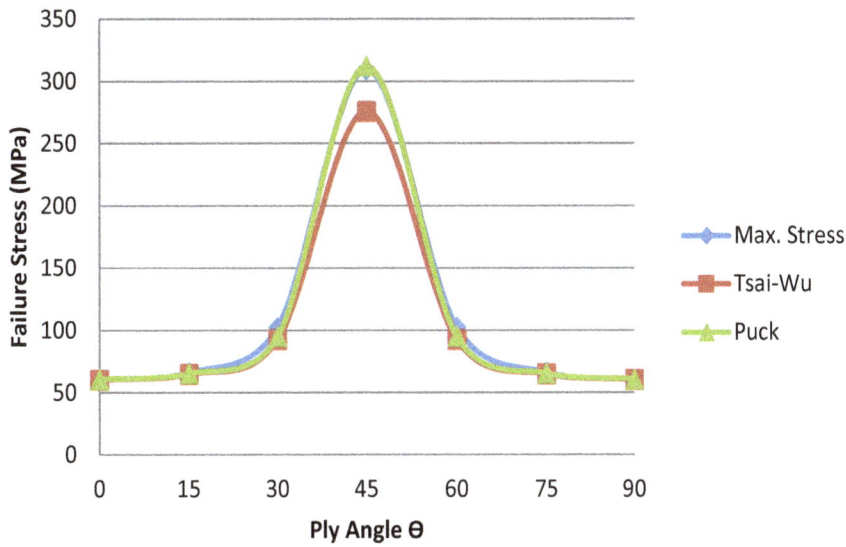

Figure 12: Variation of failure stress with ply angle ө for (ө°/-ө°/ ө°/-ө°) laminate.

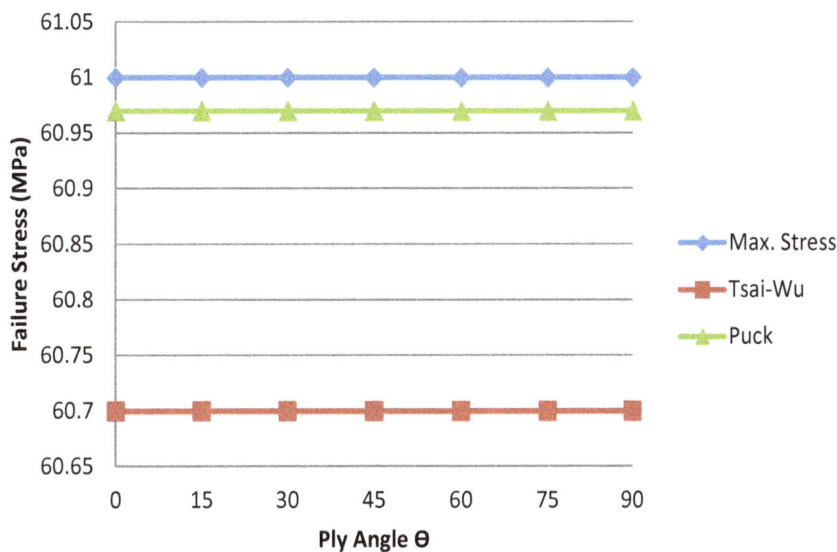

Figure 13: Variation of failure stress with ply angle ө for (ө°/ө°/ ө°/ө°) laminate.

Case 2: (Θ°/90°-Θ°/ Θ°/90°-Θ°)s: Shown in the Figure 11.

Case 3: (Θ°/-Θ°/ Θ°/-Θ°)s: For (Θ°/-Θ°/ Θ°/-Θ°)$_s$ laminate scheme maximum failure strength occurs at 45° ply angle. The obtained curve is symmetric about 45° ply angle. Minimum failure strength is obtained when Θ is 0° or 90°.

Case 4: (Θ°/Θ°/ Θ°/Θ°)s: Shown in the Figure 13.

Conclusions

1. The Boron-Epoxy composition shows maximum strength and stiffness compare to Graphite-Epoxy and Aramid-Epoxy compositions for all fiber volume fractions.

2. Off axis failure strength of lamina decreases hyperbolically as off axis angle increases. The effect of fiber volume fraction is significant only up to 10°. Beyond 10° failure strength is same for all volume fractions.

3. Under uniaxial loading condition (Θ°/-Θ°/ Θ°/-Θ°) Laminate have higher failure strength than (Θ°/ Θ°/ Θ°/Θ°) Laminates. Hence it is better to use angle ply instead of unidirectional plies under uniaxial loading conditions.

4. For biaxial loading conditions cross ply and angle ply laminates with ply angle 45° exhibits highest failure strength than all other combinations.

References

1. Huang ZM (1999) Micromechanical strength formulae for unidirectional composites. Materials Letters. 40: 164-169

2. Tabiei A, Chen Q (2001) Micromechanics based composite material model for crashworthiness explicit finite element simulation. J of Thermoplastic Composite Materials.14: 264-288.

3. Reddy YSN, Reddy JN (1992) Linear and non-linear failure analysis of composite laminates with transverse shear. Composite Science and Technology. 44: 227-255.

4. Kam TY, Sher HF, Chao TN (1996) Predictions of deflection and first-ply failure load of thin laminated composite plates via the finite element approach. Int J Solids Structure. 33: 375-398.

5. Rahimi N, Hussain AK (2012) Capability assessment of finite element software in predicting the last Ply failure of composite laminates. Int Symposium on Robotics and Intelligent Sensors 2012 (IRIS 2012) Procedia Engineering. 41: 1647-1653.

6. Ramtekkar GS, Desai YM, Shah AH (2004) First ply failure of laminated composite plates - A mixed finite element approach. J of Reinforced Plastics and composites.

7. Pipes RB, Cole BW (1973) On the off axis strength test for anisotropic materials. J of Composite Materials. 7: 246.

8. Kam TY, Sher HF, Chao TN (1996) Predictions of deflection and first-ply failure load of thin laminated composite plates via the finite element approach. Int J Solids Structure 33: 375-398.

9. Joo SG, Hong CS (2000) Progressive failure analysis of composite laminates using 3-D finite element method key engineering materials. 183-187: 535-540.

10. Kaw AK (1997) Mechanics of composite materials. CRC Press New York: 149-184.

11. Kumar YVS, Srivastava A (2003) First ply failure analysis of laminated stiffened plates composite structures. 60: 307-315.

12. Liu PF, Zheng JY (2010) Recent development on damage modeling and finite element analysis for composite laminates: A review material and design. 31: 3825-3834.

13. Pal P, Bhattacharyya SK (2007) Progressive failure analysis of cross-ply laminated composite plates by finite element method. J Reinforced Plastic and Composites 26: 465-477.

14. Rahimi MA, Rahimi N, Hussain AK, Mahmud J, Musa M (2012) Parametric study on failure analysis of composite laminate under uniaxial tensile loading. IEEE Colloquium on Humanities Science and Engineering Research Sabah Malaysia.

15. Tolson S, Zabras N (1991) Finite element analysis of progressive failure in laminated composite plates composite & structures 38: 361-376.

32

Effect of Sub-Zero Treatment on the Wear Resistance of P/M Tool Steels

Sobotova J*, Kuřík M, Krum S and Lacza J

Faculty of Mechanical Engineering, Czech Technical University in Prague, Karlovo nám 13, 121 35 Prague 2, Czech Republic

Abstract

Sub-zero treatment of tool steels is included within the cycle of conventional heat treatment. This kind of heat treatment has been reported to improve wear resistance of tools. The improvement is attributed to precipitation of fine carbide particles, but is depending on a number of the other factors as well. This paper follows the previous works, in which the effect of sub-zero treatment on the mechanical and structural properties of P/M tool steels was evaluated. Two types of P/M cold work tool steels were used in the research: Vanadis 6 and high speed steel Vanadis 30. They were austenitized, nitrogen gas quenched and tempered. The 4 hours long sub-zero period of dwell at -196°C was also incorporated between quenching and tempering. Wear evaluation was carried out using a pin-on-disk method. The observed values of wear resistance have been collated with the values of hardness and the bending strength. The results of the work are supplemented by a detailed analysis of the carbide particles of both monitored conditions.

Keywords: Sub-zero treatment; Deep cryogenic treatment; P/M tool steel; Wear resistance; SEM; EDS

Abbreviations: CHT: Conventional Heat Treatment; DCT: Deep Cryogenic Treatment; C/M: Conventional Metallurgy; P/M: Power Metallurgy; HSS: High Speed Steel; SEM: Scanning Electron Microscopy; EDS: Energy Dispersive Spectroscopy; HRC: Hardness by Rockwell

Introduction

Sub-zero treatment (also referred to as cryogenic treatment) is a way of treatment at low temperatures (lower than -80°C) added to the Conventional Heat Treatment (CHT). Therefore, essentially it is the extension of standard heat treatment processes. In case of tool steels, cryogenic treatment at temperatures ranging from -140°C to -196°C, referred to as Deep Cryogenic Treatment (DCT), included between the quenching and tempering is the most commonly used.

The influence of CHT on changes of structural and material characteristics of tool steels is well known [1,2]. The fact that the presence of high alloying elements content decreases the temperature of martensitic transformation start and finish, thereby increasing the fraction of retained austenite in the structure after hardening, is also known. To reduce this undesirable amount of the retained austenite, the cryogenic treatment can be used. Additionally, the authors of summary articles [1-4] and our own work [5,6] published in 2005-2015 addressing this issue agree on a finding that cryogenic processing of tool steels causes precipitation of very fine carbides in the structure. Superposition of both effects leads to a change of properties such as hardness, toughness, wear resistance and dimensional stability.

Based on the known results it can be stated that the effect of the cryogenic treatment on the functional properties of tool steels depends on a number of factors including the chemical composition of the processed material, method of its primary production and its application, the parameters of the cryogenic processing, but the experimental methods of functional properties evaluation as well. From a materials perspective it is logical that the effects of these various factors are mutually intertwined. Also, it is not possible to assume the same degree of sensitivity to these factors for different materials. Generally, it can be said that in the available literature it is possible to clearly find more publications dedicated to monitoring the effect of cryogenic treatment on the functional characteristics of tool steels produced by conventional metallurgy (C/M) when compared with publications addressing the same issue for the steels produced by powder metallurgy (P/M). It is known that in case of P/M tool steels, finer and more homogeneous structure is achieved in comparison with the C/M steels. That is inter alia reflected in higher toughness of the P/M steels and their wear resistance at higher cutting speeds.

The aim of the paper is to analyze influence of the DCT on precipitation effect of two tool steels of different chemical composition in connection to wear resistance evaluated in laboratory conditions.

Experimental Procedure

The experimental material was the P/M steel Vanadis 6 nominally containing (in wt.%) 2.1% C, 1.0% Si, 0.4% Mn, 6.8% Cr, 1.5% Mo and 5.4% V and P/M high-speed steel Vanadis 30 containing 1.28% C, 4.2% Cr, 5.0% Mo, 6.4% W, 3.1% V and 8.5% Co. The experimental steels were selected due to their significantly divergent chemical composition. On the basis of that, the different levels of precipitation phenomena of different carbides types are expected. For Vanadis 6 P/M tool steel for cold work, the occurrence of Cr-based M_7C_3 type and V-based MC type carbides in the structure has been confirmed in the previous works [6,7]. In case of P/M HSS Vanadis 30, V-based MC type and W-based M6C type carbides are expected to be present in the structure [8,9]. Commonly used parameters of CHT were employed for both materials (nitrogen gas quenching at pressure of 5 bars). In both cases, DCT step was inserted between quenching and tempering. Heat treatment parameters of the experimental materials are given in Table 1.

5 samples of Vanadis 6 (10 mm × 10 mm × 100 mm) and 5 samples of Vanadis 30 (Ø11 mm × 100 mm) were prepared for each combination of heat treatment parameters presented in Table 1. Surface roughness Ra of all samples for three point bending test was 0.2 – 0.3 μm. Test was carried out using Instron 5582 universal tensile testing machine with following parameters: the distance between supports was 80 mm,

*Corresponding author: Sobotova J, Faculty of Mechanical Engineering, Czech Technical University in Prague, Karlovo nám 13, 121 35 Prague 2, Czech Republic
E-mail: Jana.Sobotova@fs.cvut.cz

Material	Austenitizing	DCT	Tempering
Vanadis 6	1050°C/30 min	-	2 × 530°C
Vanadis 6	1050°C/30 min	-196°C/4 h	2 × 530°C
Vanadis 30	1100°C/5 min	-	3 × 560°C
Vanadis 30	1100°C/5 min	-196°C/4 h	3 × 560°C

Table 1: Parameters of heat treatment.

loading rate of 1 mm/min and loading force was applied to the central region up to the moment of the fracture. The fragments of the samples after the three point bending test were subsequently used for hardness measurements, metallography analysis and wear resistance test using pin on disc method. Metallographic samples were prepared using a standard procedure and observed with JEOL scanning electron microscope equipped with Oxford Instruments EDS analyzer. Six SEM micrographs at a magnification of 5000 × were always evaluated in a sense of quantification of the structural changes. EDS mapping of Vanadis 30 was done using accelerating voltage of 15 kV and a standard working distance of 8 mm. Through the EDS maps the size distribution of carbides was determined using NIS ELEMENTS image analysis. Characterization of the carbide particles was carried out using point EDS analysis as well. The hardness was measured with Emco hardness tester using Rockwell C method (HRC). Five measurements were made on each specimen and the mean value and standard deviation was calculated. Wear resistance was determined using Tribometer THT-S-CE-0000 by a pin on disc method according to the standard [10]. The surface of the testing material was polished to surface roughness lower than 0.04 μm. The test was running under dry sliding conditions at ambient temperature of 22°C and relative humidity of 40% to 50%. All specimens were ultrasonically cleaned in ethanol and dried in air. Al_2O_3 ball with a diameter of 6 mm was used as a counterpart. Applied normal load was 5 N, sliding speed 0,064 m/s, total sliding distance 100 m and track radius 2 mm. Transversal sections of the groove after the pin on disc test were measured at 6 locations on each sample using Hommel T 1000 roughness tester. Illustration of the roughness tester record of transversal section measurement of the groove is in Figure 1.

From the thusly measured values volume loss was calculated. The volume loss served to determining wear rate.

Results and Discussion

The results of hardness measurement are in Figure 2. The picture shows the hardness values after CHT are similar for both steels. After DCT on the other hand, the hardness of Vanadis 30 practically does not change whereas that of Vanadis 6 is significantly lowered.

Presented fact is due to change of the tempering curve when DCT step is inserted between quenching and tempering of Vanadis 6. This phenomena regarding Vanadis 6 is described in [11].

Influence of DCT on bending strength of the both evaluated materials is in Figure 3.

These results are in good agreement with measured hardness values (Figure 2). It can be stated that the significant drop in hardness of Vanadis 6 steel after DCT causes a slight increase in its three point bending strength. Three point bending strength of Vanadis 30 steel after DCT remains almost unchanged within the standard deviation.

The results of the pin on disc test are in Figure 4. There is also a good correlation with the results already discussed. In case of Vanadis 6 steel after DCT there is a slight increase in wear rate, whereas Vanadis 30 does not show any effect of the cryogenic treatment on the wear rate.

Unexpected difference in values was observed in standard deviation

values of the wear rate. Significantly higher standard deviation of the Vanadis 6 steel wear rate should be the subject of verification within further measurements.

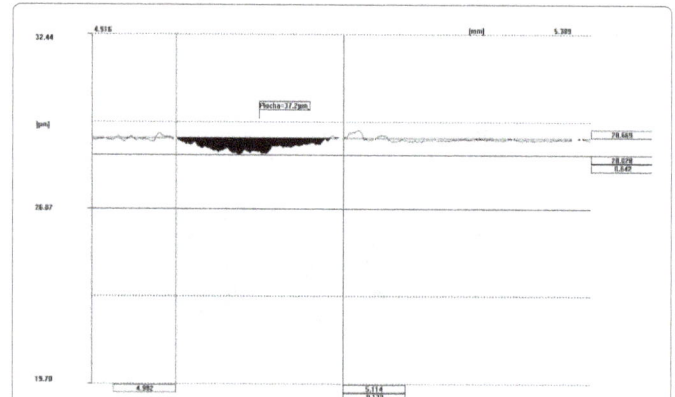

Figure 1: Illustration of the roughness tester record of transversal section measurement of the groove after the pin on disc test.

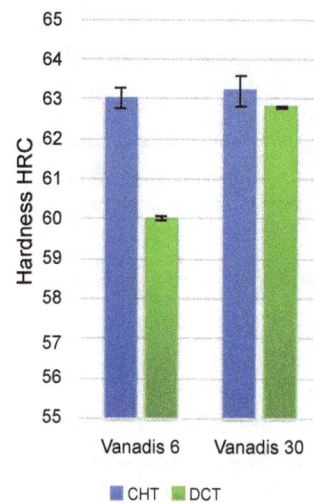

Figure 2: Influence of DCT on hardness of Vanadis 6 and Vanadis 30 steels.

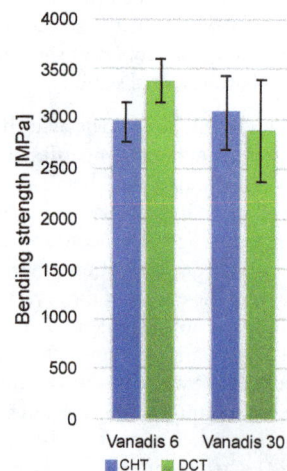

Figure 3: Influence of DCT on three point bending strength of Vanadis 6 and Vanadis 30 steels.

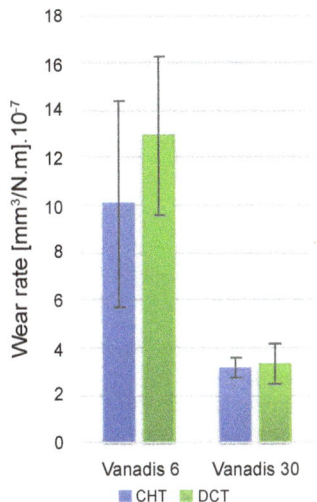

Figure 4: Influence of the DCT on wear rate of Vanadis 6 and Vanadis 30 steels.

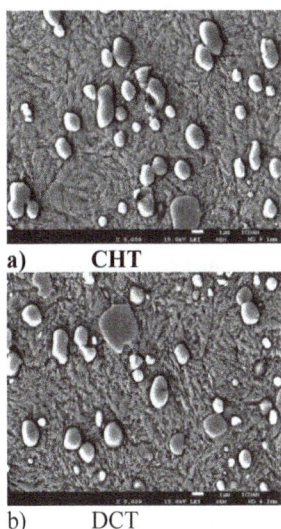

Figure 5: Influence of DCT on microstructure of Vanadis 6 steel, austenitizing temperature of 1050°C (SEM).

Microstructure comparison of Vanadis 6 after CHT and DCT is in Figure 5. In both cases, the microstructure consists of tempered martensite matrix containing uniformly distributed carbides. Carbides are of two types. Eutectic ones contain predominantly vanadium and secondary ones contain mainly chromium. In comparison with Figure 5a, the microstructure in Figure 5b additionally shows, the presence of small globular carbides already identified in previous work for Vanadis 6 after DCT, but only in case of lower austenitizing temperature of 1025°C [7,12].

Even though the detailed quantitative analysis of carbide particles after DCT was not carried out within this work, it is possible to state that the presence of the fine globular carbides in the microstructure was confirmed. These carbides should ultimately cause greater wear resistance compared to the CHT which was confirmed in but not in this paper. It is necessary to note that in [7] the used pin on disc ball was made from bearing steel 100Cr6 and the utilized method of evaluation

was different from the one used in this paper. In addition, based on previously obtained results of the authors and study of the available literature it is possible to assume that using of a higher loading forces during the pin on disc test would be more suitable [13].

Microstructure comparison of Vanadis 30 after CHT and DCT is in Figure 6. In both cases, the microstructure consists of tempered martensite matrix containing uniformly distributed lighter and darker carbides.

Detailed analysis of the carbide particles was carried out using EDS mapping. Figure 7a shows the reference image of Vanadis 30 steel after DCT. EDS maps of vanadium, tungsten, molybdenum and iron content are in Figures 7b-7e.

From the reference image of Vanadis 30 and corresponding EDS

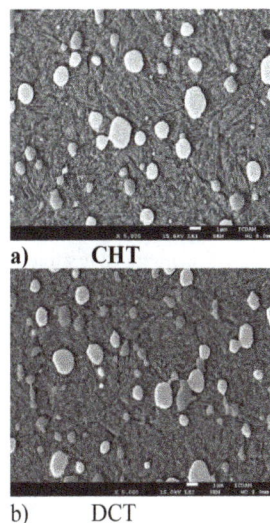

Figure 6: Influence of DCT on microstructure of Vanadis 30 steel, austenitizing temperature of 1100°C (SEM).

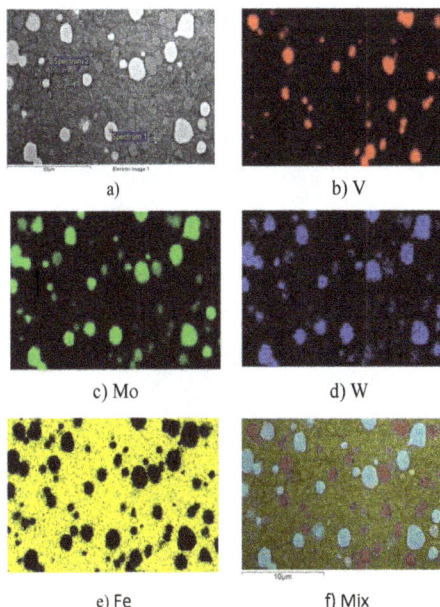

Figure 7: Metallographical analysis of Vanadis 30 after DCT (SEM, EDS).

maps of individual elements it is evident that the darker carbides contain predominantly vanadium and the lighter ones are tungsten and molybdenum based. Figure 7f shows an image composed of individual element maps. Matrix color proves the occurrence of all elements in the matrix whereas the colour of the larger carbides points at increased content of tungsten and molybdenum. The smaller carbides are predominantly vanadium based. These results were also confirmed by the point EDS analysis of the carbides marked in Figure 7a. EDS spectra of point analysis are in Figure 8. Spectrum 1 confirmed the carbides based on tungsten and molybdenum (hereinafter designated as W-carbides). The spectrum contains a relatively high iron peak. Spectrum 2 characterizes vanadium based carbides (designated as V-carbides). Molybdenum and tungsten are observed in this case as well, although their lower content is detected. The results of analysis of carbide particles is in good agreement with the results of the work [8], where authors characterized carbides in P/M HSS steel of chemical composition app. similar to Vanadis 30.

EDS maps in Figures 7b-7d were further used for an evaluation of size distribution of the carbides after CHT and DCT in Vanadis 30 using image analysis. As mentioned before, DCT should cause an increase in number of the small carbides in the microstructure compared with the microstructure after CHT. Therefore, only the numbers of carbides in the size class up to 1 μm^2 are presented in Table 2.

From the Table 2 it is evident that after DCT of Vanadis 30, number of carbides based on vanadium slightly decreases while the amount of carbides based on tungsten doubles. However in general, it can be stated that influence of DCT on fine carbides precipitation was not confirmed. Truth is that the result is in good accord with the results of mechanical tests and wear test of pin on disc, which has not shown any different values when compared with the state after CHT either.

However, it is imperative to say that the decisive factor regarding the usage of cryogenic treatment will be the results of the durability tests of real tools.

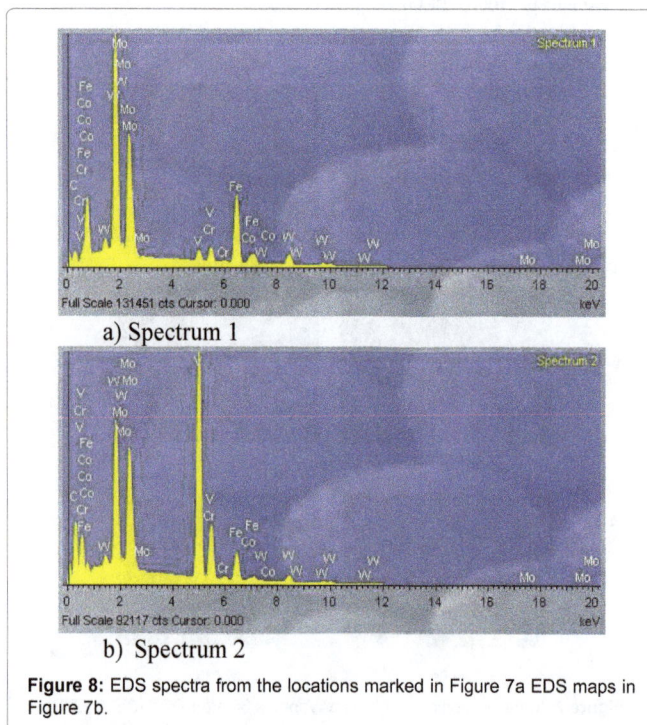

a) Spectrum 1

b) Spectrum 2

Figure 8: EDS spectra from the locations marked in Figure 7a EDS maps in Figure 7b.

Number of carbide particles of the size up to 1 μm^2			
Vanadis 30	V-Carbide	W-Carbide	Total
CHT	166	27	193
DCT	158	57	215

Table 2: Influence of DCT on number of carbide particles of the size up to 1 μm^2 in Vanadis 30 P/M HSS steel.

Conclusions

If DCT step is inserted between quenching and tempering of both Vanadis 6 P/M tool steel for cold work and Vanadis 30 P/M HSS steel, it is possible to state the following:

1. Significant drop in hardness due to DCT of Vanadis 6 causes a slight increase of three point bending strength. Three point bending strength of Vanadis 30 steel after DCT remains almost unchanged within the standard deviation.

2. In case of Vanadis 6 steel after DCT there is a slight increase in the wear rate, whereas Vanadis 30 does not show any effect of the cryogenic treatment on wear rate.

3. Metallographic analysis of Vanadis 6 after DCT confirmed higher amount of small globular carbides in comparison with state after DCT while the austenitizing temperature is 1050°C.

4. Analysis of carbide particles in Vanadis 30 confirmed the presence of carbides based on vanadium and carbides based on tungsten and molybdenum.

5. After DCT of Vanadis 30, number of carbides based on vanadium slightly decreases while the amount of carbides based on tungsten increased.

6. DCT does not seem to cause any significant effect on fine carbides precipitation in Vanadis 30.

Acknowledgement

This work was supported by the Ministry of Education, Youth and Sport of the Czech Republic, program NPU1, project No. LO1207 and by the Grant Agency of the Czech Technical University in Prague, grant No. SGS15/149/OHK2/2T/12.

References

1. Akincioglu S, Gokkaya H, Uygur AI (2015) Review of cryogenic treatment on cutting tools. The International Journal of Advanced Manufacturing Technology 78: 1609-1627.

2. Gill S, Singh J, Singh R, Singh H (2011) Metallurgical principles of cryogenically treated tool steels-A review on the current state of science. The International Journal of Advanced Manufacturing Technology 54: 59-82.

3. Kalsi N, Sehgal R, Sharma V (2010) Cryogenic treatment of tool materials: A review. Materials and Manufacturing Processes 25: 1077-1100.

4. Das D, Dutta A, Ray K (2010) Sub-zero treatments of AISI D2 steel: Part I. Microstructure and hardness. Materials Science and Engineering: A 527: 2182-2193.

5. Jurci P, Sobotova J, Cejp J, Salabova P (2010) Effect of sub-zero treatment on mechanical properties of Vanadis 6 PM Ledeburitic tool steel. In: 19th International Conference on Metallurgy and Materials, Conference Proceedings. Ostrava: Tanger Ltd., pp: 518-523.

6. Bilek P, Sobotova J, Jurci P (2011) Evaluation of the microstructural changes in Cr-V ledeburitic steels depending on the austenitization temperature. Materials and technology 45: 489-493.

7. Sobotova J, Jurci P, Dlouhy I (2016) The effect of sub-zero treatment on microstructure, fracture toughness and wear resistance of Vanadis 6 tool steel. Materials Science and Engineering: A 652: 192-204.

8. Godec M, Batic B, Mandrino D, Nagode A (2010) Characterization of the carbides and the martensite phase in powder-metallurgy high-speed steel. Materials Characterization. 61: 452-458.

9. Sobotova J, Kueik M, Cejp J (2015) Influence of heat treatment conditions on properties of high-speed P/M Steel Vanadis 30. Key Engineering Materials 647: 17-22.

10. American Society for Testing and Materials (2010) Standard test method for wear testing with a pin-on-disk apparatus: G99-05. Reapproved. Philadelphia, USA: ASTM International.

11. Jurci P, Kusy M, Ptacinova J, Kuracina V (2014) Long-term sub-zero treatment OD P/M Vanadis 6 Ledeburitic tool Steel-A preliminary study. In: International Conference on Heat treatment. Praha: Asociace pro tepelné zpracování kovů-ECOSOND, S.R.O., 2014.

12. Jurci P, Dománkova M, Sobotova J (2015) Microstructure and hardness of sub-zero treated and no tempered P/M Vanadis 6 ledeburitic tool steel. Vacuum. 111: 92-101.

13. Das D, Dutta A, Ray K (2010) Sub-zero treatments of AISI D2 steel: Part II. Wear behavior. Materials Science and Engineering: A. 527: 2194-2206.

33

Finite Element Analysis of Internal Door Panel of a Car by Considering Bamboo Fiber Reinforced Epoxy Composite

Eskezia E*, Abera A and Daniel Tilahun

School of Mechanical and Industrial Engineering, Addis-Ababa University, King George VI Street -385, Ethiopia

Abstract

In this paper work the dynamic structural Finite Element Analysis of internal door panel of a vehicle by considering bamboo fiber reinforced epoxy composite (BFREC) materials was conducted. The objective of the paper is to develop a suitable model of internal door panel for Toyota DX car, to conduct a transient dynamic structural analysis (stress and displacement analysis) of internal door panel by finite element method, to compare the performance of BFREC material with previously recommended materials of internal door panel. The door panel of Toyota Corolla DX model vehicle was used to develop the geometric model of the internal door panel by CATIA V5 R20 modeling software. This 3-D geometric model was imported to using ANSYS Workbench 15.0. The transient dynamic structural FEA was done after assigning loading and boundary conditions. The applied load considered for this analysis is the self-inertial weight of the panel due to the acceleration field produced while the door is closing. The equivalent stress and the displacement are noted and investigated to compare with the literatures revised. The result shows that, bamboo fiber reinforced epoxy composite panel has the smallest mass and equivalent stress values, as compared with the ligno-cellulosic composite and polypropylene one. Based on these realities, it is recommended that bamboo fiber reinforced with epoxy composite materials are suitable for internal structural automotive panel applications.

Keywords: Bamboo fiber; Epoxy; Composite; Door panel; Transient structural analysis; Equivalent stress; Displacement; Finite element analysis

Introduction

In order to conserve natural resources and economize energy, weight reduction has been the main focus of automobile designers and manufacturers in the present scenario. Weight reduction can be achieved primarily by the research of better material, design optimization and better manufacturing processes. Due to rise in demand of lightweight and more efficient vehicles and better mechanical performance of materials in automotive applications, different material combinations such as composites, plastic and light weight metals are implemented on different structural parts of vehicles. Applications of composite materials in automotive industries already include some structural parts, such as dashboard, roof, floor, front and back bumper, passenger safety cell, and door panels [1-5].

The internal door panel of an automobile is typically made of different materials. Unlike the materials used on the exterior side of the vehicle door, the material on the interior side serves a greater purpose other than just aesthetic appeal. The internal door panel of an automobile contributes to the overall functionality and ergonomics of the ride, such as: armrests, various switches, lights, electronic systems like the window controls and locking mechanism; etc. [6-9].

Composite materials made of natural fibers and polymer matrix provides synergistic properties, improving their strength and durability. These materials are suitable for achieving automotive interior components, where in addition to their low weight have also high rigidity and good thermal and sound insulation. The most important internal vehicle elements include car internal door panels [10-13].

Materials and Methods

In this analysis, the bamboo fiber/epoxy composite materials with a considerable composition are used as the materials of internal door panel of an automotive.

During selecting the material, the characteristics of the composite, the ways of fiber extraction, the ease of the manufacturing process, the types of the matrix used for the composite and some other criteria are taken into consideration. In addition to this, the selection of the material for this specific research work is basically focusing on the Ethiopian bamboo fiber and the researches which have previously done on it. Based on these measures, the composite with 25% bamboo fiber and 75% epoxy resin was selected [14] (Table 1).

Modeling and Analysis

The following steps are used in the solution procedure using ANSYS Workbench software for transient structural Finite Element Analysis of a mechanical problem [15] (Figures 1 and 2).

1. Imported the geometry of the panel from modeling software to the ANSYS workbench.

2. The material type and its properties are specified.

3. Meshing the imported panel model.

4. The boundary conditions and external loads are applied.

5. The solution is generated based on these input parameters.

6. Finally, the solution can be displayed.

*Corresponding author: Eskezia E, School of Mechanical and Industrial Engineering, Addis-Ababa University, King George VI Street-385, Ethiopia
E-mail: arsame2008@gmail.com

Material Property	Values
Density (ρ)	1.12 g/cm³
Tensile Strength	187.73 MPa
Flexural Strength	190.32 MPa
Compressive Strength	114.13 MPa
Shear Strength	81.18 MPa
Young's Modulus	3852
Shear Modulus	1580

Table: 1: Mechanical properties of Bamboo fiber reinforced epoxy composite [10].

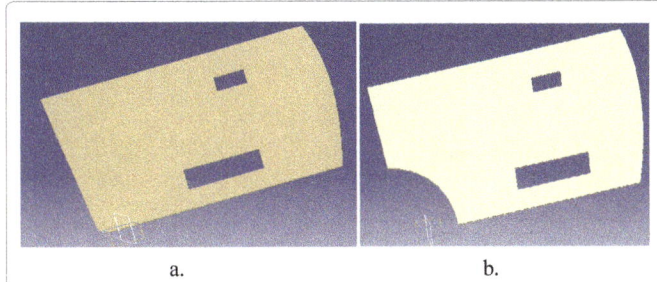

Figure 1: The 3D modelling of the internal door panel; a). Front left door panel; b). Rear left door panel.

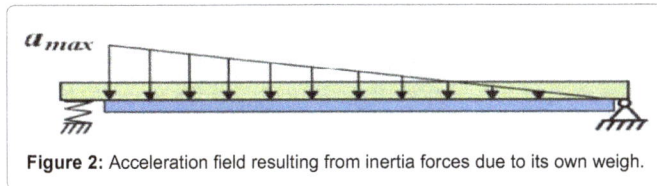

Figure 2: Acceleration field resulting from inertia forces due to its own weigh.

The internal door panel has several areas where constraints are applied as follows:

- Upper part rests on a metal door structure, thus blocking shifting on the y direction (U_2).

- In screw mounting areas shifting are blocked on all three directions (U_1, U2 and U3).

- On clips systems panel mounting metal structure areas, shifting are blocked on all the three directions (U_1, U_2 and U_3).

Where, U_1, U_2 and U_3 are the displacements in x, y and z directions, respectively.

Results and Discussion

The transient structural dynamic analysis determines characteristics of the stress and deformation of the structures (the panel) caused by the applied loading systems and boundary conditions (Figures 3-10).

Discussion

This transient structural dynamic analysis of the internal door panel of a vehicle using BFREC was performed for self-weight inertial load intensity of the shock produced while closing the door.

Comparing the results obtained by FEA of the BFREC panel with the panels of other previously recommended materials (ligno-cellulosic composite and polypropylene plastic composite materials) is important to see the improved achievement. The comparison is carried out by making everything the same, except the material properties; i.e. at the same acceleration field (350 m/s²) in the same model and the same method of FEM analysis (Figure 11).

Equivalent (Von-Mises) stress

The results of this analysis show that the equivalent (Von-Mises) stress of the BFREC panel is the smallest one as compared to that of the lignocellulosic composite and polypropylene panels. This implies that BFREC material is less stressed and thus, has a better performance.

Displacement

The maximum displacements of the BFREC panel are lower by about 34% than the lignocellulosic composite panel and 50% of the

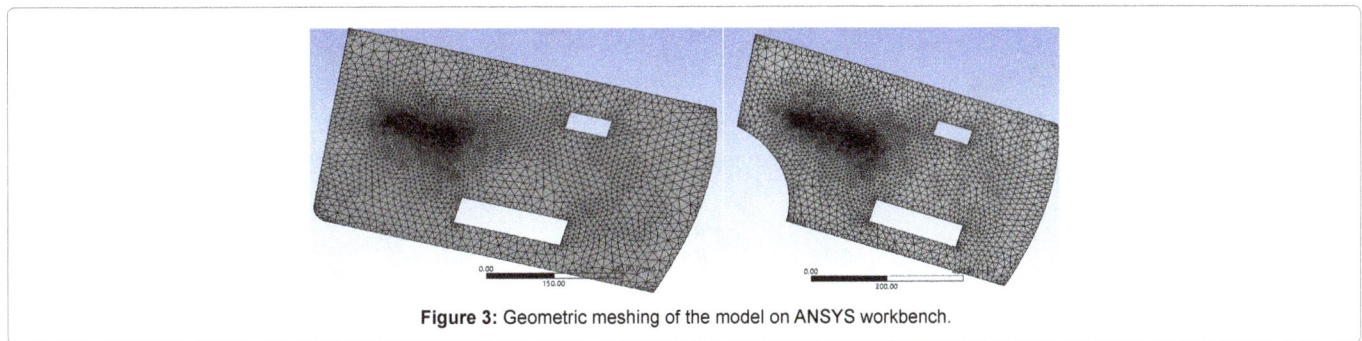

Figure 3: Geometric meshing of the model on ANSYS workbench.

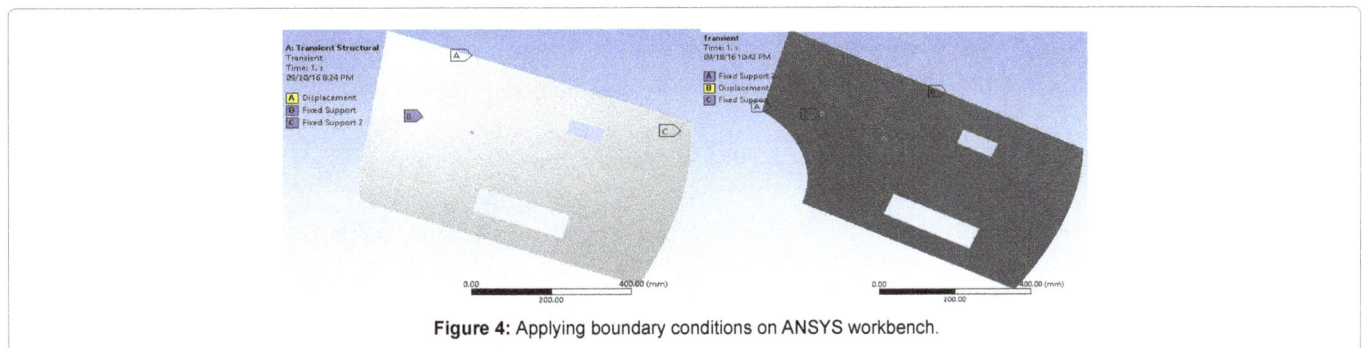

Figure 4: Applying boundary conditions on ANSYS workbench.

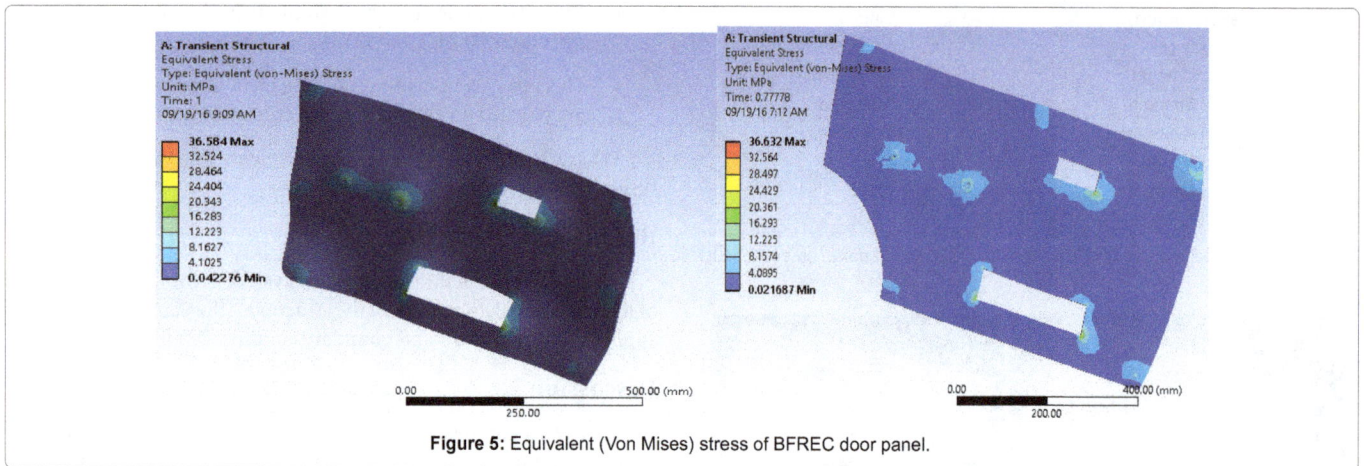

Figure 5: Equivalent (Von Mises) stress of BFREC door panel.

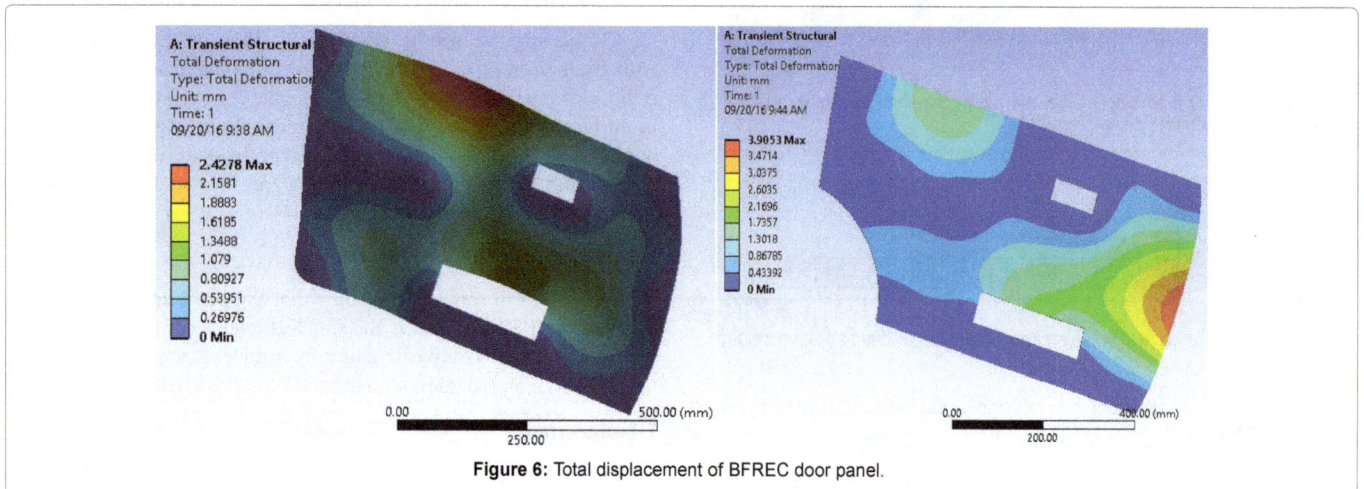

Figure 6: Total displacement of BFREC door panel.

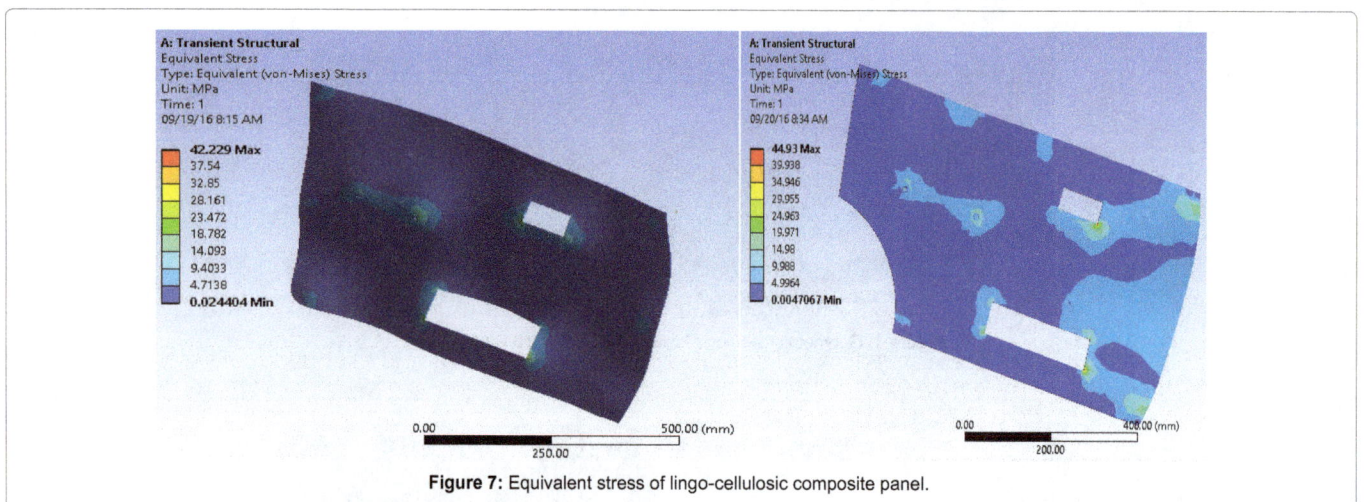

Figure 7: Equivalent stress of lingo-cellulosic composite panel.

polypropylene panel. This is due to the greater rigidity of BFREC material.

Moreover, the mass of BFREC panel is reduced by 6% and 8% than that of lignocellulosic composite and polypropylene materials respectively. The smaller mass of the BFREC panel helps to make the vehicle lightweight, so that the efficiency and fuel economy of the vehicle is improved by reducing its dead weight [16] (Tables 2 and 3).

Conclusion

In order to achieve the goal of this study, different tasks were performed and the following conclusions are drawn.

1. The transient structural dynamic analysis of the modeled panels was performed by ANSYS Workbench 15.0 analysis software.

2. Under the same applied load and boundary conditions, the

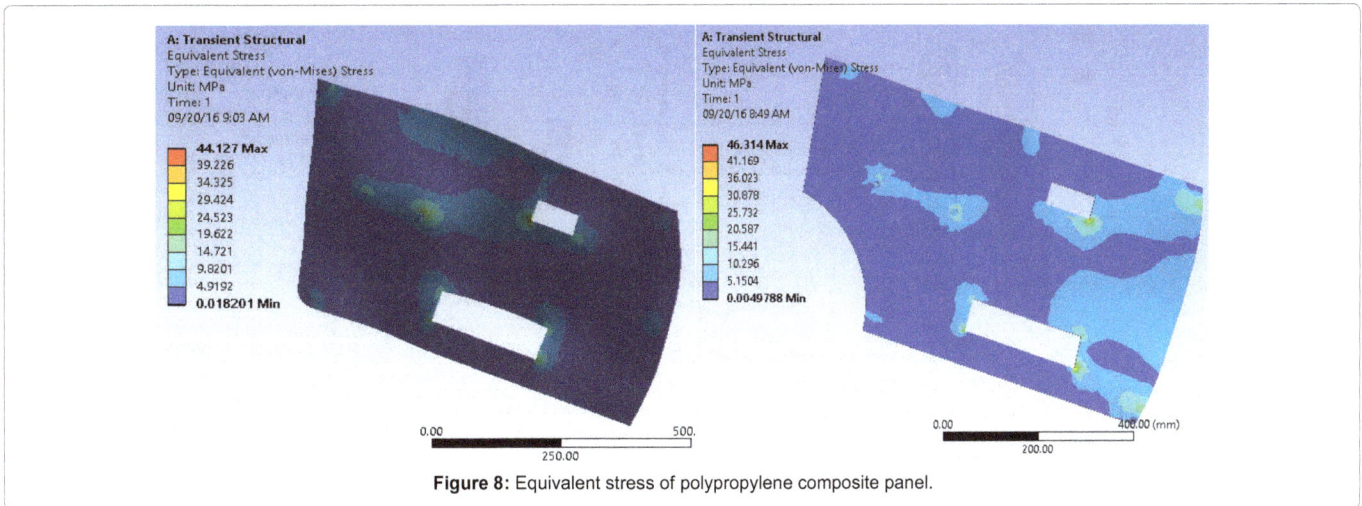

Figure 8: Equivalent stress of polypropylene composite panel.

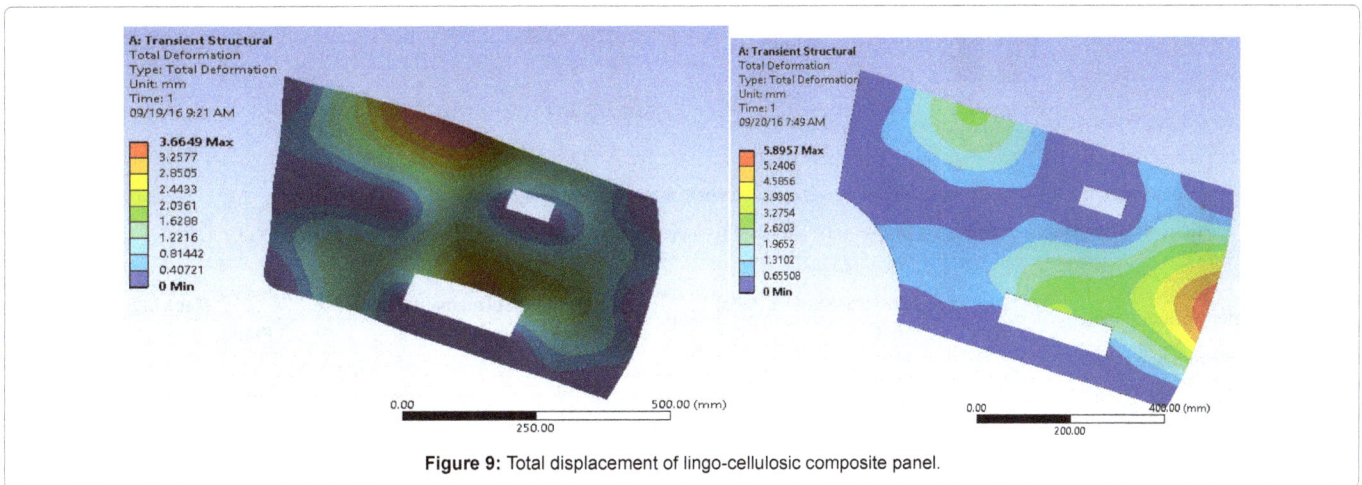

Figure 9: Total displacement of lingo-cellulosic composite panel.

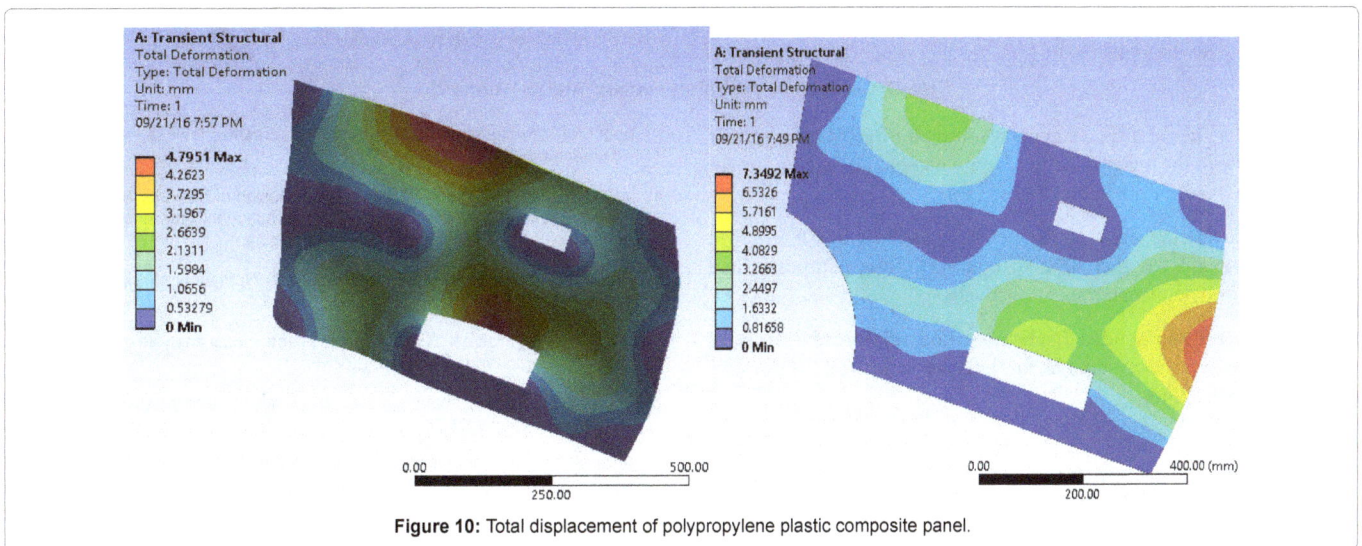

Figure 10: Total displacement of polypropylene plastic composite panel.

smallest equivalent (Von-Mises) stress is recorded on BFREC panel as compared to the ligno-cellulosic composite and polypropylene panels. Thus, BFREC panel performs in a better way at a given loading condition.

3. Displacements of BFREC materials internal door panel obtained by FEA are smaller by 34% and 50% than that of the ligno-cellulosic composite and polypropylene plastic materials respectively.

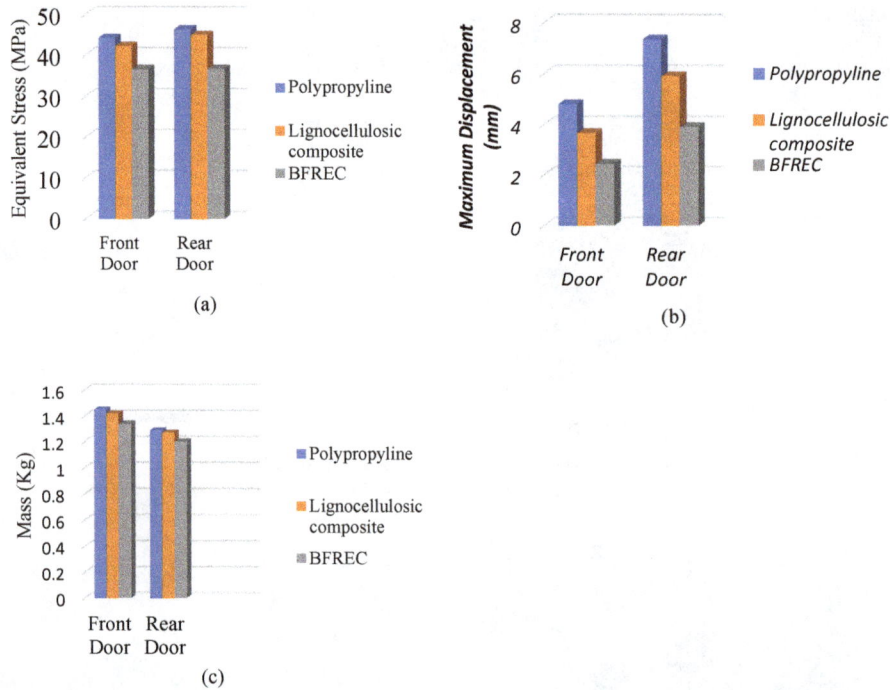

The charts plotted above will show the comparisons of these values clearly for both front and rear door panels.

Figure 11: Comparison of the results of different materials; (a) equivalent stress, (b) maximum displacement, (c) weight of the panel.

Internal door panel	Equivalent Stress (MPa)		Maximum Displacement (mm)		Mass (kg)	
	Front	Rear	Front	Rear	Front	Rear
Polypropylene panel	44.13	46.31	4.80	7.35	1.45	1.29
Ligno-cellulosic composite	42.23	44.93	3.66	5.90	1.42	1.27
BFREC panel	36.58	36.63	2.43	3.91	1.34	1.20

Table 2: Comparing the FEA results of the panels at an acceleration of 350 m/s^2.

	Front door	Rare door
Number of nodes	47136	41751
Number of elements	22785	20015

Table 3: Geometric meshing of the model on ANSYS Workbench.

4. Small values of displacements resulting for BFREC component are due to the high rigidity and low weight material component given its lower thickness of the panel.

5. Moreover, the mass of the newly modeled panel is reduced by 6% and 8% than that of ligno-cellulosic composite and polypropylene materials respectively.

Finally, it is recommended that bamboo fiber reinforced with epoxy composite materials are suitable for internal structural automotive door panel applications.

References

1. Bundele S, Bindu RS (2012) Automotive door design and structural optimization of front door for commercial vehicle with ULSAB concept for cost and weight reduction. Global J Researches in Engineering (Automotive Engineering).

2. Terciu OM, Curtu I, Cerbu C (2012) FEM modeling of an automotive door trim panel made of lignocelluloses composites in order to determine the stresses and displacements in case of a door slam simulation. The 4th International Conference of Advanced Composite Materials Engineering COMAT 2012 Brasov Romania.

3. Grujicic M (2006) Multi-disciplinary design optimization of a composite car door for structural performance, NVH, crashworthiness, durability, and manufacturability. CU-ICAR.

4. Wei T, Yan W, Ling-Yang W, Yu Z (2012) Design of the vehicle door structure based on finite element method. The 2nd International Conference on Computer Application and System Modeling Wuhan, China.

5. Darwish S, Hussein HMA, Gemeal A (2012) Numerical study of automotive doors. Int J Eng Technol IJET -IJENS 12.

6. Raghuveer M, Prakash SG (2014) Design and impact analysis of a car door. Int J Modern Engineering Research (IJMER).

7. Koricho EG (2012) Implementation of composite and plastics materials for vehicle light weight: 44-74.

8. Ademet E (2015) Experimental analysis of E-Glass /Epoxy & E-Glass /polyester composites for auto body panel. AIJRSTEM: 377-383.

9. Amada S, Untao S (2001) Fracture properties of bamboo, Composites Part B 32: 451-459.

10. Tarekegne F, Conze P (2011) Study of Ethiopia high land bamboo composite for structural materials. Commissioned by Fortune Enterprise P.L.C collaborated with GTZ on a public private partnership project (PPP) Addis Ababa, Ethiopia

11. Sreenivasu S, Reddy AC (2014) Mechanical properties evaluation of bamboo fiber reinforced composite materials. Int J Engineering Research 3: 187-194.

12. Kumar D (2014) Characterization of tensile properties of treated bamboo natural fiber polymer composite. Int J Mechanical Engineering and Technology (IJMET) 5:110-115.

13. Kelemwork S (2008) Anatomical characteristics of Ethiopian lowland bamboo (Oxy-tenanthera-abyssinica) International Center for Bamboo & Rattan, China.

14. ANSYS (2009) Finite element analysis and concepts, Guide for the use and applicability of workbench simulation tools from ANSYS Inc.

15. Barschke M (2009) Finite element modeling of composite materials using kinematic constraints. Engineering and science 5: 133-153.

16. Morgan K (2015) Shock and vibration using ANSYS mechanical. ANSYS Inc.

Permissions

List of Contributors

Yang-Hee Kwon and Sung-Gul Hong
Department of Architecture and Architectural Engineering, Seoul National University, 1 Gwanak-Ro, Gwanak-Gu, Seoul 08826, Republic of Korea

Soo-Hyung Chung
SENSE B/D, 6 Beodeunaru-ro 19-gil, Youngdeungpo-gu, Seoul 07226, Republic of Korea

Taparia N, Ritesh Kumar C, Kanwar L and Verma D
Department of Mechanical engineering, SRM University, India

Saldaña HA
School of Chemical Sciences and Engineering is also from the Autonomous University of the State of Morelos

Márquez Aguilar PA and Molina OA
Research Center for Engineering and Autonomous University of the State of Morelos Av. University # 1001, Col. Chamilpa. Cp. 62209. Cuernavaca Morelos, Mexico

Wang C, Zhi-Peng Z and Kuan-Fang H
College of Mechanical and Electrical Engineering, Hunan University of Science and Technology, Xiantan 411201, PR China
Hunan Provincial Key Laboratory of Health Maintenance for Mechanical Equipment, Hunan University of Science and Technology, Xiantan 411201, PR China

Si-Wen X
College of Mechanical and Electrical Engineering, Hunan University of Science and Technology, Xiantan 411201, PR China
Hunan Provincial Key Laboratory of High efficiency precision cutting for Difficult-to-cut Material, Hunan University of Science and Technology, Xiangtan 411201, PR China

Amasaka K
Department of Mechanical Engineering, Aoyama Gakuin University, Japan

Sung-Hoon Kang, Sung-Gul Hong and Yang-Hee Kwon
Department of Architecture and Architectural Engineering, Seoul National University, 1 Gwanak-Ro, Gwanak-Gu, Seoul 08826, Republic of Korea

Latif U and Naeem Shah A
Department of Mechanical Engineering, University of Engineering and Technology, Lahore 54000, Pakistan

Wu W
State Key Laboratory of Automotive Safety and Energy, Tsinghua University, Beijing 100084, PR China

Qian G
Paul Scherrer Institute, Nuclear Energy and Safety Department, Laboratory for Nuclear Materials, 5232 Villigen PSI, Switzerland

Cui X
Sustainable Energy Systems Group, Lawrence Berkeley National Laboratory, 1 Cyclotron Road MS 90R2002, Berkeley CA 94720, USA

Chun Xu C and Muk Choa H
Division of Mechanical and Automotive Engineering, Kongju National University 275, Budae-dong, Cheonan-si, Chungcheongnam-Do 331-717, South Korea

Bogis Haitham
Center of Excellence for Industrial Design and Manufacturing Research (CEIDM) Mechanical Engineering, King Abdulaziz University, Jeddah, Saudi Arabia

El-Gizawy A. Sherif
Center of Excellence for Industrial Design and Manufacturing Research (CEIDM) Mechanical Engineering, King Abdulaziz University, Jeddah, Saudi Arabia
Industrial and Technological Development Center, Mechanical and Aerospace Engineering, University of Missouri-Columbia Columbia, Missouri 65211, USA

Chitti Babu S
Industrial and Technological Development Center, Mechanical and Aerospace Engineering, University of Missouri-Columbia Columbia, Missouri 65211, USA

Ameen MM
Department of Physics, College of Education, University of Salahaddin, Erbil, Kurdistan Region, Iraq

Mawlud SQ and Ahmed KF
Department of Physics, College of Education, University of Salahaddin, Erbil, Kurdistan Region, Iraq
Advanced Optical Material Research Group, Department of Physics, Faculty of Science, University of Technology Malaysia, Skudai and Johor, Malaysia

Md. Sahar R
Advanced Optical Material Research Group, Department of Physics, Faculty of Science, University of Technology Malaysia, Skudai and Johor, Malaysia

Pengyun Xu, Haiyong Jiang and Xiaoshun Zhao
Mechanical and Electronical Engineering College, Agriculture University of Hebei, Baoding 071001, P.R. China

Drakakaki Arg, Diamantogiannis G and Apostolopoulos Ch
University of Patras, Panepistimioupoli Patron 265 04, Greece

Apostolopoulos Alk
University of Ioannina, Epirus, Greece

Krasikov E
National Research Centre, Kurchatov Institute, 123182, Moscow, Russia

Ghani AO, Agelin-chaab M and Barari A
Department of Automotive, Mechanical and Manufacturing Engineering, University of Ontario Institute of Technology, Oshawa, Canada

Hasan AO
Department of Mechanical Engineering, Al-Hussein Bin Talal University, Maan, P.O. Box-20, Jordan

Abu-jrai A
Department of Environmental Engineering, Al-Hussein Bin Talal University, Maan, P.O. Box-20, Jordan

Suzuki K
Department of Intelligent Mechanical Systems Engineering, Kagawa University 2217-20, Hayashi-cho, Takamatsu-city, Kagawa 761-0396, Japan

Xuan-Linh N and Huan-Xin L
Key Laboratory of Pressurized Systems and Safety, Ministry of Education, East China University of Science and Technology, Shanghai 200237, P.R. China

Regaiguia B
Département de métallurgie et génie des matériaux, Laboratoire de métallurgie et génie des matériaux, Université Badjimokhtar 23000 Annaba, Algeria

Abderrazak D
Département de métallurgie et génie des matériaux, Laboratoire de mise en forme, Université Badjimokhtar 23000 Annaba, Algeria

Bahwini T, Zhong Y and Gu C
School of Engineering, RMIT University, Melbourne, Australia

Smith J
Department of Surgery, School of Clinical Sciences at Monash Health, Monash University, Clayton, Australia

Nagaral M
Aircraft Research and Design centre, HAL, Bangalore-560037, Karnataka, India

Auradi V
R&D Centre, Department of Mechanical Engineering, SIT, Tumkur-572103, Karnataka, India

Parashivamurthy KI
Department of Mechanical Engineering, Government Engineering College, Chamarajnagar-571313, Karnataka, India

Kori SA
Department of Mechanical Engineering, Basaveshwara Engineering College, Bagalkot-587102, Karnataka, India

Bostan M and Akhtari AA
Department of Civil Engineering, College of Engineering, Razi University, Kermanshah, Iran

El Messiry M and El Deeb R
Textile Engineering Department, Faculty of Engineering, Alexandria University, Alexandria, Egypt

Chen Y, Enearu OL and Sutharssan T
School of Engineering and Technology, University of Hertfordshire, College Lane, Hatfield, Hertfordshire, AL10 9AB, UK

Montalvao D
Bournemouth University, Department of Design and Engineering, Faculty of Science and Technology, Poole BH12 5BB, UK

Xiao Q
Department of Naval Architecture, Ocean and Marine Engineering, University of Strathclyde, Glasgow G4 0LZ, UK

Massoud EZ
Department of Naval Architecture, Ocean and Marine Engineering, University of Strathclyde, Glasgow G4 0LZ, UK
Mechanical Engineering Department, College of Engineering and Technology, Arab Academy for Science, Technology and Maritime Transport, P.O. Box 1029, Abu Qir, Alexandria, Egypt

Teamah MA and Saqr KM
Mechanical Engineering Department, College of Engineering and Technology, Arab Academy for Science, Technology and Maritime Transport, P.O. Box 1029, Abu Qir, Alexandria, Egypt

Badreddine R and Abderrazak D
Departement de metallurgie et génie des materiaux, laboratoire de metallurgie et genie des materiaux, universite badjimokhtar 23000 Annaba

Kheireddine S
Research Center in Industrial Technologies, CRTI P. O. Box 64, Cheraga 16014 Algiers, Algeria

Cote P and Dumas G
LMFN Lab, Mechanical Engineering Department, Laval University, Quebec, QC G1V 0A6, Canada

Kakade IN, Sanghavi L, Sharma P and Mishra V
Centre for Nano-Science and Engineering, Indian Institute of Science, Bangalore, Karnataka, India

Altunok M, Kureli I, Doganay S and Onduran A
Faculty of Technology, Gazi University, Turkey

Afshari D and Afrabandpey A
Department of Mechanical Engineering, University of Zanjan, Iran

Uniyal P, Gunwant D and Misra A
Department of Mechanical Engineering, Gbpuat Pantnagar, India

Sobotova J, Kuřík M, Krum S and Lacza J
Faculty of Mechanical Engineering, Czech Technical University in Prague, Karlovo nám 13, 121 35 Prague 2, Czech Republic

Eskezia E, Abera A and Daniel Tilahun
School of Mechanical and Industrial Engineering, Addis-Ababa University, King Jeorge VI Street -385, Ethiopia

Index

9 781682 855706